Mechanics

Berkeley Physics Course   Volume 1, Second Edition

This book was set in Laurel by York Graphic Services, Inc. The editors were Jack L. Farnsworth, Eva Marie Strock, and Ida Abrams Wolfson; the designer was Michael A. Rogondino; and the production supervisor was Adam Jacobs. The drawings were done by Ayxa Art.
The printer was Halliday Lithograph Corporation; the binder, The Book Press, Inc.

Front cover: NGC 4594 spiral galaxy in Virgo, seen on edge; 200-in. photograph. The dark band is due to absorption by a ring of matter surrounding the bright central core. (*Photograph courtesy of the Hale Observatories.*)

Back cover: Hydrogen bubble chamber picture of the production of an anti-Ξ in the reaction $K^+ + p$. (*Photograph courtesy of the Lawrence Berkeley Laboratory.*)

MECHANICS

Library of Congress Cataloging in Publication Data

Kittel, Charles.
Mechanics.

(Berkeley physics course, v. 1)
1. Mechanics. I. Knight, Walter D., joint author. II. Ruderman, Malvin A., joint author. III. Helmholz, A. Carl, ed. IV. Moyer, Burton J., ed. V. Title. VI. Series.
QC1.B375 vol. 1 [QC125.2] 530′.08s [531] 72-7444
ISBN 0-07-004880-0

1234567890 HDBP 79876543

The first edition of the Berkeley Physics Course MECHANICS, Vol. 1 copyright © 1963, 1964, 1965 by Educational Development Center was supported by a grant from the National Science Foundation to EDC. This material is available to publishers and authors on a royalty-free basis by applying to the Educational Development Center.

Contents

One of the urgent problems confronting universities today is that of undergraduate teaching. As research has become more and more absorbing to the faculty, a "subtle discounting of the teaching process" (to quote philosopher Sidney Hook) has too often come into operation. Additionally, in many fields the changing content and structure of knowledge growing out of research have created great need for curriculum revision. This is particularly true, of course, in the physical sciences.

It is a pleasure, therefore, to contribute a foreword to the Berkeley Physics Course and Laboratory, which is a major curriculum improvement program at the undergraduate level designed to reflect the tremendous revolutions in physics of the last hundred years. The course has enlisted the efforts of many physicists working in forefront areas of research and has been fortunate to have the support of the National Science Foundation, through a grant to Educational Services Incorporated. It has been tested successfully in lower division physics classes at the University of California, Berkeley, over a period of several semesters. The course represents a marked educational advance, and I hope it will be very widely used.

The University of California is happy to act as host to the inter-university group responsible for developing this new course and laboratory and pleased that a number of Berkeley students volunteered to help in testing the course. The financial support of the National Science Foundation and the cooperation of Educational Services Incorporated are much appreciated. Most gratifying of all, perhaps, is the lively interest in undergraduate teaching evinced by the substantial number of University of California faculty members participating in the curriculum improvement program. The scholar-teacher tradition is an old and honorable one; the work devoted to this new physics course and laboratory shows that the tradition is still honored at the University of California.

Clark Kerr

Foreword

Table of Values

Value and Units	Item	Symbol or Abbreviation	Derivation of Value
Gases			
22.4×10^3 cm^3/mol	Molar volume at STP	V_0	
2.69×10^{19} cm^{-3}	Loschmidt's number	n_0	N_0/V_0
6.0222×10^{23} mol^{-1}	Avogadro's number	N_0	
8.314×10^7 ergs mol^{-1} deg^{-1}	Gas constant	R	
1.381×10^{-16} erg/K	Boltzmann's constant	k	R/N
1.01×10^6 dyn/cm^2	Atmospheric pressure		
$\approx 10^{-5}$ cm	Mean free path at STP		
3.32×10^4 cm/s	Speed of sound in air at STP		
Atomic			
6.6262×10^{-27} erg-s	Planck's constant	h	
1.0546×10^{-27} erg-s	Planck's constant/2π	$\hbar$	$h/2\pi$
13.6 electron volts	Energy associated with 1 Rydberg	Ry	
1.6022×10^{-12} erg	Energy associated with 1 electron volt	eV	
1.2398×10^{-4} cm	Wavelength associated with 1 electron volt		hc^2/e
2.4180×10^{14} s^{-1}	Frequency associated with 1 electron volt		
0.5292×10^{-8} cm	Bohr radius of the ground state of hydrogen	a_0	$\hbar^2/me^2$
$\approx 10^{-8}$ cm	Radius of an atom		
0.9274×10^{-20} erg/G	Bohr magneton	μ_B	$e\hbar/2mc$
137.036	Reciprocal of fine-structure constant	α^{-1}	$\hbar c/e^2$
Particles			
1.67265×10^{-24} g	Proton rest mass	M_p	
1.67496×10^{-24} g	Neutron rest mass	M_n	
1.66057×10^{-24} g	1 unified atomic mass unit ($\equiv \frac{1}{12}$ mass of C^{12})	u	
0.910954×10^{-27} g	Electron rest mass	m	
0.93828×10^9 eV	Energy equivalent to proton rest mass	E_p	M_pc^2
0.511004×10^6 eV	Energy equivalent to electron rest mass		mc^2
0.93150×10^9 eV	Energy equivalent to 1 atomic mass unit		
1836	Proton mass/electron mass		M_p/m
2.818×10^{-13} cm	Classical radius of the electron	r_0	e^2/mc^2
4.80325×10^{-10} esu	Charge on proton	e	
1.60219×10^{-19} C	Charge on proton	e	
2.423×10^{-10} cm	Electron Compton wavelength	λ_C	h/mc

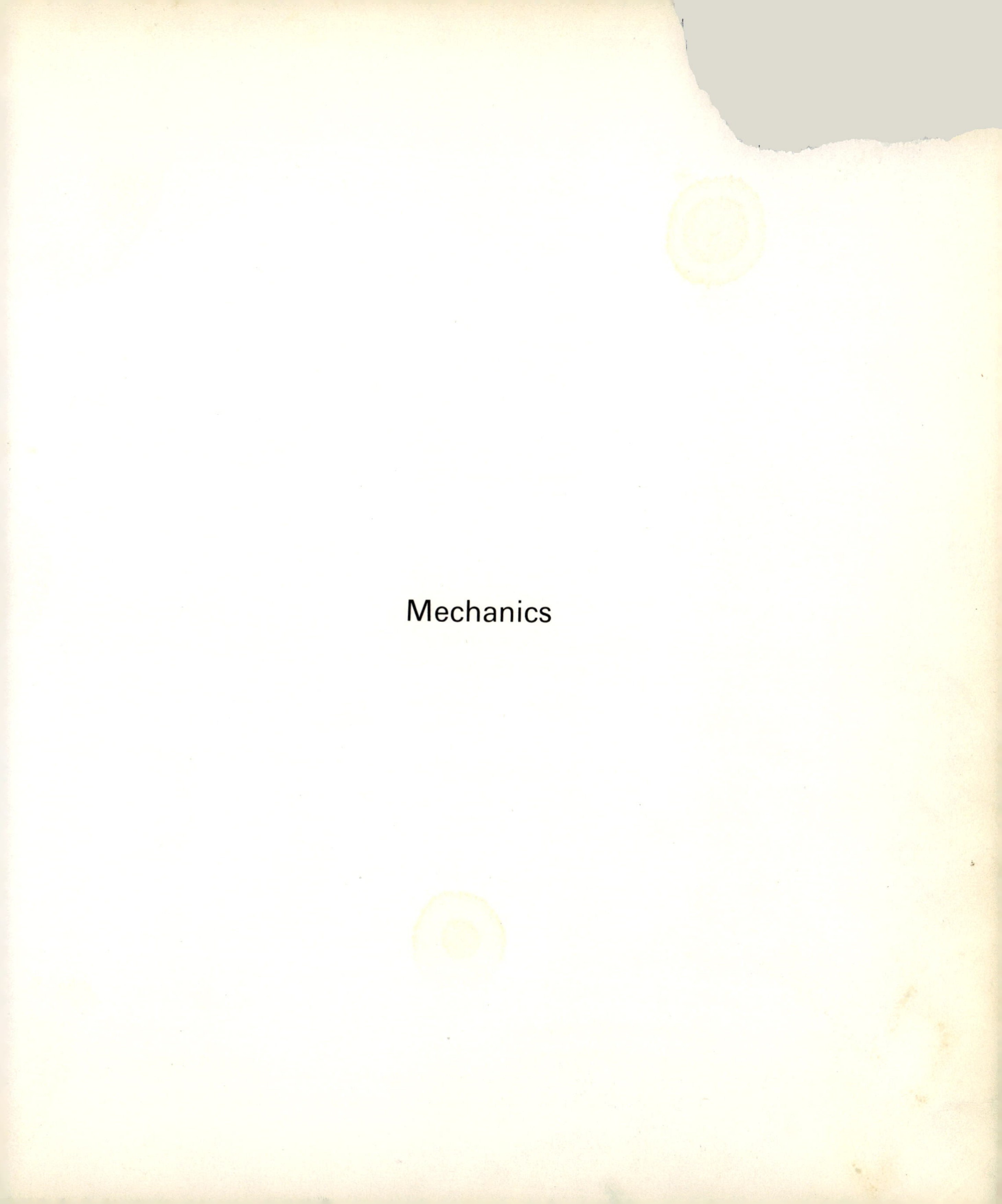

Mechanics

Charles Kittel

Professor of Physics
University of California
Berkeley

Walter D. Knight

Professor of Physics
University of California
Berkeley

Malvin A. Ruderman

Professor of Physics
New York University

Revised by

A. Carl Helmholz

Professor of Physics
University of California
Berkeley

Burton J. Moyer

Dean of the College of Liberal Arts
University of Oregon
Eugene

McGRAW-HILL BOOK COMPANY New York St. Louis San Francisco Düsseldorf Johannesburg Kuala Lumpur London Mexico Montreal New Delhi Panama Rio de Janeiro Singapore Sydney Toronto

Volume 1 of the Berkeley Physics Course has been in use in its bound form for about seven years. Several years ago it seemed appropriate to consider a revision. At this point each of us had taught the course in Berkeley several times, and on the basis of our experience and talks with colleagues, both in Berkeley and at other institutions, we had developed and considered changes to make a more "teachable" text for an introductory course for engineering and physical science students. Thus we proceeded to such a revision.

We have tried to keep the fresh approach that was characteristic of the whole Berkeley Physics Course, the use of examples drawn from research laboratories, and the presentation of interesting topics often previously judged to be too advanced for an introductory course. We have removed some of the Advanced Topics from Vol. 1 and have removed Chap. 15, Particles of Modern Physics, in the belief that they are not often used in a course at this level. The most substantial change has been the complete rewriting of Chap. 8 on Rigid Body Motion. Although this chapter is certainly more mundane now, it is more suited to the level of the students. The order of presentation of topics remains the same except that Chaps. 3 and 4 have been interchanged in the hope that some familiarity with the ordinary applications of Newton's Laws of Motion will provide the student with background for a better understanding of the somewhat more advanced concept of galilean transformations. Finally, because students have encountered substantial difficulties with mathematics, particularly differential equations, we have added a number of Mathematical Notes.

The Teaching Notes that follow give some detail of the philosophy of using this book as a text. There is still a good deal more material than can be comfortably used in a one-quarter or a one-semester course. An instructor should make conscious choices of the material that he wishes to use. In recent years the change to the quarter system at Berkeley has unfortunately made it necessary to separate laboratory work from the first quarter covering the subject of mechanics. An introductory course should

Preface to the Second Edition of Volume 1

be tied to the laboratory, and the revision of the Berkeley Physics Laboratory by Alan Portis and Hugh Young provides accompanying laboratory work valuable for any introduction to mechanics.

We have benefited from the help and criticisms of many colleagues. The help of Miss Miriam Machlis in preparing this revision has been particularly extensive.

A. Carl Helmholz
Burton J. Moyer

This is a two-year elementary college physics course for students majoring in science and engineering. The intention of the writers has been to present elementary physics as far as possible in the way in which it is used by physicists working on the forefront of their field. We have sought to make a course that would vigorously emphasize the foundations of physics. Our specific objectives were to introduce coherently into an elementary curriculum the ideas of special relativity, of quantum physics, and of statistical physics.

This course is intended for any student who has had a physics course in high school. A mathematics course including the calculus should be taken at the same time as this course.

There are several new college physics courses under development in the United States at this time. The idea of making a new course has come to many physicists, affected by the needs both of the advancement of science and engineering and of the increasing emphasis on science in elementary schools and in high schools. Our own course was conceived in a conversation between Philip Morrison of Cornell University and Charles Kittel late in 1961. We were encouraged by John Mays and his colleagues of the National Science Foundation and by Walter C. Michels, then the Chairman of the Commission on College Physics. An informal committee was formed to guide the course through the initial stages. The committee consisted originally of Luis Alvarez, William B. Fretter, Charles Kittel, Walter D. Knight, Philip Morrison, Edward M. Purcell, Malvin A. Ruderman, and Jerrold R. Zacharias. The committee met first in May 1962, in Berkeley; at that time it drew up a provisional outline of an entirely new physics course. Because of heavy obligations of several of the original members, the committee was partially reconstituted in January 1964 and now consists of the undersigned. Contributions of others are acknowledged in the prefaces to the individual volumes.

The provisional outline and its associated spirit were a powerful influence on the course material finally produced. The outline covered in detail the topics and attitudes that we believed should and could be taught

Original Preface to the Berkeley Physics Course

to beginning college students of science and engineering. It was never our intention to develop a course limited to honors students or to students with advanced standing. We have sought to present the principles of physics from fresh and unified viewpoints, and parts of the course may therefore seem almost as new to the instructor as to the students.

The five volumes of the course as planned will include:

I. Mechanics (Kittel, Knight, Ruderman)
II. Electricity and Magnetism (Purcell)
III. Waves and Oscillations (Crawford)
IV. Quantum Physics (Wichmann)
V. Statistical Physics (Reif)

The authors of each volume have been free to choose that style and method of presentation which seemed to them appropriate to their subject.

The initial course activity led Alan M. Portis to devise a new elementary physics laboratory, now known as the Berkeley Physics Laboratory. Because the course emphasizes the principles of physics, some teachers may feel that it does not deal sufficiently with experimental physics. The laboratory is rich in important experiments and is designed to balance the course.

The financial support of the course development was provided by the National Science Foundation, with considerable indirect support by the University of California. The funds were administered by Educational Services Incorporated, a nonprofit organization established to administer curriculum improvement programs. We are particularly indebted to Gilbert Oakley, James Aldrich, and William Jones, all of ESI, for their sympathetic and vigorous support. ESI established in Berkeley an office under the very competent direction of Mrs. Mary R. Maloney to assist in the development of the course and the laboratory. The University of California has no official connection with our program, but it has aided us in important ways. For this help we thank in particular two successive Chairmen of the Department of Physics, August C. Helmholz and Burton J. Moyer; the faculty and nonacademic staff of the Department; Donald Coney, and many others in the University. Abraham Olshen gave much help with the early organizational problems.

Your corrections and suggestions will always be welcome.

Eugene D. Commins
Frank S. Crawford, Jr.
Walter D. Knight
Philip Morrison
Alan M. Portis
Edward M. Purcell
Frederick Reif
Malvin A. Ruderman
Eyvind H. Wichmann
Charles Kittel, *Chairman*

Berkeley, California
January 1965

This volume is obviously intended for use as a text. The level is that of students who have had some calculus and are taking more and who have had a high school physics course. At the University of California in Berkeley, students in the physical sciences and engineering start calculus in the first quarter of their freshman year and take a course such as this along with calculus in their second quarter. They have had differential calculus by the start of the physics course and reach integration at least by the middle of the quarter. Such a tight scheduling does require fairly close cooperation with those giving the mathematics course. Of course they have not studied differential equations by this time, and so some material about the solution of simple kinds of differential equations is included in the Mathematical Notes at the ends of Chaps. 3 and 7. There are few enough types to be solved in this kind of a mechanics course so that we believe a student can learn each one of the types.

The teacher will find that the Film Lists have been put all together at the end of the book rather than at the end of each chapter. The Commission on College Physics Resource Letter is a very complete list of films. Special ones have been singled out that seemed especially suitable for the subject of mechanics. In recent years a great many film loops have been made. Some of these are very helpful as short illustrations of special topics; each instructor will find through his own use those that are well suited to his teaching.

Although the problems that have been added in this revision are mostly easier than the ones they have replaced, we have not included very simple problems and plug-in problems. Some of these are valuable in giving the student a little confidence. But we believe that each instructor can make these up for himself or at least find them in other books. No two teachers will want to give a mechanics course in exactly the same way, and the use of special problems gives them a good opportunity for diversity. There are also now several problem books that are useful. Some of them as well as other books on mechanics at this level are listed in the Appendix.

Teaching Notes

There are of course several ways to use this book as a text. One of the ways in which the first edition has apparently rarely been used, but for which we believe there might be a very good use for the entire book, is for a course in mechanics following a one-year noncalculus course, such as one might find in smaller institutions that do not have the facilities for both a calculus and a noncalculus introductory course. For such a course, which might be given to second- or third-year college students, the whole book could well be covered since many of the topics would have been included in less advanced form in the first year.

For the regular introductory section of a general physics course, this book contains too much material, and we urge the instructor to abstain from trying to cover everything. Many introductory courses do not include special relativity, so that the first nine chapters make up a coherent introduction to classical mechanics. But even this much material, if one tries to cover it all, is too great for a nine- or ten-week quarter course or the fraction of a semester that is usually devoted to mechanics. Therefore we give some suggestions below for minimum coverage of chapters. Sometimes it is not desirable to include any electrical or magnetic problems in the beginning course. We believe that the text can be used in this fashion, but it is true that many students find the electrical problems very interesting. Many instructors find it difficult to be ruthless in cutting material. Our own experience is that it is better to cover some material well than to cover more material less well. The advanced sections and the Advanced Topics should give the talented students something with which to stretch their abilities and the students who go on in physics a reference work that can be used in connection with later studies.

With these comments we proceed to the details of the several chapters.

Chapter 1. As in the first edition, this chapter is not an essential part of the study of mechanics, but it may provide interesting reading for those with broader interests. For instructors who wish to assign the reading, it may provide a good place to illustrate the concept of *order of magnitude.*

Chapter 2. Vectors introduce the student to the language that is very useful in physics. As pointed out in the text, the vector product can be omitted here along with the examples of magnetic forces in which $\mathbf{v}$ and $\mathbf{B}$ are not perpendicular. One can proceed to Chap. 6 without needing the vector product and return to it at that time. The scalar product is used often in finding magnitudes and in Chap. 5 on work and energy, so it is highly desirable to introduce it here. In addition it provides a tool for solving numbers of interesting problems. The section on vector derivatives is also useful, but the parts treating the unit vectors $\hat{r}$ and $\hat{\theta}$ can be omitted and introduced much later. Hopefully, circular motion is a good introduction of the dynamics to come.

Chapter 3. This is a long chapter with a good many applications. Newton's laws are introduced in conventional form and we proceed to applications of the Second Law. For a shortened course or one that does not include electrical and magnetic applications, the section on them can be omitted entirely or the magnetic field can be treated only for the case of velocity and magnetic field perpendicular. Conservation of momentum is then introduced through Newton's Third Law. Kinetic energy is referred to in collision problems even though it is not introduced until Chap. 5. Most students have heard of it in high school and do not find difficulty with it; but it can be omitted if desired.

Chapter 4. As pointed out in the text, this chapter is not of the conventional type. Many physicists find appeal in the introduction of galilean transformations, and for those planning to go on to special relativity, it does provide a nice introduction to transformations of coordinates. However, to nonphysics students and to those with limited time, it may be too much "frosting on the cake" and should be omitted. Some reference to accelerated frames of reference and fictitious forces should probably be included, but these can be taken from the first few pages.

Chapter 5. Work and kinetic energy are introduced, first in one dimension and then in three dimensions. The scalar product is really necessary here, but certainly the use of the line integral can be skirted. Potential energy is treated in detail. In a shorter course, the discussion of conservative fields could well be omitted as could the discussion of electrical potential. However, this is an important chapter and should not be hurried through.

Chapter 6. This chapter treats collisions again and introduces the center-of-mass system of reference. Center of mass is an important concept for rigid bodies, and although the center-of-mass system is widely used, a shortened version of a mechanics course could well omit this. The introduction of angular momentum and torque requires the use of the vector product. By this time, students have achieved a level where they can grasp and use the vector product, and if it has been omitted earlier, it can be taken up here. The conservation of angular momentum is an appealing topic to many students.

Chapter 7. Here the Mathematical Notes should be studied first if the students have had difficulty with differential equations. The mass on the spring and the pendulum provide straightforward examples of this important subject of oscillatory motion. In a shortened version, the sections on average values of kinetic and potential energy, damped motion, and forced oscillations can be omitted completely. The laboratory can provide excellent examples of this type of motion. The Advanced Topics on the Anharmonic Oscillator and the Driven Oscillator will be interesting to the more advanced student.

Chapter 8. The present authors believe that an introductory treatment of rigid bodies is valuable to all students. The ideas of torque and angular acceleration about a fixed axis are not difficult, and they provide the student connections with the real, visible world. The simple treatment of the gyro is also valuable; but the introduction of principal axes, products of inertia, and rotating coordinate systems should probably be omitted in most courses.

Chapter 9. Central-force problems are very important. Some instructors may not wish to spend so much time on evaluating the potential inside and outside spherical masses, and this of course can be omitted. They may also find the labor of integrating the r equation of motion too much, in which case they can omit it. They should enjoy the Advanced Topic. There is a good deal that can be cut from this chapter if necessary, but the work of mastering it is very rewarding. The two-body problem and the concept of reduced mass are also useful but again can be omitted in a shortened course.

Chapter 10. This chapter reviews a number of methods of determining the speed of light. For a course in mechanics, this material is not essential. We believe that students will be interested in it, but it could be assigned as outside reading. Then comes the Michelson-Morley experiment, which in a course like this is the most convincing evidence of the need for a change from the galilean transformation. The doppler effect is introduced because of the evidence that the recessional doppler effect provides for high speeds of distant stars, and the chapter closes with a section on the speed of light as the ultimate speed for material objects and the failure of the newtonian formula for kinetic energy. For those with limited time for the study of special relativity, a cursory reading of the chapter might be sufficient.

Chapter 11. In this chapter the Lorentz transformation equations are derived and applied to the most common characteristics of special relativity, length contraction, and time dilation. The velocity transformations are introduced and some examples given. This chapter is the basis for the following chapters, and consequently ample time should be allowed for the study of it.

Chapter 12. The results of Chap. 11 are used to show the need for a change in the definition of momentum, and of relativistic energy, and finally to show the origin of $E = mc^2$. The relation to experiments with high-energy particles and to high-energy nuclear physics needs to be emphasized. At this stage students may be only vaguely aware of, for example, nuclear physics; but the examples are so pertinent to the public today that it should be easy to teach. Finally the subject of particles with zero rest mass will answer the questions of many alert students.

Chapter 13. A number of examples of the subjects developed in the previous chapter are treated here. The center-of-mass system is brought in and its advantages pointed out. In a shortened course all this can be omitted. Good students will be interested in it, and it can be referred to as outside reading in other physics courses treating special relativity.

Chapter 14. In recent years the study of general relativity has become quite popular, and this chapter could provide a bridge to reading in general relativity. It is, of course, not central to the subject of special relativity in the usual sense, but many students may be interested in the difference between gravitational and inertial mass, and almost all will have heard about the tests of general relativity.

The beginning year of college physics is usually the most difficult. In the first year many more new ideas, concepts, and methods are developed than in advanced undergraduate or graduate courses. A student who understands clearly the basic physics developed in this first volume, even if he may not yet be able to apply it easily to complex situations, has put behind him many of the real difficulties in learning physics.

What should a student do who has difficulty in understanding parts of the course and in working problems, even after reading and rereading the text? First he should go back and reread the relevant parts of a high school physics book. "Physics," the PSSC text, is particularly recommended. "Harvard Project Physics" is also very good. Then he should consult and study one of the many physics books at the introductory college level. Many of these are noncalculus texts and so the difficulties introduced by the mathematics will be minimized. The exercises, particularly worked-out exercises, will probably be very helpful. Finally, when he understands these more elementary books, he can go to some of the other books at this level that are referred to in the Appendix. Of course, he should remember that his instructors are the best source for answering his questions and clearing up his misunderstandings.

Many students have difficulty with mathematics. In addition to your regular calculus book, many paperbacks may be helpful. An excellent review of the elements of calculus is available as a short manual of self-instruction: "Quick Calculus," by Daniel Kleppner and Norman Ramsey (John Wiley & Sons, Inc., New York, 1965).

Note to the Student

Units

Every mature field of science and engineering has its own special units for quantities which occur frequently. The *acre-foot* is a natural unit of volume to an irrigation engineer, a rancher, or an attorney in the western United States. The *MeV* or *million electron volts* is a natural unit of energy to a nuclear physicist; the *kilocalorie* is the chemist's unit of energy, and the *kilowatt-hour* is the power engineer's unit of energy. The theoretical physicist will often simply say: Choose units such that the speed of light is equal to unity. A working scientist does not spend much of his time converting from one system of units to another; he spends more time in keeping track of factors of 2 and of plus or minus signs in his calculations. Nor will he spend much time arguing about units, because no good science has ever come out of such an argument.

Physics is carried out and published chiefly in the gaussian cgs and the SI or mks units. Every scientist and engineer who wishes to have easy access to the literature of physics will need to be familiar with these systems.

The text is written in the gaussian cgs system; but a number of references are made to the SI units (Système Internationale), which until recently were more commonly called mks or mksa units. The transformation from cgs to SI units in mechanical problems is easy, as will be explained in the text. However, when one comes to problems in electricity and magnetism there is difficulty. In the text, explanation is given of both systems, and some examples are worked in both systems. It is not clear whether the change to the SI units that began more than twenty years ago will continue. In the current physics literature there still seem to be more papers in the cgs system, which is the reason for retaining it in this volume. In a course such as this, we want to make it as easy as possible for both sceintists and engineers to read the journals, particularly physics journals.

Notation

Physical Constants

Approximate values of physical constants and useful numerical quantities are printed inside the front and back covers of this volume. More precise values of physical constants are tabulated in E. K. Cohen and J. W. M. DuMond, *Rev. Mod. Phys.*, 37:537 (1965) and B. N. Taylor, W. H. Parker, and D. N. Langenberg, *Rev. Mod. Phys.*, 41:375 (1969).

Signs and Symbols

In general we have tried to adhere to the symbols and unit abbreviations that are used in the physics literature—that are, for the most part, agreed upon by international convention.

We summarize here several signs which are used freely throughout the book.

$=$ is equal to

$\approx$ is approximately equal to; is roughly equal to

$\equiv$ is identical with

$\sim$ is of the order of magnitude of

$\propto$ is proportional to

Usage of the signs $\approx$, $\simeq$, and $\sim$ is not standardized, but the definitions we have given are employed fairly widely by physicists. The American Institute of Physics encourages use of the sign $\approx$ where others might write either $\approx$ or $\simeq$. (*Style Manual*, American Institute of Physics, rev. ed., November 1970)

The sign $\sum_{j=1}^{n}$ or $\sum_{j}^{n}$ denotes summation over what stands to the right of Σ over all entries between $j = 1$ and $j = N$. The notation $\sum_{i,j}$ denotes double summation over the two indices i and j. The notation $\sum_{i,j}{}'$ or $\sum_{\substack{i,j \\ i \neq j}}$ denotes summation over all values of i and j except $i = j$.

Order of Magnitude

By this phrase we usually mean "within a factor of 10 or so." Free and bold estimation of the order of magnitude of a quantity characterizes the physicist's work and his mode of speech. It is an exceptionally valuable professional habit, although it often troubles beginning students enormously. We say, for example, that 10^4 is the order of magnitude of the numbers 5500 and 25,000. In cgs units the order of magnitude of the mass of the electron is 10^{-27} g; the accurate value is $(0.910954 \pm 0.000005) \times 10^{-27}$ g.

We say sometimes that a solution includes (is accurate to) terms of order x^2 or E, whatever the quantity may be. This is also written as $O(x^2)$ or $O(E)$. The language implies that terms in the exact solution which involve higher powers (such as x^3 or E^2) of the quantity may be neglected for certain purposes in comparison with the terms retained in the approximate solution.

Prefixes

The following tabulation shows the abbreviation and numerical significance of some frequently used prefixes:

10^{12}	T	tera-	10^{-3}	m	milli-
10^{9}	G	giga-	10^{-6}	μ	micro-
10^{6}	M	mega-	10^{-9}	n	nano-
10^{3}	k	kilo-	10^{-12}	p	pico-

GREEK ALPHABET

A	α	alpha
B	β	beta
Γ	γ	gamma
Δ	δ	delta
E	ϵ	epsilon
Z	ζ	zeta
H	η	eta
Θ	θ	theta
I	ι	iota
K	κ	kappa
Λ	λ	lambda
M	μ	mu
N	ν	nu
Ξ	ξ	xi
O	o	omicron
Π	π	pi
P	ρ	rho
Σ	σ	sigma
T	τ	tau
Υ	υ	upsilon
Φ	ϕ φ	phi
X	χ	chi
Ψ	ψ	psi
Ω	ω	omega

Characters not often used as symbols are shaded; for the most part they are too close in form to roman characters to be of value as independent symbols.

Mechanics

CONTENTS

Introduction

THE NATURAL WORLD

To every man the natural world seems immense and complex, the stage for a startling diversity of appearances and events. These impressions are supported by estimates of the general order of magnitude of the values of interesting quantities concerning the natural world. At this stage we shall not enter into the arguments and measurements that lead to the figures given. The most remarkable thing about these numbers is that we know them at all; it is not of pressing importance that some of them are known only approximately.

The universe is immense. From astronomical observations we infer the value 10^{28} centimeters (cm) or 10^{10} light years (yr) for a characteristic dimension loosely called the *radius* of the universe. The value is uncertain by perhaps a factor of 3. For comparison, the distance of the earth from the sun is 1.5×10^{13} cm and the radius of the earth is 6.4×10^{8} cm.

The number of atoms in the universe is very large. The total number of protons and neutrons in the universe, with an uncertainty perhaps of a factor of 100, is believed to be of the order of 10^{80}. Those in the sun number 1×10^{57}, and those in the earth 4×10^{51}. The total in the universe would provide about $10^{80}/10^{57}$ (or 10^{23}) stars equal in mass to our sun. [For comparison, the number of atoms in an atomic weight (Avogadro's number) is 6×10^{23}.] Most of the mass of the universe is believed to lie in stars, and all known stars have masses between 0.01 and 100 times that of our sun.

Life appears to be the most complex phenomenon in the universe. Man, one of the more complex forms of life, is composed of about 10^{16} cells. A cell is an elementary physiological unit that contains about 10^{12} to 10^{14} atoms. Every cell of every variety of living matter is believed to contain at least one long molecular strand of DNA (deoxyribonucleic acid) or of its close relative RNA (ribonucleic acid). The DNA strands in a cell hold all the chemical instructions, or genetic information, needed to construct a complete man, bird, etc. In a DNA molecule, which may be composed of 10^{8} to 10^{10} atoms, the precise arrangement of the atoms may vary from individual to individual; the arrangement always varies from species to species.[1] More than 10^{6} species have been described and named on our planet.

Inanimate matter also appears in many forms. Protons, neutrons, and electrons combine to form about one-hundred

[1] The term *species* is defined roughly by the statement that two populations are different species if some describable difference(s) can be found between them and if they do not interbreed in a state of nature.

different chemical elements and about 10^3 identified isotopes. The individual elements have been combined in various proportions to form perhaps 10^6 or more identified, differentiated chemical compounds, and to this number may be added a vast number of liquid and solid solutions and alloys of various compositions having distinctive physical properties.

Through experimental science we have been able to learn all these facts about the natural world, to classify the stars and to estimate their masses, compositions, distances, and velocities; to classify living species and to unravel their genetic relations; to synthesize inorganic crystals, biochemicals, and new chemical elements; to measure the spectral emission lines of atoms and molecules over a frequency range from 100 to 10^{20} cycles per second (cps);[1] and to create new fundamental particles in the laboratory.

These great accomplishments of experimental science were achieved by men of many types: patient, persistent, intuitive, inventive, energetic, lazy, lucky, narrow, and with skilled hands. Some preferred to use only simple apparatus; others invented or built instruments of great refinement, size, or complexity. Most of these men had in common only a few things: They were honest and actually made the observations they recorded, and they published the results of their work in a form permitting others to duplicate the experiment or observation.

THE RÔLE OF THEORY

The description we have given of the natural universe as immense and complex is not the whole story, for theoretical understanding makes several parts of the world picture look much simpler. We have gained a remarkable understanding of some central and important aspects of the universe. The areas that we believe we understand (summarized below), together with the theories of relativity and of statistical mechanics, are among the great intellectual achievements of mankind.

1 The laws of classical mechanics and gravitation (Volume 1), which allow us to predict with remarkable accuracy the motions of the several parts of the solar system (including comets and asteroids), have led to the prediction and discovery of new planets. These laws suggest possible mechanisms for the formation of stars and galaxies, and, together

[1]The approved unit for cycles per second has become Hertz (Hz), and so this phrase could have been written "from 100 to 10^{20} Hz."

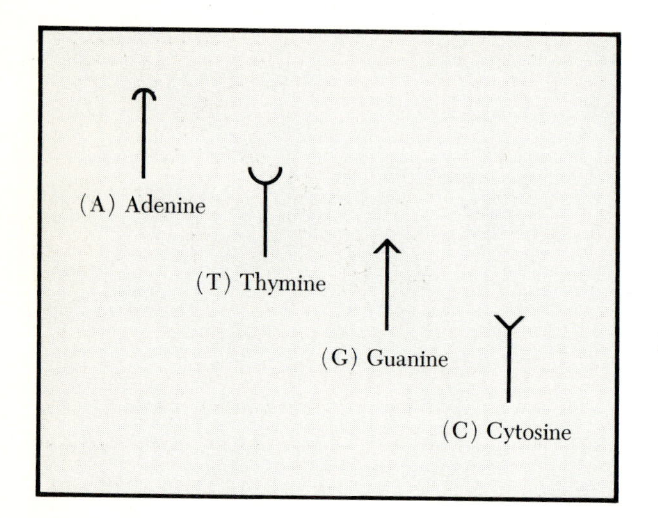

FIG. 1.1 (*a*) Schematic representation of the four nucleotide bases from which the DNA molecule is derived.

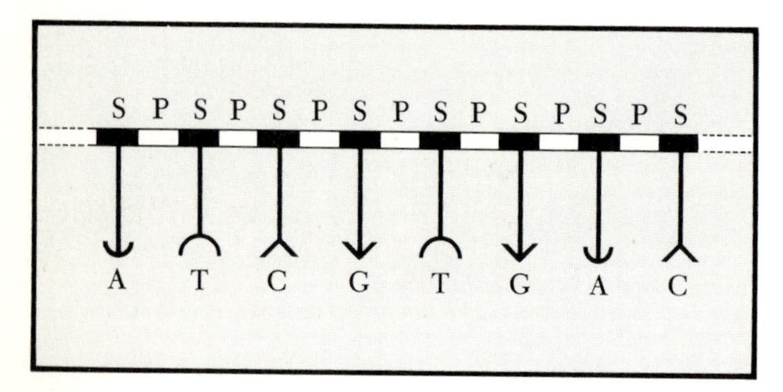

(*b*) The nucleotides are connected to sugar groups S which, in turn, are bound to phosphate groups P to form a chain.

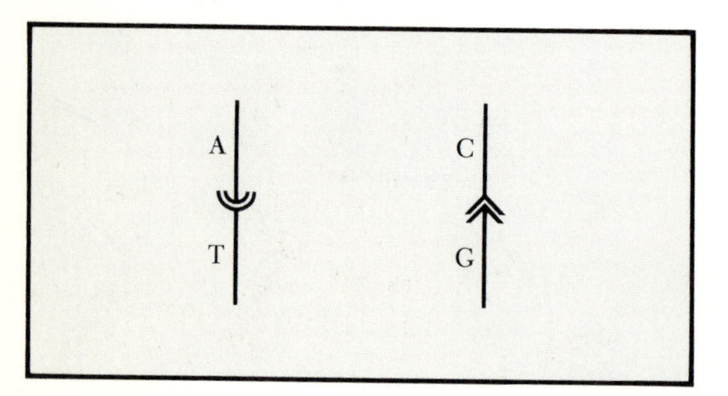

(*c*) The complete DNA molecule is composed of a double chain in the form of a helix. The two strands are connected by hydrogen bonds between adenine and thymine groups or between guanine and cytosine groups.

with the laws of radiation, they give a good account of the observed connection between the mass and luminosity of stars. The astronomical applications of the laws of classical mechanics are the most beautiful but not the only successful applications. We use the laws constantly in everyday life and in the engineering sciences. Our contemporary ventures into space and the use of satellites are based upon refined applications of the laws of classical mechanics and gravitation.

2 The laws of quantum mechanics (Volume 4) give a very good account of atomic phenomena. For simple atoms predictions have been made that agree with experiment to 1 part in 10^5 or better. When applied to large-scale terrestrial and celestial events, the laws of quantum mechanics result in predictions indistinguishable from the laws of classical mechanics. Quantum mechanics provides, in principle, a precise theoretical basis for all of chemistry and metallurgy and for much of physics, but often we cannot handle the equations on existing or foreseeable computers. In some fields nearly all the problems seem too difficult for a direct theoretical attack based on first principles.

3 The laws of classical electrodynamics, which give an excellent account of all electric and magnetic effects, except on the atomic scale, are the basis of the electrical engineering and communications industries. Electric and magnetic effects on the atomic scale are described exactly by the theory of quantum electrodynamics. Classical electrodynamics is the subject of Volumes 2 and 3; some aspects of quantum electrodynamics are touched on in Volume 4, but a complete discussion of the field must be deferred until a later course.

4 At another, narrower level, the principle of operation of the genetic code is understood—in particular, the mechanism of storage of genetic information. We find that the information storage of the cell of a simple organism exceeds that of the best present-day commercial computers. In nearly all life on our planet the complete coding of genetic information is carried in the DNA molecule by a double linear sequence (possessing 10^6 to 10^9 entries, depending on the organism) of only four different molecular groups, with specific but simple rules governing the pairing of members opposite each other in the double sequence (see Fig. 1.1). These matters are a part of the subject of molecular biology.

The physical laws and theoretical understanding mentioned in the above summaries are different in character from the direct results of experimental observations. The laws, which summarize the essential parts of a large number of observations, allow us to make successfully certain types of predictions, limited in practice by the complexity of the system. Often the laws suggest new and unusual types of experiments. Although the laws can usually be stated in compact form,[1] their application may sometimes require lengthy mathematical analysis and computation.

There is another aspect of the fundamental laws of physics: Those laws of physics that we have come to understand have an attractive simplicity and beauty.[2] This does not mean that everyone should stop doing experiments, for the laws of physics have generally been discovered only after painstaking and ingenious experiments. The statement does mean that we shall be greatly surprised if future statements of physical theory contain ugly and clumsy elements. The aesthetic quality of the discovered laws of physics colors our expectations about the laws still unknown. We tend to call a hypothesis attractive when its simplicity and elegance single it out among the large number of conceivable theories.

In this course we shall make an effort to state some of the laws of physics from viewpoints that emphasize the features of simplicity and elegance. This requires that we make considerable use of mathematical formulations, although at the present level of study this use normally will not exceed the bounds of introductory calculus. As we go along, we shall try also to give some of the flavor of good experimental physics, although this is very hard to do in a textbook. The research laboratory is the natural training ground in experimental physics.

GEOMETRY AND PHYSICS

Mathematics, which permits the attractive simplicity and compactness of expression necessary for a reasonable discussion of

[1]The first sentence of a short paperback is: "These lectures will cover all of physics." R. Feynman, "Theory of Fundamental Processes," W. A. Benjamin, Inc., New York, 1961.

[2]"It seems that if one is working from the point of view of getting beauty in one's equations, and if one has really a sound insight, one is on a sure line of progress." P. A. M. Dirac, *Scientific American*, 208 (5):45–53 (1963). But most physicists feel the real world is too subtle for such bold attacks except by the greatest minds of the time, such as Einstein or Dirac or a dozen others. In the hands of a thousand others this approach has been limited by the inadequate distribution among men of "a sound insight."

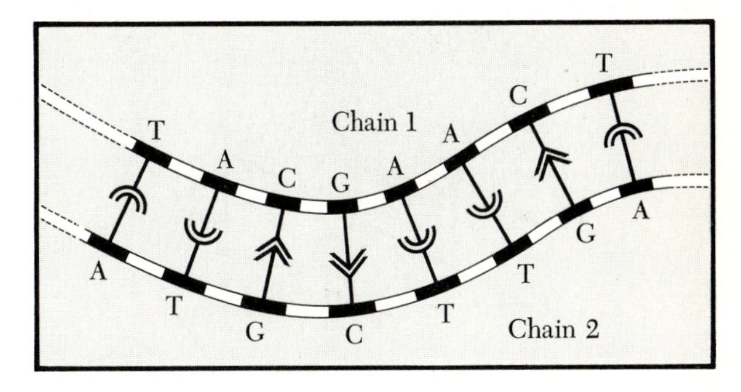

(*d*) All genetic information in the cell is contained in the order in which the nucleotide bases occur.

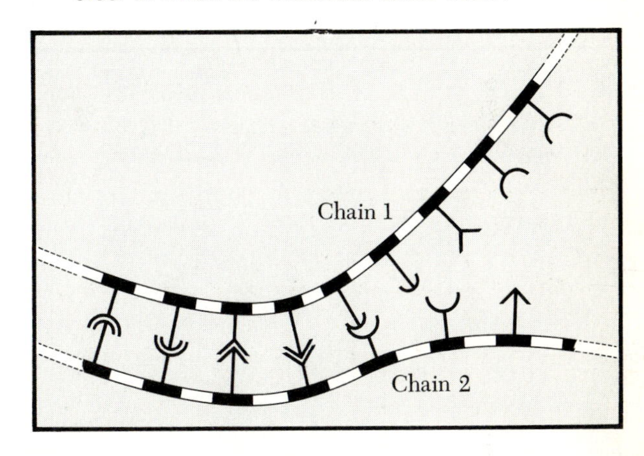

(*e*) When the cell reproduces, each DNA molecule splits into two separate chains.

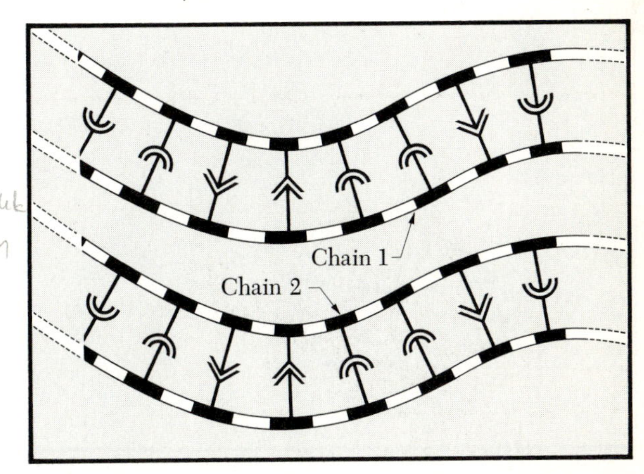

(*f*) Each free chain then forms its complement from existing cell material to produce two *identical* new DNA molecules.

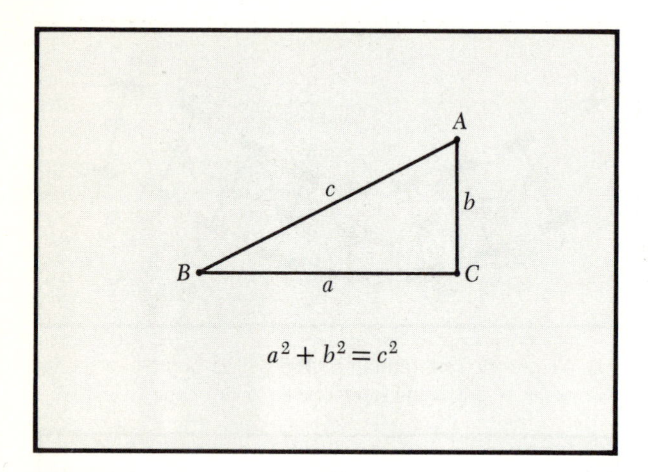

FIG. 1.2 Do the axioms of euclidean geometry, from which the pythagorean theorem is logically derived, accurately describe the physical world? Only experiment can decide.

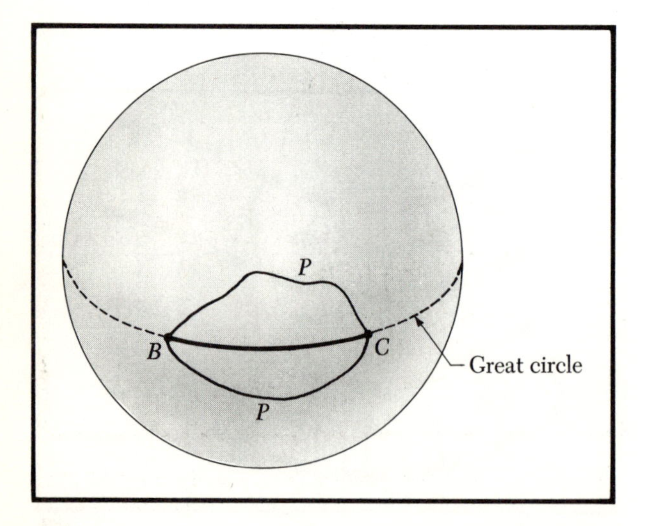

FIG. 1.3 The shortest, "straight-line" distance between points B and C on a sphere lies along the great circle through these points and not along any other paths P.

the laws of physics and their consequences, is the language of physics. It is a language with special rules. If the rules are obeyed, only correct statements can be made: The square root of 2 is 1.414 . . . , or $\sin 2a = 2 \sin a \cos a$.

We must be careful not to confuse such truths with exact statements about the physical world. It is a question of experiment, rather than contemplation, to see whether the measured ratio of the circumference to the diameter of a physical circle really is 3.14159. . . . Geometrical measurement is basic to physics, and we must decide such questions before proceeding to use euclidean or any other geometry in the description of nature. Here certainly is a question about the universe: Can we assume for physical measurements the truth of the axioms and theorems of Euclid?

We can say only a few simple things about the experimental properties of space without becoming involved in difficult mathematics. The most famous theorem in all mathematics is that attributed to Pythagoras: *For a right-angled triangle the square of the hypotenuse equals the sum of the squares of the adjacent sides* (Fig. 1.2). Does this mathematical truth, which assumes the validity of euclidean geometry, also hold true in the physical world? Could it be otherwise? Contemplation of the question is insufficient, and we must appeal to experiment for an answer. We give arguments that are somewhat incomplete because here we are not able to use the mathematics of curved three-dimensional space.

Consider first the situation of two-dimensional beings who live in a universe that is the surface of a sphere. Their mathematicians have described to them the properties of spaces of three dimensions or even more, but they have as much difficulty in developing an intuitive feeling about such matters as we have in picturing four-dimensional space. How can they determine whether they live on a curved surface? One way is to test the axioms of plane geometry by trying to confirm experimentally some of the theorems in Euclid. They may construct a straight line as the shortest path between any two points B and C on the surface of a sphere; we would describe such a path as a *great circle*, as shown in Fig. 1.3. They can go on to construct triangles and to test the pythagorean theorem. For a very small triangle, each of whose sides is small in comparison with the radius of the sphere, the theorem would hold with great but not perfect accuracy; for a large triangle striking deviations would become apparent (see Figs. 1.4 to 1.6).

If B and C are points on the equator of the sphere, the "straight line" connecting them is the section of the equator from B to C. The shortest path from C on the equator to the north pole A is the line of fixed longitude that meets the equator BC at a right angle. The shortest path from A to B is a path of fixed longitude that also meets the equator BC at a right angle. Here we have a right triangle with $b = c$. The pythagorean theorem is clearly invalid on the sphere because c^2 cannot now be equal to $b^2 + a^2$; further, the sum of the interior angles of the triangle ABC is always greater than 180°. Measurements made on the curved surface by its two-dimensional inhabitants enable them to demonstrate for themselves that the surface is indeed curved.

It is always possible for the inhabitants to say that the laws of plane geometry adequately describe their world, but the trouble lies with the meter sticks used to measure the shortest path and thus define the straight line. The inhabitants could say that the meter sticks are not constant in length but stretch and shrink as they are moved to different places on the surface. Only when it is determined by continued measurements in different ways that the same results always hold does it become evident that the simplest description of why euclidean geometry fails lies in the curvature of the surface.

The axioms of plane geometry are not self-evident truths in this curved two-dimensional world; they are not truths at all. We see that the actual geometry of the universe is a branch of physics that must be explored by experiment. We do not customarily question the validity of euclidean geometry to describe measurements made in our own three-dimensional world because euclidean geometry is such a good approximation to the geometry of the universe that any deviations from it do not show up in practical measurements. This does not mean that the applicability of euclidean geometry is self-evident or even exact. It was suggested by the great nineteenth-century mathematician Carl Friedrich Gauss that the euclidean flatness of three-dimensional space should be tested by measuring the sum of the interior angles of a large triangle; he realized that if three-dimensional space is curved, the sum of the angles of a *large enough* triangle might be significantly different from 180°.

Gauss[1] used surveying equipment (1821–1823) to measure accurately the triangle in Germany formed by Brocken, Hohehagen, and Inselberg (Fig. 1.7). The longest side of the triangle

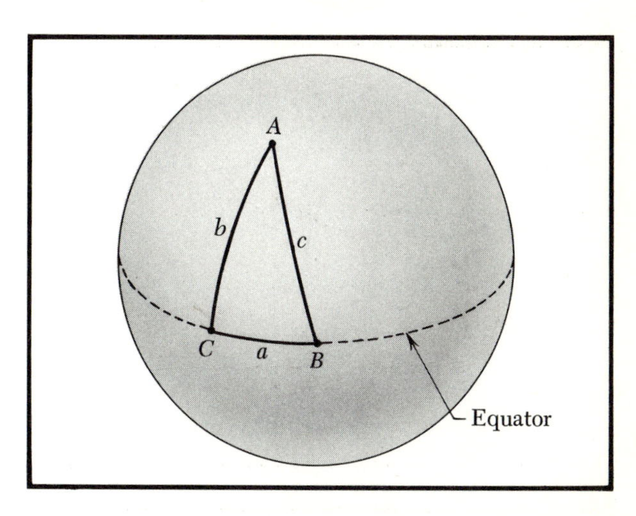

FIG. 1.4 Given three points ABC, the two-dimensional beings could construct triangles with "straight lines" as sides. They would find that for small right triangles $a^2 + b^2 \approx c^2$ and the sum of the angles of the triangle is slightly greater than 180°.

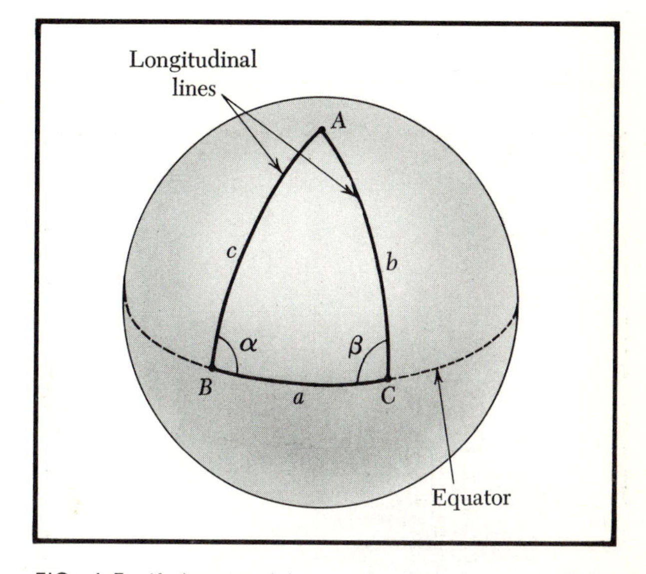

FIG. 1.5 If they used larger triangles, the sum of the angles would become increasingly greater than 180°. Here, with B and C on the equator, and A on the pole, α and β are *both* right angles. Obviously $a^2 + b^2 \neq c^2$, because b is equal to c.

[1] C. F. Gauss, "Werke," vol. 9, B. G. Teubner, Leipzig, 1903; see especially pp. 299, 300, 314, and 319. The collected works of Gauss are a remarkable example of how much a gifted man can accomplish in a lifetime.

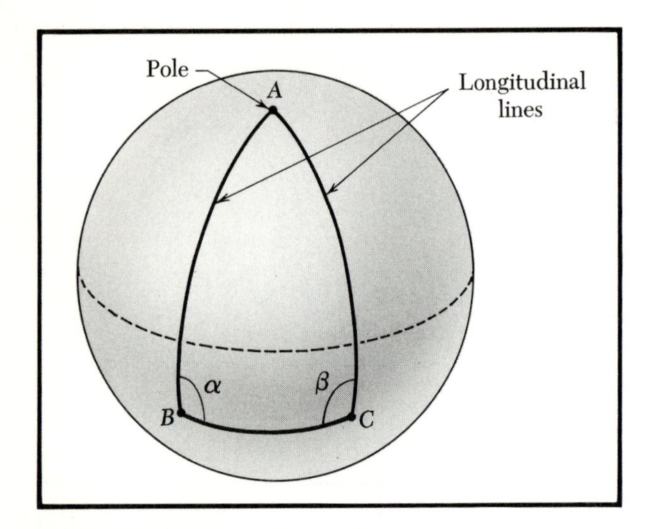

FIG. 1.6 For this triangle, with B and C below the equator, $\alpha + \beta > 180°$, which can only happen because the two-dimensional "space" of the spherical surface is curved. A similar argument can be applied to three-dimensional space. The radius of curvature of the two-dimensional space shown here is just the radius of the sphere.

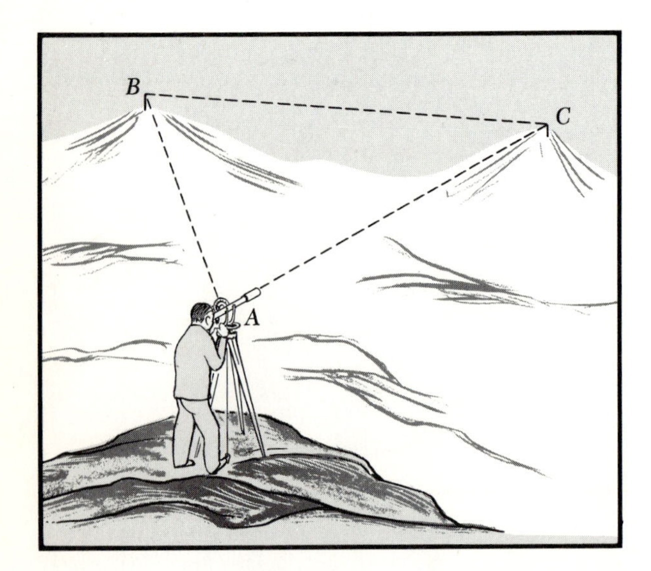

FIG. 1.7 Gauss measured the angles of a triangle with vertices on three mountain tops and found no deviation from 180° within the accuracy of his measurements.

was about 100 kilometers (km). The measured interior angles were

$$\begin{array}{rr} & 86°13'58.366'' \\ & 53°\ \ 6'45.642'' \\ & 40°39'30.165'' \\ \hline \text{Sum} & 180°00'14.173'' \end{array}$$

(We have not found a statement about the estimated accuracy of these values; it is likely that the last two decimal places are not significant.) Because the surveying instruments were set up *locally* horizontal at all three vertices, the three horizontal planes were not parallel. A calculated correction called the *spherical excess*, which amounts to 14.853″ of arc, must be subtracted from the sum of the angles. The sum thus corrected,

$$179°59'59.320''$$

differs by 0.680″ of arc from 180°. Gauss believed this to lie within the observational error, and he concluded that space was euclidean within the accuracy of these observations.

We saw in the earlier example that euclidean geometry adequately described a small triangle on the two-dimensional sphere but departures became more evident as the scale increased. To see if our own space is indeed flat we need to measure very large triangles whose vertices are formed by the earth and distant stars or even galaxies. But we are faced with a problem: Our position is fixed by that of the earth and we are not yet free to wander through space with instruments to measure astronomical triangles. How can we test the validity of euclidean geometry to describe measurements in space?

Estimates of the Curvature of Space

Planetary Predictions A first lower limit of about 5×10^{17} cm to the radius of curvature of our own universe is implied by the consistency of astronomical observations within the solar system. For example, the positions of the planets Neptune and Pluto were inferred by calculation before their visual confirmation by telescopic observation. Small perturbations of the orbits of the known planets led to this discovery of Neptune and Pluto very close to the positions calculated for them. The outermost planet in the solar system is Pluto, and we can easily believe that a slight error in the laws of geometry would have destroyed this coincidence. The average radius of the orbit of Pluto is 6×10^{14} cm; the closeness of the coincidence between the predicted and observed positions implies a radius of curvature of space of at least 5×10^{17} cm. An infinite radius of curvature (flat space) is not incompatible with the data. It would take us too far from our present purpose to discuss the numerical details of how the estimate of 5×10^{17} cm is arrived at or to define precisely what is meant by the radius of curvature of a three-dimensional space. The two-dimensional analog of the surface of a sphere can be used in this emergency as a useful crutch.

Trigonometrical Parallax Another argument was suggested by Schwarzschild.[1] In two observations taken 6 months apart, the position of the earth relative to the sun has changed by 3×10^{13} cm, the diameter of the earth's orbit. Suppose that at these two times we observe a star and measure the angles α and β as in Fig. 1.8. (Here α and β are the Greek characters alpha and beta.) If space is flat, the sum of the angles $\alpha + \beta$ is always less than 180° and the sum approaches this value as the star becomes infinitely distant. One-half of the deviation of $\alpha + \beta$ from 180° is called the *parallax*. But in a curved space it is not necessarily true that $\alpha + \beta$ is always less than 180°. An example is shown in Fig. 1.6.

We return to our hypothetical situation of two-dimensional astronomers living on the surface of a sphere to see how they discover that their space is curved from a measurement of the sum $\alpha + \beta$. From our previous discussion of the triangle ABC we see that when the star is a quarter of a circumference away, $\alpha + \beta = 180°$; when the star is nearer, $\alpha + \beta < 180°$; and when it is farther away, $\alpha + \beta > 180°$. The astronomer need merely look at stars more and more distant and measure $\alpha + \beta$ to see when the sum begins to exceed 180°. The same argument is valid within our three-dimensional space.

There is no observational evidence that $\alpha + \beta$ as measured by astronomers is ever greater than 180°, after an appropriate correction is made for the motion of the sun relative to the center of our galaxy. Values of $\alpha + \beta$ less than 180° are used to determine by triangulation the distances of nearby stars. Values less than 180° can be observed out to about 3×10^{20} cm,† the limit of angle measurement with present telescopes. It cannot be inferred directly from this argument

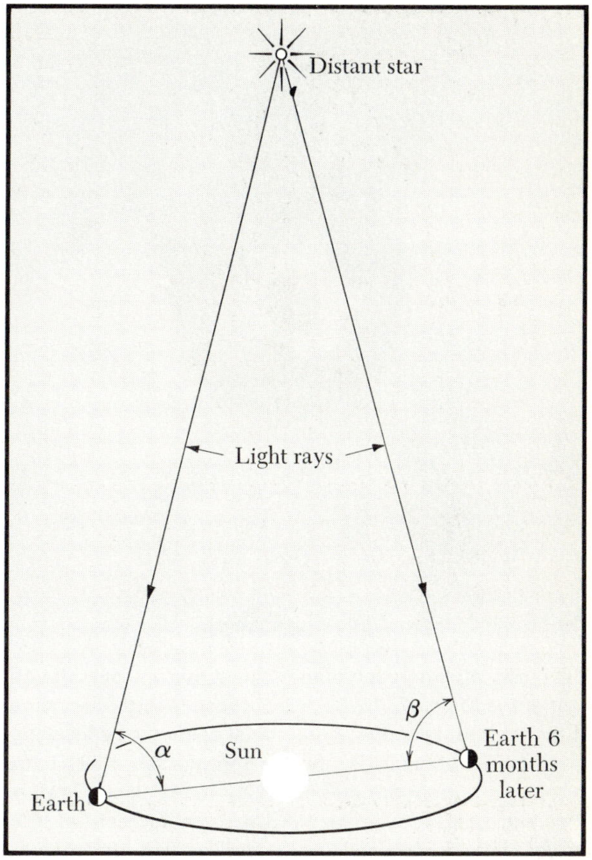

FIG. 1.8 Schwarzschild's demonstration that on a flat surface $\alpha + \beta < 180°$. The *parallax* of a star is defined as $\frac{1}{2}(180° - \alpha - \beta)$.

[1] K. Schwarzschild, *Vierteljahrsschrift der astronomischen Ges.*, **35**:337 (1900).

† It may be objected that the distance measurements themselves assume that euclidean geometry is applicable. Other methods of estimating distance are available, however, and are discussed in modern texts on astronomy.

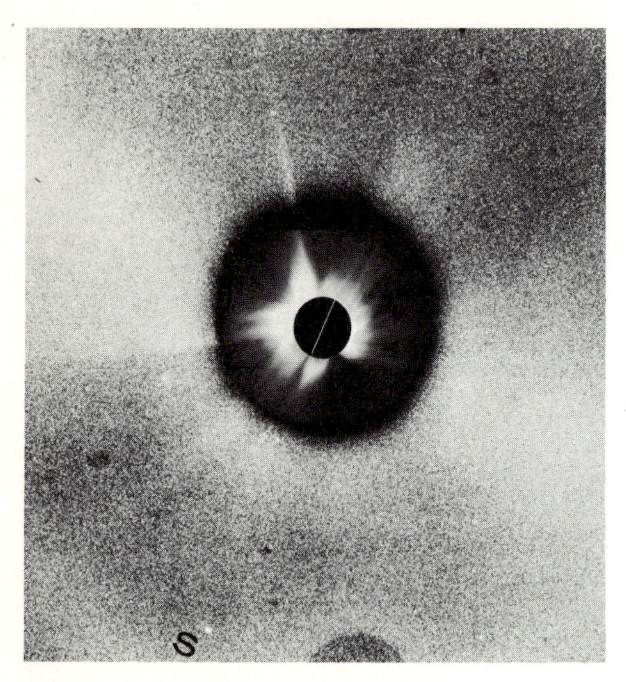

FIG. 1.9 A photograph of the solar corona in near infrared light at the March 7, 1970, solar eclipse records the image of the fourth-magnitude M star, ϕ Acquarii (just above and to right of S), about 11 sun's radii from the sun. Half circles at top and bottom are pressure plate marks. Insert in dark occulting disk is Gordon Newkirk's photograph of the eclipse, which has been used to orient this photograph. *(Photograph by the courtesy of Carl Lilliequist and Ed Schmahl—partial financial support of this experiment from the Department of Astrogeophysics, University of Colorado)*

that the radius of curvature of space must be larger than 3×10^{20} cm; for some types of curved space other arguments are needed. The answers come out finally that the radius of curvature (as determined by triangulation) must be larger than 6×10^{19} cm.

At the beginning of this chapter we said that a characteristic dimension associated with the universe is inferred to have a value of the order of 10^{28} cm or 10^{10} light yr. This number corresponds, for example, to the distance light would travel in a time equal to the age of the universe as inferred from observations that would be too lengthy to present here.[1] The most elementary interpretation of this length calls it the radius of the universe; another possible interpretation calls it the radius of curvature of space. Which is it? This is a cosmological question. (An excellent introduction to the speculative science of cosmology is given in the book by Bondi cited in the Further Reading section at the end of this chapter.) We summarize our belief about the radius of curvature of space by the statements that it is not smaller than 10^{28} cm and that we do not know that space on a large scale is not flat.

The foregoing observations bear upon the average radius of curvature of space and are not sensitive to bumps that are believed to exist in the immediate neighborhood of individual stars and that contribute a local roughness to the otherwise flat, or slightly curved, space. Experimental data that bear upon this question are extremely hard to acquire, even for the neighborhood of our sun. By careful and difficult observations of stars visible near the edge of the sun during a solar eclipse, it has been established that light rays are slightly curved when they pass near the edge of the sun and, by inference, close to any similarly massive star (see Figs. 1.9 and 1.10). For a grazing ray the angle of bend is very slight, amounting to only 1.75″. Thus as the sun moves through the sky the stars that are almost eclipsed, if we could see them in the daytime, would appear to spread out very slightly from their normal positions. This observation merely says that the light moves in a curved path near the sun; it does not by itself insist upon the unique interpretation that the space around the sun is curved. Only with accurate measurements by various measuring instruments close to the sun's surface could we establish directly that a curved space is the most efficient and natural description. One other kind of observation bears upon the possibility of a curved space. The orbit of Mercury, the planet nearest the sun, differs very

[1] One evidence for this is mentioned in Chap. 10 (page 319).

slightly from that predicted by application of Newton's laws of universal gravitation and motion (see Fig. 14.9). Could this be an effect of curved space near the sun? To answer such a question we would have to know how a possible curvature would affect the equations of motion for Mercury, and this involves more than just geometry. [These topics are discussed further (but briefly) in Chap. 14.]

In a remarkable and beautiful series of papers, Einstein [A. Einstein, *Berl. Ber.*, 778, 799, 844 (1915); *Ann. d. Phys.* **49:** 769 (1916)] described a theory of gravitation and geometry, the general theory of relativity, which predicted, in quantitative agreement with the observations, just the two effects described above. There are still few confirmations of the geometric predictions of the theory. However, despite the meager evidence, the essential simplicity of the general theory has made it widely accepted, although in recent years there has been considerable research in this field (see Chap. 14).

Geometry on a Smaller Scale From astronomical measurements we concluded that euclidean geometry gives an extraordinarily good description of measurements of lengths, areas, and angles, at least until we reach the enormous lengths of the order of 10^{28} cm. But so far nothing has been said about the use of euclidean geometry to describe very small configurations comparable in size to the 10^{-8} cm of an atom or the 10^{-12} cm of a nucleus. The question of the validity of euclidean geometry ultimately must be phrased as follows: Can we make sense of the subatomic world, can we make a successful physical theory to describe it, while assuming that euclidean geometry is valid? If we can, then there is no reason at present to question euclidean geometry as a successful approximation. We shall see in Volume 4 that the theory of atomic and subatomic phenomena does not seem to lead to any paradoxes that have thus far blocked our understanding of them. Many facts are not understood, but none appear to lead to contradictions. In this sense euclidean geometry stands the test of experiment down at least to 10^{-13} cm.

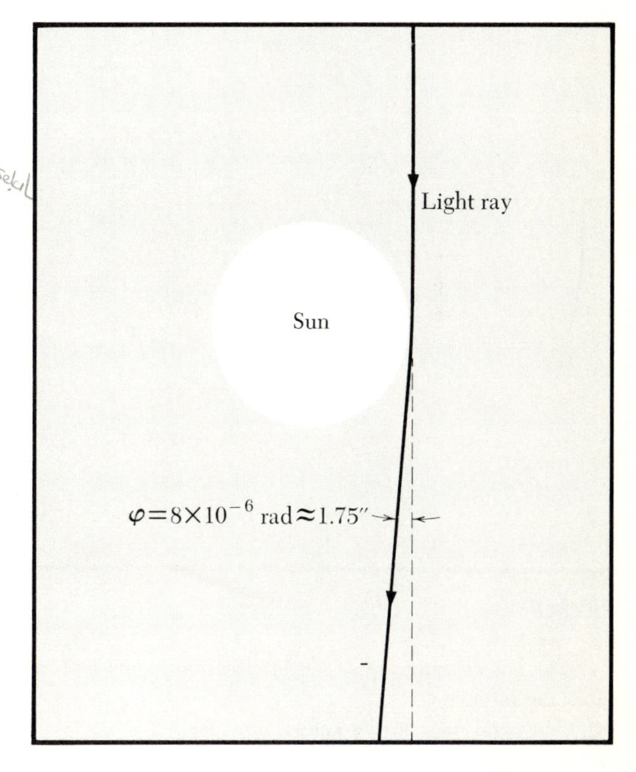

FIG. 1.10 The bending of light by the sun was predicted by Einstein in 1915 and verified by observation shortly afterward.

INVARIANCE

We shall summarize some of the consequences of the experimental validity of euclidean geometry for empty space. The *homogeneity* and *isotropy* of euclidean space can be expressed by two invariance principles, which, in turn, imply two fundamental conservation principles.

Invariance under Translation By this principle we mean that empty space is homogeneous, i.e., that it is not different from point to point. If figures are moved without rotation from one location to another, there is no change in their size or geometric properties. We assume also that the physical properties of an object, such as its inertia or the forces between its constituent particles, do not change merely upon displacing the object to another region of empty space. Thus the natural frequency of a tuning fork or the characteristic spectrum lines of an atom are not altered by such displacement.

Invariance under Rotation By experiment it is known that empty space is isotropic to high precision, so that all directions are equivalent. Geometric and physical properties are unaltered by the reorientation in direction of an object in empty space. It is possible to imagine a space that is not isotropic; for example, the speed of light in some direction could be greater than its value in another direction at right angles to the first. There is no evidence in free space for an effect of this kind; within a crystal, however, many such *anisotropic* effects are encountered. In regions of space close to massive stars and other strong sources of gravitation, effects can be observed that may be interpreted as slight departures from homogeneity and isotropy of space. (We have alluded to two such effects in the preceding section, and there are others.)

The property of invariance under translation leads to the *conservation of linear momentum;* invariance under rotation leads to the *conservation of angular momentum.* These conservation principles are developed in Chaps. 4 and 6, and the concept of invariance is developed in Chaps. 2 and 4.

The foregoing lengthy discussion about geometry and physics is an example of the types of questions that physicists must ask about the basic character of our universe. But we shall not treat such matters further at this level of our study.

PROBLEMS

1. *The known universe.* Using information in the text, estimate the following:

(*a*) The total mass in the known universe.

Ans. $\approx 10^{56}$ g.

(*b*) The average density of matter in the universe.

Ans. $\sim 10^{-29}$ g/cm^3, equivalent to 10 hydrogen atoms/m^3.

(*c*) The ratio of the radius of the known universe to that of a proton. Take the radius of the proton to be 1×10^{-13} cm and the mass of the proton to be 1.7×10^{-24} g.

2. *Signals across a proton.* Estimate the time required for a signal traveling with the speed of light to move a distance equal to the diameter of a proton. Take the diameter of the

proton to be 2×10^{-13} cm. (This time is a convenient reference interval in the physics of elementary particles and nuclei.)

3. *Distance of Sirius.* The parallax of a star is one-half the angle subtended at the star by the extreme points in the earth's orbit around the sun. The parallax of Sirius is 0.371″. Find its distance in centimeters, light years, and parsecs. One parsec is the distance to a star whose parallax is 1″. (See the table of values inside the front and back covers.)

Ans. 8.3×10^{18} cm; 8.8 light yr; 2.7 parsecs.

4. *Size of atoms.* Using the value of Avogadro's number given in the table inside the back cover of the book and your estimate of an average density for common solids, estimate roughly the diameter of an average atom, that is, the dimension of the cubical space filled by the atom.

5. *Angle subtended by moon.* Obtain a millimeter scale and, when viewing conditions are favorable, try the following experiment: Hold the scale at arm's length and measure the diameter of the moon; measure the distance from the scale to your eye. (The radius of the moon's orbit is 3.8×10^{10} cm, and the radius of the moon itself is 1.7×10^{8} cm.)

(*a*) If you were able to try the measurement, what was the result?

(*b*) If the measurement could not be made, from the data given above calculate the angle subtended by the moon at the earth. *Ans.* 9×10^{-3} radians (rad).

(*c*) What is the angle subtended at the moon by the earth? (see p. 52, Chap. 2.) *Ans.* 3.3×10^{-2} rad.

6. *Age of the universe.* Assuming the radius of the universe given on page 4, find the age of the universe from the assumption that a star now on the radius has been traveling outward from the center since the beginning at $0.6c = 1.8 \times 10^{10}$ cm/s (c = speed of light in free space).

Ans. $\approx 2 \times 10^{10}$ yr.

7. *Angles in a spherical triangle.* Find the sum of the angles in the spherical triangle shown in Fig. 1.5, assuming A is at the pole and a = radius of sphere. In order to find the angle at A, consider what would be the value of a in order for the angle to be 90°.

FURTHER READING

These first two references are contemporary texts for the high school level. They are excellent for review and clarification of concepts. The second reference contains much material on history and philosophy.

Physical Science Study Committee (PSSC), "Physics," chaps. 1–4, D. C. Heath and Company, Boston, 1965.

F. J. Rutherford, G. Holton, and F. J. Watson, "Project Physics Course," Holt, Rinehart and Winston, Inc., New York, 1970. A product of Harvard Project Physics (HPP).

O. Struve, B. Lynds, and H. Pillans, "Elementary Astronomy," Oxford University Press, New York, 1959. Emphasizes the main ideas of physics in relation to the universe; an excellent book.

"Larousse Encyclopedia of Astronomy," Prometheus Press, New York, 1962. This is a beautiful and informative book.

H. Bondi, "Cosmology," 2d ed., Cambridge University Press, New York, 1960. Brief, clear, authoritative account, with emphasis on the observational evidence, but lacking in substantial recent work.

D. W. Sciama, "Modern Cosmology," Cambridge University Press, New York, 1971. This account includes recent developments.

Robert H. Haynes and Philip C. Hanawalt, "The Molecular Basis of Life," W. H. Freeman and Company, San Francisco, 1968. A collection of *Scientific American* articles with some relating text.

Gunther S. Stent, "Molecular Genetics," W. H. Freeman and Company, San Francisco, 1971. An introductory account.

Ann Roe, "The Making of a Scientist," Dodd, Mead & Company, New York, 1953; Apollo reprint, 1961. This is an excellent sociological study of a group of leading American scientists of the late 1940s. There have probably been some significant changes in the scientific population since the book was first published in 1953.

Bernice T. Eiduson, "Scientists: Their Psychological World," Basic Books, Inc., Publishers, New York, 1962.

A. Einstein, autobiographical notes in "Albert Einstein: Philosopher-Scientist," P. A. Schilpp (ed.), Library of Living Philosophers, Evanston, 1949. An excellent short autobiography. It is a pity that there are so few really great biographies of outstanding scientists, such as that of Freud by Ernest Jones. There is little else comparable in depth and in honesty to the great literary biographies, such as "James Joyce," by Richard Ellman. The autobiography of Charles Darwin is a remarkable exception. Writers about scientists appear to be overly intimidated by Einstein's sentence: "For the essential of a man like myself lies precisely in *what* he thinks and *how* he thinks, not in what he does or suffers."

L. P. Wheeler, "Josiah Willard Gibbs; The History of a Great Mind," Yale University Press, New Haven, Conn., 1962.

E. Segrè, "Enrico Fermi, Physicist," The University of Chicago Press, Chicago, 1971.

Experimental Tools of Physics. The photographs on this and the following pages show some of the instruments and machines that are contributing actively to the advancement of the physical sciences.

A nuclear magnetic resonance laboratory for chemical structure studies. (*ASUC Photography*)

Study of nuclear magnetic resonance spectra: a sample is shown spinning rapidly between the polepieces of an electromagnet to average out magnetic field variations. (*Esso Research*)

Operator in a nuclear magnetic resonance laboratory ready to place a sample in the probe in the variable temperature controller in which the sample is spun. (*Esso Research*)

A magnet constructed of superconducting wire, for operation at low temperature. The coils shown are rated to produce a magnetic field of 54,000 gauss. Such apparatus is the heart of a modern low-temperature laboratory. (*Varian Associates*)

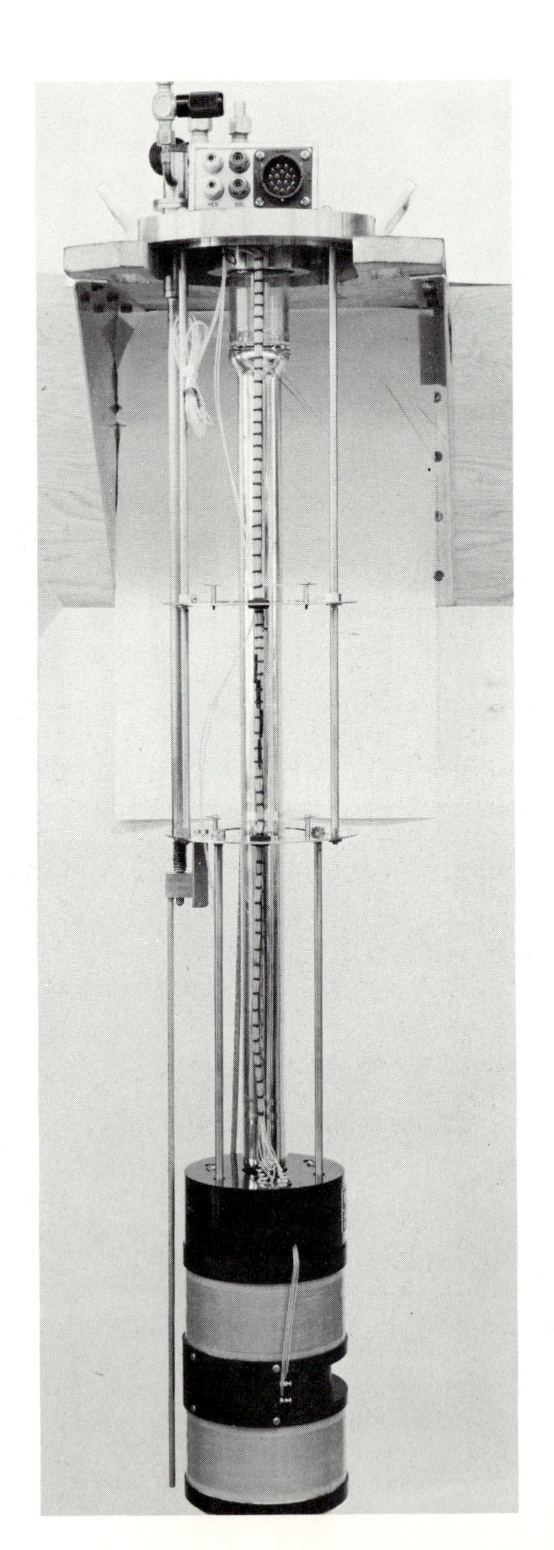

The large radio telescope in Australia. The dish is 210 ft in diameter. It stands in a quiet valley 200 mi west of Sydney, New South Wales. In this remote location, there is a minimum of electrical interference. (*Australian News and Information Bureau*)

A high-energy particle accelerator: the Bevatron at Berkeley. Protons are injected at the lower right (*Lawrence Berkeley Laboratory*). By this time much-higher-energy accelerators are operating at the Brookhaven Laboratory on Long Island, at the CERN Laboratory in Geneva, at Serpukhov in Russia, and at NAL near Chicago.

The 200-in. Hale telescope pointing to zenith; seen from the south. (*Photograph courtesy of the Hale Observatories*)

Reflecting surface of 200-in. mirror of Hale telescope and observer shown in prime-focus cage. (*Photograph courtesy of the Hale Observatories*)

Observer in prime-focus cage changing film in the 200-in. Hale telescope. (*Photograph courtesy of the Hale Observatories*)

NGC 4594 Spiral galaxy in Virgo, seen edge on; 200-in. photograph. (*Photograph courtesy of the Hale Observatories*)

Human red blood cells viewed by the scanning-electron microscope and magnified 15,000 times. Disklike objects are the red blood cells connected by a mesh-work of fibrin strands. Note the realistic three-dimensional character of the picture. *(Photograph courtesy of Dr. Thomas L. Hayes, Donner Laboratory, Lawrence Berkeley Laboratory, University of California, Berkeley)*

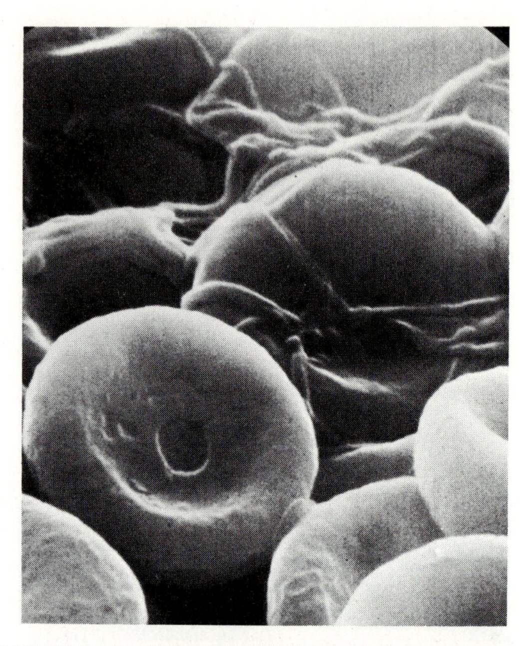

Scanning-electron-microscope installation showing electron optical column that forms probing electron beam (left) and display console containing synchronous cathode-ray-tube beam (right). Auxiliary equipment includes piezoelectric micromanipulator in column of instrument, TV frame rate display and TV tape recorder, Polaroid recording camera, and signal monitor oscilloscope. *(Photograph courtesy of Dr. Thomas L. Hayes, Donner Laboratory, Lawrence Berkeley Laboratory, University of California, Berkeley)*

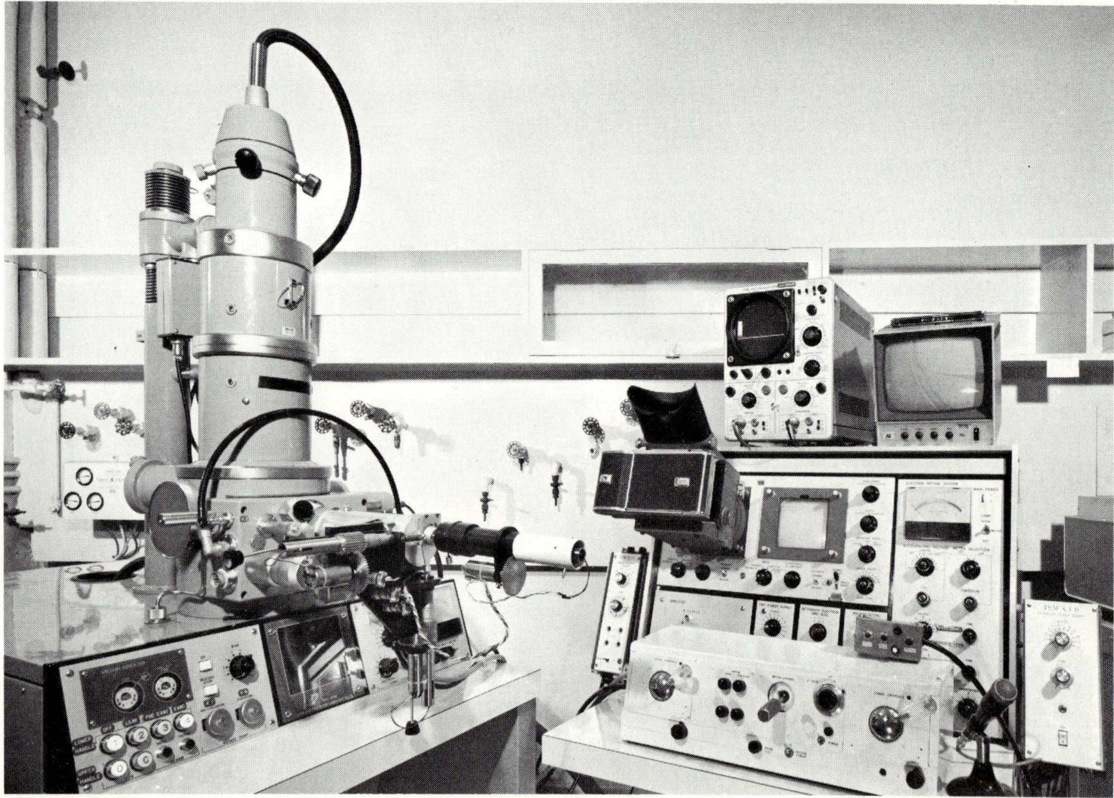

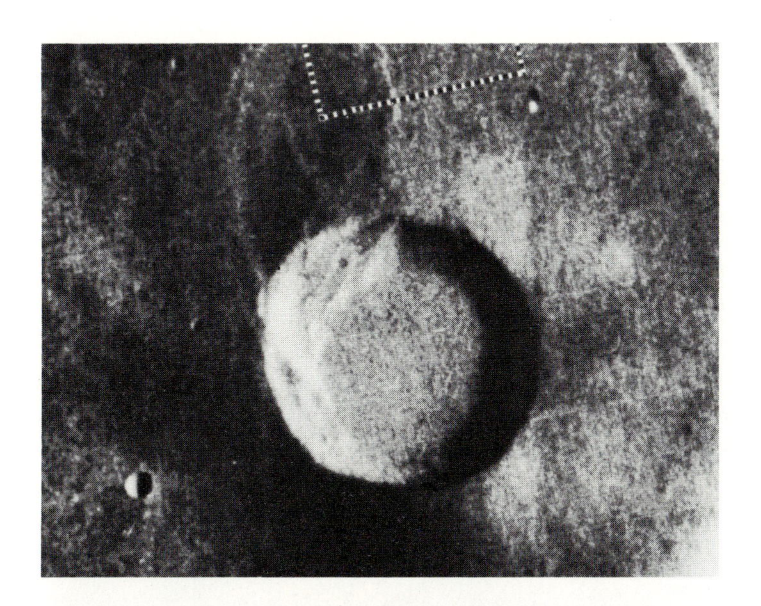

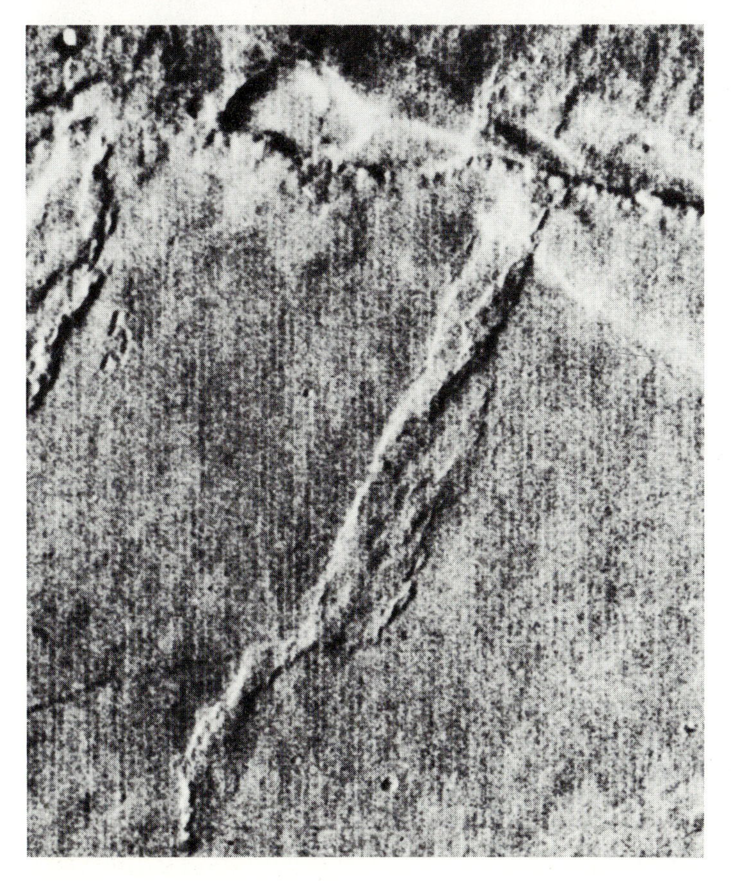

A 43-mi-wide Martian crater (top) was photographed by Mariner 9 on December 16, 1971. The sun shines from the right. The white dotted rectangle inscribes the area shown in the bottom picture taken by Mariner's high-resolution camera on December 22. The ridges, similar to lunar mare ridges, are inferred to be breaks in the crust along which extrusion of lava has taken place. Both pictures have been enhanced by computer processing. (*Photograph courtesy of the Jet Propulsion Laboratory, California Institute of Technology, NASA*)

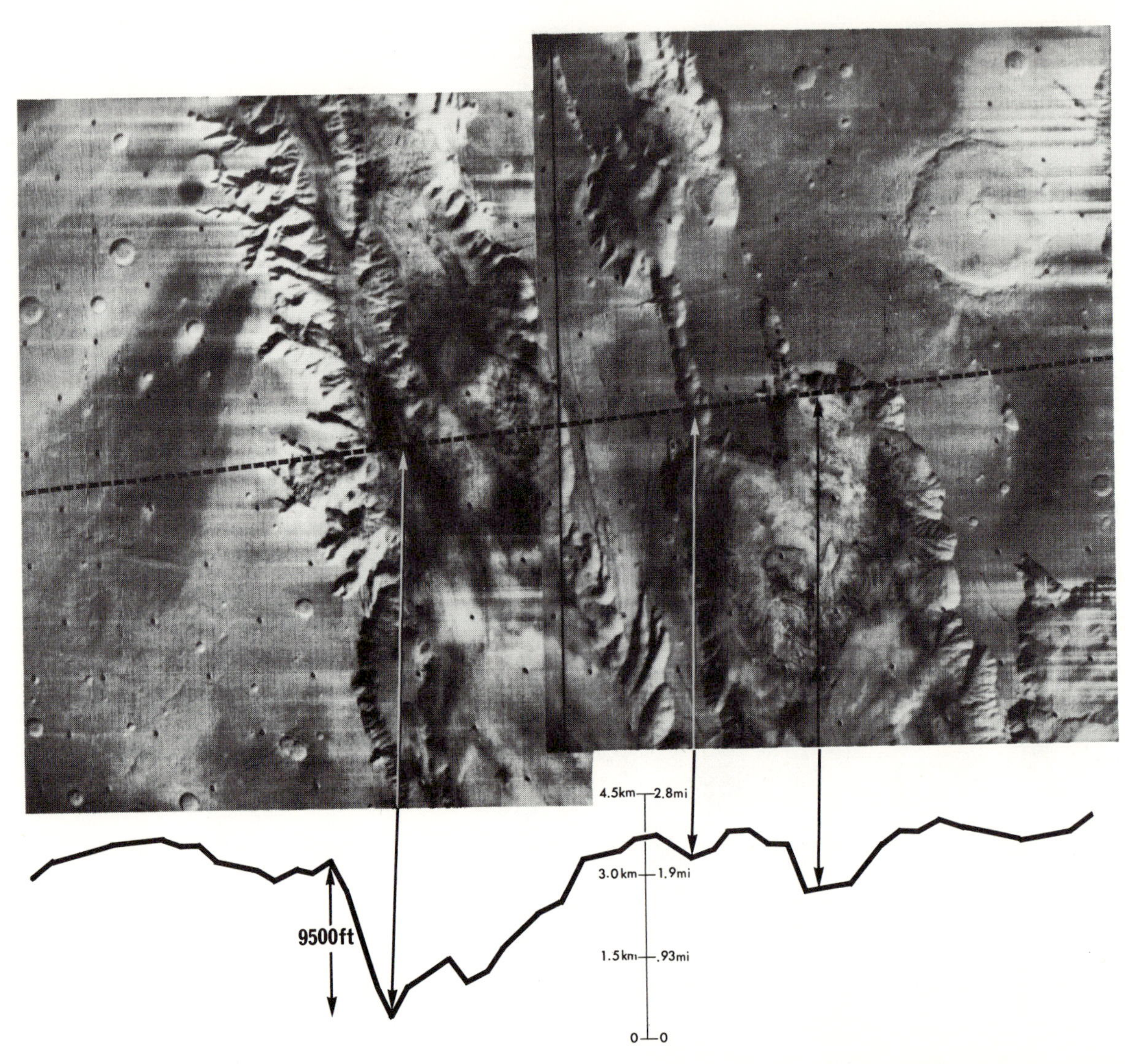

This mosaic of two photographs of the Tithonius Lacus region on Mars taken by the Mariner 9 spacecraft revealed a canyon twice as deep as the Grand Canyon in Arizona when the pictures were compared with pressure measurements taken by the ultraviolet spectrometer experiment aboard the spacecraft. The arrows connect the depths deduced from the pressure measurements taken by the ultraviolet spectrometer and the corresponding features on the photograph. The dotted line is the scan path of the spectrometer. The photographs were taken from an altitude of 1070 mi and cover an area of 400 mi across. (*Photograph courtesy of the Jet Propulsion Laboratory, California Institute of Technology, NASA*)

CONTENTS

Vectors

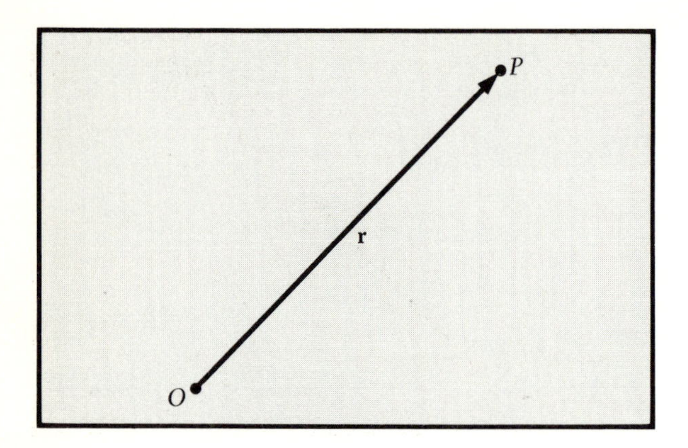

FIG. 2.1 The vector **r** represents the position of a point *P* relative to another point *O* as origin.

LANGUAGE AND CONCEPTS: VECTORS

Language is an essential ingredient of abstract thought. It is difficult to think clearly and easily about sophisticated and abstract concepts in a language that has no words appropriate to such concepts. To express new scientific concepts, new words are invented and added to languages; many such words are put together from classical Greek or Latin roots. If it satisfies the needs of the scientific community, a new word may be adopted in many modern languages. Thus *vector* in English is *vecteur* in French, *Vektor* in German, and ВЕКТОР (pronounced "vector") in Russian.

Vector is the word defining a quantity that has both *direction* and *magnitude* and that combines with other vectors according to a specific rule.[1] Throughout mechanics (and other branches of physics as well) we shall meet quantities (velocity, force, electric field, magnetic dipole moment) that have both magnitude and direction, and consequently it is important to develop the language and the techniques to deal with these quantities. Although vector analysis often ranks as a branch of mathematics, its value in physics is so great as to merit the inclusion of an introduction here.

Vector Notation

Because symbols are the components of the language of mathematics, an important part of the art of mathematical analysis is the technique of using notation well. Vector notation has two important properties:

1. Formulation of a law of physics in terms of vectors is independent of the choice of coordinate axes. Vector notation offers a language in which statements have a physical content without ever introducing a coordinate system.
2. Vector notation is concise. Many physical laws have simple and transparent forms that are disguised when these laws are written in terms of a particular coordinate system.

Although in solving problems we may wish to work in special coordinate systems, we shall state the laws of physics in vector form wherever possible. Some of the more complicated laws, which cannot be expressed in vector form, may be expressed in terms of tensors. A *tensor* is a generalization of a vector and includes a vector as a special case. The vector

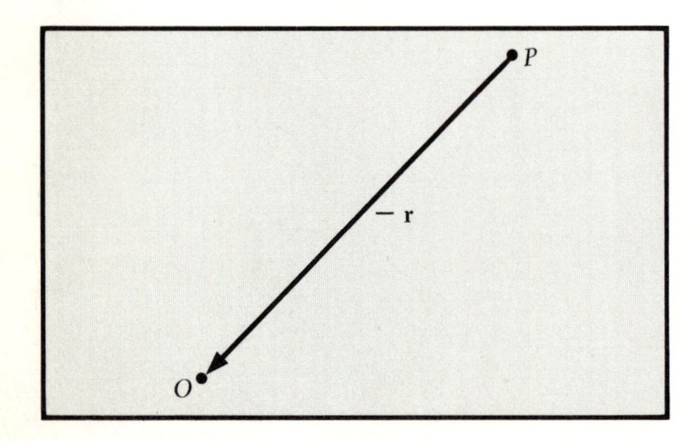

FIG. 2.2 The vector −**r** is equal in magnitude but opposite in direction to **r**.

[1] This meaning for the word *vector* is a natural extension of an earlier usage, now obsolete, in astronomy: an imaginary straight line that joins a planet moving around the focus of an ellipse to that focus. The specific rule is given on page 31.

analysis we know today is largely the result of work done toward the end of the nineteenth century by Josiah Willard Gibbs and Oliver Heaviside.

The vector notation we adopt is the following: On the blackboard a vector quantity named A is denoted by putting a wavy line under A or by putting an arrow over the letter. In print vectors always appear in boldface type. The magnitude of a vector is printed in italics: A is the magnitude of $\mathbf{A}$; A is also written as $|\mathbf{A}|$. A unit vector is a vector of unit length; a unit vector in the direction of $\mathbf{A}$ is written with a caret as $\hat{\mathbf{A}}$, which we read as "A hat," or "A caret." We summarize the notation by the identity

$$\mathbf{A} \equiv \hat{\mathbf{A}}A \equiv A\hat{\mathbf{A}}$$

Figures 2.1 to 2.4 show a vector, the negative of that vector, multiplication by a scalar, and a unit vector.

The usefulness and applicability of vectors to physical problems is largely based on euclidean geometry. Statement of a law in terms of vectors usually carries with it the assumption of euclidean geometry. If the geometry is not euclidean, addition of two vectors in a simple and unambiguous way may not be possible. For curved space there exists a more general language, metric differential geometry, which is the language of general relativity, the domain of physics in which euclidean geometry is no longer sufficiently precise.

We have considered a vector to be a quantity having direction as well as magnitude. These properties have absolutely no reference to a particular coordinate system, although we assume that the direction can be defined, for example, by reference to a laboratory room, the fixed stars, etc. We shall see, however, that there are some quantities having magnitude and direction that are not vectors, such as finite rotations (page 33). A quantity having magnitude but *not* direction is a *scalar*. The magnitude of a vector is a scalar. Temperature is a scalar and mass is a scalar. On the other hand, velocity $\mathbf{v}$ is a vector and force $\mathbf{F}$ is a vector.

Equality of Vectors Having developed the notation, we now proceed to some vector operations: addition, subtraction, and multiplication. Two vectors $\mathbf{A}$ and $\mathbf{B}$ describing similar physical quantities (e.g., forces) are defined to be equal if they have the same magnitude and direction; this is written $\mathbf{A} = \mathbf{B}$. A vector does not necessarily have location, although a vector may refer to a quantity defined at a particular point. Two vectors can be compared even though they measure a physical quantity

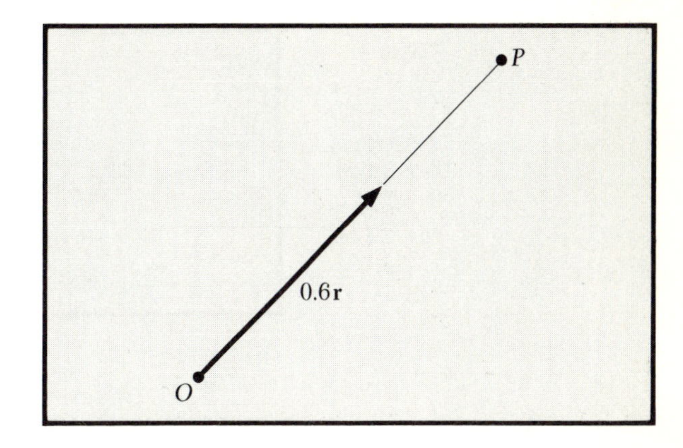

FIG. 2.3 The vector $0.6\mathbf{r}$ is in the direction of $\mathbf{r}$ and is of magnitude $0.6r$.

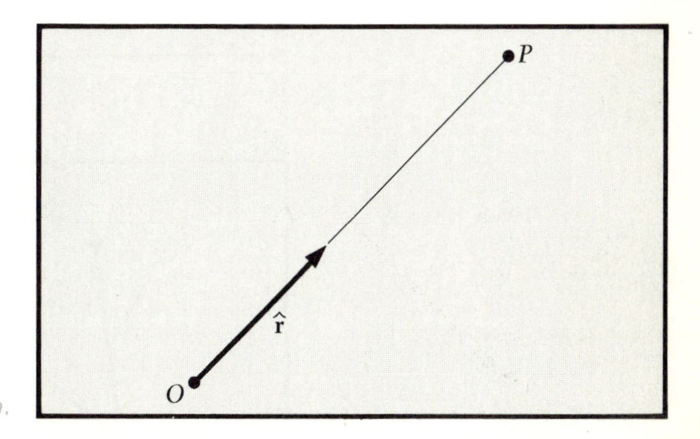

FIG. 2.4 The vector $\hat{\mathbf{r}}$ is the unit vector in the direction of $\mathbf{r}$. Note that $\mathbf{r} = r\hat{\mathbf{r}}$.

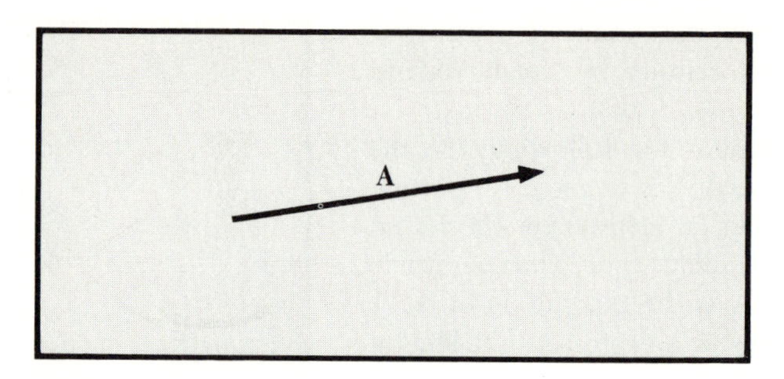

FIG. 2.5 (*a*) Vector **A**.

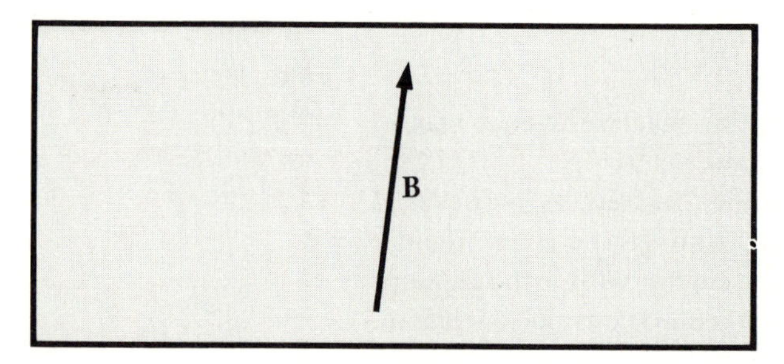

(*b*) Vector **B**.

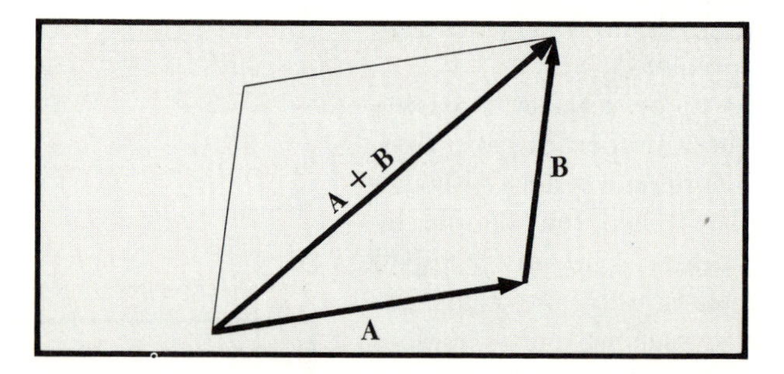

(*c*) The vector sum **A** + **B**.

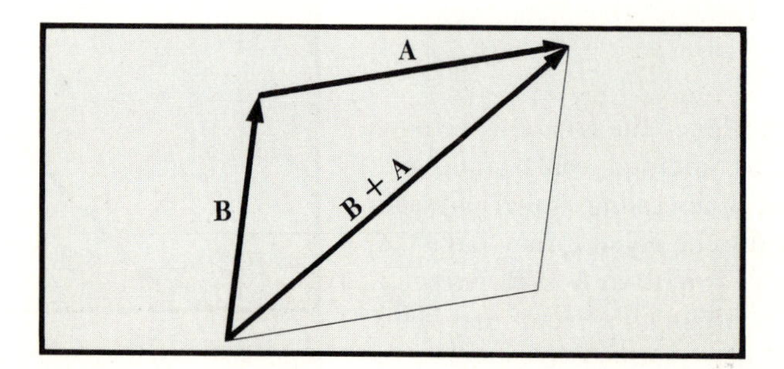

(*d*) The vector sum **B** + **A** is equal to **A** + **B**.

defined at different points of space and time. If we did not have confidence, based on experiment, that we can consider space to be flat, i.e., euclidean—except perhaps at enormous distances—then we could not unambiguously compare two vectors at different points.

VECTOR ADDITION

A vector is represented geometrically by a directed straight-line segment, or arrow, whose length in chosen scale units equals the magnitude of the vector. The sum of two vectors **A** and **B** is defined by the geometrical construction shown in Fig. 2.5*a* to *c*. This construction is often called the *parallelogram law of addition of vectors*. The sum $\mathbf{A} + \mathbf{B}$ is defined by carrying **B** parallel to itself until the tail of **B** coincides with the head of **A**. The vector drawn from the tail of **A** to the head of **B** is the sum $\mathbf{A} + \mathbf{B}$. From the figure it follows that $\mathbf{A} + \mathbf{B} = \mathbf{B} + \mathbf{A}$, so that vector addition is said to be commutative, as shown in Fig. 2.5*d*. Vector subtraction is defined by Fig. 2.6*a* and *b* with $\mathbf{B} + (-\mathbf{B}) = 0$ defining the negative vector.

Vector addition satisfies the relation $\mathbf{A} + (\mathbf{B} + \mathbf{C}) = (\mathbf{A} + \mathbf{B}) + \mathbf{C}$, so that vector addition is said to be associative (see Fig. 2.7). The sum of a finite number of vectors is independent of the order in which they are added. If $\mathbf{A} - \mathbf{B} = \mathbf{C}$, then by adding **B** to both sides we obtain $\mathbf{A} = \mathbf{B} + \mathbf{C}$. If k is a scalar,

$$k(\mathbf{A} + \mathbf{B}) = k\mathbf{A} + k\mathbf{B} \tag{2.1}$$

so that multiplication of a vector by a scalar is said to be distributive.

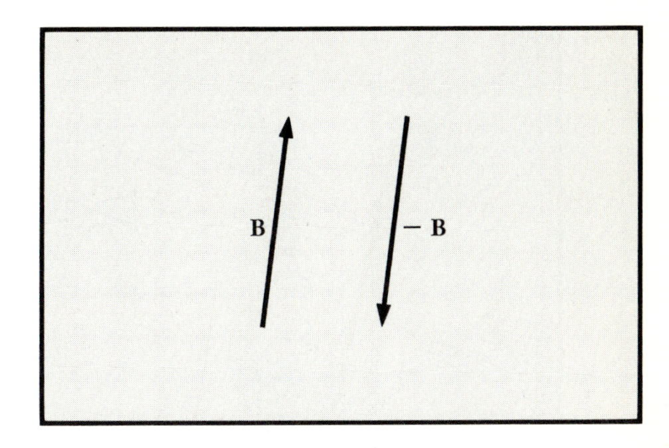

FIG. 2.6 (*a*) Vectors **B** and $-\mathbf{B}$.

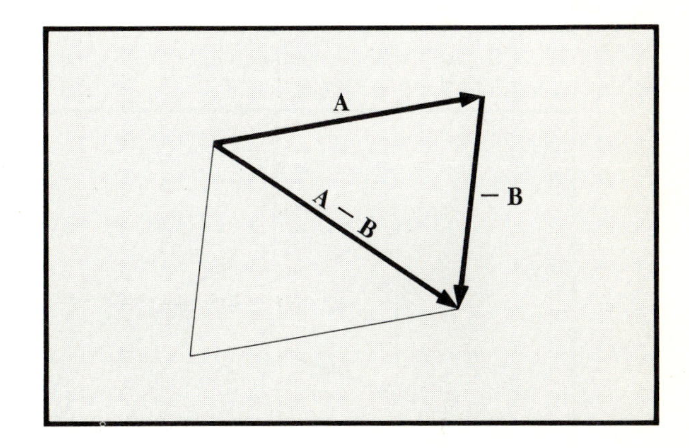

(*b*) Formation of $\mathbf{A} - \mathbf{B}$; vector subtraction.

When Is a Physical Quantity Representable by a Vector?

A displacement is a vector because it describes both the direction of the line from the initial position to the final position and the length of the line; the example of addition given above is easily recognized as applying to displacements in euclidean space. In addition to displacements, other physical quantities have the same laws of combination and invariance properties as displacements. Such quantities can also be represented by vectors. To be a vector a quantity must satisfy two conditions:

1 It must satisfy the parallelogram law of addition.

2 It must have a magnitude and a direction independent of the choice of coordinate system.

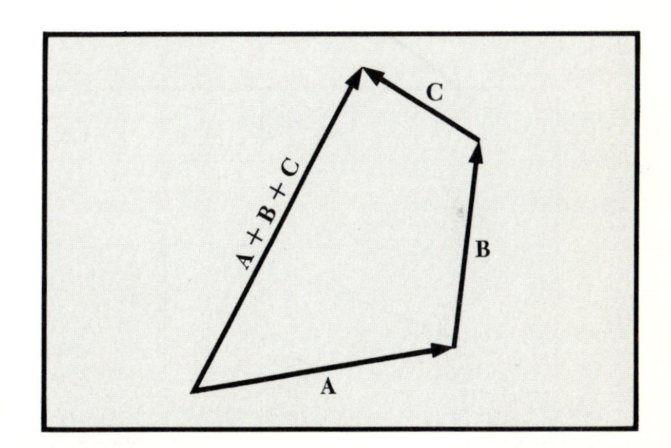

FIG. 2.7 Sum of three vectors: $\mathbf{A} + \mathbf{B} + \mathbf{C}$. Verify for yourself that this sum is equal to $\mathbf{B} + \mathbf{A} + \mathbf{C}$.

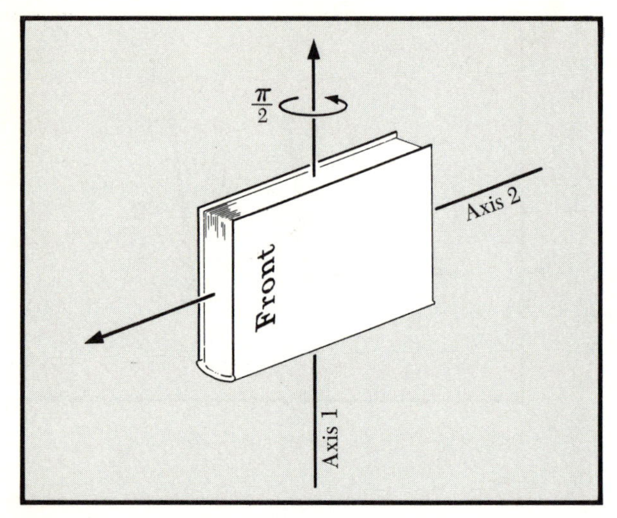

FIG. 2.8 (*a*) Original orientation of book. It is then rotated by $\pi/2$ radians (rad) about Axis 1.

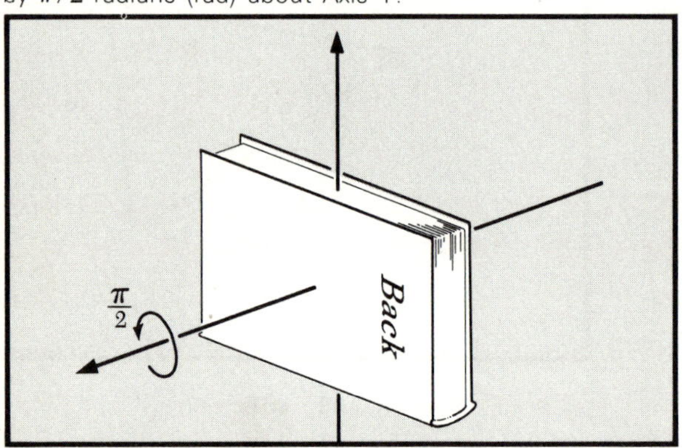

(*b*) Orientation after a rotation of $\pi/2$ rad about Axis 1.

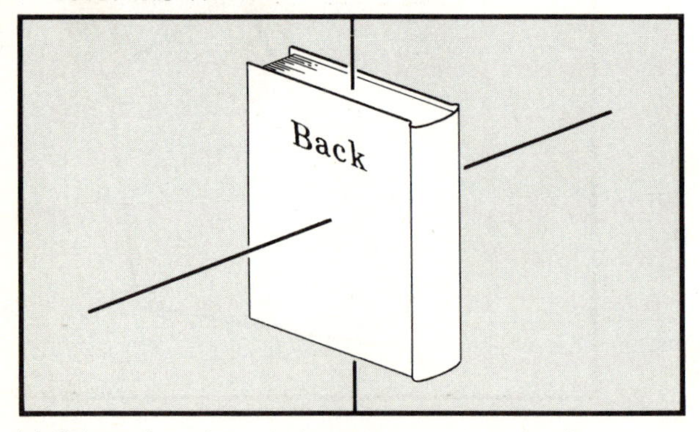

(*c*) Orientation after a subsequent rotation of $\pi/2$ rad about Axis 2.

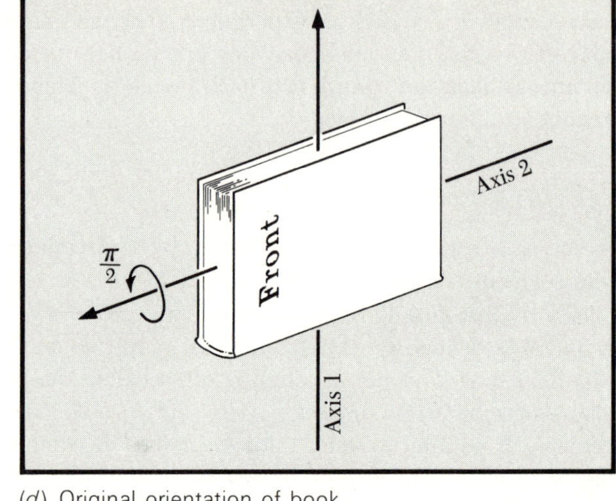

(*d*) Original orientation of book.

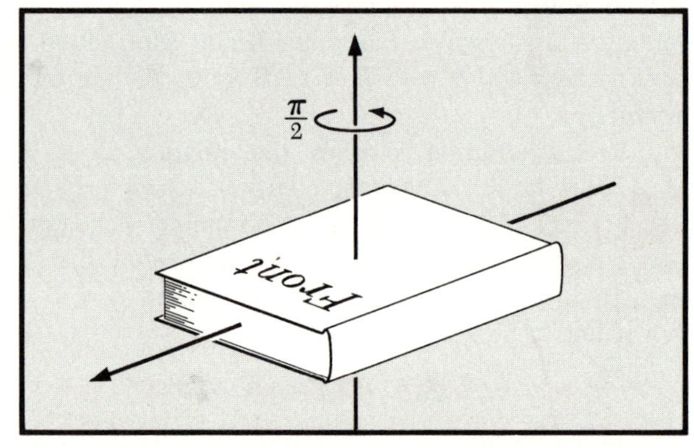

(*e*) Orientation after a rotation of $\pi/2$ rad about Axis 2.

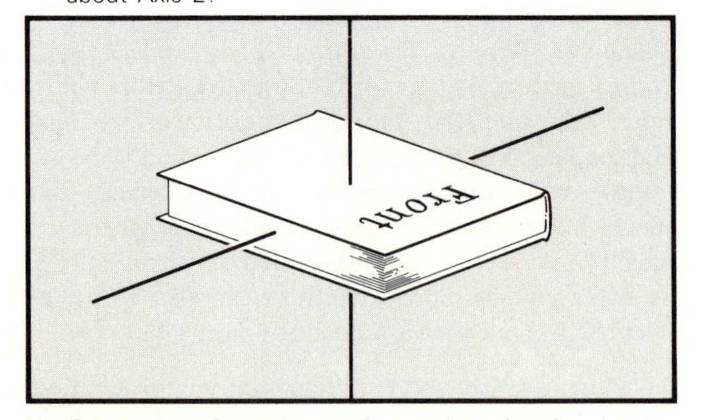

(*f*) Orientation after subsequent rotation of $\pi/2$ rad about Axis 1.

Finite Rotations Are Not Vectors Not all quantities that have magnitude and direction are necessarily vectors. For example, a rotation of a rigid body about a particular axis fixed in space has a magnitude (the angle of rotation) and a direction (the direction of the axis). But two such rotations do not combine according to the vector law of addition, unless the angles of rotation are infinitesimally small.[1] This is easily seen when the two axes are perpendicular to each other and the rotations are by $\pi/2$ rad (90°). Consider the object (a book) in Fig. 2.8*a*. The rotation (1) leaves it as in Fig. 2.8*b*, and a subsequent rotation (2) about another axis leaves the object as in Fig. 2.8*c*. But if to the object as originally oriented (Fig. 2.8*d*) we apply first the rotation (2), (Fig. 2.8*e*) and then the rotation (1), the object ends up as shown in Fig. 2.8*f*. The orientation in the sixth figure is *not* the same as in the third. Obviously the commutative law of addition is not satisfied by these rotations. Despite the fact that they have a magnitude and a direction, finite rotations cannot be represented as vectors.

PRODUCTS OF VECTORS

Although there is no reason to ask whether the sum of two vectors is a scalar or a vector, such a question has importance in reference to the product of two vectors. There are two particularly useful ways in which to define the product of two vectors. Both products satisfy the distributive law of multiplication: The product of **A** into the sum of **B** + **C** is equal to the sum of the products of **A** into **B** and **A** into **C**. One type of product is a scalar, and the other is for many purposes a vector. Both products are useful in physics. Other possible definitions of product are not useful: Why is AB not a useful definition of the product of two vectors? By AB we mean the ordinary product $|\mathbf{A}|\ |\mathbf{B}|$ of the magnitudes of **A** and **B**. We observe that if $\mathbf{D} = \mathbf{B} + \mathbf{C}$, then, in general, $AD \neq AB + AC$. This absence of the distributive property makes AB useless as a product of **A** and **B**.

Scalar Product The scalar product of **A** and **B** is defined as that number which is obtained by taking the magnitude of **A** times the magnitude of **B** times the cosine of the angle between them (see Fig. 2.9*a* to *c*). The scalar product is a scalar.

[1]Angular velocities are vectors even though finite angular rotations are not.

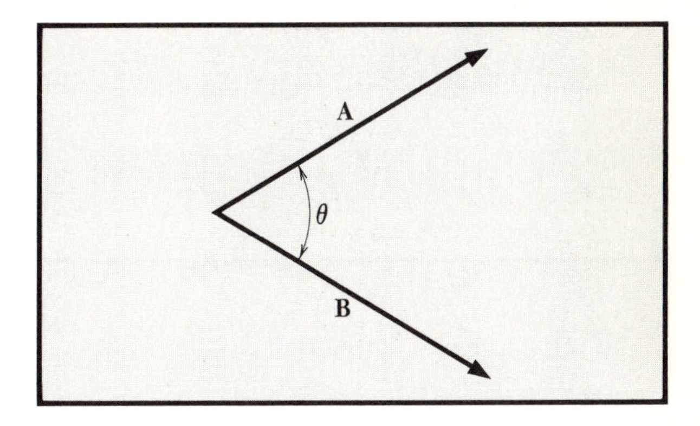

FIG. 2.9 (*a*) In forming $\mathbf{A}\cdot\mathbf{B}$, bring vectors **A** and **B** to a common origin.

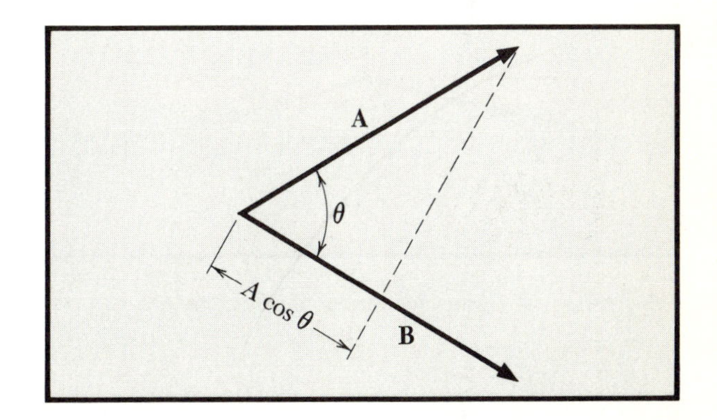

(*b*) $B(A\cos\theta) = \mathbf{A}\cdot\mathbf{B}$.

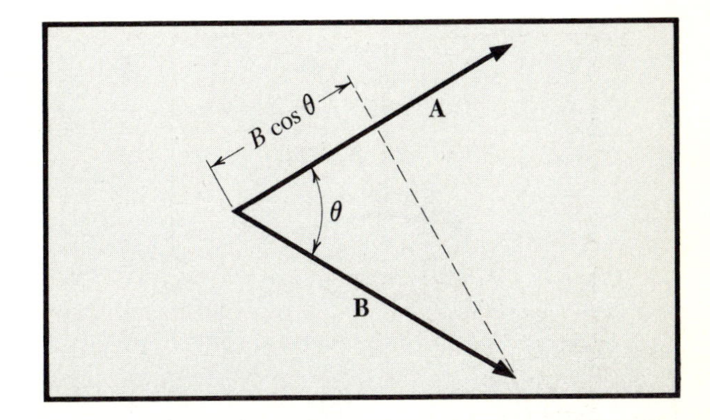

(*c*) $A(B\cos\theta) = \mathbf{A}\cdot\mathbf{B}$. Here θ, the Greek character theta, denotes the angle between **A** and **B**.

Often the scalar product is called the *dot product* because we denote it by the symbol $\mathbf{A} \cdot \mathbf{B}$ where

$$\boxed{\mathbf{A} \cdot \mathbf{B} \equiv AB \cos(\mathbf{A},\mathbf{B})} \tag{2.2}$$

Here $\cos(\mathbf{A},\mathbf{B})$ denotes the cosine of the angle between **A** and **B**. We see that no coordinate system is involved at all in the definition of scalar product. We note that $\cos(\mathbf{A},\mathbf{B}) = \cos(\mathbf{B},\mathbf{A})$, so that the scalar product is commutative:

$$\mathbf{A} \cdot \mathbf{B} = \mathbf{B} \cdot \mathbf{A} \tag{2.3}$$

We read $\mathbf{A} \cdot \mathbf{B}$ as "**A** *dot* **B**."

If the angle between **A** and **B** should lie between $\pi/2$ and $3\pi/2$, then $\cos(\mathbf{A},\mathbf{B})$ and $\mathbf{A} \cdot \mathbf{B}$ will be negative numbers. If $\mathbf{A} = \mathbf{B}$, then $\cos(\mathbf{A},\mathbf{B}) = 1$ and

$$\mathbf{A} \cdot \mathbf{B} = A^2 = |\mathbf{A}|^2$$

If $\mathbf{A} \cdot \mathbf{B} = 0$ and $A \neq 0$, $B \neq 0$, we say that **A** is *orthogonal* to **B** or perpendicular to **B**. Note that $\cos(\mathbf{A},\mathbf{B}) = \hat{\mathbf{A}} \cdot \hat{\mathbf{B}}$, so that the scalar product of two unit vectors is just the cosine of the angle between them. The magnitude of the projection of **B** on the direction of **A** is

$$B \cos(\mathbf{A},\mathbf{B}) = B\hat{\mathbf{A}} \cdot \hat{\mathbf{B}} = \mathbf{B} \cdot \hat{\mathbf{A}}$$

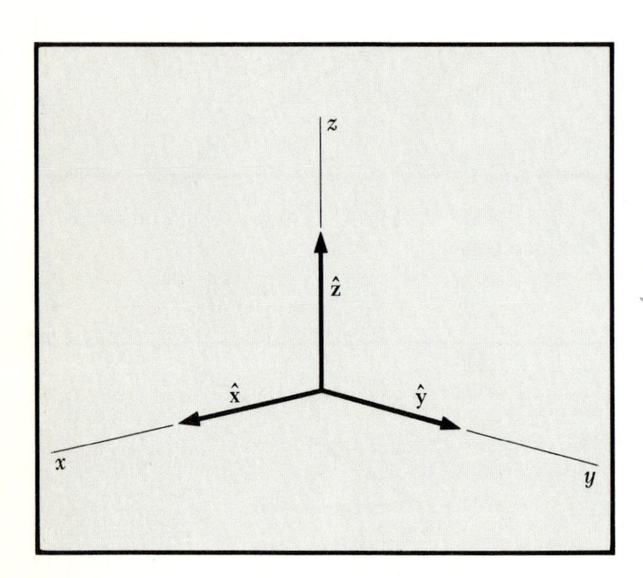

FIG. 2.10 (*a*) Cartesian orthogonal unit vectors $\hat{\mathbf{x}}$, $\hat{\mathbf{y}}$, $\hat{\mathbf{z}}$.

where $\hat{\mathbf{A}}$ is the unit vector in the direction of **A**. The projection of **A** on the direction of **B** is

$$A \cos(\mathbf{A},\mathbf{B}) = \mathbf{A} \cdot \hat{\mathbf{B}}$$

Scalar product multiplication has no inverse: if $\mathbf{A} \cdot \mathbf{X} = b$, there is no unique solution for **X**. Division by a vector is a meaningless, undefined operation.

Components, Magnitudes, and Direction Cosines Let $\hat{\mathbf{x}}$, $\hat{\mathbf{y}}$, and $\hat{\mathbf{z}}$ be three orthogonal[1] unit vectors that define a cartesian coordinate system as in Fig. 2.10*a*. An arbitrary vector **A** may be written

$$\mathbf{A} = A_x\hat{\mathbf{x}} + A_y\hat{\mathbf{y}} + A_z\hat{\mathbf{z}} \tag{2.4}$$

where A_x, A_y, and A_z are called the *components* of **A**, as illustrated in Fig. 2.10*b*. It is readily seen that $A_x = \mathbf{A} \cdot \hat{\mathbf{x}}$ since

$$\mathbf{A} \cdot \hat{\mathbf{x}} = A_x\hat{\mathbf{x}} \cdot \hat{\mathbf{x}} + A_y\hat{\mathbf{y}} \cdot \hat{\mathbf{x}} + A_z\hat{\mathbf{z}} \cdot \hat{\mathbf{x}} = A_x$$

and

$$\hat{\mathbf{y}} \cdot \hat{\mathbf{x}} = 0 = \hat{\mathbf{z}} \cdot \hat{\mathbf{x}}$$
$$\hat{\mathbf{x}} \cdot \hat{\mathbf{x}} = 1$$

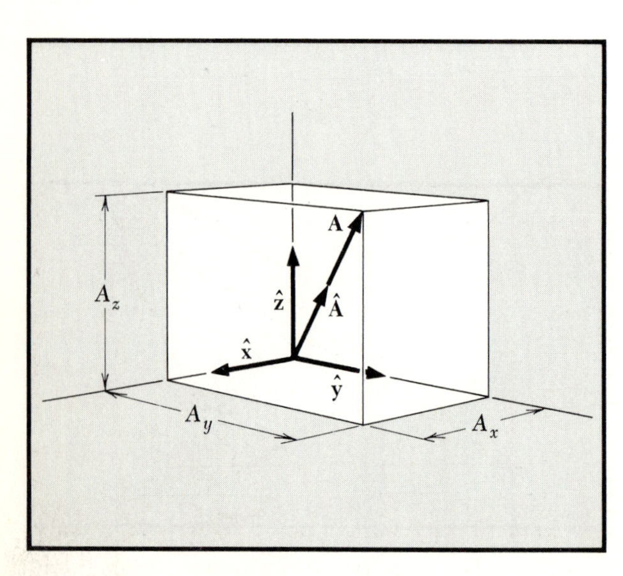

(*b*) $\mathbf{A} = \hat{\mathbf{x}}A_x + \hat{\mathbf{y}}A_y + \hat{\mathbf{z}}A_z$.

[1] Orthogonal as used here means *mutually perpendicular*.

In terms of these components A_x, A_y, and A_z, the magnitude of **A** is

$$A = \sqrt{\mathbf{A}\cdot\mathbf{A}} = \sqrt{(A_x\hat{\mathbf{x}} + A_y\hat{\mathbf{y}} + A_z\hat{\mathbf{z}})\cdot(A_x\hat{\mathbf{x}} + A_y\hat{\mathbf{y}} + A_z\hat{\mathbf{z}})}$$

$$= \sqrt{A_x^2 + A_y^2 + A_z^2} \qquad (2.5)$$

If we desire to write an expression for the unit vector $\hat{\mathbf{A}}$ (also shown in Fig. 2.10*b*), we can see that

$$\hat{\mathbf{A}} = \hat{\mathbf{x}}\frac{\hat{\mathbf{x}}\cdot\mathbf{A}}{A} + \hat{\mathbf{y}}\frac{\hat{\mathbf{y}}\cdot\mathbf{A}}{A} + \hat{\mathbf{z}}\frac{\hat{\mathbf{z}}\cdot\mathbf{A}}{A}$$

$$= \hat{\mathbf{x}}\frac{A_x}{A} + \hat{\mathbf{y}}\frac{A_y}{A} + \hat{\mathbf{z}}\frac{A_z}{A} \qquad (2.6)$$

is such an expression. From Fig. 2.11 and Eq. (2.6) we deduce that the angles that **A** makes with the x, y, and z axes have cosines A_x/A, A_y/A, and A_z/A, or $\hat{\mathbf{x}}\cdot\hat{\mathbf{A}}$, $\hat{\mathbf{y}}\cdot\hat{\mathbf{A}}$, and $\hat{\mathbf{z}}\cdot\hat{\mathbf{A}}$. These are called *direction cosines* and have the property that the sum of the squares of the three direction cosines is equal to unity, as can easily be seen with the help of Eq. (2.5).

The scalar product of two vectors **A** and **B** is conveniently remembered in terms of the components

$$\boxed{\mathbf{A}\cdot\mathbf{B} = A_xB_x + A_yB_y + A_zB_z} \qquad (2.7)$$

Applications of the Scalar Product We treat several applications of the scalar product.

1 *Law of cosines.* Let $\mathbf{A} - \mathbf{B} = \mathbf{C}$; then, on taking the scalar product of each side of this expression with itself, we have

$$(\mathbf{A} - \mathbf{B})\cdot(\mathbf{A} - \mathbf{B}) = \mathbf{C}\cdot\mathbf{C}$$

or

$$A^2 + B^2 - 2\mathbf{A}\cdot\mathbf{B} = C^2$$

which is exactly the famous trigonometric relation

$$A^2 + B^2 - 2AB\cos(\mathbf{A},\mathbf{B}) = C^2 \qquad (2.8)$$

The cosine of the angle between the directions of two vectors is given by

$$\cos(\mathbf{A},\mathbf{B}) = \cos\theta_{AB} = \frac{\mathbf{A}\cdot\mathbf{B}}{AB}$$

as in Eq. (2.2) (see Fig. 2.12*a* and *b*).

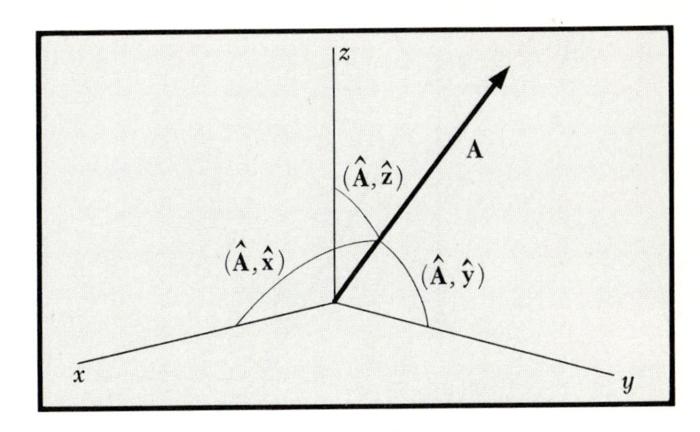

FIG. 2.11 Direction cosines refer to the angles indicated.

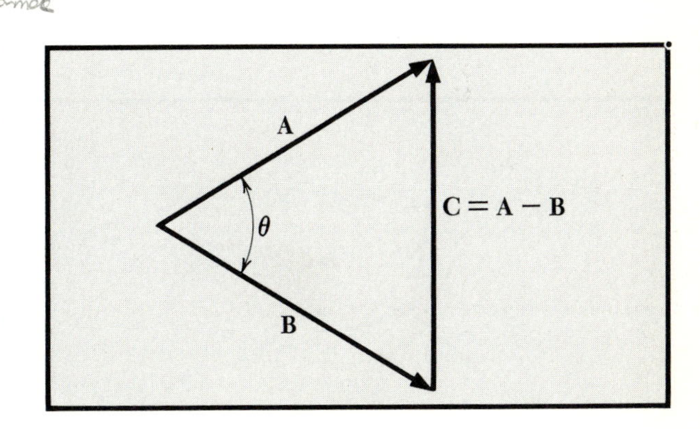

FIG. 2.12 (*a*) $\mathbf{C}\cdot\mathbf{C} = C^2 = (\mathbf{A} - \mathbf{B})\cdot(\mathbf{A} - \mathbf{B})$
$= A^2 + B^2 - 2\mathbf{A}\cdot\mathbf{B}$
$= A^2 + B^2 - 2AB\cos\theta.$

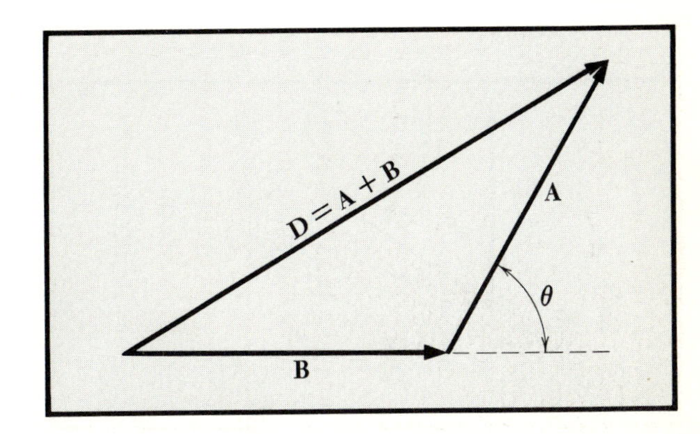

(*b*) $\mathbf{D}\cdot\mathbf{D} = D^2 = (\mathbf{A} + \mathbf{B})\cdot(\mathbf{A} + \mathbf{B})$
$= A^2 + B^2 + 2AB\cos\theta.$

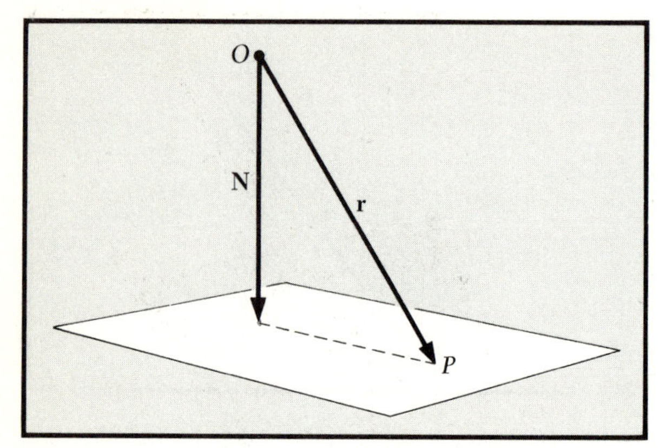

FIG. 2.13 Equation of a plane; **N** is the normal to the plane from the origin O. The equation of the plane is $\mathbf{N} \cdot \mathbf{r} = N^2$.

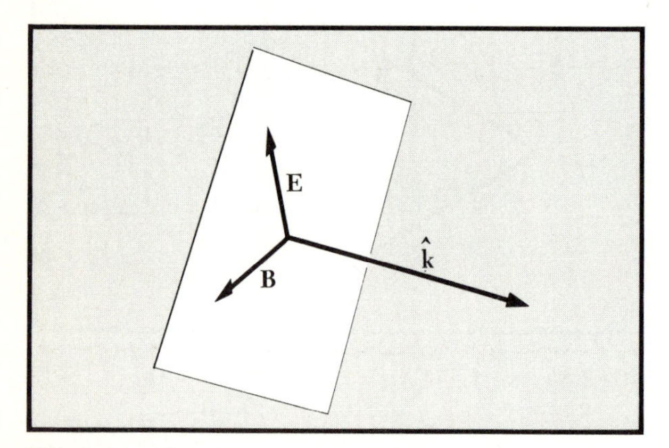

FIG. 2.14 Electric and magnetic fields in a plane electromagnetic wave in free space are perpendicular to the propagation direction $\hat{\mathbf{k}}$. Thus $\hat{\mathbf{k}} \cdot \mathbf{E} = \hat{\mathbf{k}} \cdot \mathbf{B} = 0$; $\mathbf{E} \cdot \mathbf{B} = 0$.

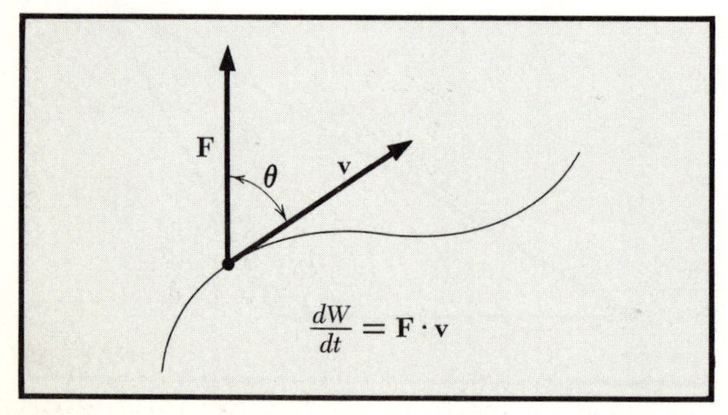

FIG. 2.15 Rate at which a force **F** does work on a particle moving with velocity **v**.

2 *Equation of a plane* (Fig. 2.13). Let **N** be a normal to the plane under consideration, which is drawn from an origin O not in the plane. Let **r** be an arbitrary vector from the origin O to any point P in the plane. The projection of **r** on **N** must be equal in magnitude to N. Thus the plane is described by the equation

$$\mathbf{r} \cdot \mathbf{N} = N^2 \tag{2.9}$$

To establish the identity of this compact expression with the usual expression in analytic geometry for the equation of a plane

$$ax + by + cz = 1$$

write **N** and **r** in terms of their components N_x, N_y, N_z and x, y, z. Now Eq. (2.9) assumes the form

$$(x\hat{\mathbf{x}} + y\hat{\mathbf{y}} + z\hat{\mathbf{z}}) \cdot (N_x\hat{\mathbf{x}} + N_y\hat{\mathbf{y}} + N_z\hat{\mathbf{z}}) = N^2$$

which reduces to

$$x\frac{N_x}{N^2} + y\frac{N_y}{N^2} + z\frac{N_z}{N^2} = 1$$

3 *Electric and magnetic vectors in an electromagnetic wave.* If $\hat{\mathbf{k}}$ is the unit vector in the direction of propagation of a plane electromagnetic wave in free space (see Fig. 2.14), then (as we shall see in Volumes 2 and 3) the electric and magnetic intensity vectors **E** and **B** must lie in a plane normal to $\hat{\mathbf{k}}$ and must be perpendicular to each other. We can express this geometric condition by the relations

$$\hat{\mathbf{k}} \cdot \mathbf{E} = 0 \qquad \hat{\mathbf{k}} \cdot \mathbf{B} = 0 \qquad \mathbf{E} \cdot \mathbf{B} = 0$$

4 *Rate of doing work.* In elementary physics (also see Chap. 5) we learned that the rate at which a force **F** does work on a particle moving with velocity **v** is equal to $Fv \cos(\mathbf{F},\mathbf{v})$. We recognize this expression as just the scalar product

$$\mathbf{F} \cdot \mathbf{v}$$

If we write generally the derivative dW/dt as a symbol for the rate of doing work, then (Fig. 2.15)

$$\frac{dW}{dt} = \mathbf{F} \cdot \mathbf{v} \tag{2.10}$$

5 *Rate at which volume is swept out.* Let **S** be a vector normal to a plane area and of magnitude equal to the area and let **v** denote the velocity at which the area is moved. The

volume per unit time traversed by the area S is a cylinder of base area S and slant height v (see Fig. 2.16), or $\mathbf{S} \cdot \mathbf{v}$. The rate at which volume is swept out is therefore

$$\frac{dV}{dt} = \mathbf{S} \cdot \mathbf{v} \tag{2.11}$$

Vector Product[1] There is a second product of two vectors that is widely used in physics. This product is vector rather than scalar in character, but it is a vector in a somewhat restricted sense. The *vector product* $\mathbf{A} \times \mathbf{B}$ is defined to be a vector normal to the plane that includes **A** and **B** with magnitude $AB|\sin(\mathbf{A},\mathbf{B})|$ as in Fig. 2.17*a*:

$$\boxed{\mathbf{C} = \mathbf{A} \times \mathbf{B} = \hat{\mathbf{C}}AB|\sin(\mathbf{A},\mathbf{B})|} \tag{2.12}$$

We read $\mathbf{A} \times \mathbf{B}$ as "**A** *cross* **B**." The sense of **C** is determined as a matter of fixed convention by the *right-hand-thread* rule: The vector **A** in the first position in the product is rotated by the smallest angle that will bring it into coincidence with the direction of **B**. The sense of **C** is that of the direction of motion of a screw with a right-hand thread (the standard thread in the United States) when the screw is rotated in the same direction as was the vector **A**, as shown in Fig. 2.17*b* on the next page.

Let us state the rule for the direction of **C** in another way: First, place together the tails of vectors **A** and **B**; this defines a plane. Vector **C** is perpendicular to this plane; that is, the vector product $\mathbf{A} \times \mathbf{B}$ is perpendicular to both **A** and **B**. Rotate **A** into **B** through the lesser of the two possible angles—curl the fingers of the right hand in the direction in which **A** was rotated, and the thumb will point in the direction of $\mathbf{C} = \mathbf{A} \times \mathbf{B}$. Note that because of this sign convention $\mathbf{B} \times \mathbf{A}$ is a vector opposite in sign to $\mathbf{A} \times \mathbf{B}$ (see Fig. 2.17*c*):

$$\mathbf{B} \times \mathbf{A} = -\mathbf{A} \times \mathbf{B} \tag{2.13}$$

Thus the vector product is not commutative. It follows from Eq. (2.12) that $\mathbf{A} \times \mathbf{A} = 0$, so that the vector product of a vector with itself is zero. The vector product does obey the distributive law:

$$\mathbf{A} \times (\mathbf{B} + \mathbf{C}) = \mathbf{A} \times \mathbf{B} + \mathbf{A} \times \mathbf{C}$$

[1] This section (pages 37 to 40) can be omitted at this time. The vector product is used in Chap. 3 (page 71), which can also be omitted; only beginning in Chap. 6 (page 185) is it essential.

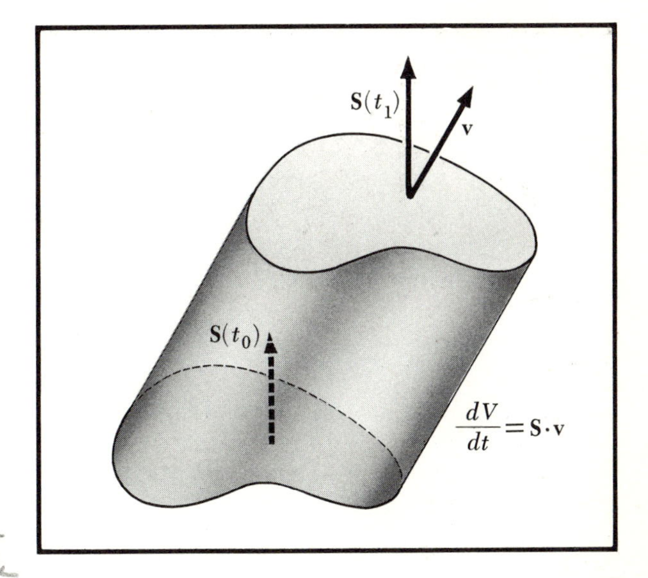

FIG. 2.16 Rate dV/dt at which area **S** moving with velocity **v** sweeps out volume.

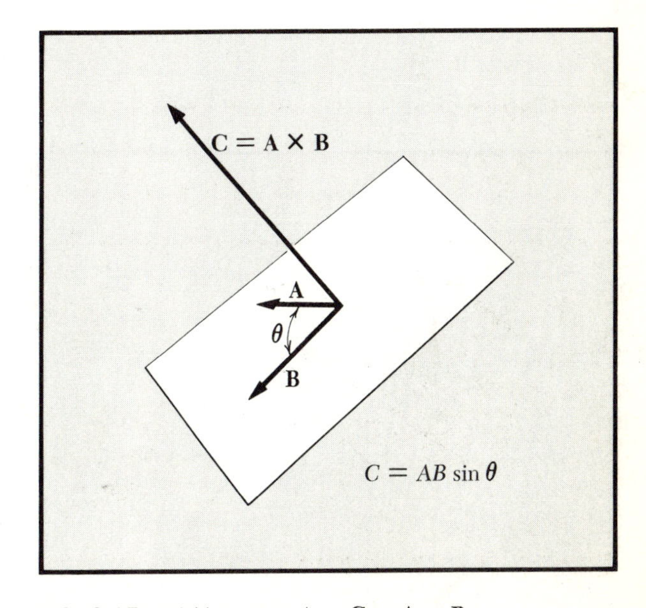

FIG. 2.17 (*a*) Vector product $\mathbf{C} = \mathbf{A} \times \mathbf{B}$.

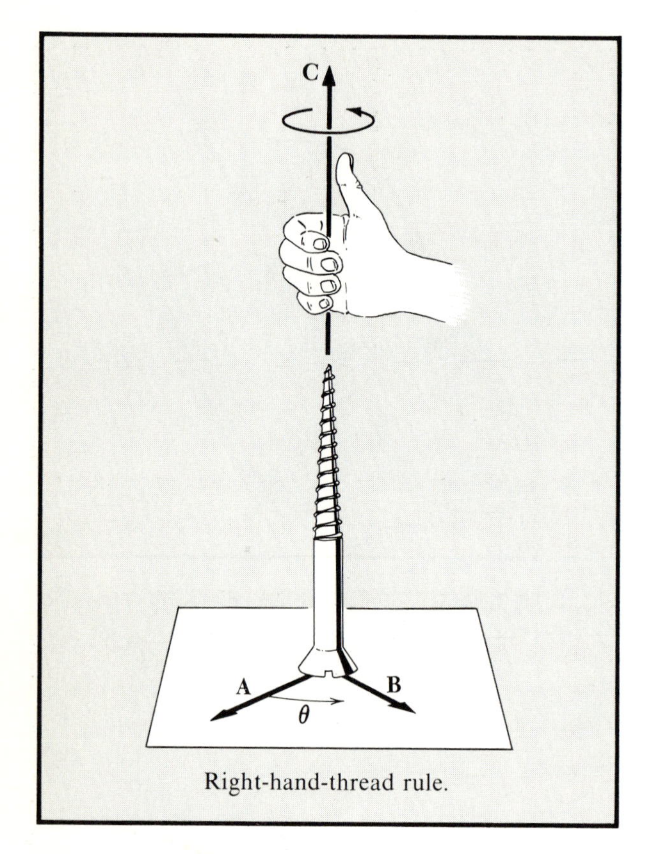

Right-hand-thread rule.

FIG. 2.17 (*cont'd.*) (*b*) Methods of determining direction of vector $\mathbf{A} \times \mathbf{B}$.

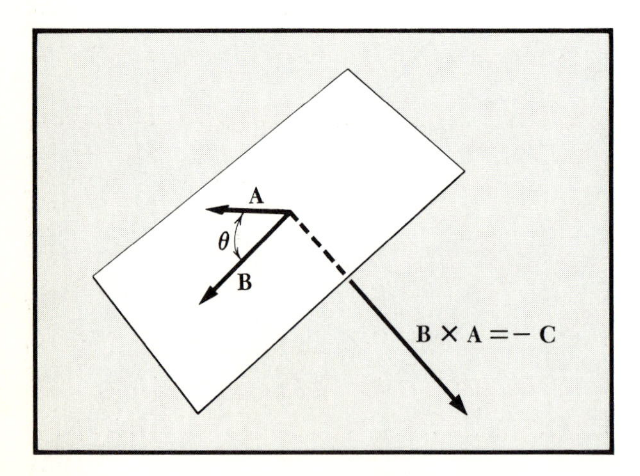

(*c*) Vector product $\mathbf{B} \times \mathbf{A}$ is opposite to $\mathbf{A} \times \mathbf{B}$.

The somewhat tedious proof may be found in any book on vector analysis.[1]

Vector Product in Cartesian Components Just as we found in Eq. (2.6) the direction cosines of a vector **A**, we can find the sines of the angles that **A** makes with the cartesian axes. This is inconvenient and the sines can more easily be found from the cosines. However, it is often useful to have the expression for the vector product of two vectors in terms of their components:

$$\begin{aligned}\mathbf{A} \times \mathbf{B} &= (A_x\hat{\mathbf{x}} + A_y\hat{\mathbf{y}} + A_z\hat{\mathbf{z}}) \times (B_x\hat{\mathbf{x}} + B_y\hat{\mathbf{y}} + B_z\hat{\mathbf{z}}) \\ &= (\hat{\mathbf{x}} \times \hat{\mathbf{y}})A_xB_y + (\hat{\mathbf{x}} \times \hat{\mathbf{z}})A_xB_z + (\hat{\mathbf{y}} \times \hat{\mathbf{z}})A_yB_z \\ &\quad + (\hat{\mathbf{y}} \times \hat{\mathbf{x}})A_yB_x + (\hat{\mathbf{z}} \times \hat{\mathbf{x}})A_zB_x + (\hat{\mathbf{z}} \times \hat{\mathbf{y}})A_zB_y\end{aligned}$$

where we have used the result $\hat{\mathbf{x}} \times \hat{\mathbf{x}} = \hat{\mathbf{y}} \times \hat{\mathbf{y}} = \hat{\mathbf{z}} \times \hat{\mathbf{z}} = 0$. The question arises: What is $\hat{\mathbf{x}} \times \hat{\mathbf{y}}$? Is it $\hat{\mathbf{z}}$ or $-\hat{\mathbf{z}}$? We make the choice that $\hat{\mathbf{x}} \times \hat{\mathbf{y}} = +\hat{\mathbf{z}}$ and construct the coordinate directions accordingly. This is called a *right-handed coordinate system*[2] and is used conventionally in physics. We agree to work only in the right-handed coordinate system, which is the kind shown in Fig. 2.10*a* and *b*.

Now $\hat{\mathbf{x}} \times \hat{\mathbf{z}} = -\hat{\mathbf{y}}$, $\hat{\mathbf{y}} \times \hat{\mathbf{z}} = \hat{\mathbf{x}}$, and so on, and we see that

$$\begin{aligned}\mathbf{A} \times \mathbf{B} &= \hat{\mathbf{x}}(A_yB_z - A_zB_y) \\ &\quad + \hat{\mathbf{y}}\,(A_zB_x - A_xB_z) + \hat{\mathbf{z}}(A_xB_y - A_yB_x)\end{aligned} \tag{2.14}$$

Note that if the indices are cyclic with xyz, the term enters the vector product with a positive sign; otherwise the sign is negative. If you are familiar with determinants, you can confirm readily that the representation

[1] For example, see C. E. Weatherburn, "Elementary Vector Analysis," p. 57, G. Bell & Sons, Ltd., London, 1928; J. G. Coffin, "Vector Analysis," p. 35, John Wiley & Sons, Inc., New York, 1911.

[2] How would we communicate our definition of right-handed to a creature in another solar system in our galaxy? We can do this by using circularly polarized radio waves. The signal carries a message that tells the remote observer in which sense we defined the waves to be polarized. The remote observer will have constructed two receivers, one with the correct sense and one with the incorrect sense, in terms of signal strength. Any method requires clear instructions: In the original analysis of the spectroscopic Zeeman effect, its discoverer incorrectly associated a positive sign to the oscillating charges in atoms because he misunderstood the sense of circularly polarized radiation. [See P. Zeeman, *Philosophical Magazine*, (5)**43**: 55 and 226 (1897). In a similar connection the first Telstar transmission on July 11, 1962, was poorly received in Great Britain because "of the reversal of a small component in the aerial feed which arose from an ambiguity in the accepted definition of the sense of rotation of radio waves." *Times* (*London*), July 13, 1962, p. 11.]

$$\mathbf{A} \times \mathbf{B} = \begin{vmatrix} \hat{\mathbf{x}} & \hat{\mathbf{y}} & \hat{\mathbf{z}} \\ A_x & A_y & A_z \\ B_x & B_y & B_z \end{vmatrix} \tag{2.15}$$

is equivalent to Eq. (2.14) and it is easier to remember.

Applications of the Vector Product In the following paragraphs we treat several applications of the vector product.

1 *Area of a parallelogram.* The magnitude

$$|\mathbf{A} \times \mathbf{B}| = AB|\sin(\mathbf{A},\mathbf{B})|$$

is the area of the parallelogram with sides **A** and **B** (or twice the area of the triangle with sides **A** and **B**)(see Fig. 2.18*a*). The direction of $\mathbf{A} \times \mathbf{B}$ is normal to the plane of the parallelogram; thus we may think of $\mathbf{A} \times \mathbf{B}$ as the *vector area* of the parallelogram. Because we have given signs to the sides **A** and **B**, the vector area comes endowed with a direction. There are physical applications where it is convenient to be able to give a direction to an area [see Eq. (2.11)].

2 *Volume of a parallelepiped.* The scalar

$$|(\mathbf{A} \times \mathbf{B}) \cdot \mathbf{C}| = V$$

is the volume of the parallelepiped of which $\mathbf{A} \times \mathbf{B}$ is the area of the base and **C** the slant height or edge (Fig. 2.18*b*). If the three vectors **A**, **B**, and **C** lie in the same plane, the volume will be zero; thus three vectors are coplanar if and only if $(\mathbf{A} \times \mathbf{B}) \cdot \mathbf{C} = 0$.

We note from inspection of the figure that

$$\mathbf{A} \cdot (\mathbf{B} \times \mathbf{C}) = (\mathbf{A} \times \mathbf{B}) \cdot \mathbf{C}$$

so that *the dot and the cross in the scalar triple product may be interchanged without altering the value of the product.* However,

$$\mathbf{A} \cdot (\mathbf{B} \times \mathbf{C}) = -\mathbf{A} \cdot (\mathbf{C} \times \mathbf{B})$$

A scalar triple product is not altered by permuting cyclically the order of the vectors, but it is reversed in sign if the cyclic order is changed. (Cyclic permutations of *ABC* are *BCA* and *CAB*; noncylic orderings of *ABC* are *BAC*, *ACB*, and *CBA*.)

3 *Law of sines.* Consider the triangle defined by $\mathbf{C} = \mathbf{A} + \mathbf{B}$ (Fig. 2.18*c*), and take the vector product of both sides of the equation with **A**:

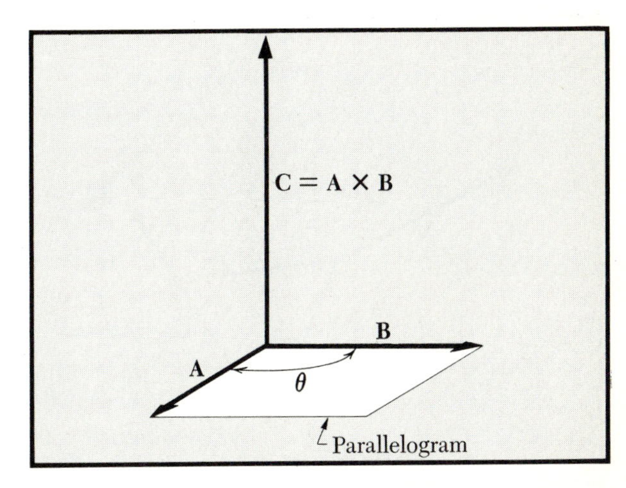

FIG. 2.18 (*a*) Vector area of parallelogram is $\mathbf{C} = \mathbf{A} \times \mathbf{B} = AB|\sin\theta|\,\hat{\mathbf{C}}$.

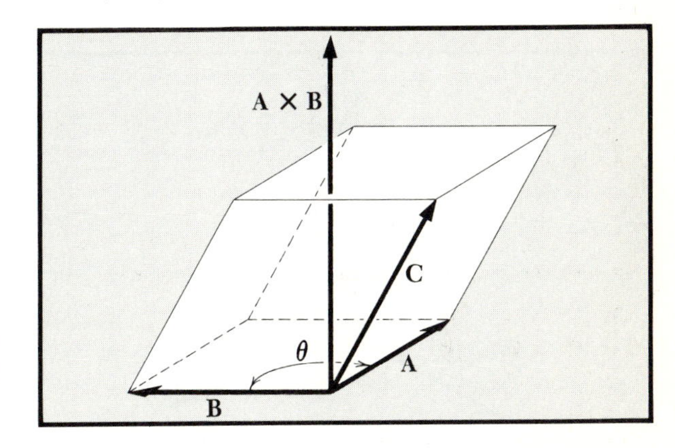

(*b*) $\mathbf{A} \times \mathbf{B} \cdot \mathbf{C}$ = base area × height = volume of parallelepiped.

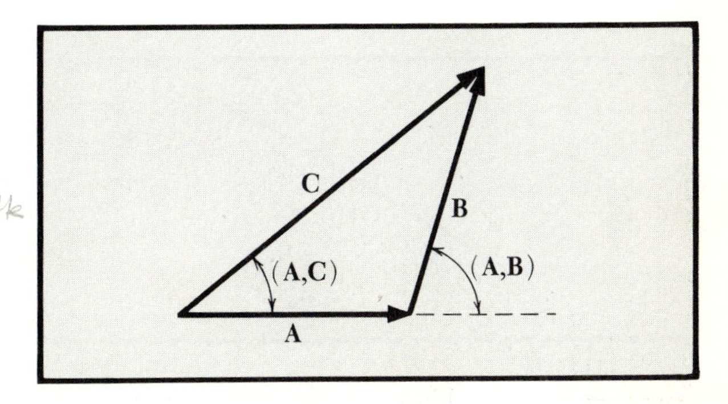

(*c*) Law of sines of triangle. Note: $\sin(\mathbf{A},\mathbf{B}) = \sin[\pi - (\mathbf{A},\mathbf{B})]$.

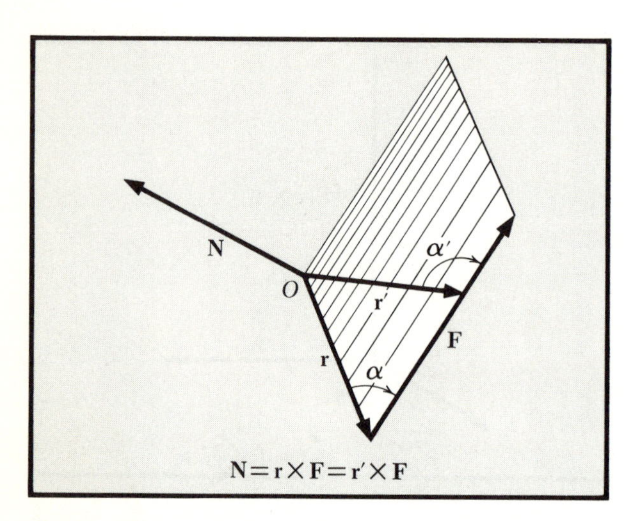

FIG. 2.18 (*cont'd.*) (*d*) Torque as a vector product.

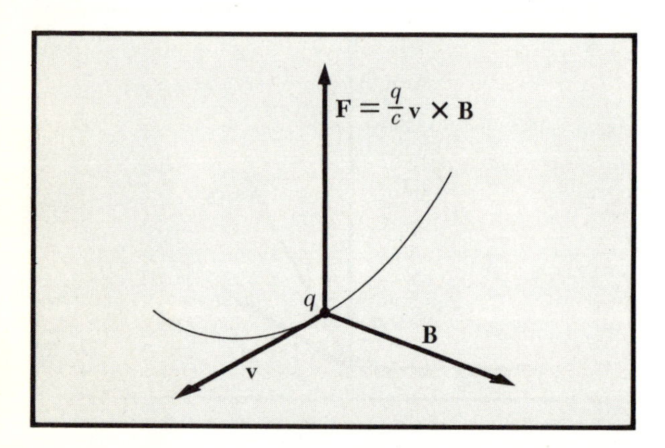

(*e*) Force on positive charge moving in a magnetic field.

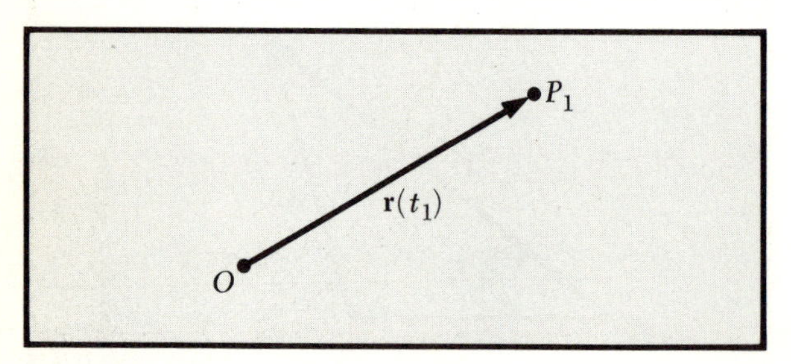

FIG. 2.19 (*a*) The position P_1 of a particle at time t_1 is specified by the vector $\mathbf{r}(t_1)$ relative to the fixed origin at point O.

$$\mathbf{A} \times \mathbf{C} = \mathbf{A} \times \mathbf{A} + \mathbf{A} \times \mathbf{B}$$

Now $\mathbf{A} \times \mathbf{A} = 0$, and the magnitudes of both sides must be equal so that

$$AC \sin(\mathbf{A},\mathbf{C}) = AB \sin(\mathbf{A},\mathbf{B})$$

or

$$\frac{\sin(\mathbf{A},\mathbf{C})}{B} = \frac{\sin(\mathbf{A},\mathbf{B})}{C} \tag{2.16}$$

This is known as the *law of sines of a triangle.*

4 *Torque.* The idea of torque is familiar from most introductory courses in physics. It is particularly important in the motion of rigid bodies discussed in Chap. 8. The torque is referred to a point and has a convenient expression in terms of vectors

$$\mathbf{N} = \mathbf{r} \times \mathbf{F} \tag{2.17}$$

where $\mathbf{r}$ is a vector from the point to the vector $\mathbf{F}$. From Fig. 2.18*d* we see that the torque has a direction perpendicular to $\mathbf{r}$ and to $\mathbf{F}$. Note that the magnitude of $\mathbf{N}$ is $rF \sin \alpha$ and $r \sin \alpha$ is the length of the perpendicular from the point (O in the figure) to $\mathbf{F}$. In the figure $r \sin \alpha = r' \sin \alpha'$. Hence the torque both in direction and in magnitude is independent of the point along $\mathbf{F}$ to which $\mathbf{r}$ is drawn.

5 *Force on a particle in a magnetic field.* The force on a point electric charge moving with velocity $\mathbf{v}$ in a magnetic field $\mathbf{B}$ is proportional to v times the perpendicular component of B; in terms of the vector product (see Fig. 2.18*e*).

$$\mathbf{F} = \frac{q}{c}\mathbf{v} \times \mathbf{B} \qquad \text{(gaussian units)}$$

$$\mathbf{F} = q\mathbf{v} \times \mathbf{B} \qquad \text{(mks units)} \tag{2.18}$$

Here q is the charge on the particle and c is the speed of light. This force law is developed in detail in Volume 2 and is used in Chap. 3 (page 70).

VECTOR DERIVATIVES

The velocity $\mathbf{v}$ of a particle is a vector; the acceleration $\mathbf{a}$ is also a vector. The *velocity* is the time rate of change of the position of a particle. The position of a particle at any time

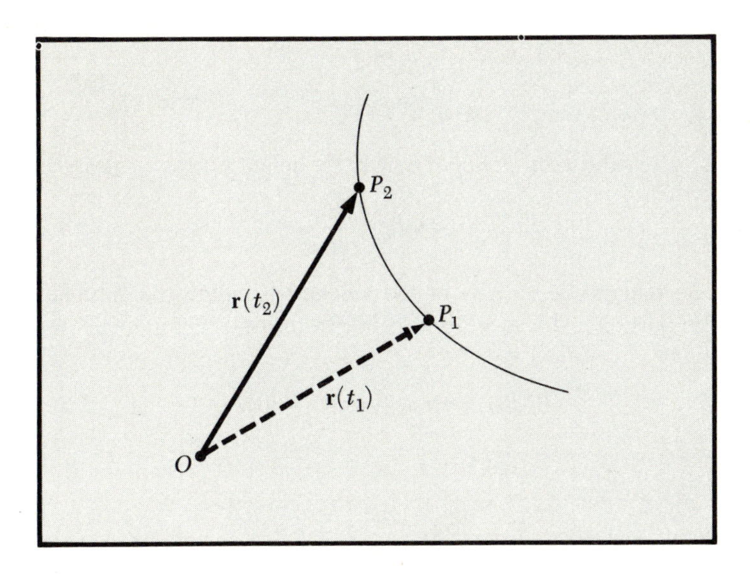

(b) The particle has advanced to P_2 at time t_2.

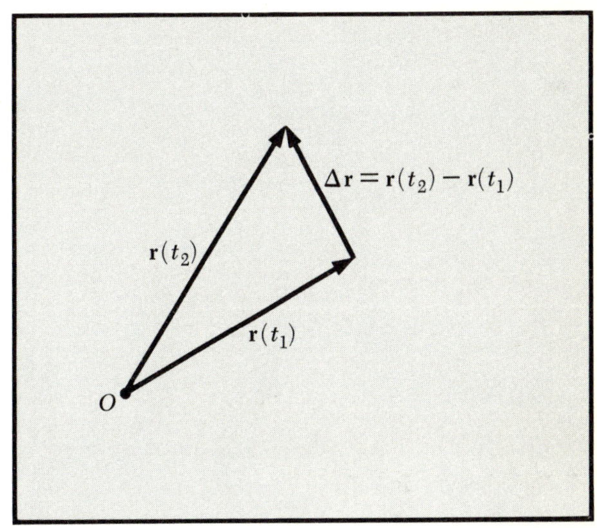

(c) The vector $\Delta\mathbf{r}$ is the difference between $\mathbf{r}(t_2)$ and $\mathbf{r}(t_1)$.

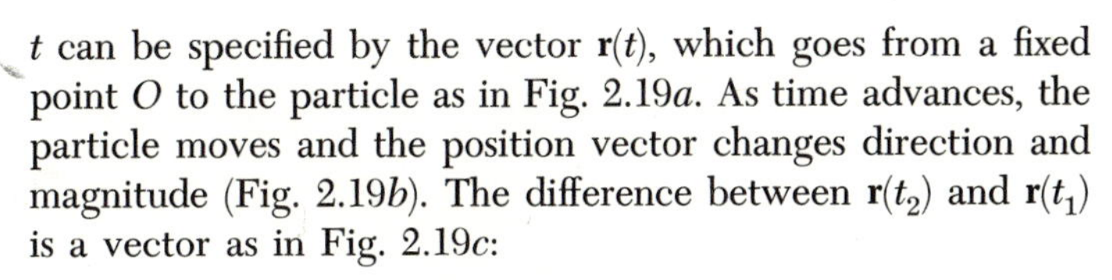

t can be specified by the vector $\mathbf{r}(t)$, which goes from a fixed point O to the particle as in Fig. 2.19a. As time advances, the particle moves and the position vector changes direction and magnitude (Fig. 2.19b). The difference between $\mathbf{r}(t_2)$ and $\mathbf{r}(t_1)$ is a vector as in Fig. 2.19c:

$$\Delta\mathbf{r} = \mathbf{r}(t_2) - \mathbf{r}(t_1)$$

If the vector $\mathbf{r}$ can be regarded as a function (a vector function) of the single scalar variable t, the value of $\Delta\mathbf{r}$ will be completely determined when the two values t_1 and t_2 are known. Thus in Fig. 2.19d $\Delta\mathbf{r}$ is the chord P_1P_2. The ratio

$$\frac{\Delta\mathbf{r}}{\Delta t}$$

is a vector collinear with the chord P_1P_2 but magnified in the ratio $1/\Delta t$. As Δt approaches zero, P_2 approaches P_1 and the chord P_1P_2 approaches the tangent at P_1. Then the vector

$$\frac{\Delta\mathbf{r}}{\Delta t} \quad \text{will approach} \quad \frac{d\mathbf{r}}{dt}$$

which is a vector tangent to the curve at P_1 directed in the sense in which the variable t increases along the curve (Fig. 2.19e).

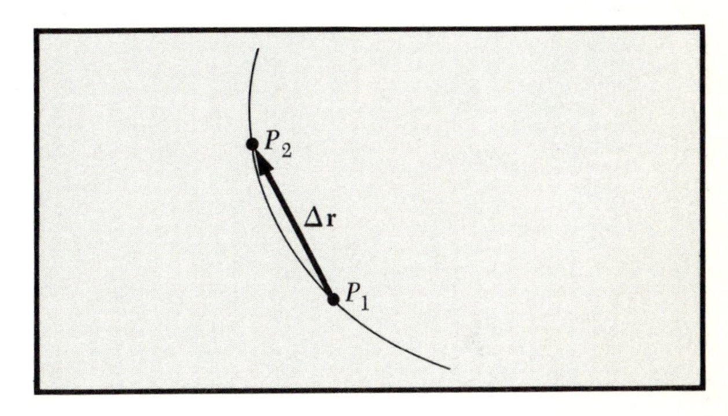

(d) $\Delta\mathbf{r}$ is the chord between the points P_1 and P_2 on the trajectory of the particle.

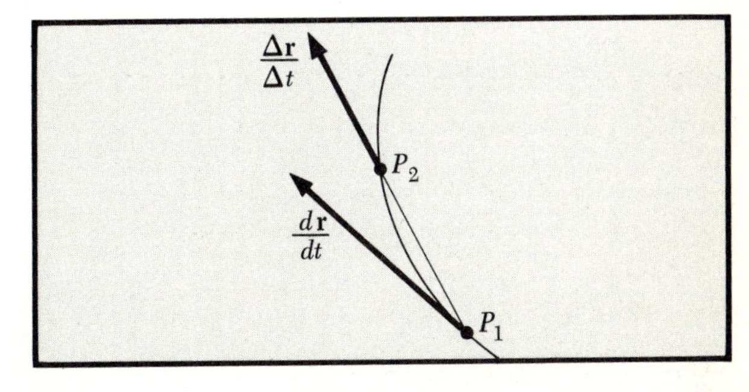

(e) As $\Delta t = t_2 - t_1 \rightarrow 0$, the vector $\Delta\mathbf{r}/\Delta t$ collinear with the chord approaches the velocity vector $d\mathbf{r}/dt$ collinear with the tangent to the trajectory at point P_1.

Velocity The vector

$$\frac{d\mathbf{r}}{dt} = \lim_{\Delta t \to 0} \frac{\Delta \mathbf{r}}{\Delta t}$$

is called the *time derivative* of **r**. By definition the velocity is

$$\mathbf{v}(t) \equiv \frac{d\mathbf{r}}{dt} \tag{2.19}$$

The magnitude $v = |\mathbf{v}|$ of the velocity is called the *speed* of the particle. The speed is a scalar. In terms of components we write

$$\mathbf{r}(t) = x(t)\hat{\mathbf{x}} + y(t)\hat{\mathbf{y}} + z(t)\hat{\mathbf{z}} \tag{2.20}$$

and

$$\boxed{\frac{d\mathbf{r}}{dt} = \mathbf{v} = \frac{dx}{dt}\hat{\mathbf{x}} + \frac{dy}{dt}\hat{\mathbf{y}} + \frac{dz}{dt}\hat{\mathbf{z}} = v_x\hat{\mathbf{x}} + v_y\hat{\mathbf{y}} + v_z\hat{\mathbf{z}}} \tag{2.21}$$

$$v = |\mathbf{v}| = \sqrt{v_x^2 + v_y^2 + v_z^2}$$

where we have assumed that the unit vectors do not change with time, so that

$$\frac{d\hat{\mathbf{x}}}{dt} = 0 = \frac{d\hat{\mathbf{y}}}{dt} = \frac{d\hat{\mathbf{z}}}{dt}$$

In general, we may write, without expressing **r** in components as in Eq. (2.20),

$$\mathbf{r}(t) = r(t)\hat{\mathbf{r}}(t)$$

where the scalar $r(t)$ is the length of the vector and $\hat{\mathbf{r}}(t)$ is a vector of unit length in the direction of **r**. The derivative of $\mathbf{r}(t)$ is defined as

$$\frac{d\mathbf{r}}{dt} = \frac{d}{dt}\left[r(t)\hat{\mathbf{r}}(t)\right] = \lim_{\Delta t \to 0} \frac{r(t + \Delta t)\hat{\mathbf{r}}(t + \Delta t) - r(t)\hat{\mathbf{r}}(t)}{\Delta t} \tag{2.22}$$

We may rewrite the numerator,[1] retaining only the first two terms in the series expansions of $r(t + \Delta t)$ and $\hat{\mathbf{r}}(t + \Delta t)$:

$$\left[r(t) + \frac{dr}{dt}\Delta t\right]\left[\hat{\mathbf{r}}(t) + \frac{d\hat{\mathbf{r}}}{dt}\Delta t\right] - r(t)\hat{\mathbf{r}}(t)$$

$$= \Delta t\left(\frac{dr}{dt}\hat{\mathbf{r}} + r\frac{d\hat{\mathbf{r}}}{dt}\right) + (\Delta t)^2\left(\frac{dr}{dt}\frac{d\hat{\mathbf{r}}}{dt}\right)$$

[1] See page 53 at the end of this chapter for expansion in series.

When this is placed in Eq. (2.22), the second term in the quotient goes to 0 as $\Delta t \to 0$ and we have

$$\mathbf{v} = \frac{d\mathbf{r}}{dt} = \frac{dr}{dt}\hat{\mathbf{r}} + r\frac{d\hat{\mathbf{r}}}{dt} \tag{2.23}$$

Here $d\hat{\mathbf{r}}/dt$ represents the rate of change of direction of the unit vector $\hat{\mathbf{r}}$. This is an example of the general rule for differentiation of the product of a scalar $a(t)$ and a vector $\mathbf{b}(t)$

$$\frac{d}{dt}a\mathbf{b} = \frac{da}{dt}\mathbf{b} + a\frac{d\mathbf{b}}{dt} \tag{2.24}$$

One contribution to the velocity in Eq. (2.23) comes from the change in the direction $\hat{\mathbf{r}}$; the other contribution comes from the change in the length r.

Since we shall apply the form of Eq. (2.23) for $\mathbf{v}$ (particularly in Chap. 9 for motion in a plane), we develop here an expression of that form for $d\mathbf{r}/dt$ utilizing the unit radial vector $\hat{\mathbf{r}}$ and a unit vector perpendicular to it that we shall call $\hat{\boldsymbol{\theta}}$.

In order to make clear these unit vectors and their time derivatives, consider the motion of a point on a circular path; in this case the unit vector $\hat{\mathbf{r}}$ will change in a time interval Δt by a vector increment $\Delta\hat{\mathbf{r}}$ to become $\hat{\mathbf{r}} + \Delta\hat{\mathbf{r}}$, as shown in Fig. 2.20*a*. If Δt is chosen so small as to approach zero, then $\Delta\hat{\mathbf{r}}$ takes the direction of the transverse unit vector $\hat{\boldsymbol{\theta}}$ shown in Fig. 2.20*b*.

Furthermore, as Δt and correspondingly $\Delta\theta$ approach zero, the magnitude of $\Delta\hat{\mathbf{r}}$ becomes simply

$$|\Delta\hat{\mathbf{r}}| = |\hat{\mathbf{r}}|\Delta\theta = \Delta\theta$$

(because $|\hat{\mathbf{r}}| = 1$) and so the vector $\Delta\hat{\mathbf{r}}$ and the ratio $\Delta\hat{\mathbf{r}}/\Delta t$ become

$$\Delta\hat{\mathbf{r}} = \Delta\theta\,\hat{\boldsymbol{\theta}} \qquad \frac{\Delta\hat{\mathbf{r}}}{\Delta t} = \frac{\Delta\theta}{\Delta t}\hat{\boldsymbol{\theta}}$$

When we pass to the limit of $\Delta t \to 0$, we obtain for the $\hat{\mathbf{r}}$ unit vector time derivative

$$\frac{d\hat{\mathbf{r}}}{dt} = \frac{d\theta}{dt}\hat{\boldsymbol{\theta}} \tag{2.25}$$

By similar arguments utilizing Fig. 2.20*c* it is readily shown that the $\hat{\boldsymbol{\theta}}$ time derivative is

$$\frac{d\hat{\boldsymbol{\theta}}}{dt} = -\frac{d\theta}{dt}\hat{\mathbf{r}} \tag{2.26}$$

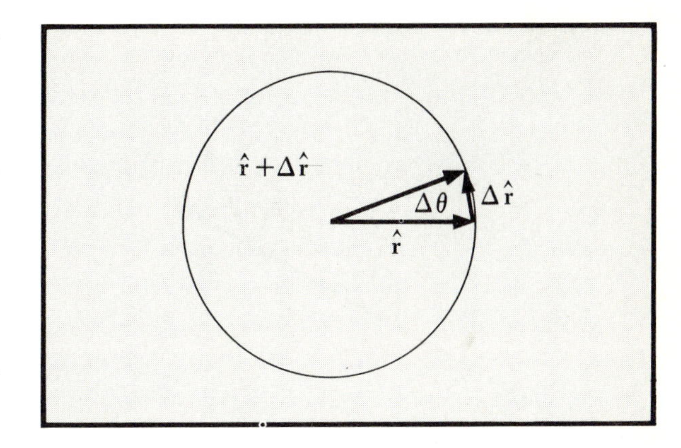

FIG. 2.20 (*a*) $\Delta\hat{\mathbf{r}}$ is the change in the unit vector $\hat{\mathbf{r}}$.

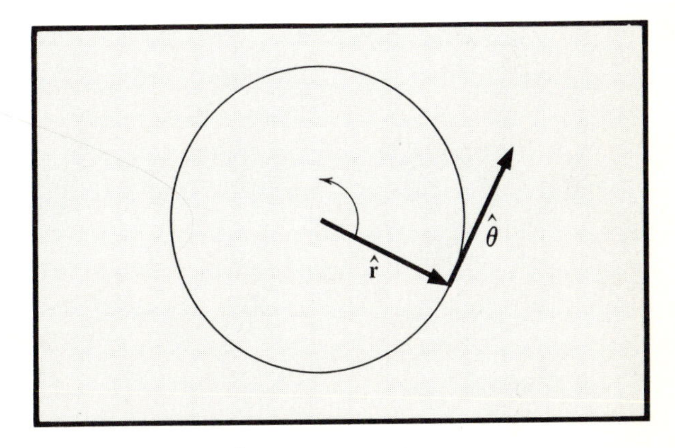

(*b*) The unit vector $\hat{\theta}$ is perpendicular to $\hat{\mathbf{r}}$ and in the direction of increasing θ.

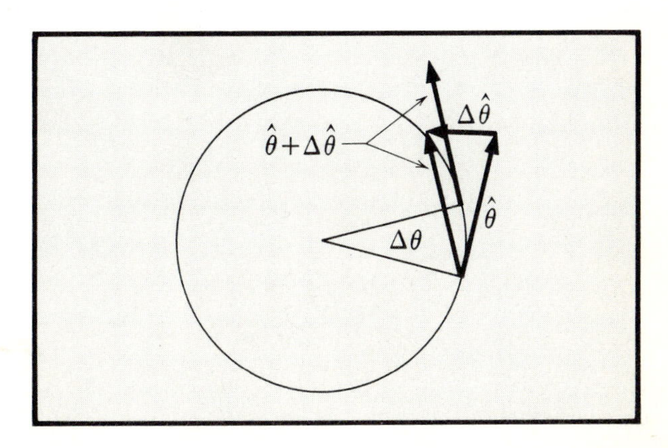

(*c*) $\Delta\hat{\theta}$ is the change in the unit vector $\hat{\theta}$.

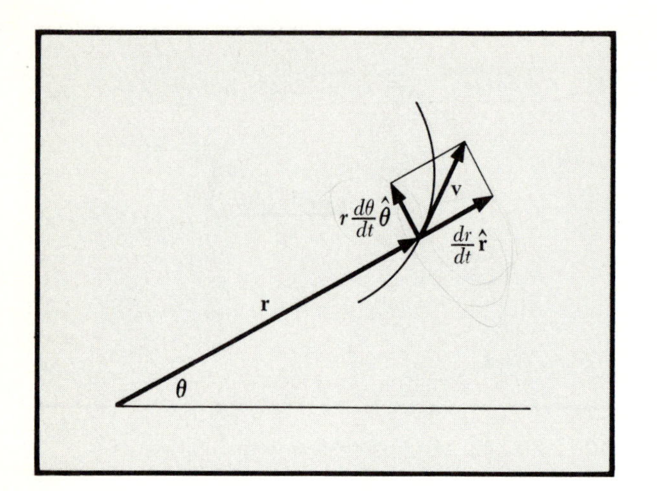

FIG. 2.21 Components of velocity vector in terms of $\hat{\mathbf{r}}$ and $\hat{\theta}$.

Now when we consider a point moving in a plane on any path, as suggested by Fig. 2.21, we recognize that the velocity vector **v** at any instant is composed of the radial component vector $dr/dt\,\hat{\mathbf{r}}$ and the transverse component vector $r\,d\hat{\mathbf{r}}/dt = r\,d\theta/dt\,\hat{\boldsymbol{\theta}}$. The latter vector utilizes Eq. (2.25). Thus the expression for **v** in the form of Eq. (2.23) is

$$\mathbf{v} = \frac{d\mathbf{r}}{dt} = \frac{dr}{dt}\hat{\mathbf{r}} + r\frac{d\theta}{dt}\hat{\boldsymbol{\theta}} \tag{2.27}$$

Acceleration Acceleration is also a vector; it is related to **v** just as **v** is related to **r**. We define acceleration as

$$\mathbf{a} \equiv \frac{d\mathbf{v}}{dt} = \frac{d^2\mathbf{r}}{dt^2} \tag{2.28}$$

Using Eq. (2.21) we obtain, in cartesian components,

$$\mathbf{a} = \frac{d\mathbf{v}}{dt} = \frac{d^2x}{dt^2}\hat{\mathbf{x}} + \frac{d^2y}{dt^2}\hat{\mathbf{y}} + \frac{d^2z}{dt^2}\hat{\mathbf{z}} \tag{2.29}$$

For the future (Chap. 9) we also need **a** in terms of r and θ; from Eq. (2.27)

$$\frac{d\mathbf{v}}{dt} = \frac{d^2r}{dt^2}\hat{\mathbf{r}} + \frac{dr}{dt}\frac{d\hat{\mathbf{r}}}{dt} + \frac{dr}{dt}\frac{d\theta}{dt}\hat{\boldsymbol{\theta}} + r\frac{d^2\theta}{dt^2}\hat{\boldsymbol{\theta}} + r\frac{d\theta}{dt}\frac{d\hat{\boldsymbol{\theta}}}{dt}$$

By reference to Eqs. (2.25) and (2.26) for $d\hat{\mathbf{r}}/dt$ and $d\hat{\boldsymbol{\theta}}/dt$, we bring this expression into the terms

$$\mathbf{a} = \frac{d\mathbf{v}}{dt} = \frac{d^2r}{dt^2}\hat{\mathbf{r}} + \frac{dr}{dt}\frac{d\theta}{dt}\hat{\boldsymbol{\theta}} + \frac{dr}{dt}\frac{d\theta}{dt}\hat{\boldsymbol{\theta}} + r\frac{d^2\theta}{dt^2}\hat{\boldsymbol{\theta}} - r\left(\frac{d\theta}{dt}\right)^2\hat{\mathbf{r}}$$

Then, by collecting terms and a little rearranging, we write this in the usual fashion:

$$\mathbf{a} = \left[\frac{d^2r}{dt^2} - r\left(\frac{d\theta}{dt}\right)^2\right]\hat{\mathbf{r}} + \frac{1}{r}\left[\frac{d}{dt}\left(r^2\frac{d\theta}{dt}\right)\right]\hat{\boldsymbol{\theta}} \tag{2.30}$$

This expression is useful in the example of motion in a circle (given below) and particularly in the study of the motion of a particle about a center of force (given in Chap. 9).

EXAMPLE

Circular Motion This example (shown in Fig. 2.22) is extremely important because of the many cases of circular motion in physics and astronomy. We want to obtain explicit expressions for the velocity and acceleration of a particle moving at constant speed in a circular orbit of constant radius r. A circular orbit can be described by

$$\mathbf{r}(t) = r\hat{\mathbf{r}}(t) \tag{2.31}$$

where r is constant and the unit vector $\hat{\mathbf{r}}$ rotates at a constant rate.

We can treat this problem in either of two ways: by using the expressions in terms of r and θ, Eqs. (2.27) and (2.30), or by using axes $\hat{\mathbf{x}}$ and $\hat{\mathbf{y}}$ fixed in space and Eqs. (2.21) and (2.29).

Method 1 Since r is constant, Eq. (2.27) gives us simply $\mathbf{v} = r\,d\theta/dt\,\hat{\boldsymbol{\theta}}$. It is customary to designate the angular velocity $d\theta/dt$ by the Greek letter ω. It is measured in radians[1] per second (rad/s) and in our present consideration is constant. Thus $\mathbf{v} = r\omega\hat{\boldsymbol{\theta}}$ and the constant speed of the particle is

$$v = \omega r \tag{2.32}$$

For the acceleration we utilize Eq. (2.30), which becomes with constant r and constant $d\theta/dt = \omega$

$$\boxed{\mathbf{a} = -r\omega^2\hat{\mathbf{r}}} \tag{2.33}$$

Thus the acceleration is constant in magnitude, and it is directed toward the center of the circular path.

Method 2 In terms of cartesian components we write the position vector of the particle at any time t in its circular motion in the form of Eq. (2.20):

$$\mathbf{r}(t) = r\cos\omega t\hat{\mathbf{x}} + r\sin\omega t\hat{\mathbf{y}} \tag{2.34}$$

The velocity vector, as given by Eq. (2.21), is then, with r constant,

$$\mathbf{v} = \frac{d\mathbf{r}}{dt} = -\omega r\sin\omega t\hat{\mathbf{x}} + \omega r\cos\omega t\hat{\mathbf{y}} \tag{2.35}$$

The speed v is the magnitude of this velocity vector

$$v = \sqrt{\mathbf{v}\cdot\mathbf{v}} = \omega r\sqrt{\sin^2\omega t + \cos^2\omega t} = \omega r \tag{2.36}$$

in agreement with Eq. (2.32). The vector $\mathbf{v}$ can be shown to be perpendicular to $\mathbf{r}$ by the fact that the scalar product of these vectors is zero.

In keeping with Eq. (2.29) we find the acceleration vector as the time derivative of $\mathbf{v}$. Differentiation of Eq. (2.35) gives

$$\begin{aligned}\mathbf{a} = \frac{d\mathbf{v}}{dt} &= -\omega^2 r\cos\omega t\hat{\mathbf{x}} - \omega^2 r\sin\omega t\hat{\mathbf{y}}\\ &= -\omega^2(r\cos\omega t\hat{\mathbf{x}} + r\sin\omega t\hat{\mathbf{y}})\\ &= -\omega^2\mathbf{r} = -\omega^2 r\hat{\mathbf{r}}\end{aligned} \tag{2.37}$$

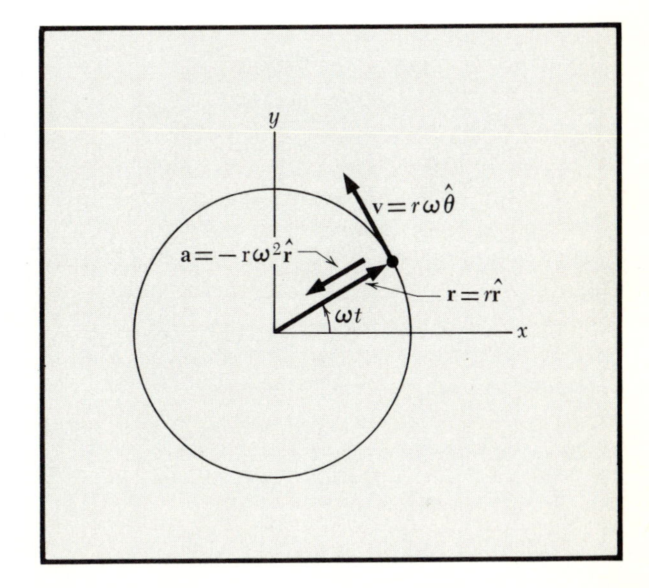

FIG. 2.22 Particle moving with constant speed in a circle of radius r. The constant angular velocity is ω. The particle velocity and acceleration are shown as derived in Eqs. (2.31) to (2.38).

[1] See page 52 at the end of this chapter for explanation of radians.

This is identical with the result obtained by Method 1 in Eq. (2.33). The acceleration has the constant magnitude $a = \omega^2 r$, and it is directed toward the center according to $-\hat{\mathbf{r}}$. By using $v = \omega r$ from Eq. (2.36) or (2.32) we may write the acceleration magnitude as

$$a = \frac{v^2}{r} \tag{2.38}$$

It is called the *centripetal* (center-seeking) acceleration, and it may be familiar to you from high school physics.

The angular velocity ω has simple connection with the ordinary frequency f. In unit time the vector $\mathbf{r}$ in Eq. (2.34) sweeps out ω rad, so that ω denotes the number of radians swept out per unit time. But the ordinary frequency f is defined as the number of complete circles swept out per unit time. Since there are 2π rad in one cycle, we must have

$$2\pi f = \omega$$

The *period* T of the motion is defined as the time to complete one cycle. We see from Eq. (2.34) that one cycle is completed in a time T such that $\omega T = 2\pi$, or

$$T = \frac{2\pi}{\omega} = \frac{1}{f}$$

For numerical orientation, suppose that the frequency f is 60 revolutions or cycles per second (60 cps). Then the period

$$T = \frac{1}{f} = \frac{1}{60} \approx 0.017 \text{ s}$$

and the angular frequency is

$$\omega = 2\pi f \approx 377 \text{ rad/s}$$

If the radius of a circular orbit is 10 cm, then the velocity is

$$v = \omega r \approx (377)(10) \approx 3.8 \times 10^3 \text{ cm/s}$$

The acceleration at any point of the orbit is

$$a = \omega^2 r \approx (377)^2(10) \approx 1.42 \times 10^6 \text{ cm/s}^2$$

In Chap. 4 a numerical example is worked out which shows that the acceleration of a point fixed on the surface of the earth at the equator due to the rotation of the earth about its own axis is about 3.4 cm/s^2.

INVARIANTS

We have mentioned (page 28) that independence of the choice of coordinate axes is an important aspect of the laws of physics and an important reason for using vector notation. Let us consider the value of the magnitude of a vector in two different coordinate systems that have a common origin but are rotated with respect to each other as in Fig. 2.23. In the two coordinate systems

$$\mathbf{A} = A_x\hat{\mathbf{x}} + A_y\hat{\mathbf{y}} + A_z\hat{\mathbf{z}}$$

and

$$\mathbf{A} = A'_x\hat{\mathbf{x}}' + A'_y\hat{\mathbf{y}}' + A'_z\hat{\mathbf{z}}' \qquad \text{(see Fig. 2.23)}$$

Since **A** has not changed, A^2 must be the same and so

$$A_x^{\,2} + A_y^{\,2} + A_z^{\,2} = A'^{\,2}_{x'} + A'^{\,2}_{y'} + A'^{\,2}_{z'}$$

In other words the magnitude of a vector is the same in all cartesian coordinate systems that differ by a rigid rotation of the coordinate axis; this is called a *form invariant*. Problem 20 (at the end of this chapter) provides a method of verifying this invariant. It is evident from its definition that the scalar product given in Eq. (2.7) is a form invariant and the magnitude of the vector product is still another form invariant. We assume that there is no change in scale; for example, the length representing one unit is unchanged by the rotation.

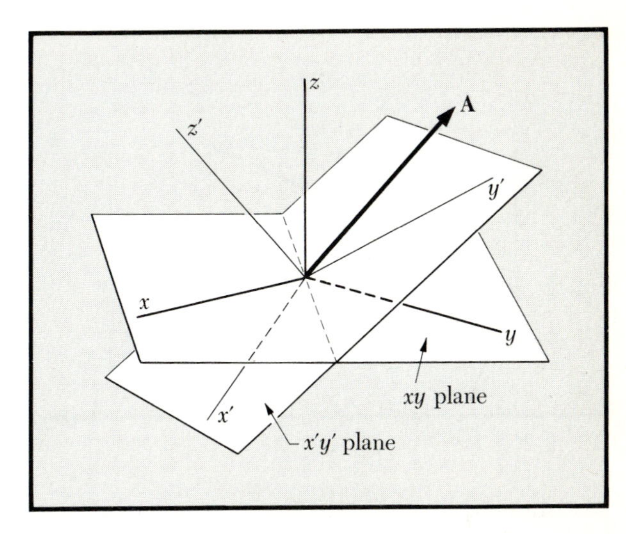

FIG. 2.23 The vector **A** can be described in coordinates xyz or coordinates $x'y'z'$ obtained from xyz by an arbitrary rotation. We say A^2 is form invariant with respect to the rotation. This means $A_x^{\,2} + A_y^{\,2} + A_z^{\,2} = A'^{\,2}_{x'} + A'^{\,2}_{y'} + A'^{\,2}_{z'}$.

We sometimes speak of a scalar function of position, such as the temperature $T(x,y,z)$ at the point (x,y,z) as a *scalar field*. Similarly, a vector whose value is a function of position, such as the velocity $\mathbf{v}(x,y,z)$ of a particle when at the point (x,y,z), is spoken of as a *vector field*. Much of the subject of vector analysis is concerned with scalar and vector fields and with differential operations on vectors, which are discussed fully in Volume 2.

Examples of Various Elementary Vector Operations We consider the vector (Fig. 2.24)

$$\mathbf{A} = 3\hat{\mathbf{x}} + \hat{\mathbf{y}} + 2\hat{\mathbf{z}}$$

(1) Find the length of **A**. We form A^2:

$$A^2 = \mathbf{A} \cdot \mathbf{A} = 3^2 + 1^2 + 2^2 = 14$$

so that $A = \sqrt{14}$ is the length of **A**.

(2) What is the length of the projection of **A** on the xy plane? The vector which is the projection of **A** on the xy plane is $3\hat{\mathbf{x}} + \hat{\mathbf{y}}$; the square of the length of this vector is $3^2 + 1^2 = 10$.

(3) Construct a vector in the xy plane and perpendicular to **A**. We want a vector of the form

$$\mathbf{B} = B_x\hat{\mathbf{x}} + B_y\hat{\mathbf{y}}$$

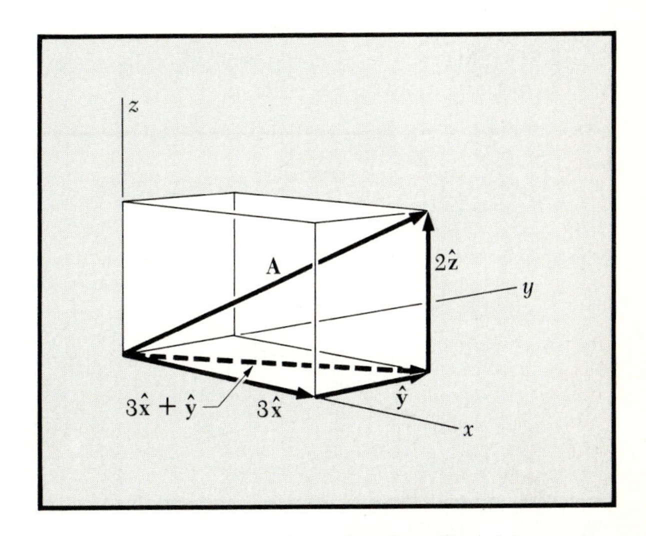

FIG. 2.24 The vector $\mathbf{A} = 3\hat{\mathbf{x}} + \hat{\mathbf{y}} + 2\hat{\mathbf{z}}$ and its projection on the xy plane.

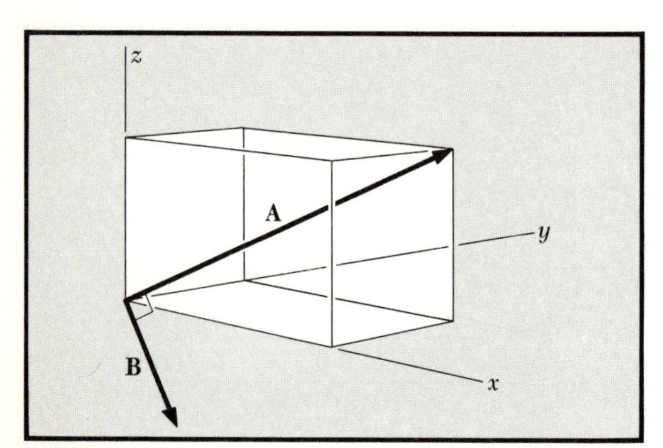

FIG. 2.25 The vector **B** is in the xy plane and perpendicular to **A**.

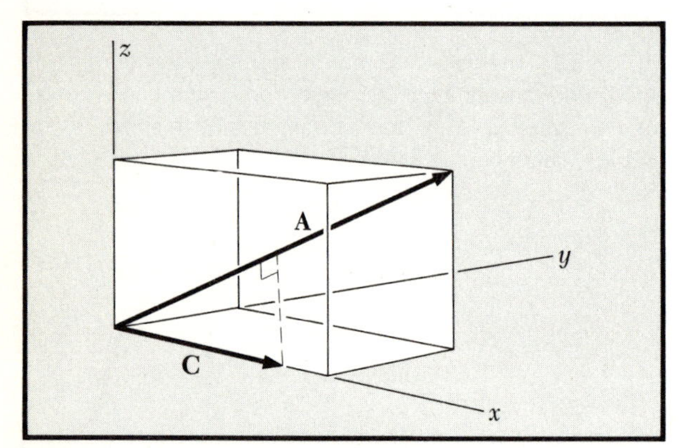

FIG. 2.26 Projection on **A** of the vector $\mathbf{C} = 2\hat{\mathbf{x}}$. $\mathbf{A} \cdot \mathbf{C}$ = (projection of **C** on **A**) times A.

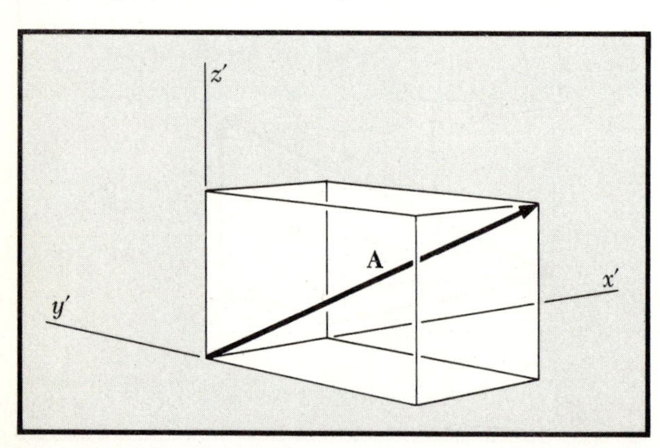

FIG. 2.27 The primed reference frame x', y', z' is generated from the unprimed system x, y, z, by a rotation of $+\pi/2$ about the z axis.

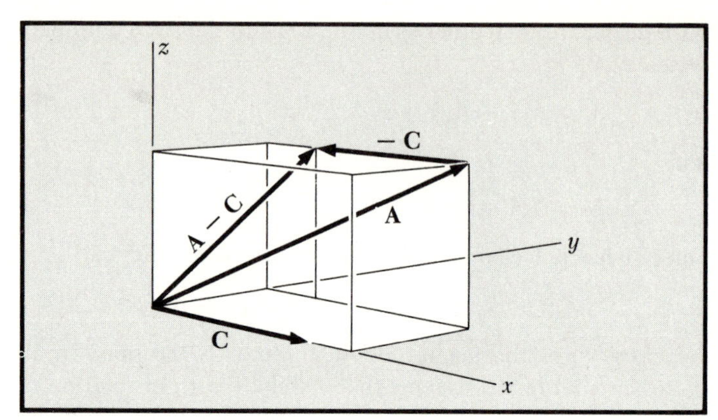

FIG. 2.28 Vector $\mathbf{A} - \mathbf{C}$.

with the property $\mathbf{A} \cdot \mathbf{B} = 0$, or

$$(3\hat{\mathbf{x}} + \hat{\mathbf{y}} + 2\hat{\mathbf{z}}) \cdot (B_x\hat{\mathbf{x}} + B_y\hat{\mathbf{y}}) = 0$$

On taking the scalar product we find

$$3B_x + B_y = 0$$

or

$$\frac{B_y}{B_x} = -3$$

The length of the vector **B** is not determined by the specification of the problem (see Fig. 2.25).

(4) Construct the unit vector $\hat{\mathbf{B}}$. We must have $\hat{B}_x^2 + \hat{B}_y^2 = 1$ or

$$\hat{B}_x^2(1^2 + 3^2) = 10\,\hat{B}_x^2 = 1$$

Thus

$$\hat{\mathbf{B}} = \sqrt{\tfrac{1}{10}}\hat{\mathbf{x}} - \sqrt{\tfrac{9}{10}}\hat{\mathbf{y}} = \frac{\hat{\mathbf{x}} - 3\hat{\mathbf{y}}}{\sqrt{10}}$$

(5) Find the scalar product with **A** of the vector $\mathbf{C} = 2\hat{\mathbf{x}}$ (see Fig. 2.26). This is seen directly to be $2 \times 3 = 6$.

(6) Find the form of **A** and **C** in a reference frame obtained from the old reference frame by a rotation of $\pi/2$ clockwise looking along the positive z axis (Fig. 2.27). The new unit vectors $\hat{\mathbf{x}}', \hat{\mathbf{y}}', \hat{\mathbf{z}}'$ are related to the old $\hat{\mathbf{x}}, \hat{\mathbf{y}}, \hat{\mathbf{z}}$ by

$$\hat{\mathbf{x}}' = \hat{\mathbf{y}} \qquad \hat{\mathbf{y}}' = -\hat{\mathbf{x}} \qquad \hat{\mathbf{z}}' = \hat{\mathbf{z}}$$

Where $\hat{\mathbf{x}}$ appeared we now have $-\hat{\mathbf{y}}'$ and where $\hat{\mathbf{y}}$ appeared we now have $\hat{\mathbf{x}}'$, so that

$$\mathbf{A} = \hat{\mathbf{x}}' - 3\hat{\mathbf{y}}' + 2\hat{\mathbf{z}}' \qquad \mathbf{C} = -2\hat{\mathbf{y}}'$$

(7) Find the scalar product $\mathbf{A} \cdot \mathbf{C}$ in the primed coordinate system. From the result of (6) we get $(-3)(-2) = 6$ exactly as in the unprimed system.

(8) Find the vector product $\mathbf{A} \times \mathbf{C}$. In the unprimed system that is

$$\begin{vmatrix} \hat{\mathbf{x}} & \hat{\mathbf{y}} & \hat{\mathbf{z}} \\ 3 & 1 & 2 \\ 2 & 0 & 0 \end{vmatrix} = 4\hat{\mathbf{y}} - 2\hat{\mathbf{z}}$$

By forming the scalar products, you may confirm that this vector is perpendicular both to $\mathbf{A}$ and to $\mathbf{C}$.

(9) Form the vector $\mathbf{A} - \mathbf{C}$. We have (see Fig. 2.28)

$$\mathbf{A} - \mathbf{C} = (3 - 2)\hat{\mathbf{x}} + \hat{\mathbf{y}} + 2\hat{\mathbf{z}} = \hat{\mathbf{x}} + \hat{\mathbf{y}} + 2\hat{\mathbf{z}}$$

PROBLEMS

1. *Position vectors.* Using the x axis as east, the y axis as north, and the z axis as up, give the vector representing the following points:
(*a*) 10 mi northeast and 2 mi up
(*b*) 5 yd southeast and 5 yd down
(*c*) 1 cm northwest and 6 cm up
Find the magnitude of each vector and the expression for the unit vector in that direction.

2. *Vector components.* Using the axes of Prob. 1, find the following:
(*a*) The components of a position vector from the origin to a point in the horizontal plane directly southeast and of length 5.0 m
(*b*) The components of a position vector to a point 15 m from the origin such that the horizontal component is 60° west from north and the vector makes an angle of 45° with the vertical

3. *Addition of vectors.* Draw the result of the following vector additions:
(*a*) Add a vector 2 cm east to one 3 cm northwest.
(*b*) Add a vector 8 cm east to one 12 cm northwest.
(*c*) Compare the results of parts (*a*) and (*b*), and frame a theorem about adding a pair of vectors that are multiples of another pair.

4. *Multiplication by a scalar.* Let $\mathbf{A} = 2.0$ cm at 70° east of north and $\mathbf{B} = 3.5$ cm at 130° east of north. Use either a protractor or polar coordinate graph paper in your solutions.
(*a*) Draw the vectors described above and two others 2.5 times as large.
(*b*) Multiply $\mathbf{A}$ by -2 and $\mathbf{B}$ by $+3$ and find the vector sum. *Ans.* 9.2 cm at 152°.
(*c*) Place a point 10 cm due north of the origin. Find multiples of $\mathbf{A}$ and $\mathbf{B}$ whose vector sum is the vector from the origin to this point.
(*d*) Work out parts (*b*) and (*c*) analytically.

5. *Scalar and vector products of two vectors.* Given two vectors $\mathbf{a} = 3\hat{\mathbf{x}} + 4\hat{\mathbf{y}} - 5\hat{\mathbf{z}}$ and $\mathbf{b} = -\hat{\mathbf{x}} + 2\hat{\mathbf{y}} + 6\hat{\mathbf{z}}$, calculate by vector methods:
(*a*) The length of each *Ans.* $a = \sqrt{50}$; $b = \sqrt{41}$.
(*b*) The scalar product $\mathbf{a} \cdot \mathbf{b}$ *Ans.* -25.
(*c*) The included angle between them *Ans.* 123.5°.
(*d*) The direction cosines for each
(*e*) The vector sum and difference $\mathbf{a} + \mathbf{b}$ and $\mathbf{a} - \mathbf{b}$ *Ans.* $\mathbf{a} + \mathbf{b} = 2\hat{\mathbf{x}} + 6\hat{\mathbf{y}} + \hat{\mathbf{z}}$.
(*f*) The vector product $\mathbf{a} \times \mathbf{b}$ *Ans.* $34\hat{\mathbf{x}} - 13\hat{\mathbf{y}} + 10\hat{\mathbf{z}}$.

6. *Vector algebra.* Given two vectors such that $\mathbf{a} + \mathbf{b} = 11\hat{\mathbf{x}} - \hat{\mathbf{y}} + 5\hat{\mathbf{z}}$ and $\mathbf{a} - \mathbf{b} = -5\hat{\mathbf{x}} + 11\hat{\mathbf{y}} + 9\hat{\mathbf{z}}$:
(*a*) Find $\mathbf{a}$ and $\mathbf{b}$.
(*b*) Find the angle included between $\mathbf{a}$ and $(\mathbf{a} + \mathbf{b})$ using vector methods.

7. *Vector addition of velocities.* In still water a man can row a boat 5 mi/h.
(*a*) If he heads straight across a stream which is flowing

2 mi/h, what will be the direction of his path and his velocity?

(*b*) In what direction must he point to travel perpendicular to the flow of the stream and what will be his speed?

8. *Composition of velocities.* The pilot of an airplane wishes to reach a point 200 mi east of his present position. A wind blows 30 mi/h from the northwest. Calculate his vector velocity with respect to the moving air mass if his schedule requires him to arrive at his destination in 40 min.

Ans. $\mathbf{v} = 279\hat{\mathbf{x}} + 21\hat{\mathbf{y}}$ mi/h; $\hat{\mathbf{x}}$ = east; $\hat{\mathbf{y}}$ = north.

9. *Vector operations; relative position vector.* Two particles are emitted from a common source and at a particular time have displacements:

$$\mathbf{r}_1 = 4\hat{\mathbf{x}} + 3\hat{\mathbf{y}} + 8\hat{\mathbf{z}} \qquad \mathbf{r}_2 = 2\hat{\mathbf{x}} + 10\hat{\mathbf{y}} + 5\hat{\mathbf{z}}$$

(*a*) Sketch the positions of the particles and write the expression for the displacement $\mathbf{r}$ of particle 2 relative to particle 1.

(*b*) Use the scalar product to find the magnitude of each vector. *Ans.* $r_1 = 9.4$; $r_2 = 11.4$; $r = 7.9$.

(*c*) Calculate the angles between all possible pairs of the three vectors.

(*d*) Calculate the projection of $\mathbf{r}$ on $\mathbf{r}_1$. *Ans.* -1.2.

(*e*) Calculate the vector product $\mathbf{r}_1 \times \mathbf{r}_2$.

Ans. $-65\hat{\mathbf{x}} - 4\hat{\mathbf{y}} + 34\hat{\mathbf{z}}$.

10. *Closest approach of two particles.* Two particles 1 and 2 travel along the x and y axes with respective velocities $\mathbf{v}_1 = 2\hat{\mathbf{x}}$ cm/s and $\mathbf{v}_2 = 3\hat{\mathbf{y}}$ cm/s. At $t = 0$ they are at

$$x_1 = -3 \text{ cm} \qquad y_1 = 0 \qquad x_2 = 0 \qquad y_2 = -3 \text{ cm}$$

(*a*) Find the vector $\mathbf{r}_2 - \mathbf{r}_1$ that represents the position of 2 relative to 1 as a function of time.

Ans. $\mathbf{r} = (3 - 2t)\hat{\mathbf{x}} + (3t - 3)\hat{\mathbf{y}}$ cm.

(*b*) When and where are these two particles closest?

Ans. $t = 1.15$ s.

11. *Body diagonals of a cube.* What is the angle between two intersecting body diagonals of a cube? (A *body diagonal* connects two corners and passes through the interior of the cube. A *face diagonal* connects two corners and runs on one face of the cube.) *Ans.* $\cos^{-1}\frac{1}{3}$.

12. *Condition for* $\mathbf{a} \perp \mathbf{b}$. Show that $\mathbf{a}$ is perpendicular to $\mathbf{b}$ if $|\mathbf{a} + \mathbf{b}| = |\mathbf{a} - \mathbf{b}|$.

13. *Parallel and perpendicular vectors.* Find x and y such that the vectors $\mathbf{B} = x\hat{\mathbf{x}} + 3\hat{\mathbf{y}}$ and $\mathbf{C} = 2\hat{\mathbf{x}} + y\hat{\mathbf{y}}$ are each perpendicular to $\mathbf{A} = 5\hat{\mathbf{x}} + 6\hat{\mathbf{y}}$. Now prove that $\mathbf{B}$ and $\mathbf{C}$ are parallel. Is it true in three dimensions that two vectors perpendicular to a third are necessarily parallel?

14. *Volume of parallelepiped.* A parallelepiped has edges described by the vectors $\hat{\mathbf{x}} + 2\hat{\mathbf{y}}$, $4\hat{\mathbf{y}}$, and $\hat{\mathbf{y}} + 3\hat{\mathbf{z}}$ from the origin. Find the volume. *Ans.* 12.

15. *Equilibrium of forces.* Three forces $\mathbf{F}_1$, $\mathbf{F}_2$, and $\mathbf{F}_3$ act simultaneously on a point particle. The resultant force $\mathbf{F}_R$ is simply the vector sum of the forces. The particle is said to be in equilibrium if $\mathbf{F}_R = 0$.

(*a*) Show that if $\mathbf{F}_R = 0$ the vectors representing three forces form a triangle.

(*b*) If $\mathbf{F}_R = 0$ as above, is it possible for any one of the vectors to lie outside the plane determined by the other two?

(*c*) A particle subject to a vertically downward force of 10 newtons (N) and suspended from a cord (tension 15 N) making an angle of 0.1 rad from the vertical cannot be in equilibrium. What third force is required to produce equilibrium?

16. *Work done by forces.* The constant forces $\mathbf{F}_1 = \hat{\mathbf{x}} + 2\hat{\mathbf{y}} + 3\hat{\mathbf{z}}$ (dynes) and $\mathbf{F}_2 = 4\hat{\mathbf{x}} - 5\hat{\mathbf{y}} - 2\hat{\mathbf{z}}$ (dynes) act together on a particle during a displacement from the point $A(20,15,0)$ (cm) to the point $B(0,0,7)$ (cm).

(*a*) What is the work done (in ergs) on the particle? The work done (Chap. 5) is given by $\mathbf{F} \cdot \mathbf{r}$, where $\mathbf{F}$ is the resultant force (here $\mathbf{F} = \mathbf{F}_1 + \mathbf{F}_2$) and $\mathbf{r}$ is the displacement.

Ans. -48 ergs.

(*b*) Calculate separately the work done by F_1 and F_2.

(*c*) Suppose the same forces were acting, but the displacement went from $\mathbf{B}$ to $\mathbf{A}$. What is the work done on the particle in this case?

17. *Torque of force about a point.* The torque or turning moment $\mathbf{N}$ of a force about a given point is given by $\mathbf{r} \times \mathbf{F}$, where $\mathbf{r}$ is the vector from the given point to the point of application of $\mathbf{F}$. Consider a force $\mathbf{F} = -3\hat{\mathbf{x}} + \hat{\mathbf{y}} + 5\hat{\mathbf{z}}$ (dynes) acting at the point $7\hat{\mathbf{x}} + 3\hat{\mathbf{y}} + \hat{\mathbf{z}}$ (cm). Remember that $\mathbf{F} \times \mathbf{r} = -\mathbf{r} \times \mathbf{F}$.

(*a*) What is the torque in dyn-cm about the origin? (Just give the result for $\mathbf{N}$ as a linear combination of $\hat{\mathbf{x}}$, $\hat{\mathbf{y}}$, and $\hat{\mathbf{z}}$.)

Ans. $14\hat{\mathbf{x}} - 38\hat{\mathbf{y}} + 16\hat{\mathbf{z}}$.

(*b*) What is the torque about the point (0,10,0)?

Ans. $-36\hat{\mathbf{x}} - 38\hat{\mathbf{y}} - 14\hat{\mathbf{z}}$.

18. *Velocity and acceleration: differentiation of vectors.* Find the velocity and acceleration of the point described by the following position vectors (t = time):

(*a*) $\mathbf{r} = 16t\hat{\mathbf{x}} + 25t^2\hat{\mathbf{y}} + 33\hat{\mathbf{z}}$

(*b*) $\mathbf{r} = 10 \sin 15t\hat{\mathbf{x}} + 35t\hat{\mathbf{y}} + e^{6t}\hat{\mathbf{z}}$

(For derivatives see the Mathematical Notes at the end of this chapter.)

19. *Random flights.* A particle follows in space a path that consists of N equal steps, each of length s. The direction in space of each step is entirely random, with no relation or correlation between any two steps. The total displacement is

$$\mathbf{S} = \sum_{i=1}^{N} \mathbf{s}_i$$

Show that the mean square displacement between initial and final positions is $\langle S^2 \rangle = Ns^2$, where $\langle\ \rangle$ denotes mean value. [*Hint:* The assumption that the direction of every step is independent of the direction of every other step means that $\langle \mathbf{s}_i \cdot \mathbf{s}_j \rangle = 0$ for all i and j, except $i = j$.]

20. *Invariance.* Consider a vector $\mathbf{A}$ in the cartesian coordinate system with unit vectors $\hat{\mathbf{x}}$, $\hat{\mathbf{y}}$, and $\hat{\mathbf{z}}$. This system is now rotated through an angle θ about the $\hat{\mathbf{z}}$ axis.

(*a*) Express the new unit vectors $\hat{\mathbf{x}}'$ and $\hat{\mathbf{y}}'$ in terms of $\hat{\mathbf{x}}$, $\hat{\mathbf{y}}$, and θ; $\hat{\mathbf{z}}' = \hat{\mathbf{z}}$.

(*b*) Express $\mathbf{A}$ in terms of $A'_{x'}$, $A'_{y'}$, $A'_{z'}$ and $\hat{\mathbf{x}}'$, $\hat{\mathbf{y}}'$, $\hat{\mathbf{z}}'$; transform to $\hat{\mathbf{x}}$, $\hat{\mathbf{y}}$, and $\hat{\mathbf{z}}$ and so find the relations between $A'_{x'}$, $A'_{y'}$, $A'_{z'}$ and A_x, A_y, A_z.

(*c*) Show that $A_x^2 + A_y^2 + A_z^2 = A'^2_{x'} + A'^2_{y'} + A'^2_{z'}$.

(This problem with an arbitrary rotation in three dimensions is complicated. One method is to use nine direction cosines among which there are six relations, three from the orthogonality of $\hat{\mathbf{x}}'$, $\hat{\mathbf{y}}'$, $\hat{\mathbf{z}}'$ and three from the fact that the sum of the squares of direction cosines is 1.)

MATHEMATICAL NOTES

Time Derivatives, Velocity, and Acceleration Dynamics involves the motion of particles and objects and consequently evolution in time; that is, some quantities describing the particles or objects are changing in time. Very often we shall use coordinates x, y, and z in our description of the physical system. Two other important types of coordinate systems, spherical polar and cylindrical, are introduced at the end of this section.

A dynamical description will consist of giving the coordinates x, y, and z as functions of the time. Figure 2.29 represents such a description with x plotted as a function of the time t. In understanding how x changes, the slope of the curve is the important characteristic. Between A and B, x increases uniformly and the slope, which is the tangent of the angle between the curve and the t axis, is constant. Between B and C the curve is parallel to the t axis and the slope is zero. We note that x does not change, and so the slope is a reflection of the x component of the velocity. Between C and D the slope becomes negative, the tangent of the angle is negative, and

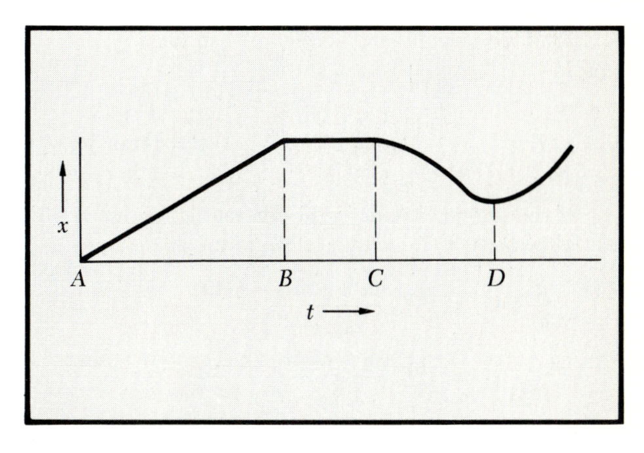

FIG. 2.29 Graph of x vs t.

x is decreasing. At D the slope becomes zero and then increases. From Eq. (2.21) we defined dx/dt as the velocity in the x direction, and this is, of course, the definition of the slope. It is important to remember that the velocity in any particular direction has a magnitude that may be either positive or negative.

It would be very time-consuming and wasteful of paper if we had to make a graph every time we wished to describe a motion. Instead we usually give a functional relation between the coordinate x, y, or z and the time t. Such a relation is $x = vt$. Since $dx/dt = v$, we see that the velocity is a constant v. Another example is $x = \frac{1}{2}at^2$; in this case $dx/dt = at = v$. We can now plot v as a function of t. What is the slope of this curve? We have already discussed it [see Eq. (2.28)] and know it is the acceleration in the x direction; thus $d^2x/dt^2 = dv/dt = a$. In the following chapters we shall refer to and use the acceleration very often.

It is worth noting here that the units of velocity are *distance* divided by *time*. There are, of course, many units of distance and many units of time. As mentioned in Chap. 1, we shall commonly use centimeters for distance and seconds for time, so that our unit of velocity is centimeters per second (cm/s). However, miles per hour (mi/h), inches per century (in./century), or kilometers per microsecond (km/μs) are perfectly feasible units. [In SI units (the mks system) meters per second (m/s) is the unit of velocity.]

In differentiating it is important to remember what is sometimes called the *chain rule*—the rule for differentiating a product of variables. The derivative of a product consists of the derivative of the first factor times the other factors, plus the derivative of the second factor times all other factors, plus the derivative of the third element times all other factors, etc.

Find the velocity and acceleration in the x, y, or z direction if:

$x = 35t$	$x = 5\cos 8t$	$x = t^2 \sin 6t$
$y = \frac{1}{2}At^2$	$y^2 = 25t$	$y = t^{\frac{1}{2}} \tan 5t$
$z = \frac{1}{2}Ct^4 + \frac{1}{4}Dt^3$	$z = 7e^{-t}$	$z = A \ln t$

If you are not familiar with differentiating sine or cosine, the following is a derivation of the formula:

$$\frac{d}{dt}\sin t = \lim_{\Delta t \to 0} \frac{\sin(t+\Delta t) - \sin t}{\Delta t}$$

$$= \lim_{\Delta t \to 0} \frac{\sin t \cos \Delta t + \cos t \sin \Delta t - \sin t}{\Delta t}$$

$$= \lim_{\Delta t \to 0} \frac{\sin t + \cos t\,\Delta t - \sin t}{\Delta t}$$

$$= \cos t \tag{2.39}$$

See Eqs. (2.44) and (2.45) for $\sin \Delta t$ and $\cos \Delta t$. Likewise

$$\frac{d}{dt}\cos t = -\sin t \tag{2.40}$$

If we wish to differentiate $\sin \omega t$, let $\omega t = z$. Then

$$\frac{d}{dt}\sin \omega t = \frac{d}{dz}\sin z \frac{dz}{dt} = \omega \cos z = \omega \cos \omega t \tag{2.41}$$

One can proceed to show

$$\frac{d}{dt}\tan t = \frac{d}{dt}\frac{\sin t}{\cos t} = \frac{\cos t}{\cos t} + \frac{\sin t \sin t}{\cos^2 t} = \frac{1}{\cos^2 t} = \sec^2 t$$

Angles In describing the position of a particle as in the case of circular motion (page 44), angles are very often useful as one element in the description. Angular velocity is the derivative of the angle with respect to time, with its vector direction parallel to the axis of rotation. There is a natural unit in the measure of an angle used throughout physics; it is the *radian* (rad). A radian is the angle subtended by an arc of a circle whose length is just equal to the radius. Since the circumference is 2π times the radius, the angle of a complete circle, which is often called 360°, is 2π rad. Dividing 360 by 2π, we get 57.3°, the degree measure of 1 rad. An angular velocity will be measured in rad/s. Knowing the angular velocity, and if the radius is constant, we can easily obtain the linear velocity by just multiplying by the radius [see Eq. (2.32)]. Notice that radians are without dimensions; that is, they are a length divided by a length, the arc divided by the radius.

The following are some problems on angular measure:

1 Find in radians the angles 90°, 240°, and 315°.

2 If $\theta = \frac{1}{5}t$, find the angular velocity. Assuming θ is in radians, find the angular velocity in degrees per second (°/s).

3 A particle moves in a circle of radius 15 cm with a speed of 5 cm/s. Find the angular velocity.

The Function e^x An interesting question from the mathematical point of view is: What function has a derivative that is equal to the function itself? If one imagines that this function can be represented by an infinite series, then one can guess such a series:

$$1 + x + \frac{x^2}{2!} + \frac{x^3}{3!} + \frac{x^4}{4!} + \frac{x^5}{5!} + \frac{x^6}{6!} + \frac{x^7}{7!} \cdots \frac{x^n}{n!} \cdots$$

If we differentiate this with respect to x, we see that the first term gives 0, but the next gives 1, the next x, the next $x^2/2!$, and so on, so that we have just what we started with. We now define this function as e^x. What is e? Setting $x = 1$, we have $e^1 = e$. Thus $e = 1 + 1 + 1/2 + 1/3! + 1/4! + 1/5! \cdots = 2.7183. \ldots$ One can also check that $e^{x+y} = e^x e^y$. We might wonder why it is that 10^x is not a function like this. In other words, where does this quantity e come from? Suppose we calculate $d(10^x)/dx$:

$$\lim_{\Delta x \to 0} \frac{10^{x+\Delta x} - 10^x}{\Delta x} = \lim_{\Delta x \to 0} \frac{10^x 10^{\Delta x} - 10^x}{\Delta x}$$

$$= \lim_{\Delta x \to 0} \frac{10^x(10^{\Delta x} - 1)}{\Delta x}$$

$$= 10^x \times 2.30 \ldots = 2.30 \ldots \times 10^x \dagger$$

From this point of view, we see that e is just such a quantity that

$$\frac{de^x}{dx} = e^x \tag{2.42}$$

One of the reasons that such a quantity is important in physics is that we very often meet with an equation $dy/dx = ky$ or the derivative of y is equal to a constant times y. We see that we can write this $dy/k\,dx = y$; and if we let kx be the independent variable z, then $dy/dz = y$. Remembering Eq. (2.42), we see that $y = e^z = e^{kx}$ is a function that satisfies our equation. Therefore we have "found a solution" to the equation $dy/dx = ky$.

Some properties of e^x are: $e^0 = 1$; $e^{-\infty} = 0$; $e^1 = e$; and $e^\alpha \approx 1 + \alpha$, where α is very small. Note here that the series

† The factor 2.30 . . . is the natural logarithm of 10. Let $10^{\Delta x} = 1 + \alpha$ where both Δx and α are small quantities:

$$\log_e 10^{\Delta x} = 2.30 \ldots \log_{10} 10^{\Delta x} = 2.30 \ldots \Delta x$$
$$\log_e(1+\alpha) = \alpha$$

Therefore $\alpha = 2.30 \ldots \Delta x$. You can check this result by the use of a table of logarithms.

for e^x looks a little like the series for sine and cosine, except that in the sine and cosine the terms alternate in sign and have only odd or even powers of x, respectively. Those with experience in mathematics know that $\sqrt{-1} = i$ when raised to increasing powers alternates in sign. Let us see what $e^{i\theta}$ looks like:

$$e^{i\theta} = 1 + i\theta + \frac{(i\theta)^2}{2!} + \frac{(i\theta)^3}{3!} + \frac{(i\theta)^4}{4!} + \frac{(i\theta)^5}{5!} \tag{2.43}$$

$$= 1 - \frac{\theta^2}{2!} + \frac{\theta^4}{4!} - \frac{\theta^6}{6!} + i\theta - \frac{i\theta^3}{3!} + \frac{i\theta^5}{5!} - \frac{i\theta^7}{7!} \cdots$$

which is just $\cos\theta + i\sin\theta$. The relation

$$e^{i\theta} = \cos\theta + i\sin\theta$$

is called *De Moivre's theorem*, and we shall have occasion to use it in Chap. 7 (in the Mathematical Notes).

In the derivation of Eq. (2.39), we noted that $\sin(\theta + \Delta\theta) = \sin\theta\cos\Delta\theta + \cos\theta\sin\Delta\theta \approx \sin\theta + \cos\theta\,\Delta\theta$, where $\Delta\theta$ is a small quantity. In other words, $\sin\Delta\theta \approx \Delta\theta$ and $\cos\Delta\theta \approx 1$, where $\Delta\theta$ is a small angle. Remember that the angle is measured in radians. The reader can look up the sine and cosine in the tables and prove this for himself. These are just the first terms of the following series expressing the sine and cosine:

$$\sin\theta = \theta - \frac{\theta^3}{3!} + \frac{\theta^5}{5!} - \frac{\theta^7}{7!} \cdots \tag{2.44}$$

$$\cos\theta = 1 - \frac{\theta^2}{2!} + \frac{\theta^4}{4!} - \frac{\theta^6}{6!} \cdots \tag{2.45}$$

Note that if θ is small, say, 0.10, then the second term in the series for the sine is $\theta^3/6 = \frac{1}{6000}$ or $\frac{1}{600}$ times the first. Thus one makes only a small error by omitting the second term, and in the limit, as θ goes to zero, the approximation is strictly accurate.

Expansion in Series Very often in physics it is important to be able to calculate the value of a function at some neighboring point when one knows it at one point. For this purpose a *Taylor expansion* is suitable. In the vicinity of a point x_0, the value of a function $f(x)$ is given by:

$$f(x) = f(x_0) + (x - x_0)\left[\frac{df(x)}{dx}\right]_{x=x_0} + \frac{1}{2}(x - x_0)^2\left[\frac{d^2f(x)}{dx^2}\right]_{x=x_0} + \cdots \tag{2.46}$$

The ratio of the third term to the second is

$$\frac{\frac{1}{2}(x - x_0)^2\left[\frac{d^2f(x)}{dx^2}\right]_{x=x_0}}{(x - x_0)\left[\frac{df(x)}{dx}\right]_{x=x_0}} \approx (x - x_0)$$

unless the derivatives are unusual in behavior. Therefore if $x - x_0$ is small compared with 1, we can with a small error (one which can at least be calculated) approximate $f(x)$ by using the formula

$$f(x) = f(x_0) + (x - x_0)\left(\frac{df}{dx}\right)_{x=x_0} \tag{2.47}$$

For example, suppose $y = Ax^5$ and that we know $y_0 = Ax_0^5$ and wish to calculate y at $x = x_0 + \Delta x$. Then

$$\left(\frac{dy}{dx}\right)_{x_0} = 5Ax_0^4 \qquad (x - x_0)\left(\frac{df}{dx}\right)_{x_0} = \Delta x\, 5Ax_0^4$$

and so

$$y = Ax_0^5 + 5Ax_0^4\,\Delta x \cdots \tag{2.48}$$

With expressions involving powers, one can write the following equation:

$$(a + bx)^n = a^n\left(1 + \frac{bx}{a}\right)^n$$

and use the binomial expansion to give

$$a^n\left(1 + \frac{bx}{a}\right)^n = a^n\left[1 + n\left(\frac{bx}{a}\right) + \frac{n(n-1)}{2!}\left(\frac{bx}{a}\right)^2 + \frac{n(n-1)(n-2)}{3!}\left(\frac{bx}{a}\right)^3 \cdots\right] \tag{2.49}$$

If bx/a is small compared with 1, we can get a good approximation by dropping all the terms after $n(bx/a)$. Applying this to our problem above, we write

$$y = A(x_0 + \Delta x)^5 = Ax_0^5\left(1 + \frac{\Delta x}{x_0}\right)^5 = Ax_0^5\left(1 + 5\frac{\Delta x}{x_0} \cdots\right)$$

$$= Ax_0^5 + 5Ax_0^4\,\Delta x \cdots$$

which agrees with Eq. (2.48).

Prove the following approximations when x is small compared to 1:

$$\frac{1}{\sqrt{1-x}} = 1 + \tfrac{1}{2}x \cdots \qquad \sqrt[3]{1+x} = 1 + \tfrac{1}{3}x \cdots$$

$$\sqrt{1+x} = 1 + \tfrac{1}{2}x \cdots \qquad \frac{1}{\sqrt[3]{1+x}} = 1 - \tfrac{1}{3}x \cdots$$

$$\sqrt{1-x} = 1 - \tfrac{1}{2}x \cdots$$

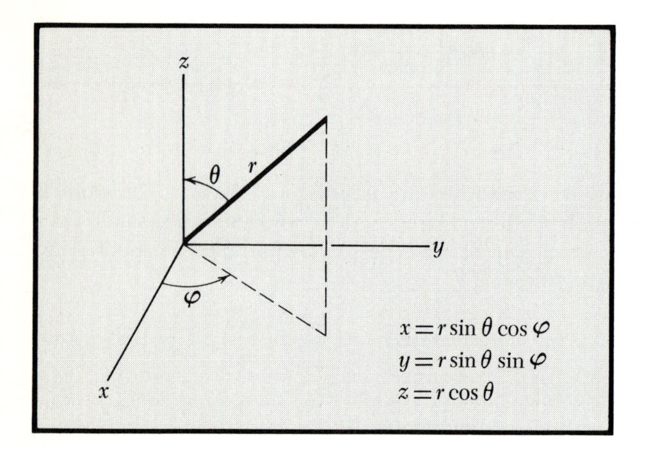

FIG. 2.30 Spherical polar coordinates.

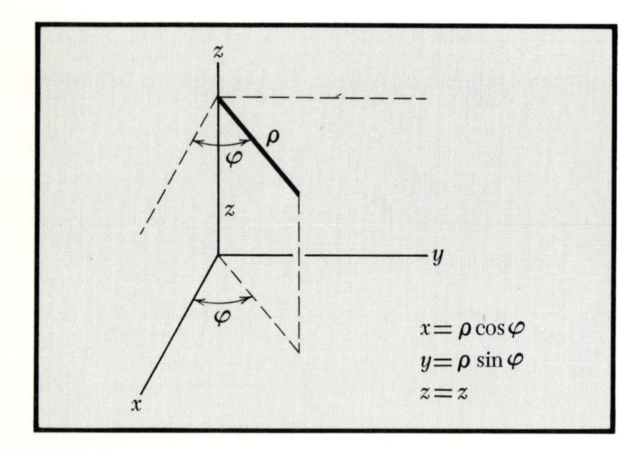

FIG. 2.31 Cylindrical polar coordinates.

Vectors and Spherical Polar Coordinates The position of a particle is expressed in spherical polar coordinates as r, θ, φ. Here r is the magnitude of the vector $\mathbf{r}$ from the origin to the particle; θ is the angle between $\mathbf{r}$ and the polar axis z; and φ is the angle between the x axis and the projection of r on the equatorial or xy plane. We take $0 \leq \theta \leq \pi$. The projection of $\mathbf{r}$ on the xy plane is of magnitude $r\sin\theta$. Note that the position in cartesian coordinates is given by

$$x = r\sin\theta\cos\varphi \quad y = r\sin\theta\sin\varphi \quad z = r\cos\theta \tag{2.50}$$

as in Fig. 2.30.

1 Let the first particle be at $\mathbf{r}_1 \equiv (r_1, \theta_1, \varphi_1)$ and a second particle at $\mathbf{r}_2 \equiv (r_2, \theta_2, \varphi_2)$. Let θ_{12} be the angle between $\mathbf{r}_1$ and $\mathbf{r}_2$. By expressing the scalar product $\hat{\mathbf{r}}_1 \cdot \hat{\mathbf{r}}_2 = \cos\theta_{12}$ in terms of $\hat{\mathbf{x}}$, $\hat{\mathbf{y}}$, $\hat{\mathbf{z}}$, show that

$$\cos\theta_{12} = \sin\theta_1\sin\theta_2\cos(\varphi_1 - \varphi_2) + \cos\theta_1\cos\theta_2 \tag{2.51}$$

where we have used the trigonometric identity

$$\cos(\varphi_1 - \varphi_2) = \cos\varphi_1\cos\varphi_2 + \sin\varphi_1\sin\varphi_2 \tag{2.52}$$

This is a good example of the power of vector methods. Try to find the result [Eq. (2.51)] otherwise!

2 Similarly, by forming the vector product $\hat{\mathbf{r}}_1 \times \hat{\mathbf{r}}_2$, find a relation for $\sin\theta_{12}$.

Cylindrical polar coordinates ρ, φ, z are an orthogonal set of coordinates defined by $x = \rho\cos\varphi$, $y = \rho\sin\varphi$, and $z = z$, as in Fig. 2.31. When used in two dimensions, the coordinates reduce to ρ and φ alone. However, we often use r and θ in place of ρ and φ. (See, for example, the formulas below.)

Formulas for Analytic Geometry

Line in xy plane	$ax + by = 1$
Line in xy plane through origin	$y = ax$
Plane	$ax + by + cz = 1$
Plane through origin	$ax + by + cz = 0$

	Cartesian coordinates	*Polar coordinates*
Circle, center at origin	$x^2 + y^2 = r_0^2$	$r = r_0$
Ellipse	$\dfrac{x^2}{a^2} + \dfrac{y^2}{b^2} = 1$	$\dfrac{1}{r} = \dfrac{1 - e\cos\theta}{a}$
	center at origin	$e < 1$; origin at focus
Parabola	$y^2 = mx$	$\dfrac{1}{r} = \dfrac{1 - \cos\theta}{a}$
	vertex at origin	origin at focus
Hyperbola	$\dfrac{x^2}{a^2} - \dfrac{y^2}{b^2} = 1$	$\dfrac{1}{r} = \dfrac{1 - e\cos\theta}{a}$
	center at origin	$e > 1$; origin at focus

Useful Vector Identities

$$\mathbf{A} \cdot \mathbf{B} = A_x B_x + A_y B_y + A_z B_z \tag{2.53}$$

$$\mathbf{A} \times \mathbf{B} = \hat{\mathbf{x}}(A_y B_z - A_z B_y) + \hat{\mathbf{y}}(A_z B_x - A_x B_z) + \hat{\mathbf{z}}(A_x B_y - A_y B_x) \tag{2.54}$$

$$(\mathbf{A} \times \mathbf{B}) \times \mathbf{C} = (\mathbf{A} \cdot \mathbf{C})\mathbf{B} - (\mathbf{B} \cdot \mathbf{C})\mathbf{A} \tag{2.55}$$

$$\mathbf{A} \times (\mathbf{B} \times \mathbf{C}) = (\mathbf{A} \cdot \mathbf{C})\mathbf{B} - (\mathbf{A} \cdot \mathbf{B})\mathbf{C} \tag{2.56}$$

$$(\mathbf{A} \times \mathbf{B}) \cdot (\mathbf{C} \times \mathbf{D}) = (\mathbf{A} \cdot \mathbf{C})(\mathbf{B} \cdot \mathbf{D}) - (\mathbf{A} \cdot \mathbf{D})(\mathbf{B} \cdot \mathbf{C}) \tag{2.57}$$

$$(\mathbf{A} \times \mathbf{B}) \times (\mathbf{C} \times \mathbf{D}) = [\mathbf{A} \cdot (\mathbf{B} \times \mathbf{D})]\mathbf{C} - [\mathbf{A} \cdot (\mathbf{B} \times \mathbf{C})]\mathbf{D} \tag{2.58}$$

$$\mathbf{A} \times [\mathbf{B} \times (\mathbf{C} \times \mathbf{D})] = (\mathbf{A} \times \mathbf{C})(\mathbf{B} \cdot \mathbf{D}) - (\mathbf{A} \times \mathbf{D})(\mathbf{B} \cdot \mathbf{C}) \tag{2.59}$$

FURTHER READING

PSSC, "Physics," chap. 6, D. C. Heath and Company, Boston, 1965.

Banesh Hoffman, "About Vectors," Prentice-Hall, Inc., Englewood Cliffs, N.J., 1966. Not a textbook, but thought-provoking to read for those with some knowledge of vectors.

G. E. Hay, "Vector and Tensor Analysis," Dover Publications, Inc., New York, 1953.

D. E. Rutherford, "Vector Methods," Oliver & Boyd Ltd., Edinburgh, or Interscience Publishers, Inc., New York, 1949.

H. B. Phillips, "Vector Analysis," John Wiley & Sons, Inc., New York, 1933. This is an old book, much used by a generation of students.

CONTENTS

3

Newton's Laws of Motion

NEWTON'S LAWS OF MOTION

This chapter is concerned chiefly with Newton's Laws of Motion. First we state the laws in their conventional forms, and then we give some applications to help the student gain confidence in using them. In Chap. 4 we treat some of the problems related to choice of frames of reference and to the galilean transformation. Although the material in Chap. 4 could be considered before that in the present chapter, some experience with straightforward applications of the laws will enhance the appreciation of the more subtle aspects presented in Chap. 4.

Newton's First Law: A body remains in a state of rest or constant velocity (zero acceleration) when no external force acts upon it; that is,

$$\mathbf{a} = 0 \qquad \text{when } \mathbf{F} = 0$$

(The philosophical questions of what is the content of the First Law, for example, whether it is entirely contained in the Second Law, are not treated here.[1])

Newton's Second Law: The rate of change of momentum of a body is proportional to the force on the body. Momentum is defined as $M\mathbf{v}$, where M is the mass and $\mathbf{v}$ is the vector velocity, so that

$$\mathbf{F} = K\frac{d}{dt}(M\mathbf{v}) = KM\frac{d\mathbf{v}}{dt} = KM\mathbf{a}$$

where we have assumed in the third and fourth terms that M is a constant. We choose our units so that $K = 1$. M is measured in grams (g), $\mathbf{a}$ in centimeters per second per second (cm/s^2); then $\mathbf{F}$ is measured in *dynes* (dyn): The dyne is thus a force that gives a mass of one gram an acceleration of one centimeter per second per second. In SI or International System of Units, M is measured in kilograms (kg), $\mathbf{a}$ in meters per second per second (m/s^2), and $\mathbf{F}$ is measured in newtons (N). One newton is the force which gives to a mass of one kilogram an acceleration of one meter per second per second.

$$1\ \text{N} = 10^3\ \text{g} \times 100\ \text{cm/s}^2 = 10^5\ \text{dyn}$$

Therefore we write

[1] See, for example, E. Mach, "The Science of Mechanics," 6th ed., p. 302ff., The Open Court Publishing Company, La Salle, Ill., 1960.

$$\boxed{\mathbf{F} = \frac{d}{dt}M\mathbf{v}} \tag{3.1}$$

and if $dM/dt = 0$

$$\boxed{\mathbf{F} = M\mathbf{a}} \tag{3.2}$$

The assumption that M is constant automatically restricts us to nonrelativistic problems with $v \ll c$. We treat special relativity in Chaps. 10 to 14 and the variation of mass with v in Chap. 12. It also restricts us in the consideration of some interesting problems such as rockets and falling chains. (We treat some of these topics in Chap. 6.) However, a rich variety of important problems with M constant is available.

Newton's Third Law: Whenever two bodies interact, the force $\mathbf{F}_{21}$† that body 1 exerts on body 2 is equal and opposite to the force $\mathbf{F}_{12}$ that body 2 exerts on body 1.

$$\boxed{\mathbf{F}_{12} = -\mathbf{F}_{21}} \tag{3.3}$$

We shall see that this law is a basis for the conservation of momentum. The finite velocity of propagation of forces (special relativity) introduces difficulties in the application of this law, and we shall mention these in Chap. 4.

A point worth emphasizing here is that these two forces $\mathbf{F}_{12}$ and $\mathbf{F}_{21}$ act on different bodies, and in the application of Newton's Second Law to a particular body it is only the force on that body that must be considered. The equal and opposite force influences only the motion of the other body. (See Prob. 1 at the end of this chapter.)

We now present a number of examples of the application of Newton's laws. Those unfamiliar with the solution of differential equations should read the Mathematical Notes at the end of the chapter in connection with the material that follows.

Motion When $\mathbf{F} = 0$ This simple case is just that of Newton's First Law. By writing

$$M\frac{d\mathbf{v}}{dt} = \mathbf{F} = 0 \tag{3.4}$$

† We adopt the convention here that F_{ij} is the force on body i due to body j.

we can immediately see that **v** must be a constant. Here the vector character of **v** is important because both the direction and the magnitude of **v** must be constant. For example, a mass moving with constant speed in a circle has a continuously changing velocity direction and therefore cannot move in such a path if $\mathbf{F} = 0$.

If the constant velocity **v** is zero, the mass M remains at rest. If it is not zero but

$$\mathbf{v} = \frac{d\mathbf{r}}{dt} = \mathbf{v}_0 \tag{3.5}$$

then we can integrate this equation to obtain

$$\mathbf{r} = \mathbf{v}_0 t + \mathbf{r}_0 \tag{3.6}$$

where $\mathbf{r}_0$ is the value of **r** at $t = 0$. These equations can, of course, be put in cartesian coordinate form.

FORCES AND EQUATIONS OF MOTION

Much more important, however, are the cases with **F** not equal to zero. Under the influence of a net force **F**, a particle of constant mass undergoes acceleration according to Newton's Second Law:

$$\mathbf{F} = M\mathbf{a} = M\frac{d^2\mathbf{r}}{dt^2} \tag{3.7}$$

This mathematical expression is an *equation of motion*. By this we mean that upon successive integration of this differential equation we obtain expressions for velocity and position of the particle as functions of time.

In order to solve such an equation we need to know the force **F**, its dependence upon the position and velocity of the particle, and also its dependence upon time if it should explicitly vary with passage of time. Clearly the solving of an equation of motion can be a difficult problem if the force is complicated in its dependence upon these variables; but fortunately many important and instructive cases involve forces constant in time and also independent of velocity.

There are several important kinds of forces known in physics: the gravitational force, the electrostatic force, the magnetic force, and, among others, the strong but short-range nuclear forces. By such forces particles may interact with each other even when separated in empty space. If a particle experiences a resultant force due to gravitational interaction with other particles or bodies, we may say that it is in a *gravitational field* produced by those bodies. When an electrically charged particle experiences a resultant force due to a distribution of electric charges on other particles or bodies in its vicinity, we consider it to be in an *electric field.*

For many applied problems in mechanics we speak of *contact forces* such as the tension in the string supporting a pendulum bob or the pressure of a plane against an object resting upon it. Frequently both field forces and contact forces are present, as in the swinging in a gravitational field of a pendulum mass supported by the tension in a thread. In ultimate analysis all contact forces are field forces, for they arise from electromagnetic interactions between atomic particles. However, for our present purposes it is often most convenient to consider simply the contact force or forces. In treating the mechanics of atomic particles we obviously are concerned only with field forces; *contact* cannot be considered in its usual simple meaning in the atomic domain.

Units In this section we depart briefly from the subject of Newton's laws to discuss the question of units. Later in the chapter when electric and magnetic forces are introduced, we discuss the units to be used for electric charge and electric and magnetic fields. Here we are concerned only with mechanical units.

In order to communicate information about motions, standards of length and time are certainly necessary. Fortunately there is general worldwide agreement on a standard of time: the second (s). Originally it was defined as a definite part of a year, the year being defined in terms of astronomical observations. However, there are practical difficulties in using this definition, so that now the second is defined in terms of the number of oscillations characteristic of an atomic system, that of the element cesium. The exact definition is that one second is the time in which there occur 9,192,631,770 oscillations of the cesium atom. Apart from experimental procedures, it is exactly the same as using a grandfather clock and saying that one second is the time for so many complete swings of the pendulum.

When we come to a unit of length, there is no worldwide agreement because the English-speaking countries use one system and the rest of the world another. Scientists have found the system of units used by the rest of the world simpler and easier to use than the British system and consequently have adopted it; hopefully the English-speaking countries will soon adopt the system also. It involves a standard, the centimeter or meter, with derived substandards that are exactly the multiples of 10 times the standard. The original standard was the distance between scratches on a bar kept in Paris, the distance being by definition exactly 100 cm or 1 m. There are practical difficulties in using this length, such as, for example, the width of the scratches. Thus we now use a better standard, the wavelength of red light from Kr^{86}: one centimeter (cm) is 16,507.6373 wavelengths. The question of whether the meter or the centimeter is the fundamental length is in a sense academic because the conversion factor is exactly 100. In this book we shall use the centimeter, although we shall give references to the use of the meter, and many books use the meter as the unit of length. The difficulties with the British system come from the fact that different units of length are not simply related to one another: for example, the foot is twelve inches, the yard is three feet, and the mile is 1,760 yards.

Newton's Second Law involves two more quantities: mass and force. Do we need standards for both these quantities? The answer, as mentioned above, is *no*. We can set a standard for one of them and use Newton's Second Law as the definition of the other. Historically the unit of mass was established and the unit of force derived from it. The unit of mass is the gram or the kilogram, which is exactly one thousand grams. The standard kilogram is also kept in Paris. The comparison of masses is an easy process, so that the adoption of the mass of a certain type of atom as a standard has not been necessary.

Again the British system is complicated by the fact that the ounce, pound, and ton are not simply related. Therefore we shall adopt the centimeter, the second, and the gram as our fundamental units of length, time, and mass, respectively, and we shall use units derived from these for force, momentum, energy, power, etc. This system is called the *cgs system*. The SI system, using the meter and the kilogram in place of the centimeter and the gram, is also commonly employed. The introduction of electricity and magnetism raises questions about units, which are discussed in Volume 2, and introduces the speed of light, which will be found in the section on magnetic fields in this chapter.

Dimensions When working through a complicated calculation, it is very important to be sure that the units on one side of the resulting equation are the same as those on the other side. For example, when calculating the distance traveled by an object, one could be sure that some mistake had been made if the answer came out in grams. An analysis of this sort is usually called *dimensional analysis*. We do not need to specify the units we are using but only the dimensions of mass, length, and time as follows.

What are the dimensions of force? We use Eq. (3.7) to see that force is mass times acceleration, that acceleration is velocity divided by time, and that velocity is distance divided by time; so that, using M, L, and T to denote mass, length, and time, we obtain

$$\text{Force} = [M][\text{accel}] = [M][L][T]^{-2}$$

$$\text{Acceleration} = \frac{[L]}{[T][T]} = [L][T]^{-2}$$

$$\text{Velocity} = \frac{[L]}{[T]} = [L][T]^{-1}$$

As an example of the use of dimensional analysis, suppose one arrives at an equation force $= \frac{3}{5}\rho v^2$ where ρ is the mass per unit volume, or density, and v is speed or velocity. Dimensional analysis will never tell whether the factor $\frac{3}{5}$ is correct since it is dimensionless, that is, a pure number. However, let us see about ρv^2.

$$\rho = [M][L]^{-3} \qquad v^2 = [L]^2[T]^{-2}$$

Therefore $\rho v^2 = [M][L]^{-3}[L]^2[T]^{-2} = [M][L]^{-1}[T]^{-2}$ whereas force, as we have seen, is $[M][L][T]^{-2}$. Thus we have made some mistake in reaching our original equation. Those who are familiar with the concept of *pressure*, which is force per unit area, will see that ρv^2 has the dimensions of pressure.

MOTION OF A PARTICLE IN A UNIFORM GRAVITATIONAL FIELD

We now proceed to some applications of Newton's Second Law. If we restrict our consideration to a laboratory region whose extent is very small compared with the size of the earth, then we may in good approximation consider the gravitational force on a particle to be everywhere *downward* and constant. The downward acceleration due to this force is given by the local value of the gravitational acceleration[1] g, and so the magnitude of the force upon a particle is thus mg. This force as a vector may then be written $\mathbf{F} = -mg\hat{\mathbf{y}}$, where x and y axes have been chosen as shown in Fig. 3.1.

If we can omit other forces such as friction, the equation of motion from Newton's Second Law [Eq. (3.7)] is then

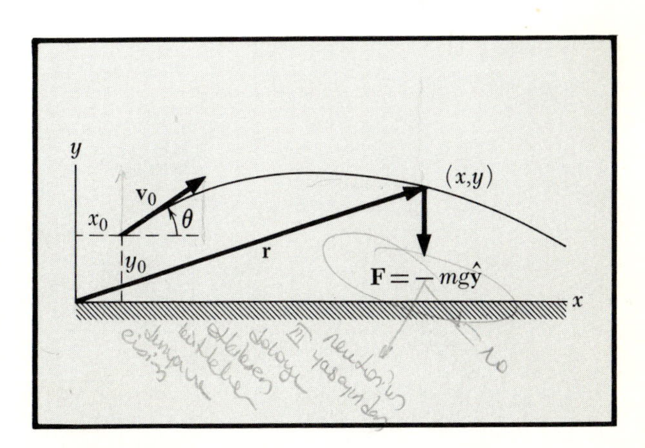

FIG. 3.1 Motion of a free particle, projected initially at (x_0, y_0) with speed v_0 at an elevation angle θ, under the influence of a uniform gravitational field. Position vector at instant pictured: $\mathbf{r} = \hat{\mathbf{x}}x + \hat{\mathbf{y}}y$. Acceleration vector: $d^2\mathbf{r}/dt^2 = (d^2x/dt^2)\hat{\mathbf{x}} + (d^2y/dt^2)\hat{\mathbf{y}} = -g\hat{\mathbf{y}}$.

[1] g is usually assumed to be 980 cm/s² = 9.80 m/s². A table of values over the surface of the earth is given in Table 4.1.

$$m\left[\hat{\mathbf{x}}\frac{d^2x}{dt^2} + \hat{\mathbf{y}}\frac{d^2y}{dt^2}\right] = -mg\hat{\mathbf{y}}$$

Since the component directions are orthogonal, we separate this into two component equations and then no longer need to retain the unit vector factors. Thus

$$\frac{d^2y}{dt^2} = -g \qquad \frac{d^2x}{dt^2} = 0 \tag{3.8}$$

The integration of these equations to obtain x and y as functions of t is taken up in the Mathematical Notes at the end of this chapter. With initial conditions as shown in Fig. 3.1, for which the initial component velocities are $v_0 \cos\theta$ and $v_0 \sin\theta$ for the x and y directions, respectively, the solutions are

$$\begin{aligned} x &= x_0 + (v_0 \cos\theta)t \\ y &= y_0 + (v_0 \sin\theta)t - \tfrac{1}{2}gt^2 \end{aligned} \tag{3.9}$$

Various special cases, such as dropping the particle from rest at initial height h, can be explored by choice of initial conditions of position and velocity and will lead to familiar results. Some cases are given in Probs. 2 to 4.

A student who is familiar with analytic geometry will recognize Eqs. (3.9) as being a parametric form, in the parameter t, for a parabola. This can be made clear by eliminating t between the two equations to obtain

$$y - \left(y_0 + \frac{v_0^2 \sin^2\theta}{2g}\right) = -\frac{g}{2v_0^2 \cos^2\theta}\left[x - \left(x_0 + \frac{v_0^2 \sin\theta \cos\theta}{g}\right)\right]^2$$

This is a standard form for a parabola with vertex at

$$x_1 = x_0 + \frac{v_0^2 \sin\theta \cos\theta}{g}$$

$$y_1 = y_0 + \frac{v_0^2 \sin^2\theta}{2g}$$

and opening downward with a vertical axis of symmetry. If air resistance were negligible, this analysis would correctly describe the motion of a projectile. In fact, it is a good approximation for objects of considerable mass moving in trajectories of limited extent with small velocities (see Prob. 20).

The parabolic representation with vertex at (x_1, y_1) as given above reveals that the maximum height attained above the launching position is

$$h = y_1 - y_0 = \frac{v_0^2 \sin^2 \theta}{2g}$$

The horizontal range, i.e., the distance at which the projectile returns to the launching elevation, is given by

$$R = 2(x_1 - x_0) = \frac{2v_0^2 \sin \theta \cos \theta}{g} = \frac{v_0^2 \sin 2\theta}{g} \tag{3.10}$$

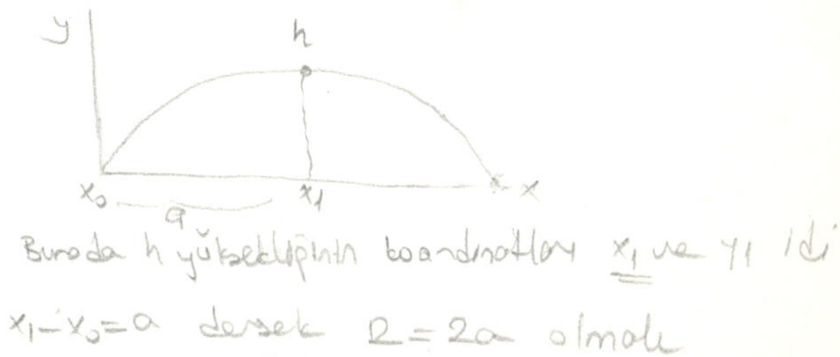

EXAMPLE

Maximum Range At what angle should an object be projected in order to make R a maximum? Before making the calculation, we can readily see that there will be a maximum for $R(\theta)$; for if θ is too small, the projectile does not stay in flight long enough to go far, while if θ is too large, the projectile just goes up and down and not far horizontally. To solve the problem analytically we can simply use the fact that for the maximum of R, $dR/d\theta = 0$. Using Eq. (3.10),

$$\frac{dR}{d\theta} = \frac{v_0^2}{g} 2 \cos 2\theta = 0$$

$$2\theta = \frac{\pi}{2}$$

$$\theta = \frac{\pi}{4} = 45°$$

NEWTONIAN LAW OF UNIVERSAL GRAVITATION

In the preceding section we treated the case of a constant gravitational field. What happens if the distance between the objects in gravitational interaction is large compared with their size? Newton's law of gravitation states:

> A particle of mass M_1 attracts any other particle of mass M_2 in the universe with a force
>
> $$\mathbf{F} = -\frac{GM_1M_2}{r^2}\hat{\mathbf{r}} \tag{3.11}$$
>
> where $\hat{\mathbf{r}}$ is the unit vector from M_1 toward M_2 and G is a constant having the value determined by experiment to be
>
> 6.67×10^{-8} dyn-cm^2/g^2 or 6.67×10^{-11} N-m^2/kg^2

Note that this is the force on M_2. The minus sign indicates that the force is attractive; it tends to decrease r.

The gravitational force is a *central* force: the force is directed along the line connecting the two point masses. The determination of the value of G is usually treated in high school texts. The classic experiment is that of Cavendish. We shall also see later (Chap. 9) that because the force depends inversely upon the square of distance, an object possessing spherical symmetry will interact as if it were a particle possessing the entire mass of the object and located at its center.

Newton himself did not know the value of G. However, he did know—in fact, he discovered—that the force law is an inverse square law, and he knew that at the surface of the earth (since the earth is essentially spherical)

$$mg = \frac{GmM_e}{R_e^2} \tag{3.12}$$

where M_e is the mass of the earth and R_e is its radius. Therefore he could find GM_e, and he could find the force at any distance r by

$$F = \frac{GmM_e}{r^2} = \frac{GmM_e}{R_e^2}\frac{R_e^2}{r^2} = mg\left(\frac{R_e}{r}\right)^2$$

Also it is known experimentally to a high degree of accuracy that the gravitational and inertial masses of a body are equal. (This will be discussed in Chap. 14.) What is meant is that the value of the mass m to be used in the gravitational force equation above is equal to the value of the mass of the same body used in Newton's Second Law $\mathbf{F} = m\mathbf{a}$. The mass in the gravitational equation is called the *gravitational mass*, and the mass in Newton's Second Law is called the *inertial mass*. The classical experiments on the equality of the two masses were carried out by Eötvös; recent and more accurate experiments are described by R. H. Dicke[1] and P. G. Roll, R. Krotkov, and R. H. Dicke.[2] The Eötvös experiment is described in Chap. 14. We have assumed the equality in Eq. (3.12).

EXAMPLE

Satellite in a Circular Orbit Consider a satellite in a circular orbit concentric and coplanar with the equator of the earth. At what radius r of the orbit will the satellite appear to remain stationary

[1] *Scientific American*, **205**:84 (1961).

[2] *Ann. Phys.* (N.Y.), **26**:442 (1964).

when viewed by observers fixed on the earth? We assume the sense of rotation of the orbit is the same as that of the earth.

In a circular orbit the gravitational attraction is equal to the mass times the centripetal acceleration:

$$\frac{GM_eM_s}{r^2} = M_s\omega^2 r \tag{3.13}$$

where M_s is the mass of the satellite. We rearrange Eq. (3.13) as

$$r^3 = \frac{GM_e}{\omega^2} = \frac{GM_eT^2}{(2\pi)^2} \tag{3.14}$$

where T is the period. We want ω for the satellite orbit to equal the angular frequency ω_e of the earth about its axis, so that the satellite will appear stationary. The angular frequency of the earth is

$$\omega_e = \frac{2\pi}{1 \text{ day}} = \frac{2\pi}{8.64 \times 10^4} = 7.3 \times 10^{-5} \text{ s}^{-1}$$

whence Eq. (3.14) becomes, with $\omega = \omega_e$,

$$r^3 \approx \frac{(6.67 \times 10^{-8})(5.98 \times 10^{27})}{(7.3 \times 10^{-5})^2} \approx 75 \times 10^{27}$$

or

$$r \approx 4.2 \times 10^9 \text{ cm}$$

The radius of the earth is 6.38×10^8 cm. This distance is roughly one-tenth of the distance to the moon and is about 6.6 times the earth's radius.

ELECTRIC AND MAGNETIC FORCES ON A CHARGED PARTICLE; UNITS

In this section we wish to consider problems involving electric and magnetic forces acting upon charged particles. In laboratory work most students will observe and measure effects of such forces on particle motions, and the subject will be treated in detail in Volume 2. At present we introduce briefly definitions of units of electric and magnetic quantities so that we can deal with the forces that occur in this important branch of mechanics.

You will remember that electric charges of like sign repel each other with forces directed along the straight line joining them. The strength of the repulsion depends inversely upon the square of the distance separating the charges, and it is directly proportional to the product of the quantities of electricity in the charges. This is Coulomb's law; it can be expressed by the statement

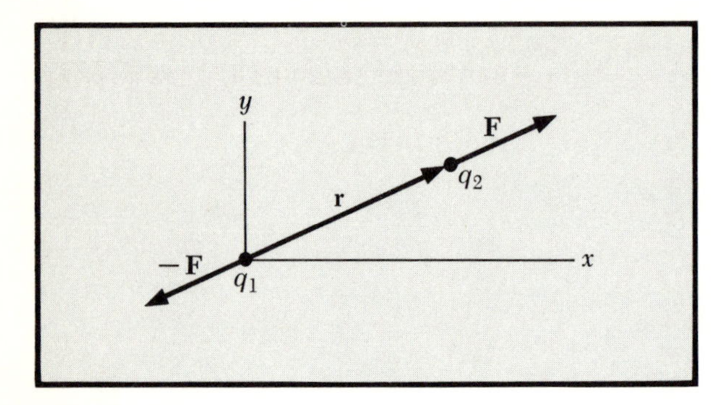

FIG. 3.2 Illustration of Coulomb's law. $\mathbf{F} = (q_1 q_2/r^2)\hat{\mathbf{r}} = (q_1 q_2/r^3)\mathbf{r}$.

$$\mathbf{F} = \frac{q_1 q_2}{r^3}\mathbf{r} = \frac{q_1 q_2}{r^2}\hat{\mathbf{r}} \tag{3.15}$$

where $\mathbf{r}$ is the vector separating point charge q_2, which experiences the force $\mathbf{F}$, from point charge q_1 taken to be at the origin. The unit vector $\hat{\mathbf{r}}$ is, of course, equal to $\mathbf{r}/r$. Figure 3.2 illustrates the situation and reminds us that if $\mathbf{F}$ acts upon q_2, then $-\mathbf{F}$ will act upon q_1.

Equation (3.15) states Coulomb's law in the gaussian system of units. In this system a unit quantity of charge q is defined as follows: Two equal point charges that repel each other with a force of 1 dyn when they are 1 cm apart are defined to be gaussian unit charges. The quantity of electricity possessed by each is said to be 1 electrostatic unit (esu), or 1 statcoulomb. The "dimensions" of electric charge are seen from Eq. (3.15) to be

$$[q] = [\text{force}]^{\frac{1}{2}}[\text{distance}] = [\text{mass}]^{\frac{1}{2}}[\text{length}]^{\frac{3}{2}}[\text{time}]^{-1}$$

Thus, in the cgs system of these quantities the dimensions of charge in the gaussian electrostatic system are

$$[q] = \text{g}^{\frac{1}{2}}\,\text{cm}^{\frac{3}{2}}\,\text{s}^{-1}$$

Clearly it is easier to use the name *esu* or *statcoulomb* than always to write this combination.

As we have mentioned, the International System of Units (SI) uses meters for distance and newtons for force. In place of the definition of the unit charge from Coulomb's law, a definition in terms of electric current, the ampere (A), is used. The quantity of charge is 1 coulomb (C) = 1 ampere-second (A-s). Then Coulomb's law must be written

$$F = k\frac{q_1 q_2}{r^2} \qquad \mathbf{F} = \frac{1}{4\pi\epsilon_0}\frac{q_1 q_2}{r^2}\hat{\mathbf{r}} \tag{3.15a}$$

where k has the dimensions

$$[\text{Force}][\text{length}]^2[\text{charge}]^{-2}$$

and ϵ_0 the inverse dimensions. In electrostatics there are many operations that introduce a factor of 4π, and so the 4π is put in the denominator of Eq. (3.15*a*) and will cancel out in such operations. The value of k is

$$k = \frac{1}{4\pi\epsilon_0} = 8.988 \times 10^9 \text{ N-m}^2/\text{C}^2$$

The charge q_p carried by the proton is the elementary charge and is almost universally designated by the symbol e. Its value in the cgs gaussian units is

$$e = +4.8022 \times 10^{-10} \text{ esu}$$

In SI units it is

$$e = +1.60210 \times 10^{-19} \text{ C}$$
$$1 \text{ C} = 2.9979 \times 10^9 \text{ esu}$$

The charge of an electron is equal to $-e$. The magnitude of the repulsive force between two protons at a separation of 10^{-12} cm is

$$F = \frac{e^2}{r^2} \approx \frac{(4.8 \times 10^{-10})(4.8 \times 10^{-10})}{10^{-12} \times 10^{-12}} \approx 2.3 \times 10^5 \text{ dyn}$$

In SI units it is

$$F = k\frac{e^2}{r^2} \approx 9.0 \times 10^9 \frac{(1.6 \times 10^{-19})^2}{(10^{-14})^2} \approx 2.3 \text{ N}$$

The force between a proton and an electron is attractive because the charges are of opposite sign.

The Electric Field When a charged particle is so situated that an electric force acts upon it, we say that it is in an *electric field*. The field and the related force on the particle receiving our attention are due to another charge or a distribution of charges in the vicinity. The field intensity **E** is defined by the relationship

$$\mathbf{F} = q\mathbf{E} \tag{3.16}$$

where q is the quantity of the "test charge" on which we observe the force **F**. The field intensity vector **E** is thus the force vector per unit positive charge at the position of the test particle.

In Fig. 3.3 we display the same situation depicted in Fig. 3.2, but here we adopt the point of view that the force **F** on q_2 is due to the field intensity **E** produced by the charge q_1 located at the origin. The vector **E** is in this case given by the following expression:

$$\mathbf{E} = \frac{q_1}{r^2}\hat{\mathbf{r}} \tag{3.17}$$

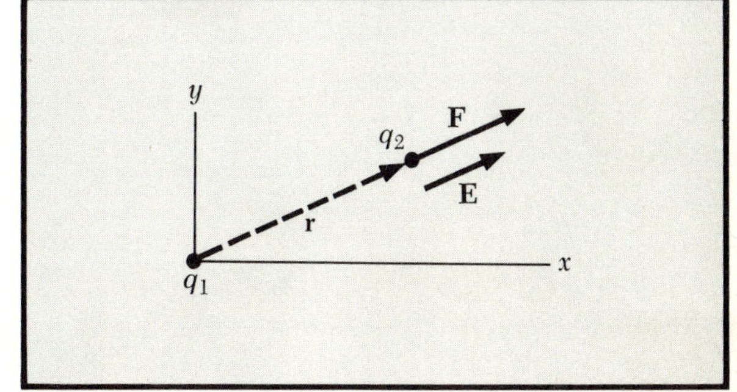

FIG. 3.3 Illustration of the concept of electric field intensity **E**. $\mathbf{E} = (q_1/r^2)\hat{\mathbf{r}}$. $\mathbf{F} = q_2\mathbf{E} = (q_1 q_2/r^2)\,\mathbf{r}$.

and the force $\mathbf{F} = q_2\mathbf{E}$ is thus the same as that expressed by Eq. (3.15). The importance of the field point of view will emerge in the study of electricity. It is particularly useful when we must deal with the electric force produced on a charged particle by distributions of electricity such as charged spheres or planes and, as explained in Volume 2, by time-varying magnetic fields.

The dimensions of field intensity clearly will be force per unit charge; thus its unit may be expressed as 1 dyn/esu. For reasons that will emerge later, field intensity is also expressed as statvolts per centimeter (statvolts/cm). The two modes of expression mean exactly the same thing:

$$1 \text{ dyn/esu} = 1 \text{ statvolt/cm}$$

the latter emphasizing the *work* involved in displacing a unit charge by unit distance in the field, and the former emphasizing the *force* acting upon a unit charge.

In the International System, Eq. (3.16) is also the defining equation for the electric field **E**, and E will have the units newtons per coulomb (N/C). In place of Eq. (3.17) for the electric field due to a charge q_1, we have

$$\mathbf{E} = k\frac{q_1}{r^2}\hat{\mathbf{r}} \tag{3.17a}$$

Exactly as in the cgs system, E can also be expressed as volts per meter (V/m) and

$$1 \text{ N/C} = 1 \text{ V/m}$$

The conversion factor from statvolts per centimeter to volts per meter is

$$2.9979 \times 10^4 \text{ V/m} = 1 \text{ statvolt/cm}$$

$$1 \text{ V/m} = \frac{1}{2.9979 \times 10^4} \text{ statvolt/cm}$$

$$\approx \frac{1}{3 \times 10^4} \text{ statvolt/cm}$$

The Magnetic Field and the Lorentz Force Up to this point we have considered only the *static* situation, where the charged particles are not moving with respect to each other or relative to the observer, and we have expressed the electrostatic force on a particle of charge q as $\mathbf{F}_{el} = q\mathbf{E}$. But if q is moving relative to the observer, it is an experimental fact that an additional force may be present in a direction perpendicular to its velocity; this is the *magnetic force*. A region in which such a velocity-dependent force exists is said to possess a *magnetic field;* and from experiment we know that a magnetic field intensity vector **B** can be related to the magnetic force by the formulation[1]

$$\mathbf{F}_{\text{mag}} = \frac{q}{c}\mathbf{v} \times \mathbf{B} \tag{3.18}$$

where c is the speed of light in empty space and $\mathbf{v}$ is the charged particle velocity, using the cgs gaussian system of units. The vector product gives $\mathbf{F}_{\text{mag}}$ perpendicular to $\mathbf{v}$ as experimentally required, and it defines the magnetic field vector **B** also to be perpendicular to $\mathbf{F}_{\text{mag}}$. Figure 3.4 illustrates these relationships for a case where $\mathbf{v}$ and **B** are 90° apart. If a wire carrying a current replaces the moving charge along the direction of $\mathbf{v}$, the direction of the force on the wire is the same as that shown in Fig. 3.4.

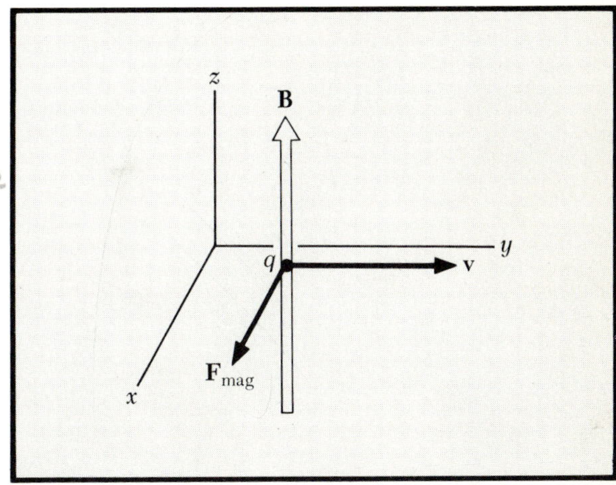

FIG. 3.4 The magnetic force $\mathbf{F}_{\text{mag}} = (q/c)\mathbf{v} \times \mathbf{B}$.

The dimensions of B as defined here are the same as those for E because the ratio v/c is dimensionless. The unit is given the name *gauss* (G), with F in dynes and q in electrostatic units. Thus if an electron moves with a velocity one-tenth the speed of light in a direction perpendicular to a magnetic field of 10,000-G intensity, the magnetic force strength will be

$$\begin{aligned} F &= (4.8 \times 10^{-10}\ \text{esu})\tfrac{1}{10}(10^4\ \text{G}) \\ &= 4.8 \times 10^{-7}\ \text{dyn} \end{aligned}$$

In SI units

$$\mathbf{F}_{\text{mag}} = q\mathbf{v} \times \mathbf{B} \tag{3.18a}$$

where q is in coulombs, v in meters per second, and F in newtons. The equation defines B with dimensions $[\text{N}][\text{s}][\text{C}]^{-1}[\text{m}]^{-1}$. In recent years a special name, the *tesla* (T), has been given to this unit, previously denoted by webers per square meter (Wb/m²).† Fortunately,

$$1\ \text{T} = 10^4\ \text{G}$$

although it must be remembered that the gauss and the tesla do not have the same dimensions; we should more properly say 1 T corresponds to 10^4 G.

[1] If the vector product in Chap. 2 has been omitted, this section can be condensed to the case of **v** and **B** perpendicular.

† W. E. Weber (1807–1891) was a German physicist and N. Tesla (1856–1943) an American inventor.

In SI units the preceding problem has $B = 1$ T, $v = 3 \times 10^7$ m/s, $q = e = 1.6 \times 10^{-19}$, and

$$F = (1.6 \times 10^{-19})(3 \times 10^7)(1.0) = 4.8 \times 10^{-12}\ \text{N}$$

The total force on a moving charged particle is the vector sum of electrostatic and magnetic forces. It is called the *Lorentz force.* (The name is sometimes applied to the magnetic force alone.) From Eqs. (3.16) and (3.18)

$$\boxed{\mathbf{F} = q\mathbf{E} + \frac{q}{c}\mathbf{v} \times \mathbf{B}} \qquad (3.19)$$

in the gaussian system, and

$$\boxed{\mathbf{F} = q\mathbf{E} + q\mathbf{v} \times \mathbf{B}} \qquad (3.19a)$$

in the SI. A great deal of physics comes out of Newton's Second Law, $\mathbf{F} = M\mathbf{a}$, used together with Eq. (3.19). Of course, an important part of the history of physics consisted of the efforts to establish these equations. [Writing Eq. (3.19) as an experimental fact here does not relieve us of the need to discuss it deeply in Volume 2.]

In this chapter we shall need the following numerical values:

The speed of light:

$$c = 2.9979 \times 10^{10}\ \text{cm/s} = 2.9979 \times 10^{8}\ \text{m/s}$$

The mass m of the electron:

$$m = 0.9108 \times 10^{-27}\ \text{g} = 0.9108 \times 10^{-30}\ \text{kg}$$

The mass M_p of the proton:

$$M_p = 1.6724 \times 10^{-24}\ \text{g} = 1.6724 \times 10^{-27}\ \text{kg}$$

In working with the Lorentz force [Eq. (3.19)] in cgs gaussian units, we express F in dynes, E in statvolts/cm, v and c in cm/s, B in gauss, and q in electrostatic units. In the SI, using Eq. (3.19a), we express F in newtons, E in V/m, v in m/s, B in teslas, and q in coulombs. The conversion factors mentioned above (and derived in Volume 2) are collected here:

$$1\ \text{m/s} = 100\ \text{cm/s}$$
$$1\ \text{statvolt/cm} = 3.0 \times 10^4\ \text{V/m}\dagger$$

† The exact values of the second and third relations are given on pages 69 and 70; these values are accurate enough for our problems.

$$1\ \text{C} = 3.0 \times 10^9 \text{ statcoulombs or esu}$$
$$1\ \text{T} \leftrightarrow 1 \times 10^4\ \text{G}$$

Motion of a Charged Particle in a Uniform Constant Electric Field The equation for the force on a charge q and mass M in an electric field $\mathbf{E}$ uniform in space and constant in time is [Eq. (3.16)]

$$\mathbf{F} = M\mathbf{a} = q\mathbf{E} \tag{3.20}$$

and so

$$\mathbf{a} = \frac{d^2\mathbf{r}}{dt^2} = \frac{q}{M}\mathbf{E}$$

is an equation for the acceleration of the charge. This result is quite similar to that for the motion of a particle in a uniform gravitational field $\mathbf{F} = -Mg\hat{\mathbf{y}}$ at the surface of the earth, where $\hat{\mathbf{y}}$ is a unit vector that points out from the center of the earth. For the gravitational problem the equation of motion is $M\mathbf{a} = -Mg\hat{\mathbf{y}}$, or $\mathbf{a} = -g\hat{\mathbf{y}}$.

Equation (3.20) can be seen by trial or by direct integration to have the general solution

$$\mathbf{r}(t) = \frac{q\mathbf{E}}{2M}t^2 + \mathbf{v}_0 t + \mathbf{r}_0 \tag{3.21}$$

where $\mathbf{r}_0$ is the position vector for the particle at $t = 0$, and $\mathbf{v}_0$ is its velocity vector at that time.

Differentiation of Eq. (3.21) will give the expression for velocity at any time, namely,

$$\mathbf{v}(t) = \frac{d\mathbf{r}}{dt} = \frac{q\mathbf{E}}{M}t + \mathbf{v}_0 \tag{3.22}$$

from which we readily see that the initial velocity (when $t = 0$) is indeed $\mathbf{v}_0$.

EXAMPLE

Longitudinal Acceleration of a Proton A proton is accelerated from rest for 1 nanosecond ($= 10^{-9}$ s) by an electric field $E_x = 1$ statvolt/cm. What is the final velocity (see Fig. 3.5)?

The velocity is given by Eq. (3.22):

$$\frac{d\mathbf{r}}{dt} = \frac{e}{M}\mathbf{E}t + \mathbf{v}_0$$

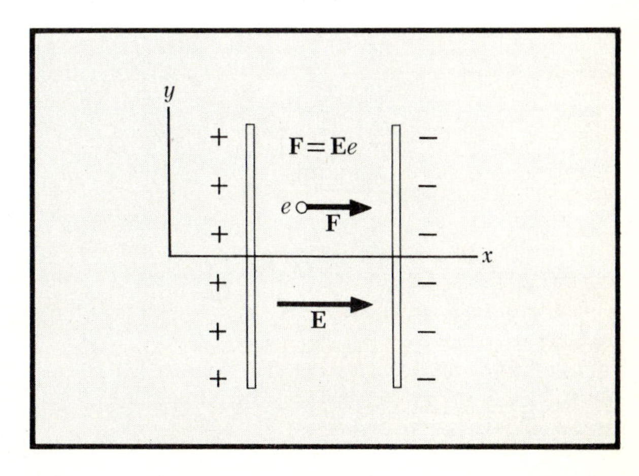

FIG. 3.5 Longitudinal acceleration of proton in electric field between charged metal plates.

For our problem this[1] reduces to

$$v_x(t) = \frac{e}{M}E_x t \qquad v_y = v_z = 0$$

because we have specified that $\mathbf{v} = 0$ at $t = 0$. Thus the final velocity at $t = 1 \times 10^{-9}$ s is, roughly,

$$v_x \approx \frac{(4.8 \times 10^{-10}\ \text{esu})(1\ \text{statvolt/cm})(1 \times 10^{-9}\ \text{s})}{(2 \times 10^{-24}\ \text{g})}$$

$$\approx 2.4 \times 10^5\ \text{cm/s}$$

Note that 1 esu $\times$ 1 statvolt/cm $\equiv$ 1 dyn $\equiv$ 1 g-cm/s^2. We have used 2×10^{-24} g for the order of magnitude of the mass of the proton.

In SI units

$$E_x = 3.0 \times 10^4\ \text{V/m}$$

$$v_x = \frac{(1.6 \times 10^{-19}\ \text{C})(3.0 \times 10^4\ \text{V/m})(1 \times 10^{-9}\ \text{s})}{2 \times 10^{-27}\ \text{kg}}$$

$$\approx 2.4 \times 10^3\ \text{m/s}$$

EXAMPLE

Longitudinal Acceleration of an Electron An electron initially at rest is accelerated through 1 cm by an electric field of 1 statvolt/cm directed in the negative sense of the x axis. What is the terminal speed?

From Eq. (3.22) we have, with $-e$ as the charge and m as the mass of the electron,

$$v_x(t) = -\frac{e}{m}E_x t \qquad x(t) = -\frac{e}{2m}E_x t^2$$

We want to eliminate t and solve for v_x in terms of x. It is convenient to form v_x^2 and then rearrange factors:

$$v_x^2 = \left(\frac{e}{m}E_x t\right)^2 = \left(\frac{2e}{m}E_x\right)\left(\frac{e}{2m}E_x t^2\right) = -\frac{2e}{m}E_x x$$

$$\approx \frac{-2 \times 5 \times 10^{-10}}{10^{-27}}(-1) \times 1 \approx 10^{-9} \times 10^{27}$$

$$\approx 10^{18}\ \text{cm}^2/\text{s}^2$$

Thus the final speed is approximately

$$|v_x| \approx 10^9\ \text{cm/s}$$

This is one-thirtieth the speed of light and is small enough so that we need not consider relativity (0.1 percent accuracy).

[1]The equation is a vector equation; with $\mathbf{E} = (E_x,0,0)$ and $\mathbf{v}_0 = 0$ it reduces to the three component equations

$$\frac{dx}{dt} = \frac{e}{M}E_x t \qquad \frac{dy}{dt} = 0 \qquad \frac{dz}{dt} = 0$$

EXAMPLE

Transverse Acceleration of an Electron After leaving the accelerating field E_x of the preceding example, the electron beam enters a region of length $L = 1$ cm in which there exists a transverse deflecting field $E_y = -0.1$ statvolt/cm, as in Fig. 3.6. What angle with the x axis does the electron beam make on leaving the deflecting region? Note that this is just like a body projected horizontally in the earth's gravitational field.

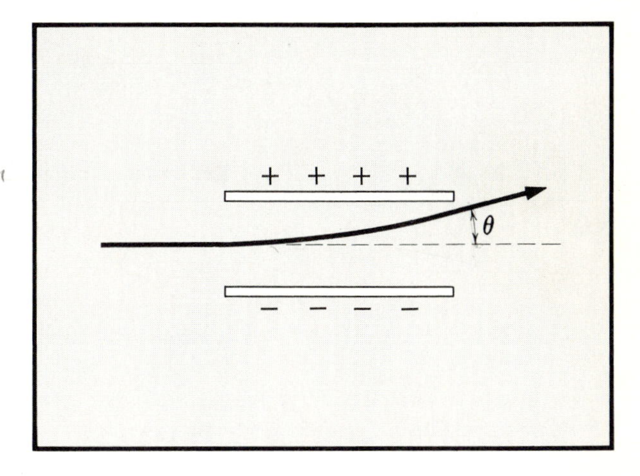

FIG. 3.6 Deflection of electron beam by a transverse electric field. The angle θ is greatly exaggerated over the value given in the example on this page.

Because there now is no x component of the field, the x component of the velocity will remain constant. The time τ spent by an electron in the deflecting region is given by

$$v_x \tau = L$$

or, if $v_x = 10^9$ cm/s,

$$\tau = \frac{L}{v_x} = \frac{1}{10^9} = 10^{-9}\ \text{s}$$

The transverse velocity acquired during this period is given by

$$v_y = -\frac{e}{m}E_y\tau \approx \frac{-5 \times 10^{-10}}{10^{-27}}(-0.1) \times 10^{-9} \approx 5 \times 10^7\ \text{cm/s}$$

The angle θ that the terminal velocity vector makes with the x axis is given by $\tan\theta = v_y/v_x$, so that

$$\theta = \tan^{-1}\frac{v_y}{v_x} \approx \tan^{-1}\frac{5 \times 10^7}{10^9} = \tan^{-1} 0.05$$

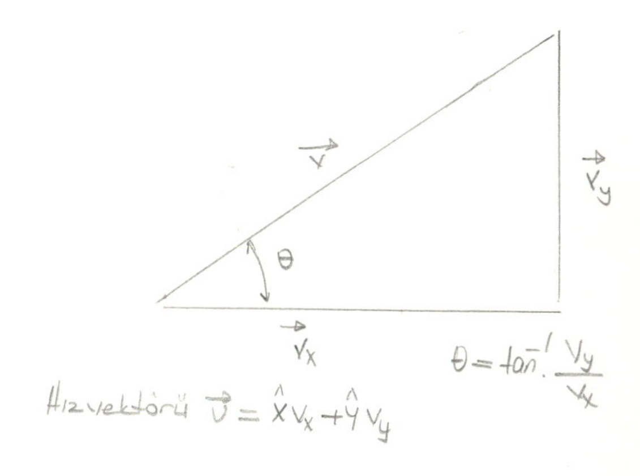

Now for a small angle we can make the approximation

$$\theta \approx \tan^{-1}\theta$$

where θ is in radians. From the above we see that $\theta \approx 0.05$ rad, which is about 3°.

By estimating the value of the next term in the series expansion of $\tan^{-1}\theta$, we can check on the error in using the approximation. Standard mathematical tables give the series expansion of trigonometric functions. Thus Dwight[1] 505.1 reads

$$\tan^{-1} x = x - \frac{x^3}{3} + \frac{x^5}{5} - \frac{x^7}{7} + \cdots \qquad \text{valid for } x^2 < 1$$

The term $x^3/3$ for $x = 0.05$ is smaller than the leading term x by the factor $x^2/3 = (0.05)^2/3 \approx 10^{-3}$, or 0.1 percent. This error may be neglected if it is less than the experimental error in the measure-

[1] For convenience nearly all references to mathematical tables will be to H. B. Dwight, "Tables of Integrals and Other Mathematical Data," 4th ed., The Macmillan Company, New York, 1961. There are a number of excellent compact collections of mathematical data; you can get along well with any one of them. You should also have a handbook of chemical and physical data, such as the "Handbook of Chemistry and Physics," Chemical Rubber Publishing Company, New York, and a 10-in. slide rule.

ment of θ. For small angles it is also true that $\sin\theta \approx \theta$ and $\cos\theta \approx 1 - \frac{1}{2}\theta^2$.

Motion of a Charged Particle in a Uniform Constant Magnetic Field[1] The equation of motion of a charged particle of mass M and charge q in a constant magnetic field $\mathbf{B}$ is, from Eq. (3.18),

$$M\frac{d^2\mathbf{r}}{dt^2} = M\frac{d\mathbf{v}}{dt} = \frac{q}{c}\mathbf{v} \times \mathbf{B} \tag{3.23}$$

Let the magnetic field be directed along the z axis:

$$\mathbf{B} = \hat{\mathbf{z}}B$$

Hence by the rule for the vector product

$$[\mathbf{v} \times \mathbf{B}]_x = v_y B \qquad [\mathbf{v} \times \mathbf{B}]_y = -v_x B \qquad [\mathbf{v} \times \mathbf{B}]_z = 0$$

Thus from Eq. (3.23)†

$$\dot{v}_x = \frac{q}{Mc}v_y B \qquad \dot{v}_y = -\frac{q}{Mc}v_x B \qquad \dot{v}_z = 0 \tag{3.24}$$

We see that the component of the velocity along the axis of the magnetic field, the z axis, is constant.

We can see directly another feature of the motion: The kinetic energy

$$K = \tfrac{1}{2}Mv^2 = \tfrac{1}{2}M\mathbf{v}\cdot\mathbf{v}$$

is constant for

$$\frac{dK}{dt} = \tfrac{1}{2}M(\dot{\mathbf{v}}\cdot\mathbf{v} + \mathbf{v}\cdot\dot{\mathbf{v}}) = M\mathbf{v}\cdot\dot{\mathbf{v}} = M\mathbf{v}\cdot\left(\frac{q}{Mc}\mathbf{v}\times\mathbf{B}\right) = 0 \tag{3.25}$$

because $\mathbf{v} \times \mathbf{B}$ is perpendicular to $\mathbf{v}$. Thus *a magnetic field does not change the kinetic energy of a free particle.*

Let us look for solutions[2] of the equations of motion of the form

[1] If the vector product in Chap. 2 has been omitted, this section can be condensed to the case of $\mathbf{v}$ and $\mathbf{B}$ perpendicular.

† We adopt here a convention common in physics: The dot over the letter means the derivative with respect to time. Thus $\dot{r} = dr/dt$, $\dot{\mathbf{A}} = d\mathbf{A}/dt$. In the same manner $\ddot{r} = d^2r/dt^2$, $\ddot{\mathbf{A}} = d^2\mathbf{A}/dt^2$.

[2] Equation (3.25) tells us that K is a constant, we must conclude that $|\mathbf{v}|$ is constant also. This result suggests that we try a solution representing a uniform circular motion, in which x and y components of the velocity are sinusoidal with $\pi/2$ phase difference. It is convenient to represent qB/Mc as a single constant having dimensions of inverse time; the dimensions are easily seen from Eq. (3.24). We expect a solution involving a rotation for which this constant is related to the angular frequency ω.

$$v_x(t) = v_1 \sin \omega t \qquad v_y(t) = v_1 \cos \omega t \qquad v_z = \text{const} \tag{3.26}$$

This motion is circular in its projection on the xy plane, with a radius that we calculate below. By differentiation of v_x and v_y in Eq. (3.26)

$$\frac{dv_x}{dt} = \omega v_1 \cos \omega t \qquad \frac{dv_y}{dt} = -\omega v_1 \sin \omega t$$

so that Eq. (3.24) becomes

$$\omega v_1 \cos \omega t = \frac{qB}{Mc} v_1 \cos \omega t \qquad -\omega v_1 \sin \omega t = -\frac{qB}{Mc} v_1 \sin \omega t$$

These equations are satisfied if

$$\omega = \frac{qB}{Mc} \equiv \omega_c \tag{3.27}$$

This relation defines the *cyclotron frequency* (or *gyrofrequency*) ω_c as the frequency of the circular motion of the particle in the magnetic field. Any value of v_1 will satisfy the equations, but we shall see that v_1 will determine the radius of the circular path.

The cyclotron frequency can also be derived by an elementary argument. The inwardly directed magnetic force qBv_1/c provides the centripetal (inward) acceleration involved in the circular motion of the particle. The magnitude of this centripetal acceleration is v_1^2/r, or $\omega_c^2 r$, because $\omega_c r = v_1$. Thus

$$\frac{qBv_1}{c} = M\omega_c^2 r = M\omega_c v_1$$

whence $\omega_c = qB/Mc$ and the radius of the circle is $r = Mcv_1/qB$ (see Fig. 3.7).

What is the complete trajectory? We have seen that it is circular with regard to x and y motion; in z it will simply progress with the constant speed v_z (which, of course, may be zero) since no force having a z component is present. By integrating Eqs. (3.26) with ω set equal to ω_c, we obtain the trajectory

$$\begin{aligned} x &= x_0 + \frac{v_1}{\omega_c} - \frac{v_1}{\omega_c} \cos \omega_c t \\ y &= y_0 + \frac{v_1}{\omega_c} \sin \omega_c t \\ z &= z_0 + v_z t \end{aligned} \tag{3.28}$$

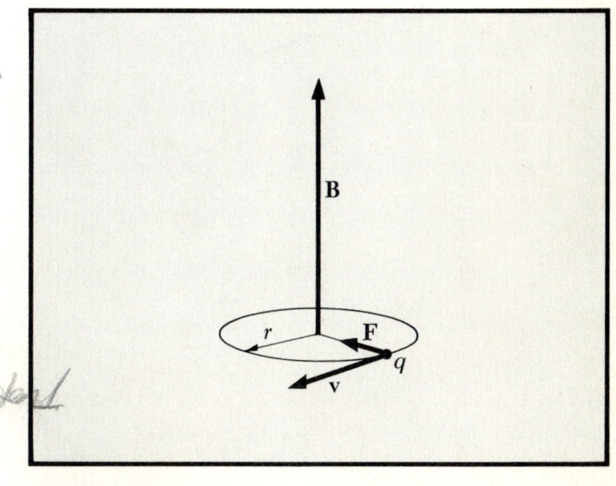

FIG. 3.7 Positive charge q, with initial velocity $\mathbf{v} \perp$ uniform $\mathbf{B}$, describes a circle with constant speed v_1, radius $r = cMv_1/qB$.

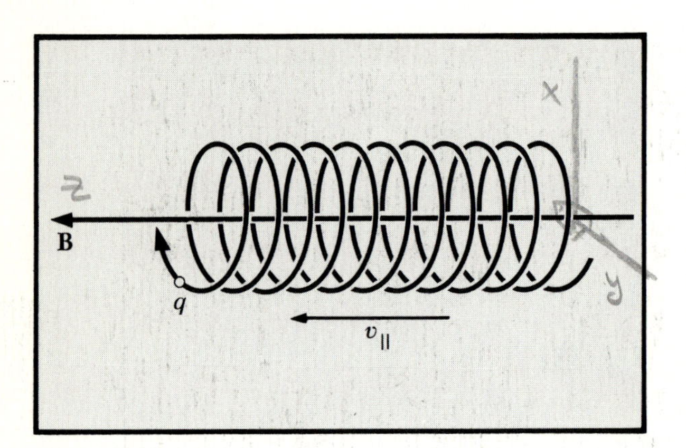

FIG. 3.8 A positive charge q describes a helix of constant pitch in a uniform magnetic field **B**. The component of velocity $v_\parallel$ parallel to **B** is a constant. If $\mathbf{B} = B_z\hat{\mathbf{z}}$, $v_\parallel = v_z$.

where we have called the constants of integration $x_0 + v_\perp/\omega_c$, y_0, and z_0 in the respective equations.

Equation (3.28) describes the x, y position of a particle moving in a circle with radius

$$r_c = \frac{v_\perp}{\omega_c} = \frac{Mcv_\perp}{qB} \tag{3.29}$$

about the center located at $(x_0 + v_\perp/\omega_c, y_0)$; and superimposed upon this uniform circular motion is a steady drift in the z direction at speed v_z, beginning at $z = z_0$ when $t = 0$. The complete motion is thus a helix whose axis is parallel to the magnetic field vector **B**, that is, along the z axis in this case. It is illustrated in Fig. 3.8. The radius r_c is frequently called the *gyroradius* or *cyclotron radius*.

We should note the product of magnetic field intensity and radius of path

$$Br_c = \frac{Mv_\perp c}{q} \tag{3.30}$$

This is an important relation; we shall see in a later chapter that it is valid in the relativistic region if the momentum, appearing here as $Mv_\perp$, is changed to its relativistic expression. The relation is used to determine the momentum of charged particles, whether at high or low velocities (see Fig. 3.9).

Check of Dimensions It is good practice always to check that the dimensions of both sides of a final equation are identical. This is an easy way to catch massive mistakes. On the right-hand side of Eq. (3.30) we have

$$\left[\frac{cMv_\perp}{q}\right] = \left[\frac{L}{T}\right]\left[M\right]\left[\frac{L}{T}\right]\left[\frac{1}{q}\right] = \left[\frac{ML^2}{qT^2}\right] \tag{3.31}$$

where we have used the notation of page 63 but kept the dimensions of q. On the left-hand side of Eq. (3.30)

$$[Br_c] = \left[\frac{F}{q}\right]\left[L\right] = \left[\frac{ML^2}{qT^2}\right] \tag{3.32}$$

because according to the Lorentz force equation (3.18) the dimensions of B in the gaussian system of units are those of force divided by charge. We see that the dimensions in Eq. (3.31) are the same as in Eq. (3.32).

In the SI units, where the force is $q\mathbf{v} \times \mathbf{B}$ instead of $(q/c)\mathbf{v} \times \mathbf{B}$, we will have

FIG. 3.9 Hydrogen bubble chamber photograph of the path of a fast electron in a magnetic field. The electron enters at the lower right. The electron shows up by losing energy by ionization of hydrogen molecules. As the electron slows up, its radius of curvature in the magnetic field decreases, hence the spiral orbit. (*Lawrence Berkeley Laboratory*)

$$\omega = \frac{qB}{M} = \omega_c \tag{3.27a}$$

$$r_c = \frac{v_1}{\omega_c} = \frac{Mv_1}{qB} \tag{3.29a}$$

and

$$Br_c = \frac{Mv_1}{q}$$

The check of these dimensions gives

$$[Br_c] = [M][L][T]^{-2}[q]^{-1}[L]^{-1}[T][L] = [M][L][T]^{-1}[q]^{-1}$$

and

$$\left[\frac{Mv_1}{q}\right] = [M][L][T]^{-1}[q]^{-1}$$

EXAMPLE

Gyrofrequency What is the gyrofrequency of an electron in a magnetic field of 10 kG, or 1×10^4 G? (A field of 10 to 15 kG is typical of ordinary laboratory iron-core electromagnets.)

We have, from Eq. (3.27),

$$\omega_c = \frac{eB}{mc} \approx \frac{(4.8 \times 10^{-10})(1 \times 10^4)}{(10^{-27})(3 \times 10^{10})} \approx 1.6 \times 10^{11}\ \mathrm{s}^{-1}$$

or in the SI units

$$\omega_c = \frac{eB}{m} \approx \frac{(1.6 \times 10^{-19})(1.0)}{10^{-30}} \approx 1.6 \times 10^{11}\ \mathrm{s}^{-1}$$

The corresponding frequency denoted by ν_c is

$$\nu_c = \frac{\omega_c}{2\pi} \approx 3 \times 10^{10}\ \mathrm{cps}$$

This is equivalent to the frequency of an electromagnetic wave in free space of wavelength

$$\lambda_c = \frac{c}{\nu_c} \approx \frac{3 \times 10^{10}}{3 \times 10^{10}} \approx 1\ \mathrm{cm}$$

The gyrofrequency $\omega_c(p)$ of a proton is lower than that of an electron in the same magnetic field in the ratio 1:1836, i.e., the ratio of the electron mass to the proton mass. For a proton in a 10-kG field

$$\omega_c(p) = \frac{m}{M_p}\omega_c(e) \approx \frac{1.6 \times 10^{11}}{1.8 \times 10^3} \approx 10^8\ \mathrm{s}^{-1}$$

The sense of rotation for the electron is opposite to that for the proton because their charges are opposite in sign.

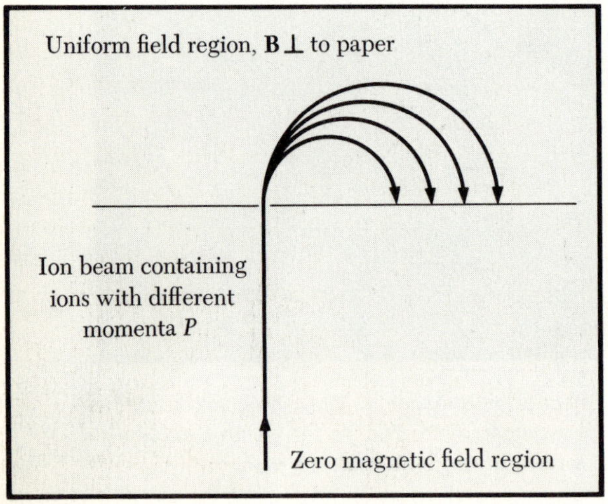

FIG. 3.10 Magnetic field as a momentum selector.

EXAMPLE

Gyroradius What is the radius of the cyclotron orbit in a 10-kG field for an electron of velocity 10^8 cm/s normal to B?

We have for the gyroradius, using Eq. (3.29),

$$r_c = \frac{v_1}{\omega_c} \approx \frac{10^8}{1.6 \times 10^{11}} \approx 6 \times 10^{-4} \text{ cm}$$

The gyroradius for a proton of the same velocity is larger in the ratio M/m:

$$r_c \approx (6 \times 10^{-4})(1.8 \times 10^3) \approx 1 \text{ cm}$$

180° Magnetic Focusing Let a beam of charged particles possessing various masses and velocities enter a region in which there is a uniform magnetic field **B** perpendicular to the beam. A particle will be deflected with a radius of curvature given by the relation $B\rho = (c/q)Mv_t$, where v_t is the velocity component in the plane normal to **B**. If we examine the beam at some point, say, after 180° of motion, we find it is spread out in the plane of the motion because the different particles with different masses and velocities have different radii of curvature, as in Fig. 3.10. By providing a narrow exit slit, the arrangement is used as a *momentum selector*, a device to obtain a beam of particles having closely equal momenta if the particles all have the same charge q. One advantage of using a deflection of 180° is that particles of equal momenta but moving through an entrance slit at slightly different angles are brought to an approximately common focus after 180°.

The accuracy of the focusing is purely a problem in geometry, and it is illustrated in Fig. 3.11*a* and *b*. Consider a trajectory that makes initially an angle θ with the ideal trajectory. The distance from the entrance slit at which it will strike the target area is given by the chord C of the circle of radius ρ. The difference in length between the diameter and the chord is

$$2\rho - C = 2\rho(1 - \cos\theta) \approx \rho\theta^2$$

where we have used for small θ the first two terms in the power series expansion of the cosine

$$\cos\theta = 1 - \frac{\theta^2}{2!} + \frac{\theta^4}{4!} - \cdots$$

as found in standard tables (Dwight 415.02). If we measure the angular focusing power by

$$\frac{2\rho - C}{2\rho} \approx \frac{1}{2}\theta^2$$

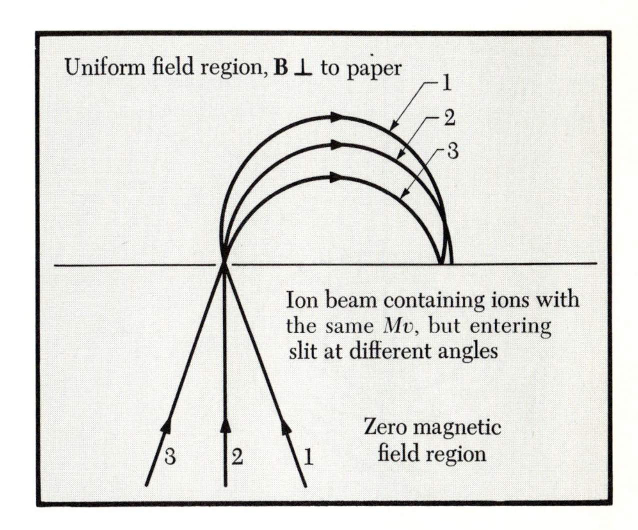

FIG. 3.11 (*a*) 180° focusing in a magnetic field. Ions of equal momenta but different directions are focused nearly together.

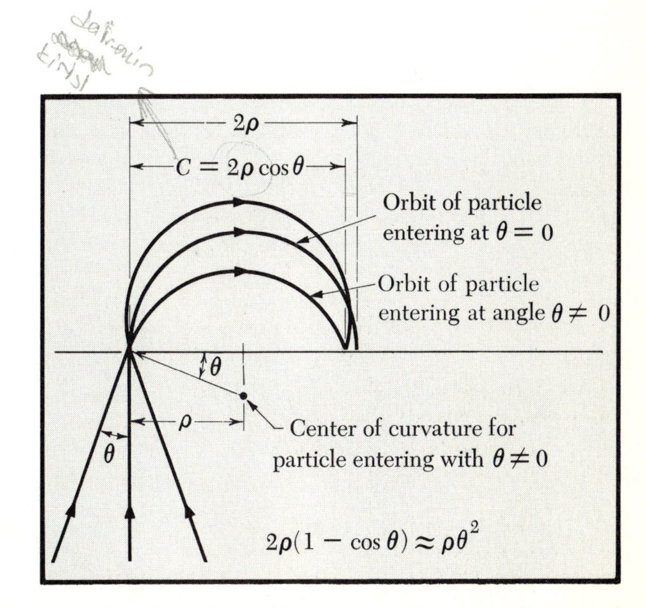

(*b*) Diagram showing details of focusing in 180° velocity selector.

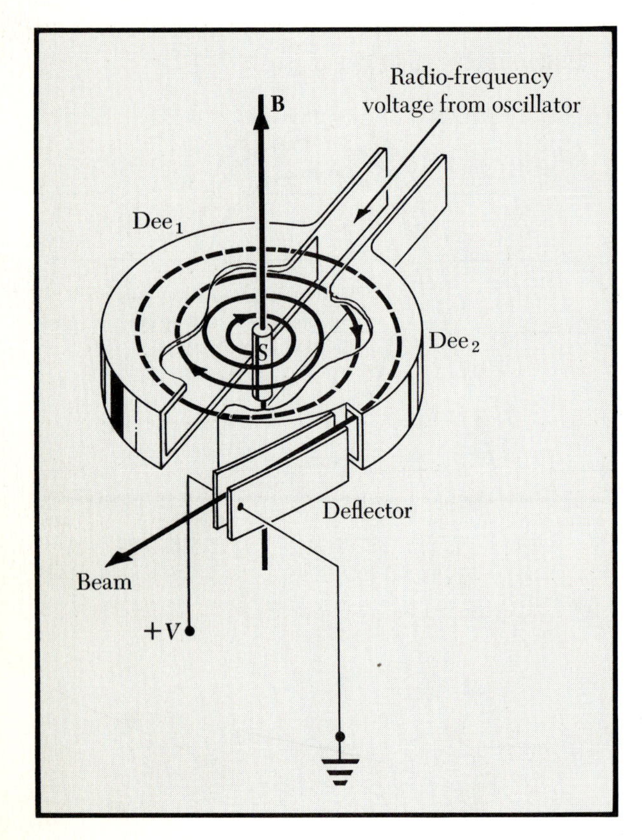

FIG. 3.12 Cutaway view of a conventional low-energy cyclotron, consisting of ion source S, hollow accelerating electrodes (Dee_1, Dee_2), and deflector. The entire apparatus is immersed in a homogeneous vertical magnetic field **B** (pointed upward). The plane of the particle orbit is horizontal and is also the median plane of the dees. The accelerating rf electric field is confined to the gap between the dees.

we have, for $\theta = 0.1$ rad, the value

$$\frac{2\rho - C}{2\rho} \approx 5 \times 10^{-3}$$

This illustrates the focusing action.

Cyclotron Acceleration Principle Charged particles in a standard cyclotron move in roughly spiral orbits in a constant magnetic field, as described in the Historical Note at the end of the chapter and as shown in Fig. 3.12. The particles are accelerated every half cycle (π rad) by an oscillating electric field. The requirement for periodic acceleration is that the frequency of the electric field should equal the cyclotron frequency of the particles.

The cyclotron frequency for protons in a magnetic field of 10 kG was shown above to be $1 \times 10^8\ \mathrm{s}^{-1}$ or $\nu_c = \omega_c/2\pi \approx 10^7$ cps ≈ 10 Mc/s. The frequency is independent of the energy of the particle so long as the velocity is nonrelativistic, i.e., small compared with the speed of light. A graph of wavelength (c/ν) against B is shown in Fig. 3.13.

In each cycle of operation the particle picks up energy from the oscillating electric field. The effective radius of the orbit increases as the kinetic energy increases, because, as shown,

$$r_c = \frac{v}{\omega_c} = \frac{\sqrt{2E/M_p}}{\omega_c}$$

where E now denotes the energy. The energy of a nonrelativistic proton in a constant magnetic field is set by the outer radius of the cyclotron: at $\omega_c = 1 \times 10^8\ \mathrm{s}^{-1}$ and $r_c = 50$ cm, we have $v = \omega_c r_c \approx 5 \times 10^9$ cm/s, or

$$E = \tfrac{1}{2}M_p v^2 \approx 10^{-24}(5 \times 10^9)^2 \approx 25 \times 10^{-6}\ \mathrm{erg}$$

In practice this velocity is sufficiently nonrelativistic for the operation of a conventional cyclotron.

CONSERVATION OF MOMENTUM

The law of conservation of momentum is probably familiar to the student from high school physics. Its importance in collision problems can hardly be overemphasized. We give here the derivation based on Newton's Third Law and reserve for Chap. 4 the discussion of an alternate derivation. The law states that:

For an isolated system, subject only to internal forces (forces between members of the system), the total linear momentum of the system is a constant; it does not change in time.

Most familiarly it is applied to collisions of two particles for which it may be stated: The sum of the momenta after the collision is equal to the sum of the momenta before the collision provided that the collision takes place in a region free from external forces:

$$\mathbf{p}_1(\text{before}) + \mathbf{p}_2(\text{before}) = \mathbf{p}_1'(\text{after}) + \mathbf{p}_2'(\text{after}) \qquad (3.33)$$

where the momentum p has been defined by

$$\mathbf{p} = M\mathbf{v} \qquad (3.34)$$

and we use primes ($\mathbf{p}'$) to indicate values after the collision. See Fig. 3.14 for a depiction of the vector momenta and Fig. 3.15 for orbits. The collision may be either elastic or inelastic. In an elastic collision all the kinetic energy of the incoming particles reappears after the collision as kinetic energy but usually divided differently between the particles. In the usual inelastic collisions, part of the kinetic energy of the incoming particles appears after the collision as some form of *internal* excitation energy (such as heat) of one or more of the particles. It is important to realize that momentum conservation applies *even* to *inelastic* collisions, in which the kinetic energy is not conserved.

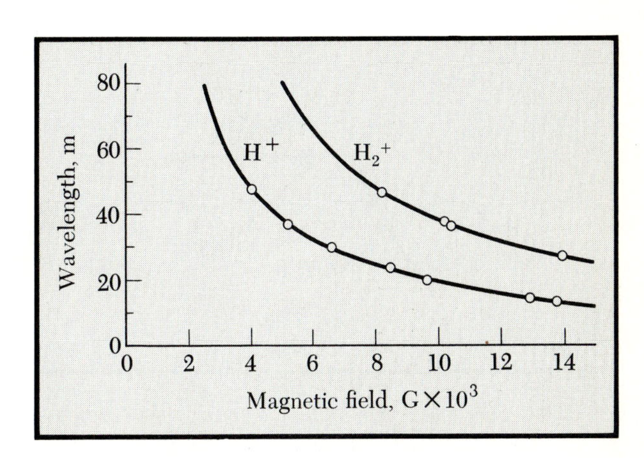

FIG. 3.13 Resonance condition in the first cyclotron (11-in. diameter). On the vertical scale is plotted the wavelength in free space of the rf power supplied to the accelerating electrodes (dees). The curves are the theoretical relations for H^+ and H_2^+ ions; the circles are the experimental observations. [Lawrence and Livingston, *Phys. Rev.*, **40**: 19 (1932)]

Derivation by Use of Newton's Third Law We assume that the bodies obey Newton's Third Law [Eq. (3.3)]. For body 1, we have

$$\mathbf{F}_{12} = \frac{d\mathbf{p}_1}{dt} = \frac{d}{dt}(M_1\mathbf{v}_1) \qquad (3.35)$$

For body 2, we have

$$\mathbf{F}_{21} = \frac{d\mathbf{p}_2}{dt} = \frac{d}{dt}(M_2\mathbf{v}_2) \qquad (3.36)$$

Adding these two, we obtain

$$\mathbf{F}_{12} + \mathbf{F}_{21} = 0 = \frac{d\mathbf{p}_1}{dt} + \frac{d\mathbf{p}_2}{dt} = \frac{d}{dt}(\mathbf{p}_1 + \mathbf{p}_2)$$

$$= \frac{d}{dt}(M_1\mathbf{v}_1 + M_2\mathbf{v}_2)$$

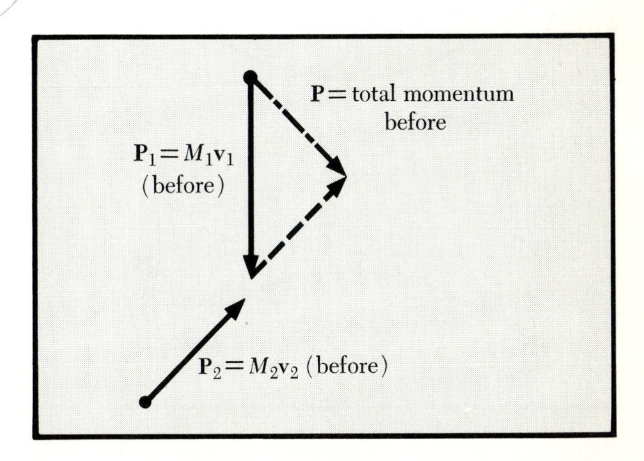

FIG. 3.14 (*a*) Before the collision the momenta $\mathbf{P}_1$ (before) and $\mathbf{P}_2$ (before) add up to $\mathbf{P}$.

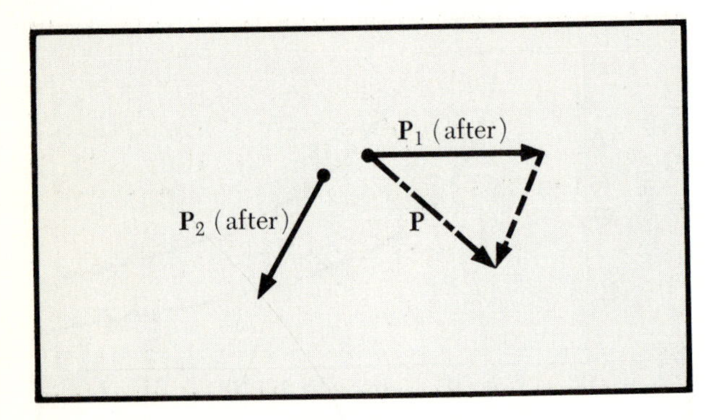

FIG. 3.14 (*cont'd.*) (*b*) While after the collision the momenta $\mathbf{P}_1$ (after) and $\mathbf{P}_2$ (after) add up to the same $\mathbf{P}$.

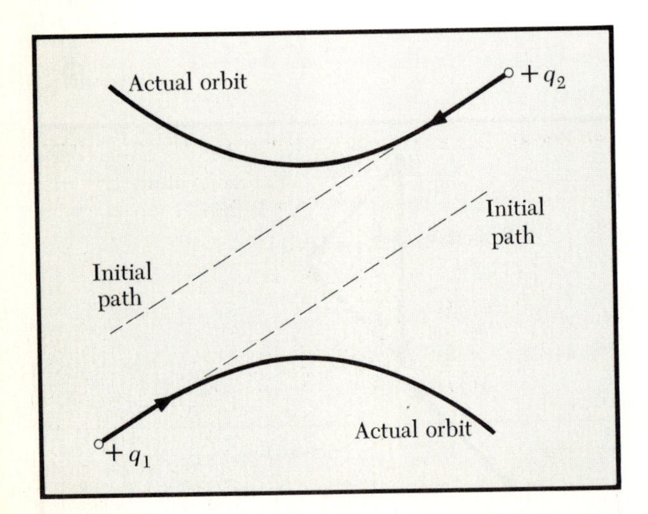

FIG. 3.15 If two moving point charges q_1 and q_2 pass close to one another, their orbits are deviated from initial straight-line paths.

Therefore

$$\mathbf{p}_1 + \mathbf{p}_2 = M_1\mathbf{v}_1 + M_2\mathbf{v}_2 = \text{const} = \mathbf{p}_1' + \mathbf{p}_2' = M_1\mathbf{v}_1' + M_2\mathbf{v}_2' \tag{3.37}$$

where again the primes indicate the values after the collision. If there are more than two bodies, the same procedure may be used with the same result applicable to any number of bodies in an isolated system.

In the following examples we discuss a number of cases of this law. Two points should be emphasized:

1 This is a vector law; and so, in the collision of two particles whose vector momenta add to define a line, the two resultant momenta add to define the same line.

2 The application of this principle alone does not enable us to solve a collision problem uniquely.

As an example of point 2, consider the problem of the collision of equal masses, one initially at rest. Only further information allows us to obtain unique answers, as illustrated in the following two cases.

(*a*) Assume that after the collision the two equal-mass particles stick together. What is their velocity? Let the original velocity of the moving body be along the x axis. Then

$$\mathbf{p}_1 = M_1 v_1 \hat{\mathbf{x}} \qquad \mathbf{p}_2 = 0$$

$$(\mathbf{p}_1' + \mathbf{p}_2') = (M_1 + M_2)\mathbf{v}' = 2M_1\mathbf{v}' = \mathbf{p}_1 = M_1 v_1 \hat{\mathbf{x}}$$

$$\mathbf{v}' = \frac{v_1}{2}\hat{\mathbf{x}}$$

(*b*) Assume that in the collision, the first particle is brought to rest. What is the velocity of the second?

$$\mathbf{p}_1' + \mathbf{p}_2' = 0 + M_2\mathbf{v}_2' = M_1 v_1 \hat{\mathbf{x}}$$
$$\mathbf{v}_2' = v_1 \hat{\mathbf{x}}$$

In order to solve a collision problem uniquely, we need information additional to the law of conservation of momentum such as that provided in one or the other of the assumptions under (*a*) or (*b*). Or the additional information may be stated in terms of elasticity or energy conservation.

EXAMPLE

Elastic Collision of Two Equal-mass Particles with One Initially at Rest We wish to prove that in this case the angle between the two momenta and velocity vectors after the collision is equal to 90°.

$$\mathbf{p}_1 + \mathbf{p}_2 = M_1\mathbf{v}_1 + 0 = M_1\mathbf{v}_1' + M_2\mathbf{v}_2'$$

Thus, since $M_2 = M_1$, and M_2 is initially at rest,

$$\mathbf{v}_1 = \mathbf{v}_1' + \mathbf{v}_2'$$

The adjective *elastic* applied to the collision means that kinetic energy $\frac{1}{2}Mv^2$ is conserved. Therefore

$$\tfrac{1}{2}M_1v_1^{\,2} = \tfrac{1}{2}M_1v_1'^{\,2} + \tfrac{1}{2}M_2v_2'^{\,2}$$

which gives

$$v_1^{\,2} = v_1'^{\,2} + v_2'^{\,2} \tag{3.38}$$

Equation (3.38) reminds us of the pythagorean theorem, and we notice from the vector diagram in Fig. 3.16 that $\mathbf{v}_1$ must be the hypotenuse of a right triangle. Therefore the angle between $\mathbf{v}_1'$ and $\mathbf{v}_2'$ must be 90°.

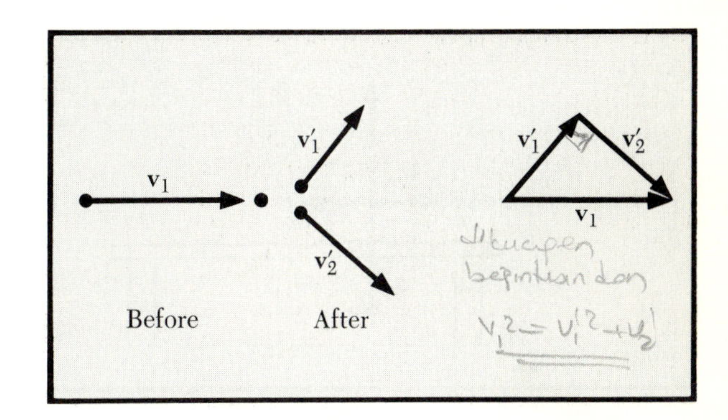

FIG. 3.16 Elastic collision of equal masses.

Further examples are given as Probs. 16 to 18, and we shall discuss collisions in greater detail in Chap. 6.

Atwood's Machine Both Newton's Second and Third Laws are utilized in the familiar *Atwood's machine* problem illustrated in Fig. 3.17. Two unequal masses are suspended by a string over a pulley assumed to be without friction and of negligible mass. Let m_2 be greater than m_1; then the acceleration will be in the direction shown for each mass and will be the same value for each because of the continuity of the string and its constancy of length. First we find the value of the acceleration.

Each mass is subject to two forces, namely, the tension of the string and the gravitational force. Newton's Third Law ensures that the tension has the same value at each body. Newton's Second Law permits us to write

$$\begin{aligned} &\text{For the motion of } m_1 \qquad T - m_1g = m_1a \\ &\text{For the motion of } m_2 \qquad m_2g - T = m_2a \end{aligned} \tag{3.39}$$

Addition of these two equations gives

$$(m_2 - m_1)g = (m_1 + m_2)a \qquad \text{or} \qquad a = \frac{m_2 - m_1}{m_2 + m_1}g \tag{3.40}$$

We can now evaluate the tension T by employing this expression for a in either one of Eqs. (3.39). This yields

$$T = \frac{2m_1m_2}{m_1 + m_2}g$$

How strong must the string be? It must not break under this tension, which means that at rest it must support a mass

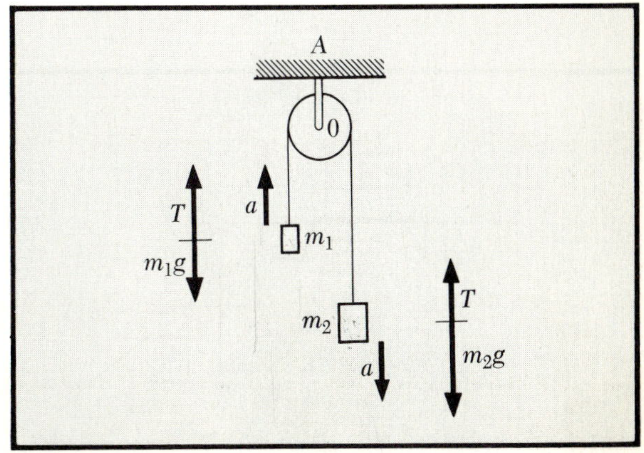

FIG. 3.17 Atwood's machine.

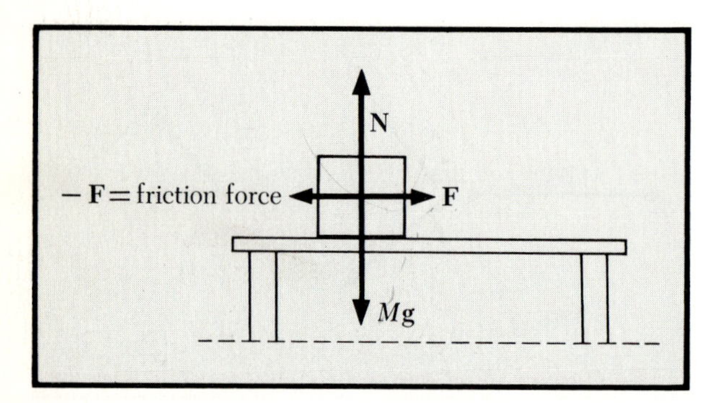

FIG. 3.18 Body on horizontal surface acted on by gravity Mg, a normal force $\mathbf{N}$, an external horizontal force $\mathbf{F}$, and a frictional force $-\mathbf{F}$.

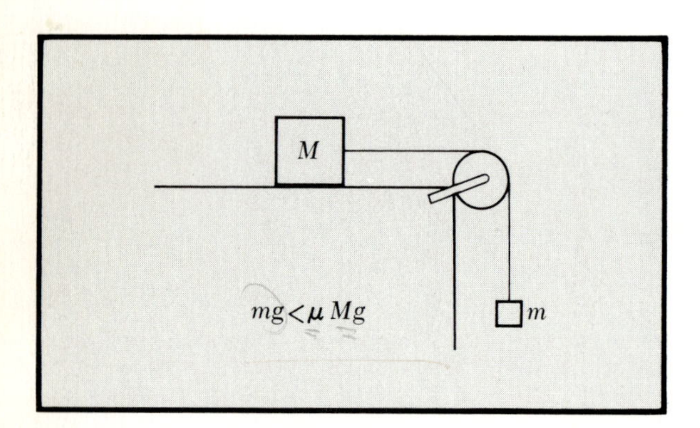

FIG. 3.19 (*a*) M far from slipping.

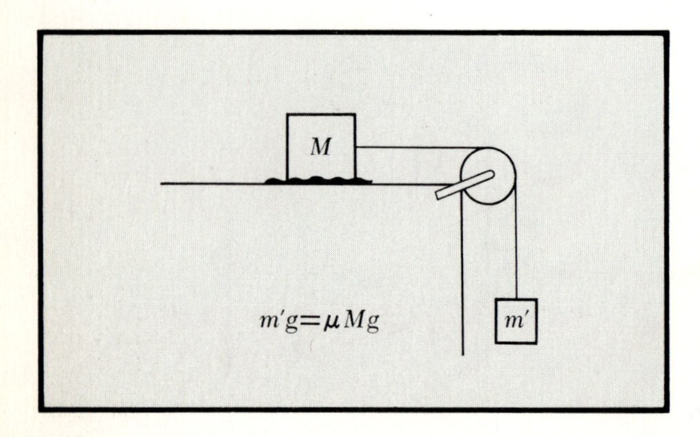

(*b*) M about to slip.

for which $mg = T$. Thus it must be at least strong enough to support the mass

$$m = \frac{2m_1m_2}{m_1 + m_2}$$

which is greater than m_1 yet smaller than m_2.

It is instructive to note that the expression for acceleration [Eq. (3.40)] can be understood in terms of $F = Ma$ for a single object by noting that the total moving mass is $m_1 + m_2$ while the net force is $(m_2 - m_1)g$. Thus, as before,

$$a = \frac{F}{M} = \frac{(m_2 - m_1)g}{m_2 + m_1}$$

CONTACT FORCES: FRICTION

In our familiar experience with ponderable objects we are often concerned with forces communicated to a body by pressures or tensions acting at the contact with another body. In the preceding section forces of this sort are exhibited in the tension of the string, and in our earlier discussion of collisions when thought of in terms of marbles or billiard balls it is assumed that contact pressures are acting briefly at the moment of collision. Another contact force that is practically very important is *friction* (see, e.g., the damping of an oscillator in Chap. 7). The force of friction may depend in a very complicated way on an object's velocity, but we treat here the simplest case—that of a constant force if the body is moving or a force just big enough to ensure equilibrium if the body is at rest.

The force of friction is parallel to the surface of contact of two objects or an object and a surface. It depends on another contact force, the normal force that a solid surface exerts on an object resting upon it. Figure 3.18 shows an object on a flat horizontal surface. Obviously the force of gravity Mg is acting vertically downward. Since the body is at rest, Newton's First Law tells us that also there must be an upward force equal to Mg. Such a force, which is normal to the surface and which prevents a body from falling through the surface, is usually denoted by N, as in Fig. 3.18. The force tending to force the body into the surface may be gravity, a component of the gravitational force, or some other force entirely, depending on the particular circumstances.

Suppose now that we exert a force $\mathbf{F}$ parallel to the surface (perhaps by attaching a string and hanging a weight on the string, as shown in Fig. 3.19) but not big enough to cause the body to slip. Again from Newton's First Law, the surface must exert an equal and opposite force $-\mathbf{F}$ on the object. This force $-\mathbf{F}$ is called the *force of friction*. It is zero until the force $\mathbf{F}$ tries to move the body.

How big can the force of friction be? We can always (except in the case of an "immovable object") exert a big enough force $\mathbf{F}$ to cause the body to slip. It is an experimental fact that

$$F_{\text{max}} = \mu N \tag{3.41}$$

where μ is a constant called the *coefficient of static friction* characteristic of the surfaces in contact. Sample values are given in Table 3.1. Remember that the force of static friction can have any value up to μN, depending on the value of the external force applied; this is shown in Fig. 3.19.

EXAMPLE

Measurement of μ The value of μ can be determined by finding the angle θ to the horizontal of an inclined plane at which a body will just slip. Referring to Fig. 3.20 and assuming that the body is about to slip, we recognize that the sum of the three forces $M\mathbf{g}$, $\mathbf{N}$, and $\mathbf{F}_{\text{fric}}$ must be zero. Taking components parallel and perpendicular to the surface, we find

$$N = Mg\cos\theta \qquad F_{\text{fric}} = Mg\sin\theta \tag{3.42}$$

Now, using the fact that $F_{\text{fric}} = \mu N$, we obtain

$$\mu = \frac{F_{\text{fric}}}{N} = \frac{Mg\sin\theta}{Mg\cos\theta} = \tan\theta \tag{3.43}$$

EXAMPLE

Slipping with Tangential Force Variable in Direction A body of mass M rests on an inclined plane with coefficient of friction $\mu > \tan\theta$. Find the magnitude of the force parallel to the plane necessary to cause the body to slip in terms of the angle from the direction straight up the plane. A variation of this problem is: If a force parallel to the plane but not necessarily up or down the plane causes the object to slip, find the direction in which it will start to move in terms of the direction of the force.

Figure 3.21 shows the forces parallel to the plane that are responsible for the equilibrium. From Fig. 3.21 we see that $\mathbf{F}_{\text{fric}}$, $Mg\sin\theta\,\hat{\mathbf{x}}$, and $\mathbf{F}$ (the external force) must add up to zero. Since the body is about to slip, we have, from the example above,

$$F_{\text{fric}} = \mu Mg\cos\theta$$

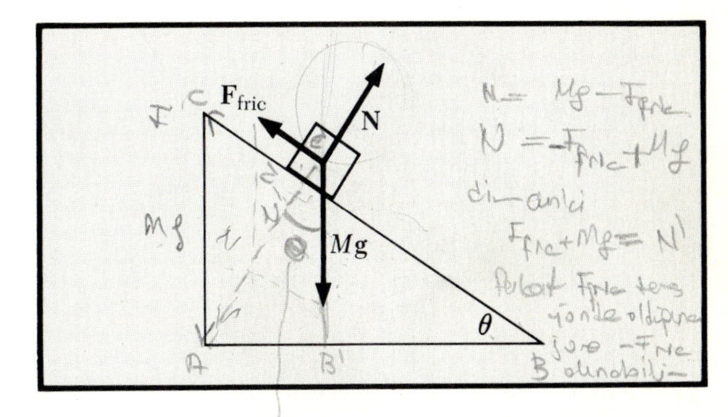

FIG. 3.20 Body about to slip down inclined plane.

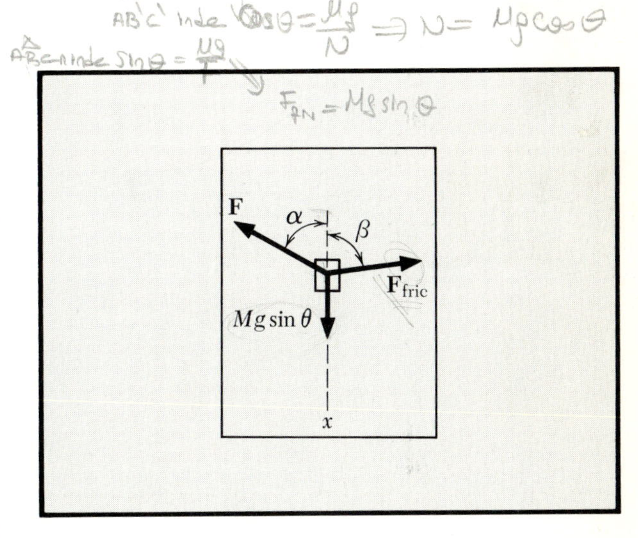

FIG. 3.21 Body about to slip on rough inclined plane under action of external force $\mathbf{F}$.

TABLE 3.1 Coefficients of Static Friction $\mu = F/N$

Material	μ
Glass on glass	0.9–1.0
Glass on metal	0.5–0.7
Graphite on graphite	0.1
Rubber on solids	1–4
Brake material on cast iron	0.4
Ice on ice	0.05–0.15
Ski wax on dry snow	0.04
Copper on copper	1.6
Steel on steel	0.58

Taking components up and down the plane,

$$F \cos \alpha + F_{\text{fric}} \cos \beta - Mg \sin \theta = 0$$

or

$$F \cos \alpha + \mu Mg \cos \theta \cos \beta = Mg \sin \theta$$

and perpendicular to this direction

$$F_{\text{fric}} \sin \beta - F \sin \alpha = 0 \qquad F \sin \alpha = \mu Mg \cos \theta \sin \beta$$

Eliminating β from these equations gives

$$\frac{F}{Mg} = \cos \alpha \sin \theta \pm \sqrt{\cos^2 \alpha \sin^2 \theta + \mu^2 \cos^2 \theta - \sin^2 \theta} \tag{3.44}$$

What is the meaning of the negative sign in Eq. (3.44)? To determine this, we note that $\mu^2 \cos^2 \theta > \sin^2 \theta$ since we have assumed that $\mu > \tan \theta$. Therefore the square-root term is greater than $\cos \alpha \sin \theta$. If we use the negative sign, F will be a negative quantity. But this is clearly an unacceptable solution since we have assumed a positive F. Therefore we must use the positive sign. Note that

$$\begin{aligned} F &= \quad Mg \sin \theta + \mu Mg \cos \theta \qquad &&\text{when } \alpha = 0 \\ &= -\, Mg \sin \theta + \mu Mg \cos \theta \qquad &&\text{when } \alpha = \pi \end{aligned}$$

which are straightforward to work out. One can also check that if $\mu = \tan \theta$ and $\alpha = \pi$, $F = 0$.

When F is just larger than the value given in Eq. (3.44), the body will slip and the direction will be just opposite to $\mathbf{F}_{\text{fric}}$. From the above equations, the value of β can be calculated giving

$$\sin \beta = \frac{\sin \alpha}{\mu}\left(\cos \alpha \tan \theta + \sqrt{\mu^2 - \tan^2 \theta \sin^2 \alpha}\,\right)$$

A check on this equation can be made by the assumption that $\beta = \pi/2$, in which case the three forces $\mathbf{F}$, $Mg \sin \theta \hat{\mathbf{x}}$, and $\mathbf{F}_{\text{fric}}$ form a right triangle.

EXAMPLE

Horizontal Motion with Constant Frictional Force Suppose the coefficient of friction between a horizontal surface and a moving body is μ. With what speed must the body be projected parallel to the surface to travel a distance D before stopping? We have a one-dimensional problem with a constant force

$$M\frac{d^2x}{dt^2} = -\mu Mg \qquad \frac{d^2x}{dt^2} = -\mu g$$

We have already worked out a solution to a similar equation in the section on gravity. See Eqs. (3.8) and (3.9). We have

$$v_x = -\mu g t + v_0 \qquad \text{and} \qquad x = -\tfrac{1}{2}\mu g t^2 + v_0 t$$

where we have let x_0 (the value of x at $t = 0$) be equal to 0. The speed desired is v_0. When the body stops, $v_x = 0$ and $t = v_0/\mu g$. Putting this value into the equation for x, we obtain, setting x equal to D,

$$D = -\tfrac{1}{2}\mu g\left(\frac{v_0}{\mu g}\right)^2 + v_0\frac{v_0}{\mu g} = \frac{1}{2}\frac{v_0^2}{\mu g}$$

or

$$v_0 = \sqrt{2D\mu g}$$

PROBLEMS

(*Note:* Always give units with numerical answers. Without units, a numerical answer is meaningless.)

1. *Newton's Third Law.* A student in elementary physics finds himself in the middle of a large ice rink with a small but finite coefficient of friction between his feet and the ice. He has been taught Newton's Third Law. Since the law says that for every action there is an equal and opposite reaction, all forces add up to zero. Therefore he assumes that there will be no force possible to accelerate him toward the side of the rink and so he must stay at the center.
(*a*) How do you tell him to get to the side?
(*b*) Once he is at the edge, what do you tell him about Newton's Second and Third Laws?

2. *Monkey and hunter.* A familiar demonstration in freshman physics lectures is illustrated by Fig. 3.22. A projectile is shot from a gun at 0 aimed at a target object located at P. The target object is released at the same instant the projectile is "fired." The projectile strikes the falling object as shown. Prove that this midair collision will result independent of muzzle velocity.

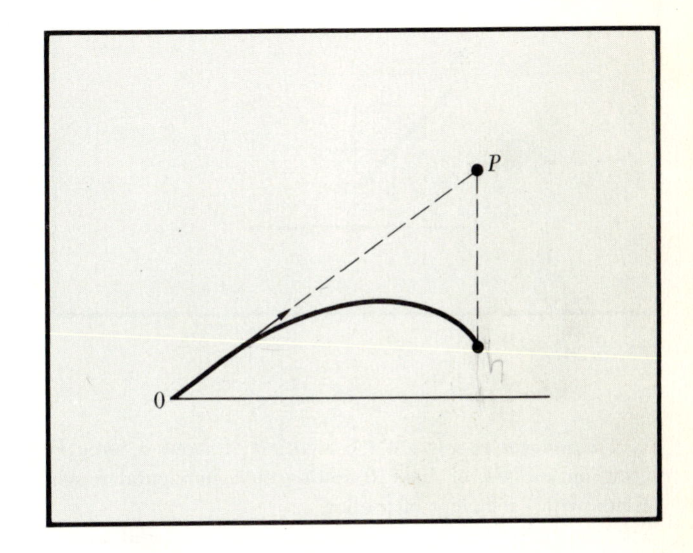

FIG. 3.22

3. *Ceiling height for a game of catch.* Two boys "play catch" with a ball in a long hallway. The ceiling height is H, and the ball is thrown and caught at shoulder height, which we call h for each boy. If the boys are capable of throwing the ball with velocity v_0, at what maximum separation can they play? *Ans.* $R = 4\sqrt{(H - h)[v_0^2/2g - (H - h)]}$.
Show that if $H - h > v_0^2/4g$, $R = v_0^2/g$. Explain the physical significance of the condition $H - h > v_0^2/4g$.

4. *Shooting upward.* The muzzle velocity of a gun is 3.0×10^3 cm/s. A man shoots one shot each second straight up into the air, which is considered frictionless.
(*a*) How many bullets will be in the air at any time?
(*b*) At what heights above the ground will they pass each other?

5. *Friction on two inclined planes.* In Fig. 3.23 planes 1 and 2 are both rough with coefficients of friction μ_1 and μ_2. Find the relation between M_1, M_2, θ_1, θ_2, μ_1, and μ_2 such that
(*a*) M_1 is about to slip down plane 1.
(*b*) M_2 is about to slip down plane 2.

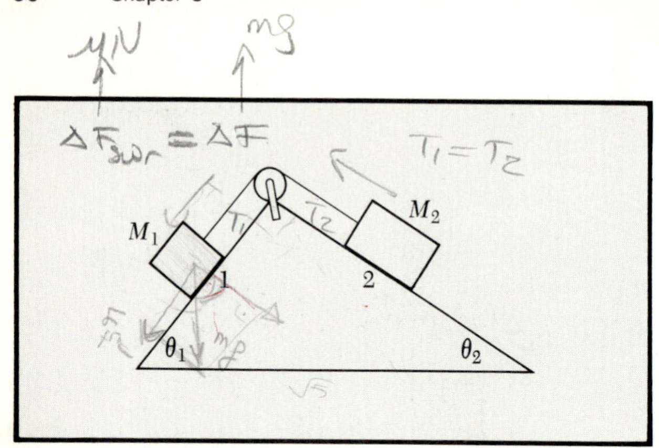

FIG. 3.23

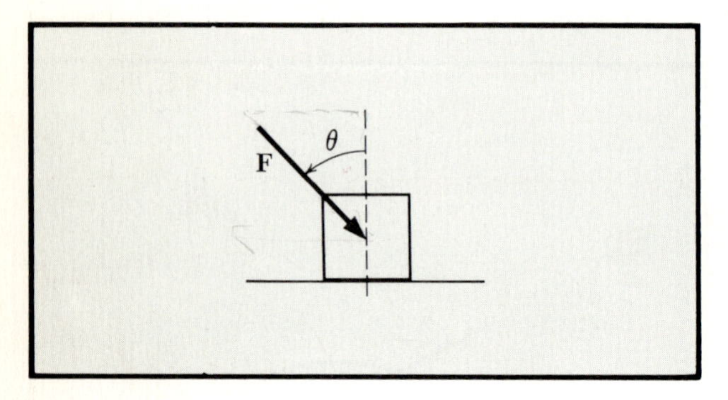

FIG. 3.24

6. *Friction not equal to μMg.* Figure 3.24 shows a force **F** acting on a block of mass M resting on a horizontal rough surface with coefficient of friction μ.

(*a*) Assuming $F \gg Mg$, find the maximum angle θ at which the force F can not make the block slip, no matter how large it is.

(*b*) Find the ratio F/Mg in terms of θ and μ such that the block will just slip. Show that the answer reduces to that of (*a*) in the limit $F \gg Mg$.

7. *Atwood's machine.* In the Atwood's machine shown in Fig. 3.17, find the tension in the string 0*A* supporting the pulley. Show that the vector sum of the three forces—this tension, m_1g, and m_2g—is equal to the rate of change of the vertical momentum.

8. *Satellite and moon.* Which travels faster, the moon or a satellite traveling around the earth at a radius just greater than the radius of the earth? What is the ratio of the speeds in terms of the ratio of the radii? What is the ratio of the periods?

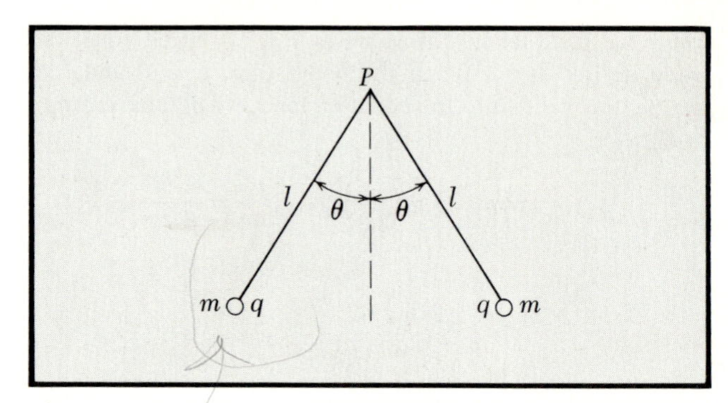

FIG. 3.25

From the facts that the moon has a period of about 27 days and a radius of orbit 240,000 mi and that the radius of the earth is 4000 mi, find the period of the satellite.

9. *Electrostatic force.* Two identical, small conducting spheres are suspended from P by threads of equal length. Initially the spheres hang in contact with each other, with $\theta \simeq 0$. They are given electric charge that is shared equally, and they then assume an equilibrium situation as shown in Fig. 3.25. Find an expression giving q in terms of m, g, ℓ, and θ. (Treat the small spheres as if they were point charges.)

10. *Proton in an electric field*

(*a*) What force (in dynes) acts upon a proton in an electric field of 100 statvolts/cm?

(*b*) If a proton were released at rest in a uniform field of this intensity, what would be its speed after 10^{-8} s?

(*c*) How far would it be from its release point after this time?

11. *Proton in a magnetic field.* A proton ($e = 4.80 \times 10^{-10}$ esu) is projected with a velocity vector $\mathbf{v} = 2 \times 10^8 \hat{\mathbf{x}}$ cm/s into a region where a uniform magnetic field exists described by $\mathbf{B} = 1000\hat{\mathbf{z}}$ G.

(*a*) Evaluate the force (in magnitude and direction) acting on the proton immediately after its projection.

(*b*) What is the radius of curvature of its subsequent path?

(*c*) Locate the position of the center of its circular path if the projection point is the origin.

12. *Ratio of electric and gravitational forces between two electrons.* The magnitude of the electrostatic force between two electrons is e^2/r^2; the magnitude of the gravitational force is Gm^2/r^2, where $G = 6.67 \times 10^{-8}$ dyn-cm^2/g^2. What is the order of magnitude of the ratio of the electrostatic to the gravitational forces between two electrons? *Ans.* 10^{42}.

13. *Crossed electric and magnetic fields.* A charged particle moves in the x direction through a region in which there is an electric field E_y and a perpendicular magnetic field B_z. What is the condition necessary to ensure that the net force on the particle will be zero? Show the **v**, **E**, and **B** vectors on a diagram. What is the condition on v_x if $E_y = 10$ statvolts/cm and $B_z = 300$ G? *Ans.* $v_x = 1 \times 10^9$ cm/s.

14. *Deflection between condenser plates.* A particle of charge q and mass M with an initial velocity $v_0\hat{\mathbf{x}}$ enters an electric field $-E\hat{\mathbf{y}}$ (see Fig. 3.26). We assume **E** is uniform, i.e., its value is constant at all points in the region between plates of length L (except for small variations near the edges of the plates, which we shall neglect).

(*a*) What forces act in the x and y directions, respectively? *Ans.* $F_x = 0$; $F_y = -qE\hat{\mathbf{y}}$.

(*b*) Will a force in the y direction influence the x component of the velocity?

(*c*) Solve for v_x and v_y as functions of time, and write the complete vector equation for $\mathbf{v}(t)$. *Ans.* $v_0\hat{\mathbf{x}} - (qE/M)t\hat{\mathbf{y}}$.

(*d*) Choose the origin at the point of entry, and write the complete vector equation for the position of the particle as a function of time while the particle is between the plates.

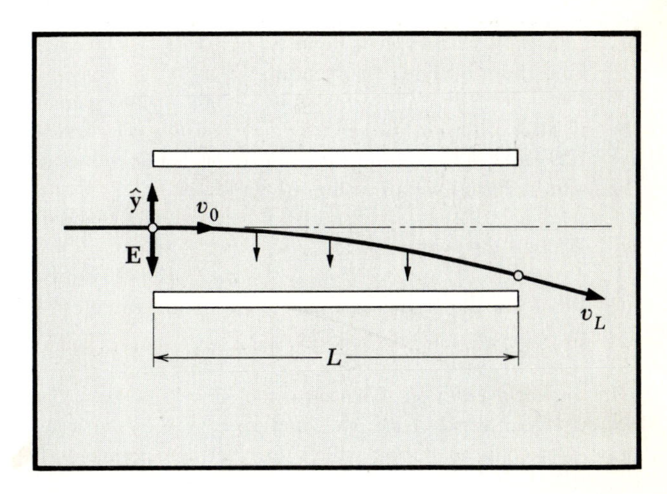

FIG. 3.26

15. *Continuation of preceding problem.* If the particle in Prob. 14 is an electron of initial kinetic energy 10^{-10} erg (kinetic energy $= \frac{1}{2}mv^2$; 1 erg is the kinetic energy of a mass of 2 g moving with speed 1 cm/s), if the electric field strength is 0.01 statvolt/cm, and if $L = 2$ cm, find:

(*a*) The vector velocity as it leaves the region between the plates.

(*b*) The angle $(\mathbf{v},\hat{\mathbf{x}})$ for the particle as it leaves the plates. *Ans.* 2.7°.

(*c*) The point of intersection between the x axis and the direction of the particle as it leaves the field. *Ans.* 1.0 cm.

16. *Collision courses.* Initially two particles are at positions $x_1 = 5$ cm, $y_1 = 0$; and $x_2 = 0$, $y_2 = 10$ cm, with $\mathbf{v}_1 = -4 \times 10^4\hat{\mathbf{x}}$ cm/s, and $\mathbf{v}_2$ is along $-\hat{\mathbf{y}}$ as in Fig. 3.27.

(*a*) What must be the value of $\mathbf{v}_2$ if they are to collide? *Ans.* $-8 \times 10^4\hat{\mathbf{y}}$ cm/s.

(*b*) What is the value of $\mathbf{v}_r$, the relative velocity? *Ans.* $4 \times 10^4\,(2\hat{\mathbf{y}} - \hat{\mathbf{x}})$ cm/s.

(*c*) Establish a general criterion for recognizing a collision course for two objects in terms of their positions $\mathbf{r}_1$, $\mathbf{r}_2$ and velocities $\mathbf{v}_1$, $\mathbf{v}_2$.

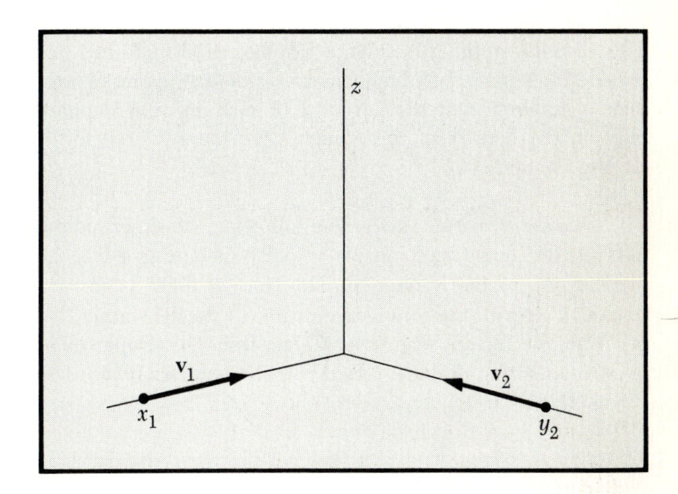

FIG. 3.27

17. *Collision kinematics.* Two masses constrained to move in a horizontal plane collide. Given initially that $M_1 = 85$ g,

$M_2 = 200$ g, $\mathbf{v}_1 = 6.4\hat{\mathbf{x}}$ cm/s, and $\mathbf{v}_2 = -6.7\hat{\mathbf{x}} - 2.0\hat{\mathbf{y}}$ cm/s:

(*a*) Find the total linear momentum.

Ans. $-796\hat{\mathbf{x}} - 400\hat{\mathbf{y}}$ g-cm/s.

(*b*) If after collision $|\mathbf{w}_1| = 9.2$ cm/s and $\mathbf{w}_2 = -4.4\hat{\mathbf{x}} + 1.9\hat{\mathbf{y}}$ cm/s, what is the direction of $\mathbf{w}_1$? (For velocities after collision we are using the symbol $\mathbf{w}$.)

Ans. $-84°$ with respect to x axis.

(*c*) What is the relative velocity $\mathbf{w}_r = \mathbf{w}_1 - \mathbf{w}_2$?

Ans. $5.4\hat{\mathbf{x}} - 11\hat{\mathbf{y}}$ cm/s.

(*d*) What are the initial and final total kinetic energies? Is the collision elastic or inelastic?

18. *Inelastic collision.* Two objects ($M_1 = 2$ g; $M_2 = 5$ g) possess velocities $\mathbf{v}_1 = 10\hat{\mathbf{x}}$ cm/s and $\mathbf{v}_2 = 3\hat{\mathbf{x}} + 5\hat{\mathbf{y}}$ cm/s just prior to a collision during which they become permanently attached to each other.

(*a*) What is their final velocity?

(*b*) What fraction of the initial kinetic energy is lost in the collision?

19. *Satellite orbit.* Consider a satellite orbit that lies just outside the equator of a homogeneous spherical planet of mass density ρ. Show that the period T of such an orbit depends only on the density of the planet. Give the equation for T. (It also contains G.)

20. *Range of mortar shells.* The following are experimental data on the range and muzzle velocity of mortar shells, all fired at 45° to the horizontal. The time of flight is also included. Compare these ranges and times with the simple theory. Can you see any regularity? (Data from U.S. Department of Army, Firing Tables FT4.2-F-1, December 1954.) Use $g = 32$ ft/s^2.

Muzzle velocity, ft/s	*Range, yd*	*Time, s*
334	1063	14.4
368	1268	15.7
400	1475	17.0
431	1683	18.2

ADVANCED TOPIC

Charged Particle in a Uniform Alternating Electric Field

Let

$$\mathbf{E} = E_x\hat{\mathbf{x}} = E_x^{\,0} \sin \omega t\hat{\mathbf{x}}$$

where $\omega = 2\pi f$ is the angular frequency and $E_x^{\,0}$ is the amplitude of the electric field vector. Often the superscript zero (^{0}) on the E is omitted if no ambiguity is introduced. The equation of motion is, from Eq. (3.20),

$$\frac{d^2x}{dt^2} = \frac{q}{M}E_x = \frac{q}{M}E_x^{\,0} \sin \omega t \tag{3.45}$$

In solving differential equations we shall often use the excellent method of trial and error, guided by physical insight. We look for a solution of the form[1]

$$x(t) = x_1 \sin \omega t + v_0 t + x_0 \tag{3.46}$$

On differentiating Eq. (3.46), we find

$$\frac{d^2x}{dt^2} = -\omega^2 x_1 \sin \omega t$$

The derivatives of the sine and cosine are given by

$$\frac{d}{d\theta}\sin\theta = \cos\theta \qquad \frac{d^2}{d\theta^2}\sin\theta = -\sin\theta$$

$$\frac{d}{d\theta}\cos\theta = -\sin\theta \qquad \frac{d^2}{d\theta^2}\cos\theta = -\cos\theta$$

Thus Eq. (3.46) is a solution of the equation of motion [Eq. (3.45)] provided that

$$-\omega^2 x_1 \sin \omega t = \frac{q}{M}E_x^{\,0} \sin \omega t \tag{3.47}$$

[1] Here we participate in one of the common occupations of a physicist: finding the solution of a differential equation subject to prescribed initial conditions. This is an art in which intuitive guessing plays an important part. Often there are strictly prescribed mathematical procedures; but often the physicist asks himself "What could happen?" or "What else would you expect?" In the end the test is to substitute the guess in the original equation to see if the solution works. If the guess is wrong, try again. Intelligent guessing saves time, but even wrong guesses illuminate the problem.

Equation (3.45) states that the acceleration of a charged particle is a sinusoidal function of the time if the applied force is sinusoidal. Because the acceleration is oscillatory, the displacement at least in part must be oscillatory. For this reason we include in Eq. (3.46) such a term as $\sin \omega t$ or $\cos \omega t$. We select $\sin \omega t$ because two successive differentiations of a sine function give a sine function. The term x_0 must be included as the initial displacement. Since we must provide also for an initial velocity, we add a term $v_0 t$ that provides for any initial velocity, including zero. The effect of the term $v_0 t$ will persist at later times as a constant velocity superimposed on the oscillating one. The form $v_0 t$ is the only possibility, a higher power of t is not consistent with Eq. (3.45).

This requires that

$$x_1 = -\frac{qE_x^{\,0}}{M\omega^2} \tag{3.48}$$

On substituting Eq. (3.48) in (3.46), we have the following result:

$$x(t) = -\frac{qE_x^{\,0}}{M\omega^2}\sin \omega t + v_0 t + x_0$$

The velocity is

$$v_x(t) = \frac{dx}{dt} = -\frac{qE_x^{\,0}}{M\omega}\cos \omega t + v_0$$

thus at $t = 0$

$$v_x(0) = -\frac{qE_x^{\,0}}{M\omega} + v_0$$

Do not confuse $v_x(0)$, which is the velocity at $t = 0$, with v_0, which is a constant to be selected to make $v_x(0)$ have the assigned value. If we choose the initial velocity to be zero we must have

$$v_0 = \frac{qE_x^{\,0}}{M\omega}$$

By substituting this in the expression for $x(t)$ above, we have

$$x(t) = -\frac{qE_x^{\,0}}{M\omega^2}\sin \omega t + \frac{qE_x^{\,0}}{M\omega}t + x_0$$

This is a somewhat unexpected result: With the boundary condition $v_x = 0$ at $t = 0$, the motion consists of an oscillation superposed on a constant drift velocity of $qE_x^{\,0}/M\omega$. This is because the particle never reverses its velocity for this special problem. The particle sidesteps continuously to the same side. Note that v_0 is *not* equal to $v_x(t = 0)$ in the present problem, but x_0 is equal to $x(t = 0)$.

The acceleration, velocity, and distance are shown as functions of the time in Fig. 3.28.

MATHEMATICAL NOTE

Differential Equations We have seen that the acceleration in cartesian coordinates is

$$\frac{d^2x}{dt^2}\hat{\mathbf{x}} + \frac{d^2y}{dt^2}\hat{\mathbf{y}} + \frac{d^2z}{dt^2}\hat{\mathbf{z}}$$

In other coordinates it will involve second and possibly first derivatives with respect to time. Newton's Second Law then reduces in one dimension to

$$M\frac{d^2x}{dt^2} = F_x \tag{3.49}$$

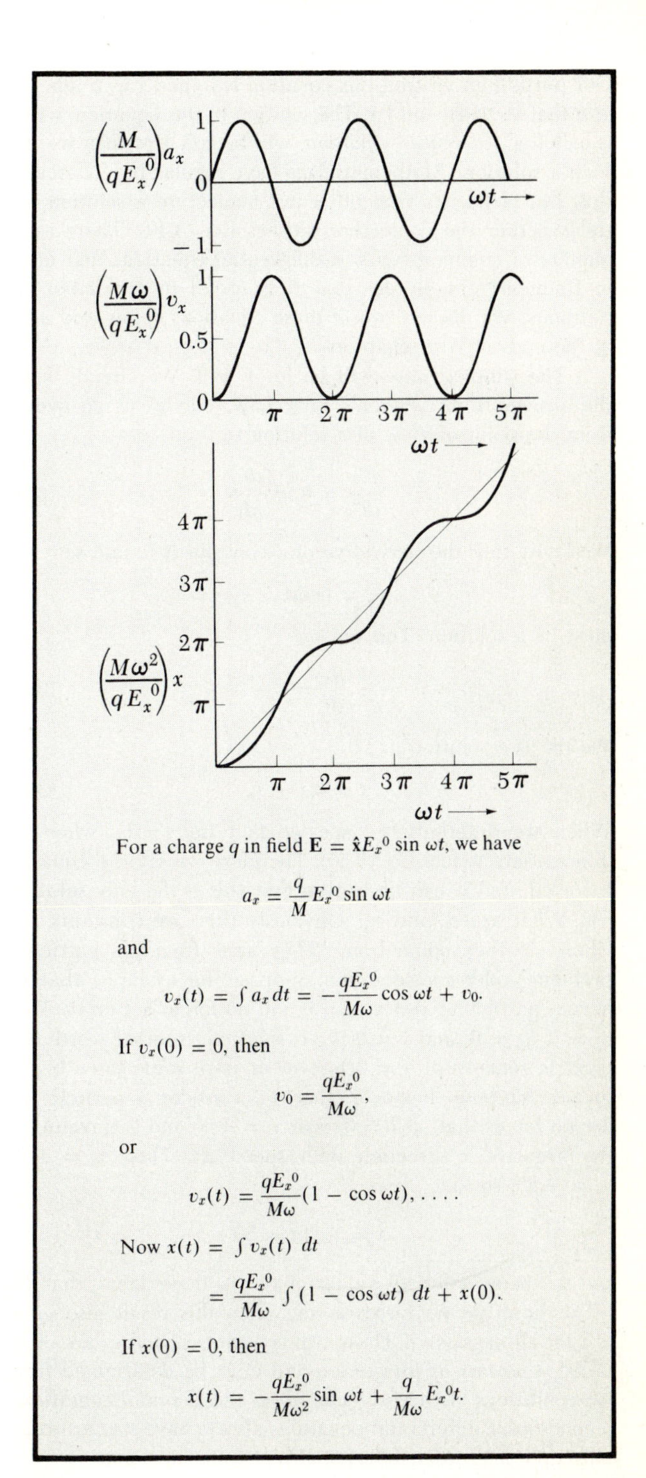

FIG. 3.28 Acceleration, velocity, and displacement plotted against ωt.

Our purpose in writing this equation is to find x as a function of t: that is, to solve for x. The solution to this equation, which is called a *differential equation*, will be $x(t)$. How can we find such a solution? Mathematicians have regular procedures for this, but physicists very often just conjecture a solution and test whether the conjecture satisfies Eq. (3.49). There are a number of common types of differential equations that occur so frequently in physics that it is useful to remember the solutions. We discuss two of these equations below and more at the ends of later chapters.

The simplest case will be for $F = 0$. We already know the answer from Newton's First Law, but let us go over it from the point of view of a solution to

$$\frac{d^2x}{dt^2} = 0 = \frac{dv_x}{dt} \tag{3.50}$$

We know that the derivative of a constant is 0, and so

$$v_x = \text{const} = v_0$$

must be a solution. This means

$$\frac{dx}{dt} = v_0 \tag{3.51}$$

We try as a solution

$$x = v_0 t + x_0 \tag{3.52}$$

When we differentiate once we find Eq. (3.51); when we differentiate twice, Eq. (3.50). Therefore we say the equation is solved, and it can be proved that this is the only solution.

What are v_0 and x_0? Obviously they are constants, but where do they come from? They arise from the particular problem under consideration. Suppose, for example, that we have a particle at rest at $x = 0$ and no forces act on it. Then $v_0 = 0$, $x_0 = 0$, and $x = 0$ is our solution: In other words, the particle remains at $x = 0$ forever or until some force begins to act. Suppose, however, that we consider a particle under no forces that, at $t = 0$, is at $x = +50$ and is traveling in the negative x direction with speed 25. Then $x_0 = +50$, $v_0 = -25$, so that

$$x = -25t + 50$$

and we know x for all values of $t > 0$. If we know that for $t < 0$ there are no forces acting, then this result also gives us x for all negative t. These two constants x_0 and v_0 are often called *constants of integration* and must be determined from the conditions of the problem, often called *initial conditions*. Second-order differential equations always have *two* arbitrary constants; first-order differential equations have *one*.

The next simplest case is $F_x = \text{const} = F_0$. We change from x to y to enable the student to correlate the solutions with the example at the beginning of the chapter:

$$\frac{d^2y}{dt^2} = \frac{F_0}{M} = a \tag{3.53}$$

where a is the constant acceleration. Let us try a solution:

$$y = \tfrac{1}{2}at^2 + v_0 t + y_0 \tag{3.54}$$

Differentiating twice gives us Eq. (3.53), and so we have a solution; v_0 and y_0 are again the arbitrary constants, or constants of integration, and must be determined for the problem under consideration. You may recognize Eq. (3.54) from your previous physics as that of a particle under the force of gravity. If y is positive upward, $a = -g$ where g is the acceleration of gravity, 980 cm/s². Below are given some examples:

1 For a body dropped from rest at $y = 10{,}000$ cm,

$$v_0 = 0 \qquad y_0 = 10{,}000$$

and so

$$y = -\tfrac{1}{2}980t^2 + 10{,}000 \qquad \text{in cm} \tag{3.55}$$

2 For a body projected upward from the origin with $v_0 =$ 980 cm/s,

$$y_0 = 0 \qquad v_0 = 980$$
$$y = -\tfrac{1}{2}980t^2 + 980t \qquad \text{in cm} \tag{3.56}$$

From these we can work out the values of y for any t. For example, in Eq. (3.56) what is the maximum height reached? We set

$$\frac{dy}{dt} = 0 = -980t + 980$$
$$t = 1 \text{ s}$$
$$y = -\tfrac{1}{2}980 \times 1^2 + 980 \times 1 = 490 \text{ cm}$$

Note that $dy/dt = 0$ means $v_y = 0$, which is the situation when the body reaches maximum height.

From the mathematical point of view the initial conditions can be understood as the values of y and dy/dt at some point, which is, in the cases above, $t = 0$. The second-order differential equation gives the curvature of y versus t, but the slope and value are not given. Therefore to find the curve uniquely it is necessary to specify both the slope and the value for some point. The student might compare the graphs of Eqs. (3.55) and (3.56).

HISTORICAL NOTE

Invention of the Cyclotron Most of the present high-energy particle accelerators are descended from the first 1-MeV proton cyclotron built by E. O. Lawrence and M. S. Livingston in LeConte Hall at Berkeley. The cyclotron was conceived by Lawrence; the conception was first published by Lawrence and Edlefsen in a talk abstracted in *Science*, **72**:376, 377 (1930). In 1932 the first results were published in a beautiful paper in the *Physical Review*, principal physics journal of the American Physical Society. Although this journal requires all papers to be accompanied by informative abstracts, few are so clear and informative as that reproduced here from the classic paper by Lawrence and Livingston. Also reproduced are two figures from the original paper. Professor Livingston is at MIT; Professor Lawrence died in 1958.

The original 11-in. magnet was almost immediately outgrown for accelerator applications; it has been rebuilt and is still used for a variety of research projects in LeConte Hall. The first successful experiments on cyclotron resonance of charge carriers in crystals were carried out with this magnet.

For an interesting account of the early history of the cyclotron, see E. O. Lawrence, "The Evolution of the Cyclotron," *Les Prix Nobel en 1951*, pp. 127–140, Imprimerie Royale, Stockholm, 1952. Picture of early cyclotron and reprint of *Phys. Rev.* article follow here.

An early cyclotron.

FURTHER READING

PSSC, "Physics," chaps. 19–21, 28 (secs. 1, 4, 6), 30 (secs. 6–8), D. C. Heath and Company, Boston, 1965.

HPP, "Project Physics Course," chaps. 2–4, 9 (secs. 2–7), 14 (secs. 3, 4, 8, 13), Holt, Rinehart and Winston, Inc., New York, 1970.

A. French, "Newtonian Mechanics," W. W. Norton & Company, Inc., New York, 1971. A complete book at this level; part of the MIT series.

Ernst Mach, "The Science of Mechanics: A Critical and Historical Account of Its Development," 6th ed., chaps. 2 and 3, The Open Court Publishing Company, La Salle, Ill., 1960. A classic account of the concepts of mechanics and their development.

Herbert Butterfield, "The Origins of Modern Science, 1300–1800," Free Press, The Macmillan Company, New York, 1965. Chapter 1 presents a historian's view of the importance of a correct understanding of motion and inertia.

L. Hopf, "Introduction to the Differential Equations of Physics," translated by W. Nef, Dover Publications, Inc., New York, 1948. A compact and pleasant introduction to differential equations that requires little mathematical preparation and is well suited for independent study.

APRIL 1, 1932 *PHYSICAL REVIEW* *VOLUME 40*

THE PRODUCTION OF HIGH SPEED LIGHT IONS WITHOUT THE USE OF HIGH VOLTAGES

By Ernest O. Lawrence and M. Stanley Livingston
University of California

(Received February 20, 1932)

Abstract

The study of the nucleus would be greatly facilitated by the development of sources of high speed ions, particularly protons and helium ions, having kinetic energies in excess of 1,000,000 volt-electrons; for it appears that such swiftly moving particles are best suited to the task of nuclear excitation. The straightforward method of accelerating ions through the requisite differences of potential presents great experimental difficulties associated with the high electric fields necessarily involved. The present paper reports the development of a method that avoids these difficulties by means of the multiple acceleration of ions to high speeds without the use of high voltages. The method is as follows: Semi-circular hollow plates, not unlike duants of an electrometer, are mounted with their diametral edges adjacent, in a vacuum and in a uniform magnetic field that is normal to the plane of the plates. High frequency oscillations are applied to the plate electrodes producing an oscillating electric field over the diametral region between them. As a result during one half cycle the electric field accelerates ions, formed in the diametral region, into the interior of one of the electrodes, where they are bent around on circular paths by the magnetic field and eventually emerge again into the region between the electrodes. The magnetic field is adjusted so that the time required for traversal of a semi-circular path within the electrodes equals a half period of the oscillations. In consequence, when the ions return to the region between the electrodes, the electric field will have reversed direction, and the ions thus receive second increments of velocity on passing into the other electrode. Because the path radii within the electrodes are proportional to the velocities of the ions, the time required for a traversal of a semi-circular path is independent of their velocities. Hence if the ions take exactly one half cycle on their first semi-circles, they do likewise on all succeeding ones and therefore spiral around in resonance with the oscillating field until they reach the periphery of the apparatus. Their final kinetic energies are as many times greater than that corresponding to the voltage applied to the electrodes as the number of times they have crossed from one electrode to the other. This method is primarily designed for the acceleration of light ions and in the present experiments particular attention has been given to the production of high speed protons because of their presumably unique utility for experimental investigations of the atomic nucleus. Using a magnet with pole faces 11 inches in diameter, a current of 10^{-9} ampere of 1,220,000 volt-protons has been produced in a tube to which the maximum applied voltage was only 4000 volts. There are two features of the developed experimental method which have contributed largely to its success. First there is the focussing action of the electric and magnetic fields which prevents serious loss of ions as they are accelerated. In consequence of this, the magnitudes of the high speed ion currents obtainable in this indirect manner are comparable with those conceivably obtainable by direct high voltage methods. Moreover, the focussing action results in the generation of very narrow beams of ions—less than 1 mm cross-sectional diameter—which are ideal for experimental studies of collision processes. Of hardly less importance is the second feature of the method which is the simple and highly effective means for the correction of the magnetic field along the paths of the ions. This makes it possible, indeed easy, to operate the tube effectively with a very high amplification factor (i.e., ratio of final equivalent voltage of accelerated ions to applied voltage). In consequence, this method in its present stage of development constitutes a highly reliable and experimentally convenient source of high speed ions requiring relatively modest laboratory equipment. Moreover, the present experiments indicate that this indirect method of multiple acceleration now makes practicable the production in the laboratory of protons having kinetic energies in excess of 10,000,000 volt-electrons. With this in mind, a magnet having pole faces 114 cm in diameter is being installed in our laboratory.

INTRODUCTION

THE classical experiments of Rutherford and his associates[1] and Pose[2] on artificial disintegration, and of Bothe and Becker[3] on excitation of nuclear radiation, substantiate the view that the nucleus is susceptible to the same general methods of investigation that have been so successful in revealing the extra-nuclear properties of the atom. Especially do the results of their work point to the great fruitfulness of studies of nuclear transitions excited artificially in the laboratory. The development of methods of nuclear excitation on an extensive scale is thus a problem of great interest; its solution is probably the key to a new world of phenomena, the world of the nucleus.

But it is as difficult as it is interesting, for the nucleus resists such experimental attacks with a formidable wall of high binding energies. Nuclear energy levels are widely separated and, in consequence, processes of nuclear excitation involve enormous amounts of energy—millions of volt-electrons.

It is therefore of interest to inquire as to the most promising modes of nuclear excitation. Two general methods present themselves; excitation by absorption of radiation (gamma radiation), and excitation by intimate nuclear collisions of high speed particles.

Of the first it may be said that recent experimental studies [4,5] of the absorption of gamma radiation in matter show, for the heavier elements, varia-

[1] See Chapter 10 of Radiations from Radioactive Substances by Rutherford, Chadwick and Ellis.

[2] H. Pose, Zeits. f. Physik **64,** 1 (1930).

[3] W. Bothe and H. Becker, Zeits. f. Physik **66,** 1289 (1930).

[4] G. Beck, Naturwiss. **18,** 896 (1930).

[5] C. Y. Chao, Phys. Rev. **36,** 1519 (1930).

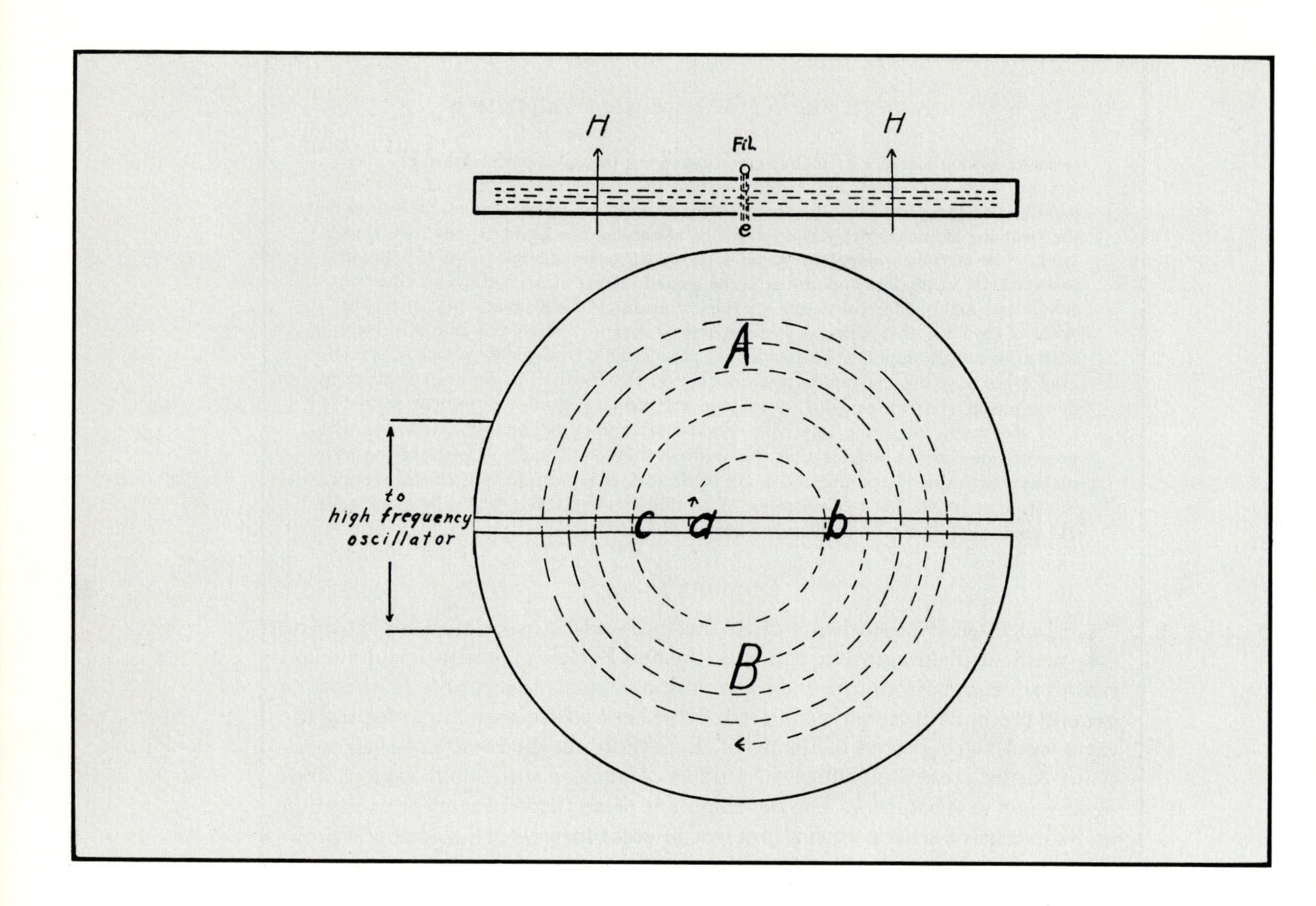
H
Fil.
H
e
A
to
high frequency
oscillator
c
a
b
B

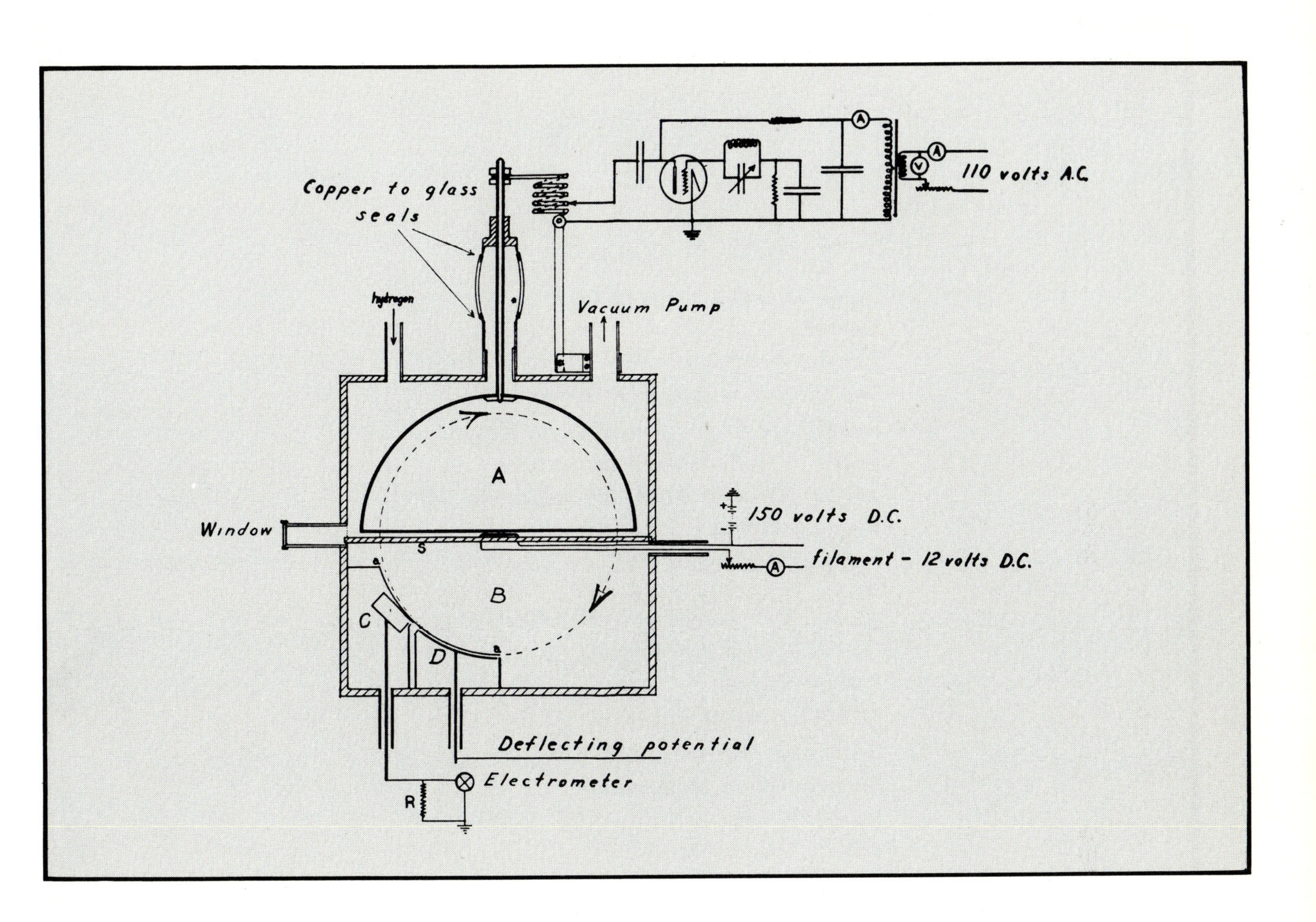
Copper to glass seals
hydrogen
Vacuum Pump
110 volts A.C.
A
150 volts D.C.
Window
S
filament - 12 volts D.C.
B
C
D
Deflecting potential
Electrometer
R

CONTENTS

Frames of Reference: Galilean Transformation

In this chapter we shall investigate some of the more subtle aspects of Newton's Second and Third Laws. The question of frames of reference was entirely omitted from Chap. 3 but is treated in some detail here. Galilean invariance and an alternative derivation of the law of conservation of momentum are the other major topics. In a sense, this chapter is not essential for proceeding to the later chapters, but it is important for a well-rounded understanding of the subject of mechanics.

INERTIAL AND ACCELERATED REFERENCE FRAMES

Newton's first two laws hold only when observed in unaccelerated reference frames. This is obvious from everyday experience. If your reference frame is at rest on a rotating merry-go-round, you do not have zero acceleration in this frame in the absence of applied forces. You can stand still on the merry-go-round platform only by pushing on some part of it, thus, according to Newton's Third Law, causing that part to exert a force $M\omega^2 r$ on your body toward the axis of rotation, where M is the mass, ω the angular velocity, and r the distance from the axis of rotation. Or suppose your reference frame is at rest in an aircraft that accelerates rapidly on takeoff. You "are pressed back against your seat by the acceleration," and you are held at rest relative to the airplane by the force exerted on you by the back of the seat.

If you were content to remain at rest or in uniform motion relative to an unaccelerated frame of reference, no force would be needed. But if you must be at rest in an accelerated reference frame, then you must experience a force such as that of the part of the merry-go-round or the back of the seat. Such forces that arise automatically in accelerated reference frames are important in physics. It is particularly important to understand forces that act in a reference frame in circular motion. It is a good idea to review now this topic, which you studied in high school. [An excellent PSSC film (PSSC MLA 0307), Frames of Reference, clarifies beautifully some of the material of this chapter. See the Film Lists for Chap. 4 at the end of the book.]

EXAMPLE

Ultracentrifuge The effects of *not* being in an inertial reference frame can be enormous! And the effects can be of great practical importance. Consider a molecule suspended in a liquid in the test chamber of an ultracentrifuge. Suppose that the molecule lies 10 cm from the axis of rotation and that the ultracentrifuge rotates at 1000 revolutions per second (60,000 rpm). Then the angular velocity is

$$\omega = (2\pi)(1 \times 10^3) \approx 6 \times 10^3 \text{ rad/s}$$

and the linear velocity is

$$v = \omega r \approx (6 \times 10^3)(10) \approx 6 \times 10^4 \text{ cm/s}$$

The magnitude of the acceleration associated with circular motion is equal to $\omega^2 r$ (see Chap. 2):

$$a = \omega^2 r \approx (6 \times 10^3)^2(10) \approx 4 \times 10^8 \text{ cm/s}^2$$

Now the acceleration g due to gravity is only 980 cm/s^2 at the surface of the earth, so that the ratio of the rotational acceleration to the gravitational acceleration is

$$\frac{a}{g} \approx \frac{4 \times 10^8}{10^3} \approx 4 \times 10^5$$

Thus the acceleration in the ultracentrifuge is 400,000 times as large as the acceleration due to gravity. (These data are characteristic of the ultracentrifuge shown in Fig. 4.1.) Suspended molecules *whose density* (*mass/volume*) *is different* from that of the surrounding liquid will experience in the ultracentrifuge cell a strong force tending to separate them from the fluid. If their density is the same as the liquid, there is no separation effect. If their density is less than that of the liquid, the differential force is inward. For example, a floating helium-filled balloon in a car rounding a curve will tend to move toward the inside of the curve.

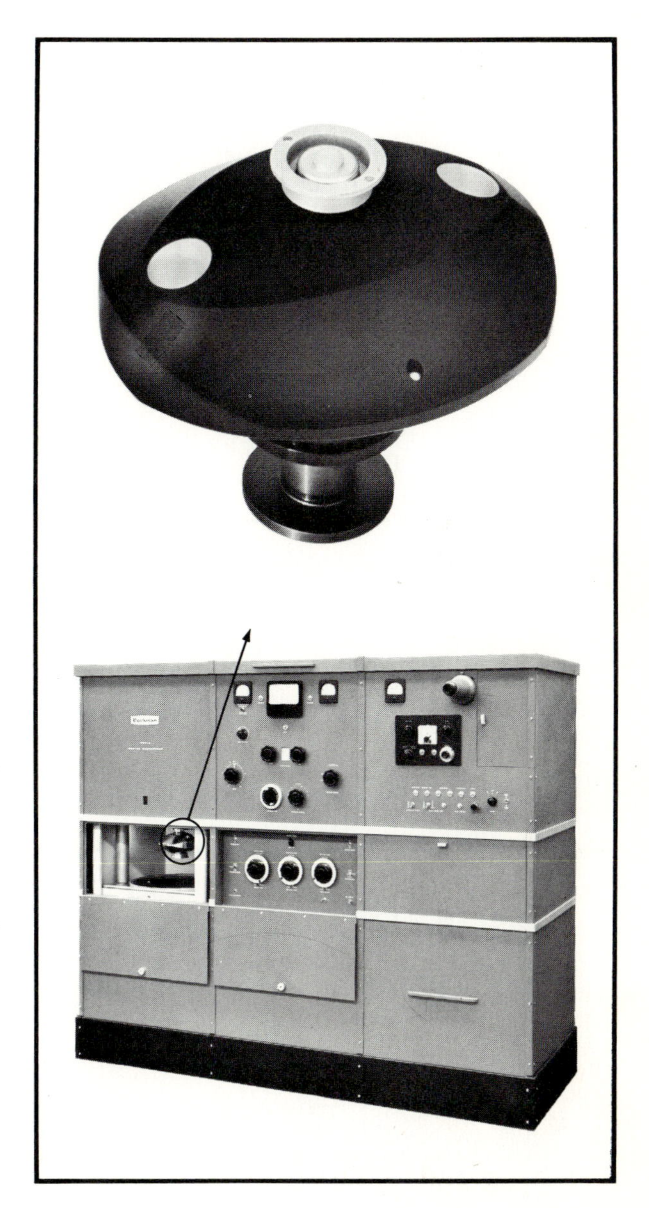

FIG. 4.1 Rotor of an ultracentrifuge. This operates at 60,000 rpm and provides a centrifugal acceleration slightly below 400,000 times the acceleration of gravity. (*Beckmann Spinco Division*)

According to Newton's First Law the suspended molecule wants to remain at rest (or moving at constant velocity in a straight line) as viewed from the laboratory. (The laboratory is a fairly good approximation to an unaccelerated reference frame.) The molecule does not want to be dragged around madly at a high angular velocity in the ultracentrifuge. To an observer at rest in the ultracentrifuge test cell the molecule will act as if there were exerted on it a force $M\omega^2 r$ tending to pull it away from the axis of rotation toward the outside of the test chamber in the centrifuge rotor. (This supposes the molecule to have greater density than the liquid.) How big is the force? Suppose the molecular weight of the molecule is 100,000, which means roughly that the mass M of the molecule is 10^5 times the mass of a proton:

$$M \approx (10^5)(1.7 \times 10^{-24}) \approx 2 \times 10^{-19} \text{ g}$$

(The mass of a proton is approximately equal to one atomic mass unit, as we see from the table of values inside the cover.) The force associated with the rotational acceleration is

$$Ma = M\omega^2 r \approx (2 \times 10^{-19})(4 \times 10^8) \approx 8 \times 10^{-11} \text{ dyn}$$

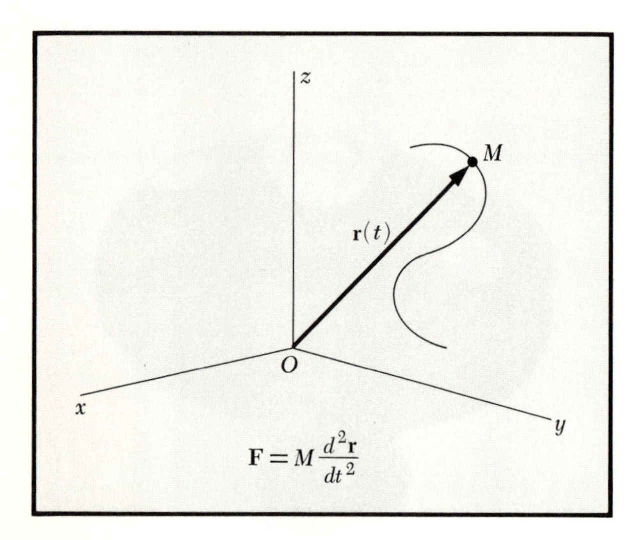

FIG. 4.2 Newton's Second Law says: Force = mass × acceleration. But, acceleration relative to what frame?

using the value of the acceleration given in the preceding paragraph.[1]

This apparent force, which seems to pull the molecule toward the outside of the test chamber, is called the *centrifugal* (center-fleeing) force. The motion toward the outside is opposed by the drag of the surrounding liquid on the molecule. Since different kinds of molecules will experience different values of the centrifugal force and different drags, they will accordingly move outward in the test cell at different speeds. In the reference frame of the ultracentrifuge cell this centrifugal force is like an artificial gravity directed outward and with intensity increasing with distance from the axis. The various species of molecules ultimately settle in this unusual *gravitational field* into a succession of layers graduated according to density. The ultracentrifuge thus provides an excellent method for the separation from one another of different kinds of molecules. The method works best on large molecules, which tend to be just the molecules of great biological interest, and so the matter of whether a molecule is at rest in an accelerated or an unaccelerated reference system turns out to be important for biological and medical research.

We return now to our discussion of inertial and accelerated reference frames. The fundamental law of classical mechanics, Newton's Second Law, states

$$\text{Force} = \frac{d}{dt}(\text{momentum}) \qquad \mathbf{F} = \frac{d}{dt}(\mathbf{p}) \tag{4.1}$$

or for the case of constant mass

$$\mathbf{F} = M\frac{d\mathbf{v}}{dt} = M\frac{d^2\mathbf{r}}{dt^2} = M\mathbf{a} \tag{4.2}$$

where $\mathbf{a}$ is the acceleration. But with respect to what sort of a frame of reference is the coordinate $\mathbf{r}$, the velocity $\mathbf{v}$, or the acceleration $\mathbf{a}$ measured? The examples above clearly indicate that the choice of frame of reference is very important, and Figs. 4.2 to 4.6 illustrate the question.

Equation (4.1) or (4.2) may be viewed as defining in a consistent way the *true force* $\mathbf{F}$ that acts on a particle or body if we can be confident that the acceleration $\mathbf{a}$ is measured with respect to a nonaccelerated reference frame. Conversely, we could say that if we happen to know the true force $\mathbf{F}$ and can find a reference frame in which the observed acceleration of the particle or body satisfies Eq. (4.2), then that reference frame is an *inertial* frame; i.e., it is without acceleration or rotation.

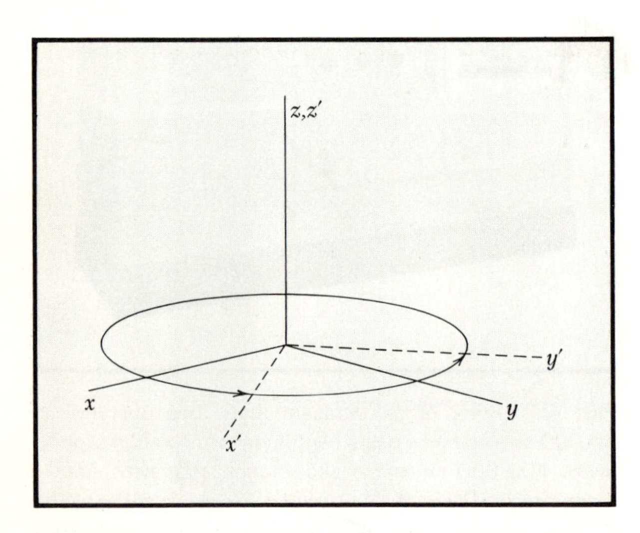

FIG. 4.3 For instance frame $S'(x',y',z')$ rotates with respect to frame $S(x,y,z)$. The acceleration of M in each of these frames is different.

[1] In SI units, $M \approx 2 \times 10^{-22}$ kg, $\omega \approx 6 \times 10^{3}$ rad/s, $r = 0.10$m, and $F \approx (2 \times 10^{-22})(4 \times 10^{6}) \approx 8 \times 10^{-16}$ N.

Our ability to say whether or not a particular reference frame is an inertial frame will depend in a strict sense upon the precision with which we can detect the effects of a small acceleration of the frame. In a practical sense, a reference frame in which no acceleration is observed for a particle believed to be free of any force and constraint is taken to be an inertial frame.

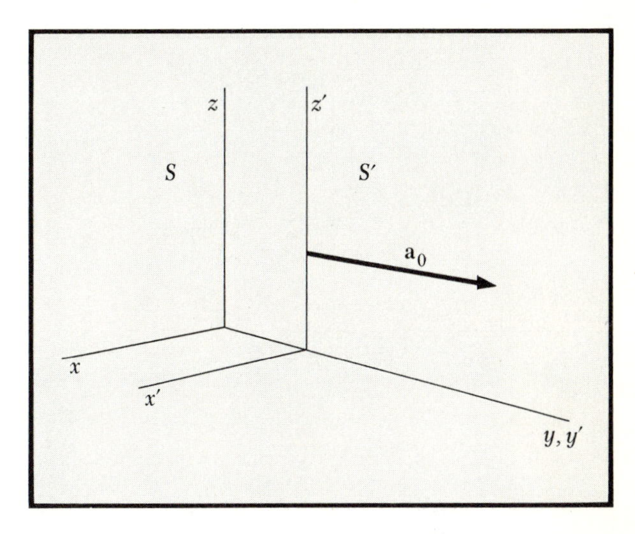

FIG. 4.4 Or, for example, frame S' has acceleration $\mathbf{a}_0$ with respect to frame S. The acceleration of M in each of these frames is different.

The Earth as a Reference Frame Does a laboratory fixed on the surface of the earth provide a good inertial reference frame? If it does not, how do we correct $\mathbf{F} = M\mathbf{a}$ to allow for the acceleration of the laboratory?

For many purposes the earth is a fairly good approximation of an inertial frame. An acceleration of a laboratory fixed on the earth results from the daily rotation of the earth about its axis. This rotation amounts to a small acceleration of the laboratory, not entirely negligible for all purposes. A point at rest on the surface of the earth at the equator must experience a centripetal acceleration given by

$$a = \frac{v^2}{R_E} = \omega^2 R_E \tag{4.3}$$

referred to the center of the earth. Here $\omega = 2\pi f$ is the angular velocity of the earth, and R_E is the radius of the earth. From Chap. 3 (page 67) we have $\omega = \approx 0.73 \times 10^{-4}\ \mathrm{s}^{-1}$. With $R_E \approx 6.4 \times 10^8$ cm, the acceleration is

$$a \approx (0.73 \times 10^{-4})^2(6.4 \times 10^8) \approx 3.4\ \mathrm{cm/s^2} \approx 0.034\ \mathrm{m/s^2}$$

The force of gravity must supply this acceleration to a mass at the equator. Therefore the force necessary to hold the mass in equilibrium against the force of gravity is less than the full force of gravity by $3.4m$ dyn, where m is the mass; or the observed acceleration of gravity is less than that at the North Pole, where a in Eq. (4.3) is zero, by 3.4 cm/s². The remainder of the large-scale gravity variation on the surface of the earth is due to the ellipsoidal shape of the earth. The total variation between the North Pole (or South Pole) and the equator is about 5.2 cm/s². Until the availability of satellites the best way to determine the flattening of the earth at the poles was by measuring the variation of gravity over the earth. Table 4.1 gives values of g at different latitudes.

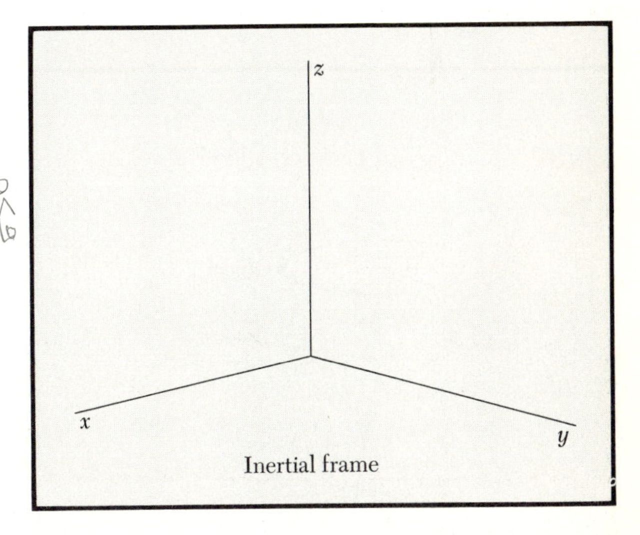

FIG. 4.5 Do there exist *inertial frames* in which we should compute $\mathbf{a}$ in the equation $\mathbf{F} = M\mathbf{a}$?

In the Advanced Topic (at the end of this chapter) we construct a more complicated form of Newton's Second Law,

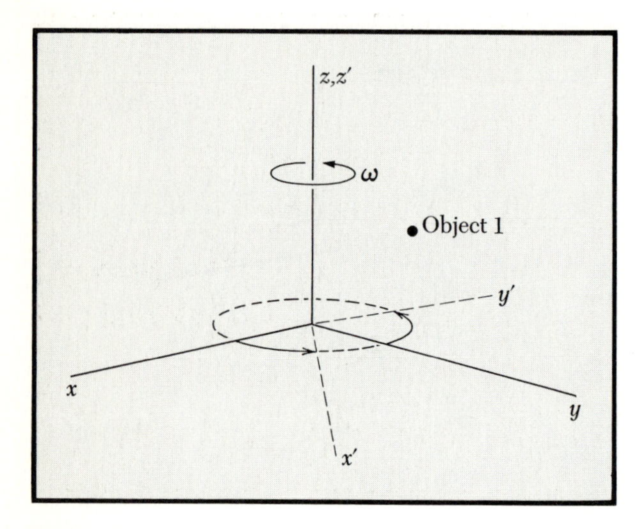

FIG. 4.6 (*a*) If $S(x,y,z)$ is such an inertial frame, then $S'(x',y',z')$, which rotates around the z axis of S, *cannot* be inertial.

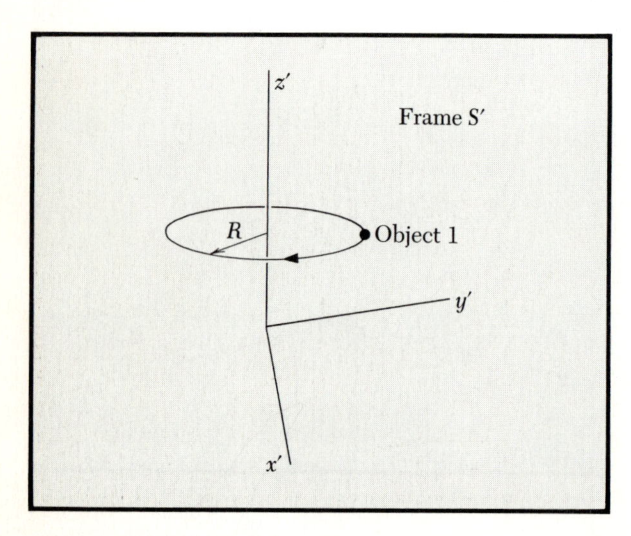

(*b*) For in frame S' object 1, although very far from all other objects, experiences acceleration. (It appears to rotate.)

TABLE 4.1 Values of g at Diverse Latitudes

Station	*Latitude*	*g, cm/s²**
North Pole	90°N	983.245
Karajak Glacier, Greenland	70°N	982.53
Reykjavik, Iceland	64°N	982.27
Leningrad	60°N	981.93
Paris	49°N	980.94
New York	41°N	980.27
San Francisco	38°N	979.96
Honolulu	21°N	978.95
Monrovia, Liberia	6°N	978.16
Batavia, Java	6°S	978.18
Melbourne, Australia	38°S	979.99

*To obtain g in m/s^2, multiply by $\frac{1}{100}$. g is approximately 9.8 m/s^2.

which is applicable in a coordinate system with axes fixed on the surface of the earth. But to obtain a valid law in the simple form of Eq. (4.1) or (4.2) *we must refer the acceleration to a reference system that is unaccelerated, an inertial or galilean frame.* In an accelerated (noninertial) reference system **F** does not equal $M\mathbf{a}$ if **a** is the acceleration as observed from the noninertial system.

Fixed Stars: An Inertial Reference Frame It is an established convention to speak of the fixed stars as a standard unaccelerated reference frame. This language contains an element of metaphysics, for the statement that the fixed stars are unaccelerated goes beyond our actual experimental knowledge. It is unlikely that our instruments could detect an acceleration of a distant star or group of stars of less than 10^{-4} cm/s² even if we made careful observations for 100 yr. For practical purposes it is convenient to refer directions in space to the stars, but also for practical purposes we shall see that we can establish by experiment a satisfactory unaccelerated reference frame. Even if the earth were surrounded continuously by a dense fog, we would be able to establish an inertial reference frame without particular difficulty.

The acceleration of the earth in its orbit around the sun is one order of magnitude smaller than the acceleration at the equator due to the rotation of the earth. Since 1 yr $\approx \pi \times 10^7$ s, the angular velocity of the earth about the sun is

$$\omega \approx \frac{2\pi}{\pi \times 10^7} \approx 2 \times 10^{-7}\ \text{s}^{-1}$$

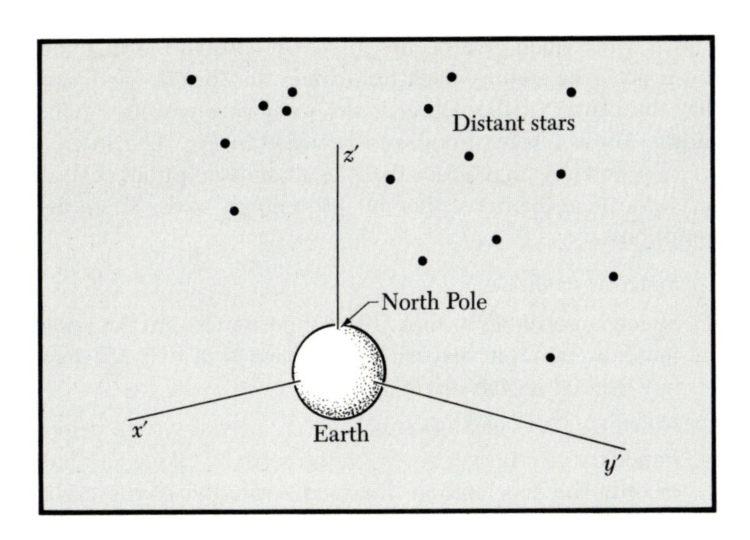

(c) For example, from frame $S'(x',y',z')$ fixed on earth, the distant stars, which are like object 1, rotate. A frame fixed on earth is not inertial because the earth spins on its axis and revolves about the sun.

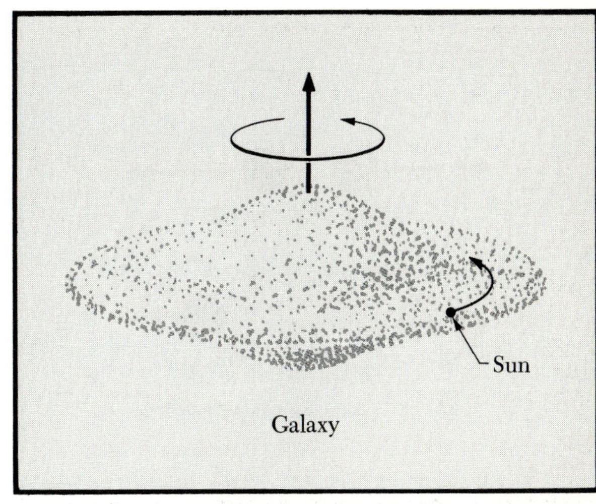

(d) Is a frame fixed on the sun inertial? Even the sun rotates around the galaxy, but this acceleration appears to be small enough to neglect.

We have, with $R \approx 1.5 \times 10^{13}$ cm, the centripetal acceleration

$$a = \omega^2 R \approx (4 \times 10^{-14})(1.5 \times 10^{13}) \approx 0.6 \text{ cm/s}^2 \quad (4.4)$$

for the acceleration of the earth in its orbit about the sun.

The acceleration of the sun toward the center of our galaxy[1] is not known experimentally. But from doppler-shift studies of spectral lines the velocity of the sun relative to the center of the galaxy is believed to be about 3×10^7 cm/s. If the sun is in a circular orbit about the center of the galaxy, which is at a distance of approximately 3×10^{22} cm from the sun, then the acceleration of the sun about the galactic center is

$$a = \omega^2 R = \frac{v^2}{R} \approx \frac{9 \times 10^{14}}{3 \times 10^{22}} \approx 3 \times 10^{-8} \text{ cm/s}^2$$

[1]Stars are not scattered randomly throughout space but are gathered into large systems widely separated from each other. Each system contains of the order of 10^{10} stars. The systems are called *galaxies;* that one which contains our own sun is known as the Galaxy. The Milky Way is part of our galaxy. The galaxies themselves are not distributed entirely at random, for there is a marked tendency to form clusters. Our galaxy belongs to a cluster of 19 members known as the *Local Group,* which forms a physical system bound by gravitational attraction.

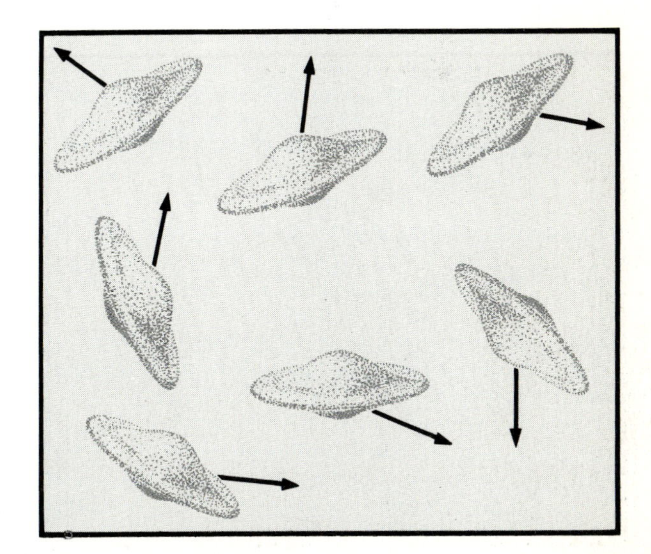

(e) Apparently we can also neglect the acceleration of our galaxy relative to others.

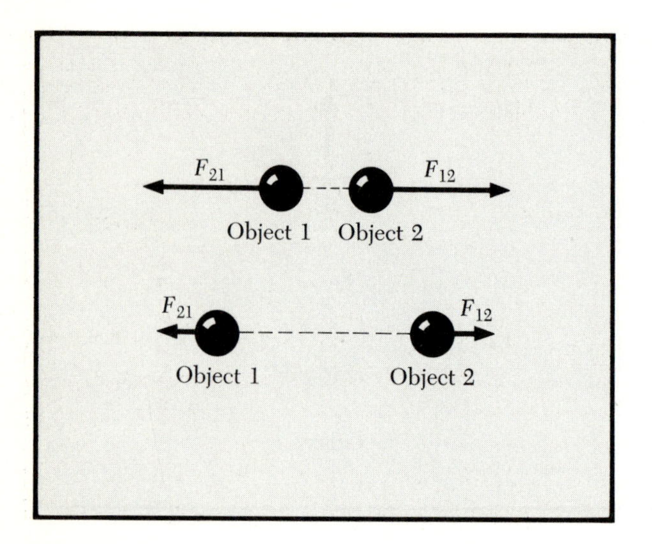

FIG. 4.7 Experimentally, the force one object exerts on another always decreases rapidly as the objects are separated to a greater distance.

This is quite small. We do not know by observation that the sun is not accelerating much faster than this, nor do we know that the center of the galaxy is not itself accelerating significantly. These accelerations are illustrated in Fig. 4.6*a* to *e*.

We do know in practice that the set of assumptions central to classical mechanics works out exceedingly well. These assumptions are:

1 Space is euclidean.

2 Space is isotropic, so that physical properties are the same in all directions in space. Thus the mass M in $\mathbf{F} = M\mathbf{a}$ does not depend on the direction of $\mathbf{a}$.

3 Newton's Laws of Motion hold in an inertial system determined for an observer at rest on the earth by taking account of only the acceleration due to the rotation of the earth about its axis and due to the motion of the earth in its orbit around the sun.

4 Newton's law of universal gravitation is valid. A brief discussion of this law was given in Chap. 3 (pages 65–67), and a more detailed discussion is given in Chap. 9.

These assumptions are difficult to test individually to great precision. The most precise tests, which relate to the motions of the planets in the solar system, usually involve the entire package of all four statements above. Two extremely precise tests of the classical package are discussed in the Historical Notes at the end of Chap. 5.

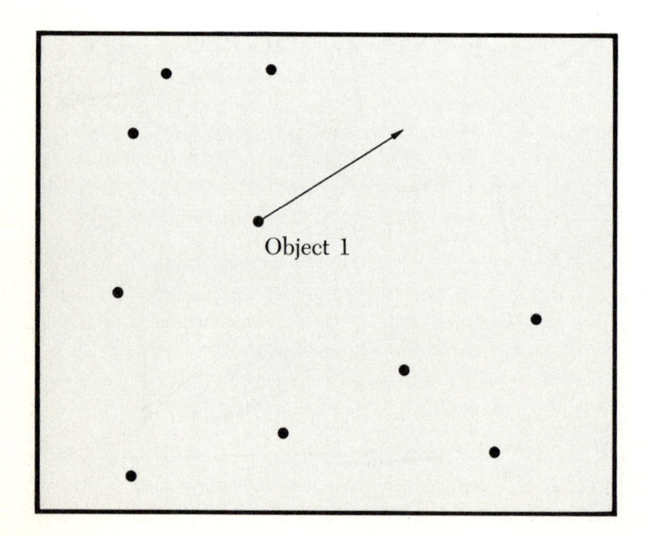

FIG. 4.8 Thus if object 1 is sufficiently far from all other objects, it will be subjected to no forces.

Forces in Inertial Reference Systems Galileo said that *a body subject to no forces has a constant velocity.*[1] We have seen that this statement is true only in an inertial reference frame—it defines an inertial frame or system.

The statement may seem ambiguous, for how do we ever know that a body has no forces acting upon it? Forces can act upon a body, not only by the direct contact of one body with another, but also when the body is isolated. Gravitational or electric forces can be important even without the very near presence of other bodies. We cannot be sure that no forces act just because no other bodies touch or are very close to a given body. But if we cannot decide a priori that some reference body is not subject to a force, we have difficulty framing laws of motion that relate forces to accelerations. We need to have an unaccelerated reference system with respect to which

[1] This is ordinarily called *Newton's First Law of Motion.*

we can measure accelerations; Galileo's way of determining such a system assumes that we have some independent way of knowing that there is no force upon it. But we don't know this because our criterion for no force is no acceleration, which demands some reference against which to measure acceleration, and so on around the circle of argument.

The situation is not hopeless, for we know the forces between two bodies must fall off fairly rapidly as the distance between the bodies grows, as in Fig. 4.7. If the forces did not fall off rapidly, we could never isolate the interactions of two bodies from among those of all other bodies in the universe. All known forces between particles fall off with distance at least as fast as that of the inverse square law. We and every other body on the earth are attracted mainly toward the center of the earth and not mainly toward some distant part of the universe. If we were not supported by the floor, we would accelerate toward the center at 980 cm/s^2. We are pulled less strongly by the sun; from Eq. (4.4), we accelerate toward it at 0.6 cm/s^2. In a reasonable description of acceleration, a body far away from all other bodies is expected to have virtually no forces exerted upon it and hence no acceleration (see Figs. 4.8 to 4.10). A typical star is at least 10^{18} cm from its nearest neighbor[1] and is expected to have only a small acceleration. We are thus led to expect that the fixed stars may define a convenient unaccelerated coordinate system to a good approximation.

A good discussion of the establishment of an unaccelerated reference system is given by P. W. Bridgman, *Am. J. Phys.*, **29**:32 (1961). Several excerpts follow:

> A system of three rigid orthogonal axes fixes a Galilean frame if three force-free massive particles projected along the three axes with arbitrary velocities continue to move along the axes with uniform velocities. Our terrestrial laboratories do not constitute such a frame, but we may construct such a frame in our laboratories by measuring how three arbitrarily projected masses deviate from the requirement . . . and incorporating these deviations as negative corrections into our specifications for the Galilean frame. There need be no reference to the stars, but the behavior of bodies can be relevantly described in terms of such immediately observable things as the rotation of the plane of the Foucault pendulum with respect to the earth or the deviation of a falling body from the perpendicular. Even if the rocket operator who is trying to put a satellite into orbit

[1]Excluding double (binary) stars, which have typical separations of the order of 10^{15} cm.

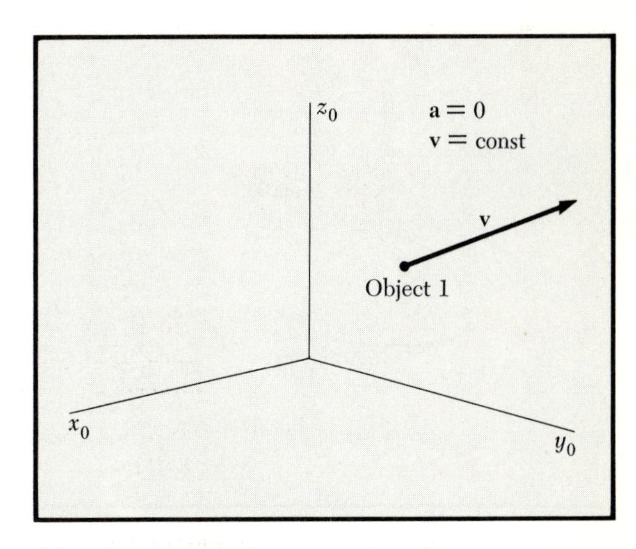

FIG. 4.9 An *inertial frame* is one in which the acceleration of an object like 1 is zero.

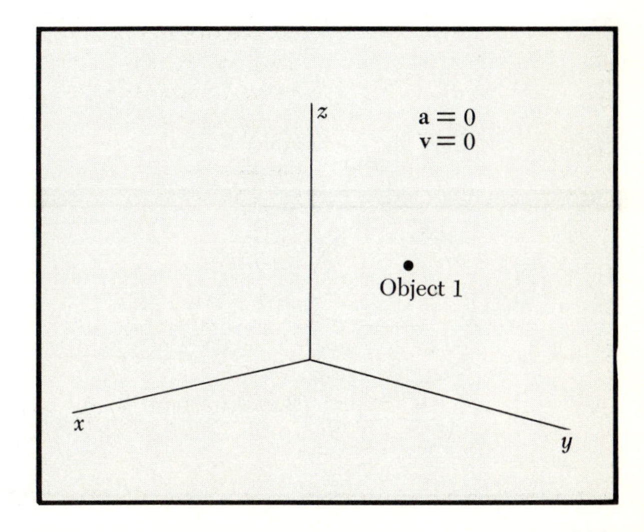

FIG. 4.10 In particular there are inertial frames in which object 1 is, and remains, at rest.

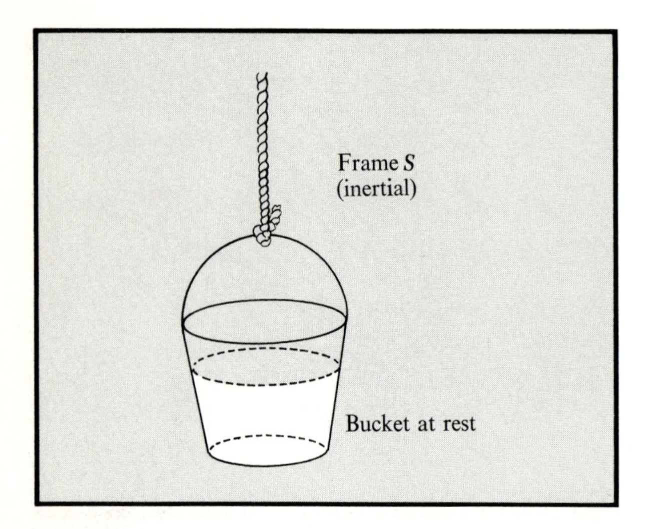

FIG. 4.11 (*a*) An example of "fictitious" forces which arise in noninertial frames: When the bucket is at rest in S, the water surface is flat. S is assumed to be unaccelerated relative to distant stars.

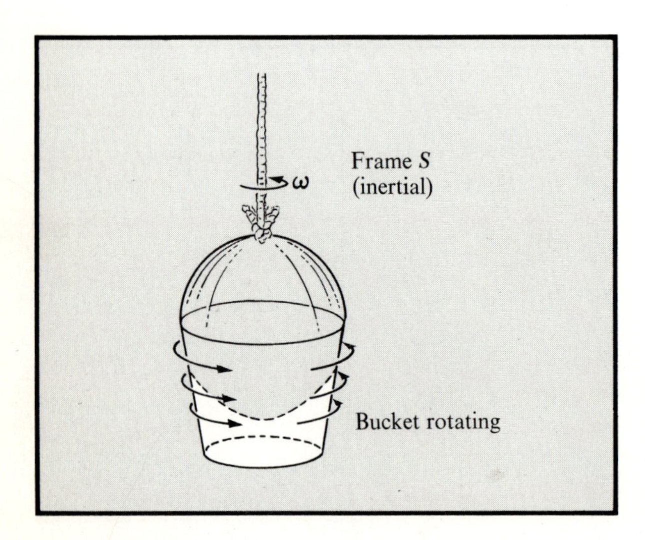

(*b*) When the bucket rotates in S, the water surface assumes a paraboloidal shape.

finds it convenient to make some of his specifications in terms of observations on the pole star, it is obvious that his apparatus must eventually be described in terrestrial terms. . . . In a Galilean frame a rotating body, after it has been set into rotation and the forces disconnected, preserves the orientation of its plane of rotation in the frame and, consequently, preserves the direction of its axis of rotation.

ABSOLUTE AND RELATIVE ACCELERATION

An inertial reference system can be found in which **F** equals $M\mathbf{a}$ to very great accuracy. This is well supported by experiment. We conclude that in an inertial reference system the forces that have been postulated to explain the motion of galaxies, stars, atoms, electrons, etc., have the common property that the force on a body does indeed decrease as it is removed further and further from its neighbors. We shall see that if we choose a noninertial reference system, there *appear* to exist forces that do not have the property of being associated with the proximity of other bodies.

The existence of an inertial reference system suggests a difficult and unanswered question: What effect does all of the other matter in the universe have upon an experiment done in a terrestrial laboratory? Suppose, for example, all the matter in the universe except that in the neighborhood of our own earth were to be given a large acceleration **a.** A particle on earth, subject to no net force originally, had zero acceleration relative to the fixed stars. When these stars are accelerated, would this particle—originally force-free but free to move—still maintain zero acceleration with respect to its nonaccelerated neighborhood, or would it experience a change in its motion relative to its immediate surroundings? Is there any difference between accelerating a particle with $+\mathbf{a}$ or the fixed stars with $-\mathbf{a}$? If only the relative acceleration is significant, the answer to the last question is *no;* if there is a meaning to absolute acceleration, the answer is *yes*. This is a fundamental unanswered question, but it is not easily susceptible to experimental investigation (see Fig. 4.11*a* to *c*).

Newton expressed this question and his own answer in a picturesque way. Consider a bucket of water. If we rotate the bucket and water relative to the stars, the water surface assumes a parabolic shape; on this everyone would agree. But suppose that instead of rotating the bucket we somehow rotated the stars about the bucket so that the relative motion was the same. Newton's belief was that the surface would be flat if we rotated the stars. This viewpoint gives a significance to absolute rotation and absolute acceleration. What we do know empirically is that all the phenomena of the rotating bucket of water can be completely described and correlated with the results of local measurements in the laboratory, with no reference whatever to the stars.

The opposite point of view, that *only acceleration relative to the fixed stars has any significance*, is a conjecture commonly called *Mach's principle*. According to this point of view the water in the bucket would adopt the parabolic form. Although there is neither experimental confirmation nor objection to this point of view, some physicists, including Einstein, have found this principle to be attractive a priori, but others have not. This is a matter for speculative cosmology.

If one believes that the average motion of the rest of the universe affects the behavior of any single particle, a number of related questions present themselves without offering any clues to the answers. Are there other relations between the properties of a single particle and the state of the rest of the universe? Would the charge on the electron, or its mass, or the interaction energy between nucleons[1] change if the number of particles in the universe or their density were somehow altered? So far the answer to this deep question, the relation of the distant universe to the properties of single particles, remains unanswered.

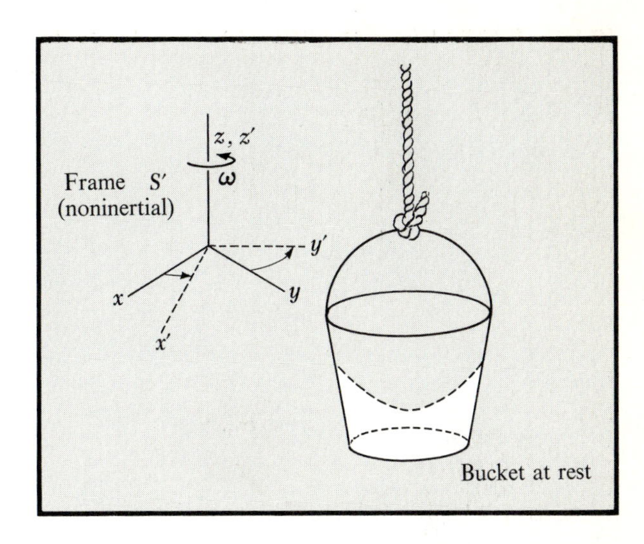

(*c*) In rotating frame S', the bucket is at rest. But the water surface is still paraboloidal! A "fictitious" centrifugal force acts on the water in the noninertial frame S'.

Fictitious Forces We give here a few examples of forces that seem to exist because a frame of reference is accelerated. We start with Newton's Second Law, at first referring our problem to an inertial frame since we know that this is valid. We then proceed to include the acceleration of a noninertial frame and relate this to the force that "seems to exist" when we refer the problem to the latter frame. Newton's Second Law states

$$\mathbf{F} = M\mathbf{a}_I \tag{4.5}$$

in which the left-hand side is the applied force and $\mathbf{a}_I$ is the acceleration as observed in the inertial frame. The mass M is assumed to be constant. The subscript I has been added to $\mathbf{a}$ to emphasize the word *inertial*. In a noninertial frame, such as on the rotating earth, we know that Eq. (4.5) is not valid as it stands. The reason is that an acceleration $\mathbf{a}_0$ that should have been included has been left out, namely, the inertial

[1] A *nucleon* is a proton or a neutron; an *antinucleon* is an antiproton or antineutron.

acceleration which the particle has because of the frame's acceleration or motion.

If $\mathbf{a}$ is the acceleration of a body as measured in the noninertial frame, we have $\mathbf{a} + \mathbf{a}_0 = \mathbf{a}_I$,† or

$$\mathbf{F} = M(\mathbf{a} + \mathbf{a}_0) \tag{4.6}$$

If we do experiments in a noninertial frame, we must always be sure to include $\mathbf{a}_0$ *in the force equation.* In working in a noninertial frame it is often convenient to think in terms of a quantity $\mathbf{F}_0$ such that Eq. (4.6) appears as

$$\mathbf{F} + \mathbf{F}_0 = M\mathbf{a} \tag{4.7}$$

where

$$\mathbf{F}_0 \equiv -M\mathbf{a}_0 \tag{4.8}$$

is called the *fictitious force,* or *pseudoforce.* The fictitious force is that quantity which must be added to the real force in order to make the sum equal to $M\mathbf{a}$, where $\mathbf{a}$ is the acceleration in the noninertial frame. If the frame has a translational acceleration $\mathbf{a}_0$, this fictitious force is just $-M\mathbf{a}_0$. Below we discuss the case of a rotating frame in which the fictitious force depends on the position in the frame. Anything fictitious in physics tends to seem confusing; you can always resolve a problem by going back to Eq. (4.6).

EXAMPLE

Accelerometer Suppose the force applied to a mass M by a spring stretched in the x direction is $F_x = -Cx$, where C is a constant. Consider a noninertial frame with the acceleration $\mathbf{a}_0 = a_0\hat{\mathbf{x}}$ in the x direction. If the mass M is at rest in this noninertial frame, then its acceleration $\mathbf{a}$ in this frame is zero and $\mathbf{F} = M(\mathbf{a} + \mathbf{a}_0)$ reduces to

$$F_x = Ma_0 = -Cx$$

so that

$$x = -\frac{Ma_0}{C} \tag{4.9}$$

Or using the fictitious force

$$\mathbf{F} + \mathbf{F}_{0x} = M\mathbf{a} = 0$$
$$F_x = -F_{0x}$$
$$-Cx = Ma_0$$

which is the same as Eq. (4.9). The displacement x is proportional

† In the first Advanced Topic, we discuss the general case of motion in a rotating frame where $\mathbf{a}_I - \mathbf{a}$ depends on the velocity and on the position in the accelerated frame.

and opposite in direction to the acceleration a_0 of the noninertial frame. The noninertial frame might be an aircraft or an automobile. We see that Eq. (4.9) describes the operation of an *accelerometer* in which a mass M is attached to a spring and constrained to move in the direction of the acceleration. The displacement x of the mass measures the acceleration a_0 of the noninertial reference frame.

EXAMPLE

Centrifugal Force and Centripetal Acceleration in a Uniformly Rotating Frame Although we discuss rotating frames in some detail in the Advanced Topics at the end of this chapter, it is worthwhile to discuss a simple and common example. Consider a point mass M at rest in a noninertial frame, so that in this frame $a = 0$. The noninertial frame rotates uniformly about an axis fixed with respect to an inertial frame. The acceleration of the point in question was seen in Chap. 2 to be

$$\mathbf{a}_0 = -\omega^2\mathbf{r} \tag{4.10}$$

with respect to the inertial frame, where $\mathbf{r}$ is directed outward to the particle from the axis and is perpendicular to the axis.

Equation (4.10) expresses the famous *centripetal acceleration.* The mass might be constrained to be at rest by a stretched spring. The specification that in the noninertial frame $\mathbf{a} = 0$ leads, by Eqs. (4.7) and (4.8), to

$$\mathbf{F} = -\mathbf{F}_0 = M\mathbf{a}_0 = -M\omega^2\mathbf{r} \tag{4.11}$$

The fictitious force $\mathbf{F}_0$ in this example is called the *centrifugal force;* it is $\mathbf{F}_0 = M\omega^2\mathbf{r}$; and it is directed away from the axis. The centrifugal force is balanced in this example by the elastic force $\mathbf{F}$ of the spring in order to produce zero acceleration (mass at rest) in the rotating noninertial frame.

If $M = 100$ g, $r = 10$ cm, and the frame rotates at 100 revolutions per second, what is the value of the centrifugal force? We have $F_0 = M\omega^2 r = (10^2)(2\pi \times 100)^2(10) \approx 4 \times 10^8$ dyn, or $(0.1)(2\pi \times 100)^2(0.1) \approx 4 \times 10^3$ N.

EXAMPLE

Experiments in a Freely Falling Elevator Let the acceleration of a noninertial frame, a freely falling elevator, be

$$\mathbf{a}_0 = -g\hat{\mathbf{y}}$$

where $\hat{\mathbf{y}}$ is measured upward from the surface of the earth and g is the acceleration of gravity. This acceleration corresponds to free fall under gravity. From Eq. (4.8) the fictitious force on a mass M in the noninertial frame is

$$\mathbf{F}_0 = -M\mathbf{a}_0 = Mg\hat{\mathbf{y}}$$

FIG. 4.12 Foucault pendulum as installed in the United Nations Headquarters in New York. The sphere, seen at left, is gold-plated and weighs 200 lb. It is suspended from the ceiling 75 ft above the floor of the lobby. A stainless steel wire holds it in such a manner as to allow it to swing freely in any plane. The sphere swings directly over the raised metal ring, which is about 6 ft in diameter. The sphere swings continuously as a pendulum, its plane shifting slowly in a clockwise direction, thus offering visual proof of the rotation of the earth. A complete cycle takes approximately 36 h and 45 min. Inscribed on it is a message by Queen Juliana of The Netherlands: "It is a privilege to live today and tomorrow." (*United Nations photograph*)

An unattached body in the elevator is acted on by the sum of the gravitational force $\mathbf{F} = -Mg\hat{\mathbf{y}}$ and the fictitious force $\mathbf{F}_0 = Mg\hat{\mathbf{y}}$ so that the total apparent force in the noninertial frame of the freely falling elevator is zero:

$$\mathbf{F} + \mathbf{F}_0 = 0$$

Thus the body is unaccelerated in the noninertial frame. This is a form of "weightlessness." The body appears to remain suspended in space if it has no initial velocity relative to the elevator.

EXAMPLE

Foucault Pendulum The Foucault pendulum demonstrates that the earth is a rotating noninertial frame (see Fig. 4.12). The experiment was first performed publicly by Foucault in 1851, under the great dome of the Pantheon in Paris, using a 28-kg mass on a wire suspension nearly 70 m long. The attachment of the upper end of the wire allows the pendulum to swing with equal freedom in any direction. The period (see Chap. 7) of a pendulum of this length is about 17 s.

Around the point on the floor directly under the point of suspension there was constructed a circular railing about 3 m in radius. On this railing was piled a ridge of sand so that a metal point extending downward from the pendulum brushed aside the sand at each swing. With successive swings it appeared that the plane of the motion of the pendulum moved in a clockwise direction as viewed from above. In 1 h the pendulum changed the plane of its swing by over 11°. A full circuit was completed in about 32 h. In one swing the plane moved by 3 mm, as measured at the circle of sand.

Why does the plane of the pendulum rotate? If the Foucault experiment were carried out at the North Pole of the earth, we could see immediately that the plane of motion of the pendulum would remain fixed in an inertial frame while the earth rotates under the pendulum once every 24 h. The rotation of the earth is counterclockwise as viewed from above the North Pole (say, from Polaris), and so to an observer on a ladder on the earth at the North Pole the plane of the pendulum appears to rotate clockwise relative to him.

The situation appears different (and more difficult to analyze) when we leave the North Pole, and the time for a full circuit is longer. Consider the relative velocities of the extreme north and south points of the circle of sand of radius r as in Fig. 4.13. The south point is farther from the axis of rotation of the earth and will therefore move in space faster than the north point. If ω denotes the angular velocity of the earth and R denotes the radius of the earth, then the center of the circle of sand moves at velocity $\omega R \cos\phi$, where ϕ is the latitude of Paris (48°51′N) as measured from the equator of the earth. The northernmost point on the ring moves at velocity

$$v_N = \omega R \cos \phi - \omega r \sin \phi$$

as we see from the figure, and the southernmost point moves at velocity

$$v_S = \omega R \cos \phi + \omega r \sin \phi$$

The difference between either velocity and that of the center of the ring is

$$\Delta v = \omega r \sin \phi$$

If the pendulum is started in the north-south plane by a push from rest at the center of the ring, the east-west component of velocity in space will be the same as that of the center of the ring. The circumference of the ring is $2\pi r$, so that the time T_0 for a full circuit is, if Δv is constant around the ring,

$$T_0 = \frac{2\pi r}{\omega r \sin \phi} = \frac{24 \text{ h}}{\sin \phi}$$

At the equator $\sin \phi = 0$ and the time becomes infinite.

What happens when the plane of the pendulum reaches the east-west plane through the center of the ring? Why should Δv remain the same here as in the north-south plane? This is hard to see without reference to a globe. Take a piece of cardboard or stiff paper and hold it out from a globe. Let it nearly touch the globe at Paris; let it be normal to the globe at this point and lie in an east-west plane. The normal direction to the surface of the globe is the line of the pendulum wire. With one hand hold the plane of the cardboard fixed while rotating the globe slowly with the other hand. Notice that one side of the line of near contact of the cardboard and the globe appears to move southward and the other side appears to move northward. Contemplation or detailed analysis gives the same value of Δv as found above: The plane of the pendulum actually turns relative to the ring on the Pantheon floor with the constant angular velocity $\omega \sin \phi$, where ω is the angular velocity of the earth and ϕ is the latitude. Many mechanics texts at the junior level have a mathematical treatment of the equation of motion of a Foucault pendulum.

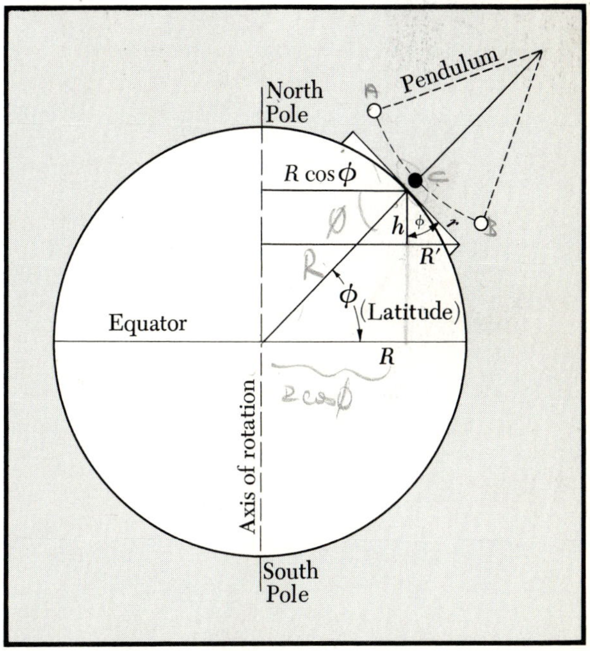

FIG. 4.13 The Foucault pendulum, greatly exaggerated in size in relation to the earth, is shown at approximately the angle of latitude ϕ of Paris. The circle of sand beneath the pendulum has a radius r. The distance from the earth's axis to the center of the pendulum's swing is $R \cos \phi$. Because of the earth's rotation the south side of the sand moves faster than the north side (relative to an inertial frame).

ABSOLUTE AND RELATIVE VELOCITY

Is there any physical meaning to absolute velocity? According to all experiments yet performed the answer is *no*. We are thus led to a fundamental hypothesis, the hypothesis of *galilean invariance:*

> The basic laws of physics are identical in all reference systems that move with uniform (unaccelerated) velocity with respect to one another.

According to this hypothesis an observer confined to a windowless box cannot tell by any experiment whether he is stationary or in uniform motion with respect to the fixed stars. Only by looking through a window, so that he can compare his motion to that of the stars, can an observer tell that he is in uniform motion with respect to them. Even then he cannot decide whether he or the stars are moving. The galilean invariance principle was one of the first to be introduced into physics. It was basic to Newton's view of the universe; it has survived repeated experiments, and it is one of the cornerstones of the theory of special relativity. It is such a remarkably simple hypothesis that it would be considered seriously even in the absence of strong evidence. The hypothesis of galilean invariance is entirely consistent with special relativity, as we shall see in Chap. 11.

What use can we make of the hypothesis? The hypothesis that absolute velocity has no meaning in physics restricts in part the form and content of all physical laws, both known and undiscovered. To two observers moving with different velocities but without relative acceleration, the laws of physics must be the same if the hypothesis is true. Suppose they both observe some particular phenomenon, such as the collision of two particles. Because of the different velocities of the observers, the observed event will be described differently by each of them. From the laws of physics we can predict what the observations of one observer will be, how the particles interact, and, finally, how they appear to the second observer.

The laws of physics of the second observer can therefore be inferred, or obtained from those of the first, by two separate lines of argument. On the one hand, they are by hypothesis the same as those of the first. Alternatively, we can predict the second observer's laws from what we predict about his observations of the phenomena described by the first observer's laws. The two methods give the same result for actual physical laws. Before proceeding, we state some empirical results on the manner in which two observers, one moving at uniform velocity with respect to the other, describe the same physical event.

GALILEAN TRANSFORMATION

If we now discuss how two observers measure a given length and time interval, we can infer how their measurements will compare for other physical quantities. Let S denote a particular inertial cartesian coordinate system, and let S′ denote another

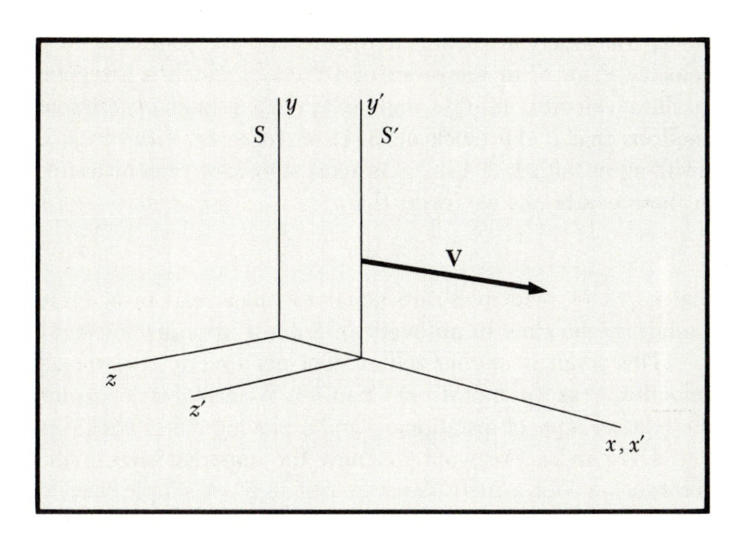

FIG. 4.14 Suppose S is an inertial frame and S' moves with constant velocity $\mathbf{V}$ relative to S. Then S' must be inertial also.

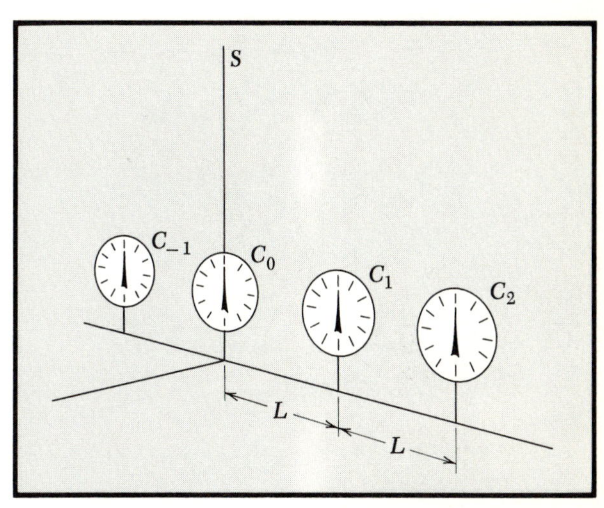

FIG. 4.15 Let us place synchronized clocks C_0, C_1, etc., at intervals of length L along x axis, at rest in S.

inertial cartesian coordinate system moving with velocity $\mathbf{V}$ with respect to the first, as shown in Fig. 4.14. The axes x', y', z' of S′ are taken to be parallel to the x, y, z axes of S. We choose $\mathbf{V}$ to be in the x direction. We wish to compare measurements of time and distance by an observer sitting in the frame S′ with those of an observer at rest in the frame S. The result of the comparison ultimately can be decided only by experiment.

If our two observers construct identical clocks, they may perform the following experiment: We suppose first that the observer on S distributes his clocks along his x axis and sets them all to read the same, for example, by looking at them to be sure they all read the same, as in Fig. 4.15.† As we shall see in Chap. 11, this is a more complicated operation than it seems; we are assuming here that the speed of light is infinite. Now we can compare the reading of clocks S′ with clocks 1, 2, 3, . . . , in S, as S′ passes by each of them (see Fig. 4.16). If such an experiment is to be done with a real macroscopic

†This procedure can be simply improved upon by correcting for the time it takes for the image of the more distant objects to reach our eyes, so that a clock that is l cm away will appear to lag behind a nearby clock by l/c s, where $c = 3 \times 10^{10}$ cm/s is the velocity of light.

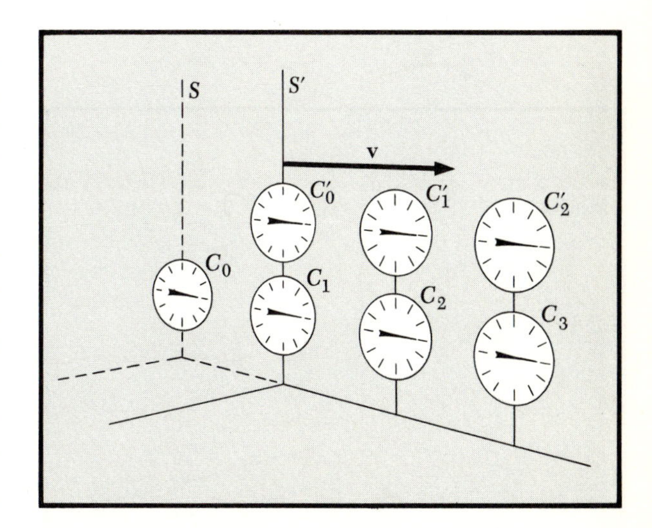

FIG. 4.16 If we place similar clocks C_0', C_1', etc., at rest in S', then to an observer in S these clocks appear synchronized with themselves *and* with C_0, C_1, etc., according to the galilean transformation.

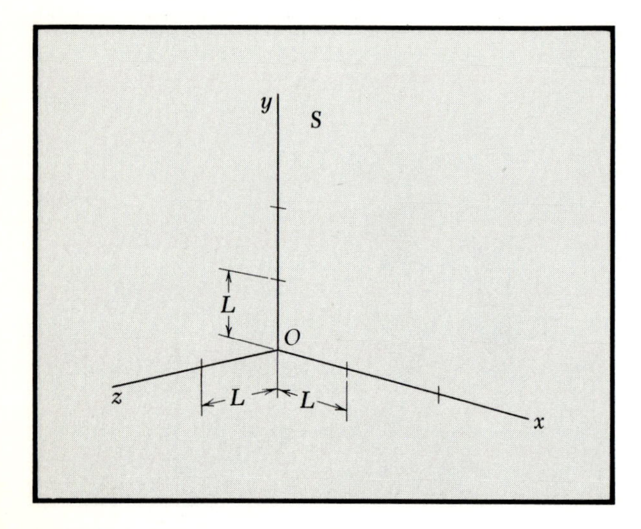

FIG. 4.17 (*a*) Let us mark off equal lengths L along the (xyz) axes of S.

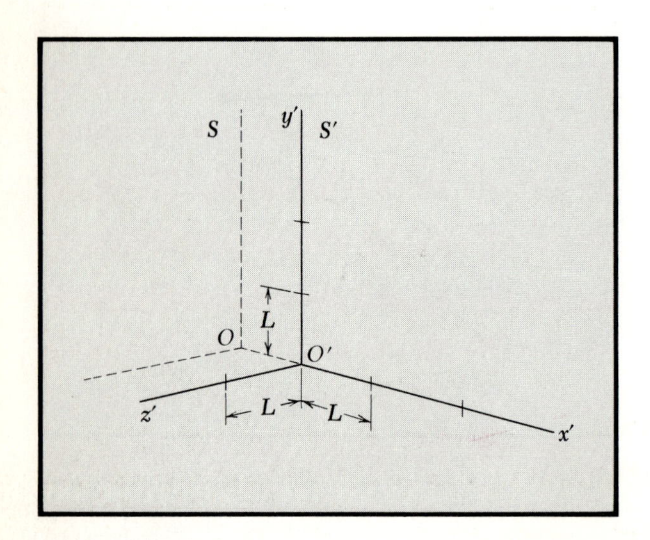

(*b*) And along the (x',y',z') axes of S'.

clock, for practical technical reasons we are restricted to a velocity V for S' of the order of 10^6 cm/s, which is a typical satellite velocity. In this regime $V/c \ll 1$, and experiment confirms that if the clock on S' is set to agree with clock 1, it will agree with 2, 3, 4. . . . As accurately as we can measure[1] in these conditions we assert that

$$t' = t \tag{4.12}$$

that is, times read in S' are equal to times read in S. Here t refers to the time of an event in S and t' to an event in S'.

This result is neither self-evident nor exactly true for all velocities V, as we shall see in Chap. 11. We can also determine the relative sizes of a stationary and a moving meter stick (see Fig. 4.17*a* to *c*). We want to know the apparent size to the observer on S of a meter stick at rest in S'. A simple way to find this is to utilize the clocks again and record the positions of both ends of the moving meter stick simultaneously, that is, when the clocks on S at the front and back read the same. We find by experiment[2] that

$$L' = L \tag{4.13}$$

provided that $V \ll c$.

We can summarize Eqs. (4.12) and (4.13) in terms of a transformation that relates the coordinates x', y', z' and time t' as measured on S' to the coordinates x, y, z and t as measured on S. The frame S' moves with velocity $V\hat{\mathbf{x}}$ as viewed from S. Suppose at $t = 0$ that $t' = 0$ and that at this time the origins O and O' coincide. If we choose identical scales for distance, we shall have the following transformation equations:

$$\boxed{t = t' \qquad x = x' + Vt' \qquad y = y' \qquad z = z'} \tag{4.14}$$

This transformation is called a *galilean transformation* and is illustrated in Fig. 4.18.

An immediate consequence of Eq. (4.14) is the *law of velocity addition:*

[1]Relativity theory predicts for a velocity $V = 10^6$ cm/s that t' and t should differ by only one part in 2×10^9, or less than 1 s in 50 yr. Although clocks with such stability can now be constructed, there was no way until the launching of satellites to keep a clock moving at 10^6 cm/s for enough time to permit a measurement. The equality $t = t'$ for $V \ll c = 3 \times 10^{10}$ cm/s is a simple extrapolation from experience and is not based to date upon very accurate measurement.

[2]Such an experiment has not been performed with great accuracy, and the belief in the equality $L = L'$ for $V \ll c$ is based mainly upon qualitative experience, the simplicity of the hypothesis, and the fact that this assumption does not lead to any paradoxes or inconsistencies.

$$v_x = \frac{dx}{dt} = \frac{dx}{dt'} = \frac{dx'}{dt'} + V = v'_x + V$$

or, in vector form,

$$\mathbf{v} = \mathbf{v}' + \mathbf{V} \tag{4.15}$$

where $\mathbf{v}'$ is the velocity measured in S′ and $\mathbf{v}$ is the velocity measured in S. The inverse transformation to Eq. (4.15) is simply $\mathbf{v}' = \mathbf{v} - \mathbf{V}$.

If the definition, Eq. (4.14), of a galilean transformation between S and S′ is combined with the fundamental postulate that the laws of physics are identical as determined by physicists on S and S′, then we can make the following statement:

> The basic laws of physics are unchanged in form in two reference frames connected by a galilean transformation.

This statement is somewhat more special than our earlier general statement that the laws of physics are identical in all reference frames that move with uniform velocity with respect to one another because it assumes $t' = t$. Except for the cases in which v^2/c^2 is not negligible compared with 1, the statement is valid. In Chap. 11 we discuss the modifications of the galilean transformation equations in the case of v comparable to c to ensure that the laws of physics are identical in all reference frames that move with uniform velocity with respect to one another.

The present assumption of invariance under Eq. (4.14) means that the laws of physics must have exactly the same appearance when written in primed and in unprimed variables, as in Eqs. (4.17) to (4.19) below. This requirement puts a definite restriction on the possible form of physical laws.

From the relation $\mathbf{v} = \mathbf{v}' + \mathbf{V}$, where $\mathbf{V}$ is the relative velocity of the two reference frames, it follows that

$$\Delta\mathbf{v} = \Delta\mathbf{v}'$$

A velocity change observed from S is equal to a velocity change observed from S′; both S and S′ are inertial frames. We recall that $\mathbf{V}$ is assumed not to change with time. Because $\Delta t = \Delta t'$ it follows that the accelerations are equal as observed from S and S′:

$$\mathbf{a} \equiv \frac{\Delta\mathbf{v}}{\Delta t} = \frac{\Delta\mathbf{v}'}{\Delta t'} \equiv \mathbf{a}' \tag{4.16}$$

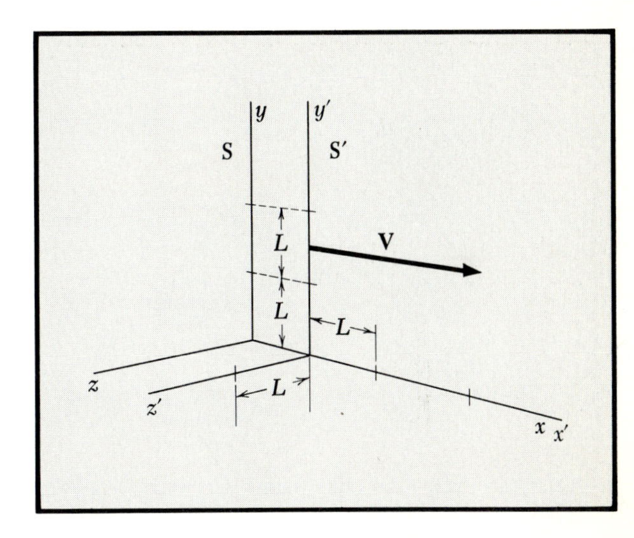

(c) Then to an observer in S, the lengths in S′ appear unaltered even though S is moving according to the galilean transformation.

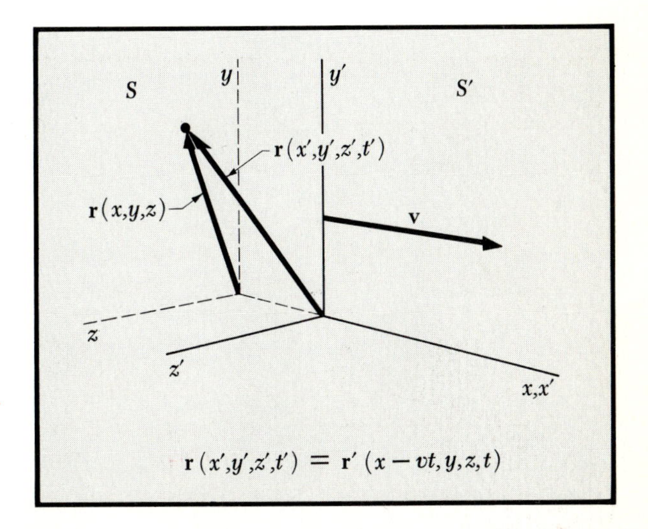

FIG. 4.18 We can thus summarize the galilean transformation from S ↔ S′: $x' = x - Vt$; $y' = y$; $z' = z$; $t' = t$.

How does the force $\mathbf{F}$ transform from S to S'? The assumption that the laws of physics are the same in primed as in unprimed variables means that

$$\mathbf{F}' = M\mathbf{a}' \tag{4.17}$$

if

$$\mathbf{F} = M\mathbf{a} \tag{4.18}$$

provided that the mass M is independent of velocity. But we have shown in Eq. (4.16) that $\mathbf{a}' = \mathbf{a}$, whence

$$\mathbf{F} = M\mathbf{a}' = \mathbf{F}' \tag{4.19}$$

and thus the forces are equal: $\mathbf{F} = \mathbf{F}'$. We conclude that if the relation $\mathbf{F} = M\mathbf{a}$ is used to define the force, observers in all inertial reference frames would agree on the magnitude and direction of the force $\mathbf{F}$ independent of the relative velocities of the reference frames.

Conservation of Momentum The law of conservation of momentum was stated in Chap. 3 (page 83). We now derive it assuming the validity of galilean invariance and conservation of energy and mass. This derivation has the advantage of not using the assumption that the forces of action and reaction are equal, which can be questioned because of the finite velocity of propagation of the force. Such problems as atomic collisions often involve radiation, and at this point we are not able to include the momentum of radiation.

We consider two free particles 1 and 2, which initially have velocities $\mathbf{v}_1$, $\mathbf{v}_2$. The initial (and final) positions are assumed to be widely separated, so that at the initial and final epochs the particles do not interact. From earlier physics (or from Chap. 5) the initial kinetic energy of the particles is known to be

$$\tfrac{1}{2}M_1v_1^2 + \tfrac{1}{2}M_2v_2^2$$

Now let the particles collide; it is *not* necessary that the collision be elastic. Momentum is conserved even if the collision is inelastic. The kinetic energy after the collision is

$$\tfrac{1}{2}M_1w_1^2 + \tfrac{1}{2}M_2w_2^2$$

where $\mathbf{w}_1$ and $\mathbf{w}_2$ are the velocities after the collision,[1] long enough after so that the particles no longer interact. The law of conservation of energy tells us that

[1]Note that we use w instead of the v' that was used in Chap. 3 for the velocity after the collision. The primed symbols here are reserved to refer to frame of reference S'.

$$\tfrac{1}{2}M_1v_1^2 + \tfrac{1}{2}M_2v_2^2 = \tfrac{1}{2}M_1w_1^2 + \tfrac{1}{2}M_2w_2^2 + \Delta\epsilon \quad (4.20)$$

where $\Delta\epsilon$ (which can be either positive or negative) is the change in internal excitation energy of the particles consequent to the collision. We must exclude from this present consideration those collisions where sound or light are given off since we are not yet prepared to include their momenta in our calculations.

The internal excitation might be a rotation or an internal vibration; it might be the excitation of a bound electron from a low energy state to a high energy state. In an elastic collision $\Delta\epsilon = 0$, but we need not restrict the derivation to elastic collisions.[1] We have assumed here that the masses M_1, M_2 of the particles are unchanged in the collision.

Now view the same collision from the primed reference frame moving with the uniform velocity $\mathbf{V}$ with respect to the original unprimed frame. In the primed reference frame the initial velocities are $\mathbf{v}_1'$, $\mathbf{v}_2'$ and the final velocities are $\mathbf{w}_1'$, $\mathbf{w}_2'$. We have

$$\mathbf{v}_1' = \mathbf{v}_1 - \mathbf{V} \qquad \mathbf{v}_2' = \mathbf{v}_2 - \mathbf{V}$$
$$\mathbf{w}_1' = \mathbf{w}_1 - \mathbf{V} \qquad \mathbf{w}_2' = \mathbf{w}_2 - \mathbf{V} \quad (4.21)$$

The statement of the law of conservation of energy in the primed frame is

$$\tfrac{1}{2}M_1(v_1')^2 + \tfrac{1}{2}M_2(v_2')^2 = \tfrac{1}{2}M_1(w_1')^2 + \tfrac{1}{2}M_2(w_2')^2 + \Delta\epsilon \quad (4.22)$$

We have assumed that the excitation energy $\Delta\epsilon$ is unchanged on changing reference frames. This agrees with experiment.

If the law of conservation of energy is to be invariant under a galilean transformation, then in both the primed and unprimed systems the initial kinetic energy must be equal to the final kinetic energy plus $\Delta\epsilon$, the internal excitation energy. That is, both Eqs. (4.20) and (4.22) must hold. Conservation of energy in the primed system can also be expressed by substituting the transformation equation (4.21) in (4.22) and noting that $(v_1')^2 = v_1^2 - 2\mathbf{v}_1 \cdot \mathbf{V} + V^2$, etc., so that Eq. (4.22) becomes

$$\tfrac{1}{2}M_1(v_1^2 - 2\mathbf{v}_1 \cdot \mathbf{V} + V^2) + \tfrac{1}{2}M_2(v_2^2 - 2\mathbf{v}_2 \cdot \mathbf{V} + V^2)$$
$$= \tfrac{1}{2}M_1(w_1^2 - 2\mathbf{w}_1 \cdot \mathbf{V} + V^2)$$
$$+ \tfrac{1}{2}M_2(w_2^2 - 2\mathbf{w}_2 \cdot \mathbf{V} + V^2) + \Delta\epsilon \quad (4.23)$$

[1] In an inelastic collision there is no violation of the principle of conservation of energy. What happens is that kinetic energy lost or gained from the motion of the bodies appears as rotational, vibrational, or other excitational motion of the interior of the bodies. Such internal motion may often be called *heat*, or thermal, motion (Volume 5).

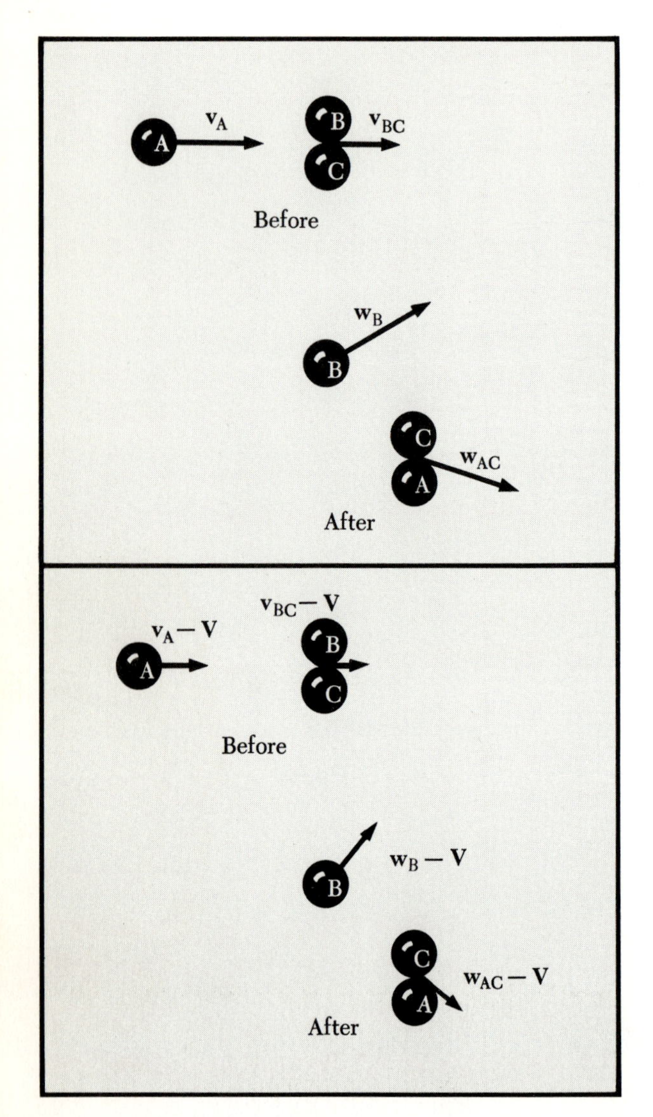

FIG. 4.19 A collision between atom A and molecule BC resulting in atom B and molecule AC. The collision is viewed from two different reference frames.

Notice that the terms in V^2 cancel between the right and left sides. This expression is identical with the law of energy conservation [Eq. (4.20)] in the unprimed system, provided the scalar products cancel in Eq. (4.23):

$$(M_1\mathbf{v}_1 + M_2\mathbf{v}_2)\cdot\mathbf{V} = (M_1\mathbf{w}_1 + M_2\mathbf{w}_2)\cdot\mathbf{V} \tag{4.24}$$

Equation (4.24) must hold for any value of **V**. Hence the general solution of Eq. (4.24) is

$$\boxed{M_1\mathbf{v}_1 + M_2\mathbf{v}_2 = M_1\mathbf{w}_1 + M_2\mathbf{w}_2}$$

This is precisely the *law of conservation of linear momentum.*

To review what we have done: We have assumed the conservation of energy and conservation of mass in a collision, and we have further assumed that these laws are valid in any inertial reference frame. That is, we assumed galilean invariance. We found that the laws can be valid in different inertial frames *only* if linear momentum is conserved in the collision. We have not used the law of conservation of mass in its fullest generality. If the collision involved an exchange of some mass, so that after the collision M_1 became $\overline{M}_1$ and M_2 became $\overline{M}_2$ but with $M_1 + M_2 = \overline{M}_1 + \overline{M}_2$, the same steps used in the above derivation could be used to derive the conservation of momentum. This is demonstrated in the second example below.

EXAMPLE

Inelastic Collision of Equal Masses As an example of these ideas, let us look at an inelastic collision of equal masses from two different frames of reference, the first frame being that in which one particle is initially at rest and the second frame being that in which the two masses initially approach each other with equal and opposite velocities. After collision the two masses stick together.

We have worked the problem in the first frame (see Chap. 3, page 84) and found that the velocity of the two masses after the collision is $v_1/2$, where v_1 is the velocity of the moving mass before collision. The loss in kinetic energy is

$$\Delta\epsilon = \tfrac{1}{2}m_1v_1^2 - \tfrac{1}{2}\cdot 2m_1\left(\frac{v_1}{2}\right)^2 = \tfrac{1}{4}m_1v_1^2$$

For the second frame, the total momentum is zero; *this frame* is often called the *center-of-mass frame.* The velocity of motion of the center-of-mass frame is $v_1/2$, so that $v_1' = v_1 - v_1/2 = v_1/2$ and $v_2' = -v_1/2$. After the collision $w_1' = w_2' = 0$ and the loss in kinetic energy is

$$\Delta\epsilon = \frac{1}{2}m_1\left(\frac{v_1}{2}\right)^2 + \frac{1}{2}m_1\left(\frac{v_1}{2}\right)^2 - 0 = \tfrac{1}{4}m_1v_1^2$$

If you are concerned about the equality of $\Delta\epsilon$ in the two frames of reference, you can work out other examples.

EXAMPLE

Chemical Reactions We show that momentum is conserved in a chemical reaction in which the atoms of the reactants are rearranged or exchanged while conserving the total mass. We assume that there are no external forces (see Fig. 4.19).

Let the reaction be represented by

$$\mathrm{A + BC \rightarrow B + AC}$$

where BC means a molecule consisting of atoms B and C. In the reaction, atom C attaches itself to atom A to form AC. In one inertial frame the law of conservation of energy may be written as

$$\tfrac{1}{2}M_{\mathrm{A}}v_{\mathrm{A}}^2 + \tfrac{1}{2}(M_{\mathrm{B}} + M_{\mathrm{C}})v_{\mathrm{BC}}^2 = \tfrac{1}{2}M_{\mathrm{B}}w_{\mathrm{B}}^2 + \tfrac{1}{2}(M_{\mathrm{A}} + M_{\mathrm{C}})w_{\mathrm{AC}}^2 + \Delta\epsilon \qquad (4.25)$$

Here $\Delta\epsilon$ represents changes in the binding energy of the molecules taking part in the reaction. In a second inertial frame moving with velocity $\mathbf{V}$ with respect to the first, the law of conservation of energy may be written as, with $\mathbf{v}_{\mathrm{A}}$ replaced by $\mathbf{v}_{\mathrm{A}} - \mathbf{V}$, etc.,

$$\tfrac{1}{2}M_{\mathrm{A}}(\mathbf{v}_{\mathrm{A}} - \mathbf{V})^2 + \tfrac{1}{2}(M_{\mathrm{B}} + M_{\mathrm{C}})(\mathbf{v}_{\mathrm{BC}} - \mathbf{V})^2 = \tfrac{1}{2}M_{\mathrm{B}}(\mathbf{w}_{\mathrm{B}} - \mathbf{V})^2 + \tfrac{1}{2}(M_{\mathrm{A}} + M_{\mathrm{C}})(\mathbf{w}_{\mathrm{AC}} - \mathbf{V})^2 + \Delta\epsilon \qquad (4.26)$$

On writing out the squares of the quantities in parentheses, we see that Eqs. (4.25) and (4.26) are consistent if

$$M_{\mathrm{A}}\mathbf{v}_{\mathrm{A}} + (M_{\mathrm{B}} + M_{\mathrm{C}})\mathbf{v}_{\mathrm{BC}} = M_{\mathrm{B}}\mathbf{w}_{\mathrm{B}} + (M_{\mathrm{A}} + M_{\mathrm{C}})\mathbf{w}_{\mathrm{AC}}$$

which is exactly a statement of the law of conservation of linear momentum.

EXAMPLE

Collision of a Heavy Particle with a Light Particle A heavy particle of mass M collides elastically with a light particle of mass m (see Fig. 4.20). The light particle is initially at rest. The initial velocity of the heavy particle is $\mathbf{v}_h = v_h\hat{\mathbf{x}}$; the final velocity is $\mathbf{w}_h$. If the particular collision is such that the light particle goes off in the forward $(+\hat{\mathbf{x}})$ direction, what is its velocity $\mathbf{w}_l$? What fraction of the energy of the heavy particle is lost in this collision?

From momentum conservation there can be no $\hat{\mathbf{y}}$ component to the final velocity of the heavy particle in this particular collision, so that

$$Mv_h\hat{\mathbf{x}} = Mw_h\hat{\mathbf{x}} + mw_l\hat{\mathbf{x}}$$

or

$$Mv_h = Mw_h + mw_l \qquad (4.27)$$

From energy conservation we have (with $\Delta\epsilon = 0$ for an elastic collision)

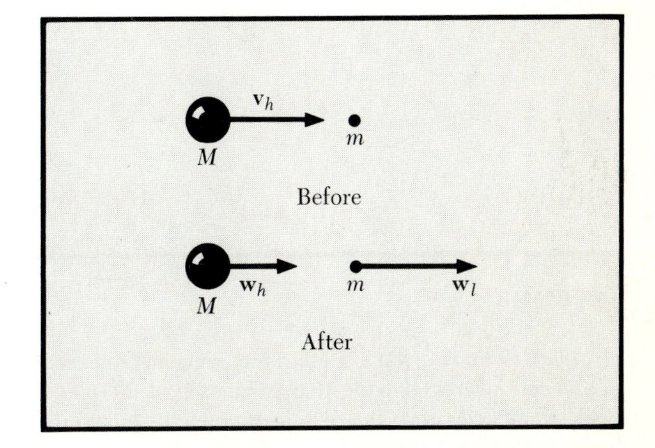

FIG. 4.20 Collision of heavy particle with light particle. (Note that the vectors represent velocities, not momenta.)

$$\tfrac{1}{2}Mv_h^2 = \tfrac{1}{2}Mw_h^2 + \tfrac{1}{2}mw_l^2$$

This may be written with the help of Eq. (4.27) as

$$\frac{1}{2}M\left(w_h^2 + \frac{2m}{M}w_hw_l + \frac{m^2}{M^2}w_l^2\right) = \tfrac{1}{2}Mw_h^2 + \tfrac{1}{2}mw_l^2 \tag{4.28}$$

If $m \ll M$, it is convenient to neglect the term of order m/M, so that Eq. (4.28) reduces to

$$mw_hw_l \approx \tfrac{1}{2}mw_l^2$$

or

$$w_l \approx 2w_h \tag{4.29}$$

Thus the light particle goes off at approximately twice the velocity of the heavy particle. It follows further, from substituting Eq. (4.29) in (4.27), that

$$Mv_h \approx Mw_h + 2mw_h$$

or

$$\frac{\Delta v_h}{v_h} = \frac{v_h - w_h}{v_h} \approx \frac{2m}{M + 2m} \approx \frac{2m}{M} \tag{4.30}$$

The fractional energy loss of the heavy particle is

$$\frac{\tfrac{1}{2}Mv_h^2 - \tfrac{1}{2}Mw_h^2}{\tfrac{1}{2}Mv_h^2} = \frac{v_h^2 - w_h^2}{v_h^2} = 1 - \left(\frac{M}{M + 2m}\right)^2 \approx \frac{4m}{M} \tag{4.31}$$

using Eq. (4.30).† Note that in both Eqs. (4.30) and (4.31) we have neglected terms in m/M compared with 1.

Other examples of the application of the conservation of linear momentum are treated in Chap. 6.

†Another way of writing this, using Δ as an operator, is

$$\frac{\Delta(\tfrac{1}{2}Mv_h^2)}{\tfrac{1}{2}Mv_h^2} = \frac{Mv_h\Delta v_h}{\tfrac{1}{2}Mv_h^2} = \frac{2\Delta v_h}{v_h} \approx \frac{4m}{M}$$

PROBLEMS

1. *Block on rotating table.* A block is to remain at rest relative to a rough horizontal table that is rotating at 20 rpm. The block is 150 cm from the axis of rotation that is vertical. How big must the coefficient of friction be? Show the centrifugal and the frictional forces on a diagram.

2. *Moving frame.* In a railroad car traveling at 500 cm/s along a straight track, a head-on collision takes place between a 100-g mass moving with velocity 100 cm/s in the same direction as the train and a 50-g mass moving in the opposite direction at 500 cm/s. Both velocities are relative to the train.

After the collision, in the car the 50-g mass is at rest; what is the velocity of the 100-g mass? How much kinetic energy has been lost? *Ans.* -150 cm/s.

Now describe the collision from the point of view of an observer at rest by the track. Is momentum conserved? How much kinetic energy is lost in this frame?

3. *Acceleration in circular motion.* An object moves in a circular path with a constant speed v of 50 cm/s. The velocity vector $\mathbf{v}$ changes direction by 30° in 2 s.
(*a*) Find the magnitude of the change in velocity $\Delta\mathbf{v}$.
(*b*) Find the magnitude of the average acceleration during the interval. *Ans.* 12.95 cm/s^2.
(*c*) What is the centripetal acceleration of the uniform circular motion? *Ans.* 13.16 cm/s^2.

4. *Effective force due to rotation.* An object fixed with respect to the surface of a planet identical in mass and radius to the earth experiences zero gravitational acceleration at the equator. What is the length of a day on that planet? *Ans.* 1.4 h.

5. *Motion in a noninertial reference frame.* Consider an inertial frame S on the surface of the earth and a noninertial frame S′ at rest in a freely falling elevator.
(*a*) What is the equation of motion in S′ of a freely falling particle in S?
(*b*) What are the applied and fictitious forces in S and S′ on the particle in (*a*)?
(*c*) What are the equations of motion in S′ of a particle moving in a horizontal circle in S? Assume $y = y' = 0$ at $t = 0$, and y is vertical.

6. *Pendulum in accelerated car.* A pendulum hangs vertically in a car at rest. At what angle will it hang when the acceleration of the car moving on a horizontal plane is 100 cm/s^2?

7. *Centrifuge for humans.* In aviation medicine studies, centrifuges, which are horizontal shafts rotating about a vertical axis and carrying an experimental subject at one end of the shaft, are used. If the distance of the subject from the center of rotation is 700 cm, how fast must the centrifuge rotate in order to subject the rider to 5 g? g = accel. of gravity.

8. *Accelerated frame.* A frame of reference has an upward acceleration of 300 cm/s^2. At $t = 0$, its origin is at rest and coincident with that of an inertial frame on the surface of the earth. (Neglect the rotation of the earth.)
(*a*) Assuming y is up and x is horizontal, find $x(t)$ and $y(t)$ in both frames for an object that is projected horizontally with a speed of 1000 cm/s at $t = 0$, neglecting gravity.
(*b*) Work (*a*) including gravity.

9. *Collision kinematics; center of mass.* Two particles of mass $M_1 = 100$ g and $M_2 = 40$ g have initial velocities $\mathbf{v}_1 = 2.8\hat{\mathbf{x}} - 3.0\hat{\mathbf{y}}$ cm/s and $\mathbf{v}_2 = 7.5\hat{\mathbf{y}}$ cm/s. They collide, and after the collision the velocities are $\mathbf{v}_1' = 1.2\hat{\mathbf{x}} - 2.0\hat{\mathbf{y}}$ cm/s and $\mathbf{v}_2' = 4.0\hat{\mathbf{x}} + 5.0\hat{\mathbf{y}}$ cm/s.
(*a*) Find the total momentum.
(*b*) Find the velocity of a reference frame in which the total momentum (before collision) is zero. This is called the *center-of-mass frame.*
(*c*) Show that the momentum is zero in this frame after the collision.
(*d*) What fraction of the initial kinetic energy is not present as kinetic energy after the collision? Is the collision elastic?

10. *Unequal-mass collision.* In the collision of two particles, the reference frame in which one is initially at rest and the other moving with velocity v is called the *laboratory frame.* Suppose the moving mass is m and the stationary one is $2m$.
(*a*) What is the velocity of the center-of-mass frame (see Prob. 9) with respect to the laboratory frame?
(*b*) How much kinetic energy is lost in both the laboratory and center-of-mass frames in a completely inelastic collision, i.e., one in which the particles stick together?
(*c*) If the collision is elastic, the velocities of the particles in the center-of-mass frame are changed in direction but not in magnitude. Find an expression relating the angle of deviation (usually called angle of scattering) of the mass m in the laboratory and in the center-of-mass frames.

Note that in the center-of-mass frame, the angle of the second particle is always 180° from the angle of the first. In the equal-mass collision, $\theta_{\text{lab}} = \theta_{\text{c.m.}}/2$. Vector diagrams are instructive.

11. *Acceleration and magnetic deflection of electrons.* (This problem and Probs. 12 to 14 are reviews of material in Chap. 3.) Suppose electrons are liberated at rest at point 0 on a metallic plane (see Fig. 4.21) and are accelerated toward a parallel plane 0.25 cm away by an electric field. A tiny hole at P permits a beam of electrons to escape into a region free of electric fields (all in a high vacuum, of course). The electric field is produced by applying voltages of -300 and 0 V to the metallic planes, as indicated. It is desired to bend the beam through a 90° angle in a circular path of radius 0.5 cm by a magnetic field B of circular outline as shown. Calculate the field strength required; also state its direction. (*Note:* An electric field intensity of 1 statvolt/cm is equal to 300 V/cm.)

12. *Transit time of ions.* A pulse of singly charged cesium ions Cs^+ is accelerated from rest by an electric field of 1 statvolt/cm acting for 0.33 cm and afterward travels 1 mm

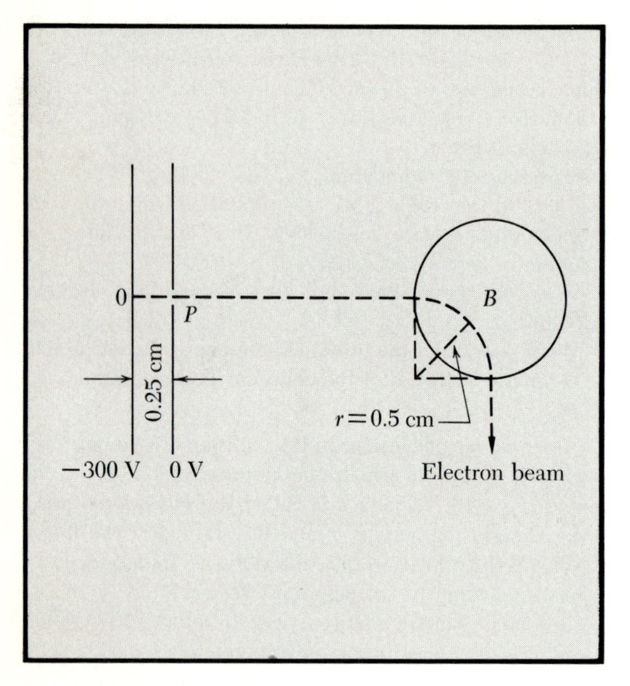

FIG. 4.21

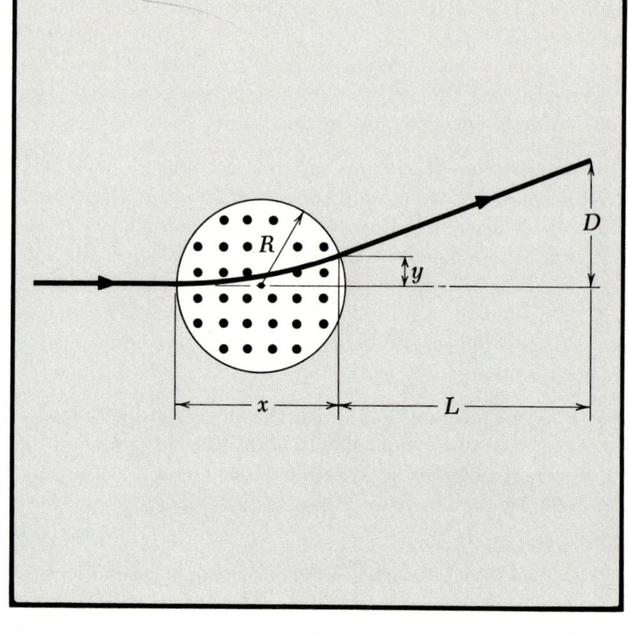

FIG. 4.22

in 87×10^{-9} s in an evacuated field-free space.

(*a*) Derive from these data a value of the atomic mass of Cs^+. *Ans.* 2.4×10^{-22} g.

Compare with the value you will find in tables, handbooks, or chemistry textbooks.

(*b*) What would be the time for protons to transit the 1-mm region? *Ans.* 7.2×10^{-9} s.

13. *Magnetic deflection of electron beam.* Deflection of an electron beam in a cathode-ray tube may be accomplished by magnetic as well as by electrostatic means. A beam of electrons of energy W enters a region of transverse uniform magnetic field of strength B. (Neglect fringe effects. See Fig. 4.22.)

(*a*) If x is the distance from the point at which the electron entered the field to the point at which it leaves it, show that

$$y = r\left[1 - \sqrt{1 - \left(\frac{x}{r}\right)^2}\right]$$

where r is the radius of curvature of the electron in the transverse magnetic field. The radius of curvature is the radius of the circle that will match (coincide with) the curved portion of the path.

(*b*) If R is the radius of the magnet poles, then $x \approx 2R$ when $r \gg R$. Use the binomial expansion to show $y \approx 2R^2/r$.

14. *Acceleration in a cyclotron.* Suppose in a cyclotron that $\mathbf{B} = \hat{\mathbf{z}}B$ and

$$E_x = E \cos \omega_c t \qquad E_y = -E \sin \omega_c t \qquad E_z = 0$$

with E constant. (In an actual cyclotron the electric field is not uniform in space.) We see that the electric field intensity vector sweeps around a circle with angular frequency ω_c. Show that the displacement of a particle is described by

$$x(t) = \frac{qE}{M\omega_c^2}(\omega_c t \sin \omega_c t + \cos \omega_c t - 1)$$

$$y(t) = \frac{qE}{M\omega_c^2}(\omega_c t \cos \omega_c t - \sin \omega_c t)$$

where at $t = 0$ the particle is at rest at the origin. Sketch the first few cycles of the displacement.

ADVANCED TOPICS

Velocity and Acceleration in Rotating Coordinate Systems

We now consider a noninertial reference frame that is rotating with constant angular velocity ω about the z axis of an inertial frame. We restrict the discussion to the case of relative coordinate rotation about a common z axis. (The formulas for the general case are derived in books on intermediate mechanics and are given at the end of this section.) The importance of this problem lies in the fact that the earth is a rotating frame of reference. In the analysis, besides the centripetal acceleration, we pick up the Coriolis acceleration, which is important in the large-scale motion of sea and air currents.

The coordinates (x_R, y_R, z_R) of a point P viewed from the rotating frame may be related simply to the coordinates (x_I, y_I, z_I) of the same point as viewed from the inertial frame. We see on studying the geometry in Figs. 4.23 and 4.24 that

$$
\begin{aligned}
x_I &= x_R \cos \omega t - y_R \sin \omega t \\
y_I &= x_R \sin \omega t + y_R \cos \omega t \qquad (4.32)\\
z_I &= z_R
\end{aligned}
$$

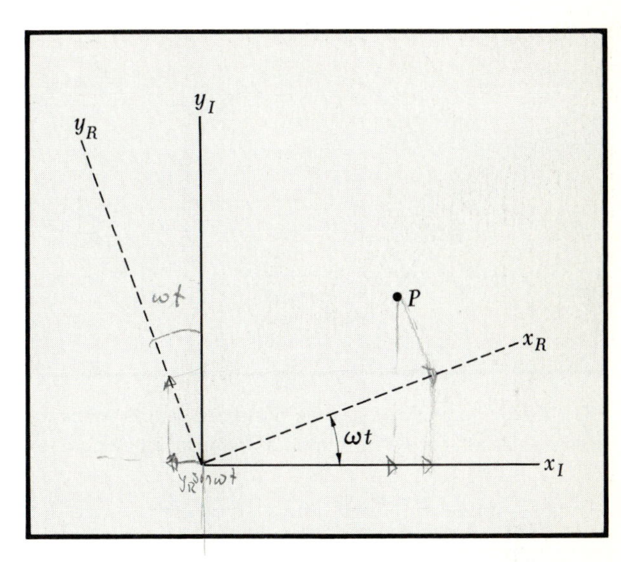

FIG. 4.23 Point P in the xy plane of an inertial frame (x_I, y_I) and of a rotating frame (x_R, y_R). The axes coincide at $t = 0$ and the rotation is with angular velocity ω about the z axis.

The relations between the velocity components in the two frames are found on differentiating Eq. (4.32) with respect to time. (For compactness we use the dot over a quantity to denote differentiation with respect to the time as introduced on page 76, Chap. 3. Thus $\dot{x} \equiv dx/dt \equiv v_x$ and $\ddot{x} \equiv d^2x/dt^2 \equiv \dot{v}_x \equiv dv_x/dt$.) We have

$$
\begin{aligned}
\dot{x}_I &= \dot{x}_R \cos \omega t - \omega x_R \sin \omega t - \dot{y}_R \sin \omega t - \omega y_R \cos \omega t \\
\dot{y}_I &= \dot{x}_R \sin \omega t + \omega x_R \cos \omega t + \dot{y}_R \cos \omega t - \omega y_R \sin \omega t \qquad (4.33)\\
\dot{z}_I &= \dot{z}_R
\end{aligned}
$$

We have taken ω to be constant, in the interest of simplicity. Notice for a particle at rest $(\dot{x}_R = \dot{y}_R = \dot{z}_R = 0)$ in the rotating frame that Eq. (4.33) reduces to

$$
\begin{aligned}
\dot{x}_I &= -\omega x_R \sin \omega t - \omega y_R \cos \omega t \\
\dot{y}_I &= \omega x_R \cos \omega t - \omega y_R \sin \omega t
\end{aligned}
$$

Similarly, for a particle at rest in the inertial frame $(\dot{x}_I = \dot{y}_I = \dot{z}_I = 0)$, we have (following some algebra)

$$
\dot{x}_R - \omega y_R = 0 \qquad \dot{y}_R + \omega x_R = 0 \qquad \dot{z}_R = 0
$$

from Eq. (4.33).

The acceleration components are found on differentiating Eq. (4.33) with respect to the time:

$$
\begin{aligned}
\ddot{x}_I &= \ddot{x}_R \cos \omega t - 2\omega \dot{x}_R \sin \omega t - \omega^2 x_R \cos \omega t \\
&\qquad - \ddot{y}_R \sin \omega t - 2\omega \dot{y}_R \cos \omega t + \omega^2 y_R \sin \omega t \\
\ddot{y}_I &= \ddot{x}_R \sin \omega t + 2\omega \dot{x}_R \cos \omega t - \omega^2 x_R \sin \omega t \qquad (4.34)\\
&\qquad + \ddot{y}_R \cos \omega t - 2\omega \dot{y}_R \sin \omega t - \omega^2 y_R \cos \omega t \\
\ddot{z}_I &= \ddot{z}_R
\end{aligned}
$$

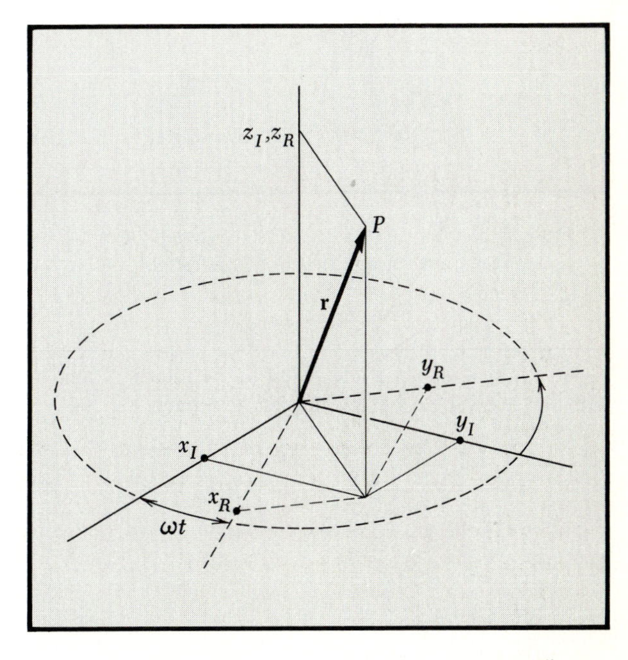

FIG. 4.24 Point P can be described with coordinates $x_I y_I z_I$ of the inertial frame or with coordinates $x_R y_R z_R$ of the rotating frame. The rotation is about the z axis.

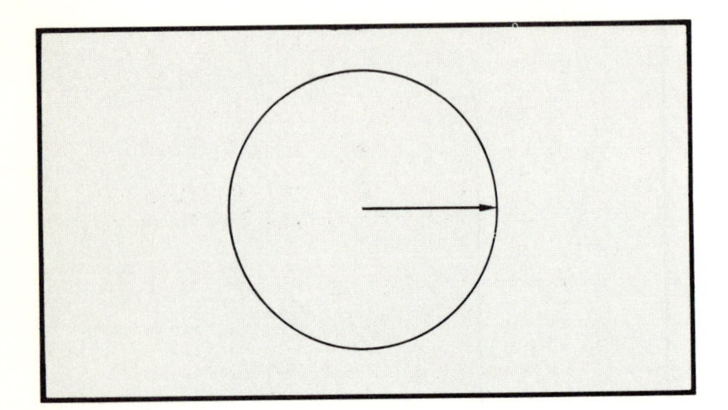

FIG. 4.25 (*a*) Path of particle projected radially outward from center as viewed from inertial frame of reference.

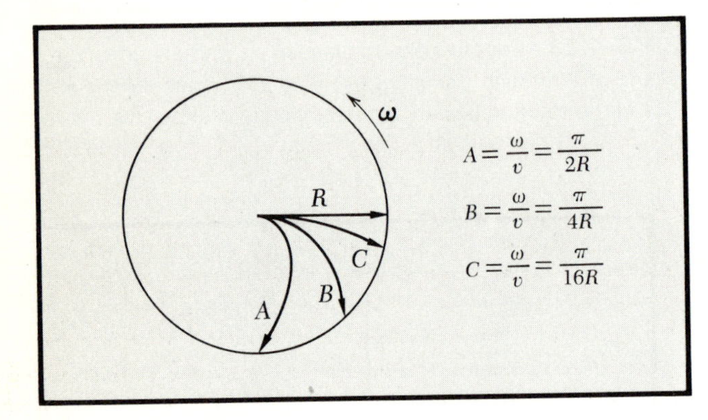

(*b*) Path of particle projected radially outward from center as viewed from rotating frame of reference.

We notice for a particle at rest in the rotating frame that, with the help of Eqs. (4.32), Eq. (4.34) reduces to

$$\ddot{x}_I = -\omega^2(x_R \cos \omega t - y_R \sin \omega t) = -\omega^2 x_I \tag{4.35}$$

$$\ddot{y}_I = -\omega^2(x_R \sin \omega t + y_R \cos \omega t) = -\omega^2 y_I \tag{4.36}$$

Equations (4.35) and (4.36) may be combined in vector form as

$$\mathbf{a}_I = -\omega^2 \mathbf{r}_I \tag{4.37}$$

where $\mathbf{a}_I \equiv \ddot{\mathbf{r}}_I$ is the acceleration of the particle relative to the inertial frame and $\mathbf{r}_I = x_I\hat{\mathbf{x}}_I + y_I\hat{\mathbf{y}}_I$, as in Eq. (4.10). Equation (4.37) is the expression of the usual centripetal acceleration.

The first terms in Eqs. (4.34) are just the acceleration in the rotating coordinates system ($\ddot{x}_R$ and $\ddot{y}_R$) projected upon the inertial coordinate axes. The second terms, however, depend on the velocity in the rotating system ($\dot{x}_R$ and $\dot{y}_R$) and will be zero if $\dot{x}_R = \dot{y}_R = 0$. They can be understood by considering a particle projected radially outward with no real forces acting. Its real path will be a straight radial line, as in Fig. 4.25*a*, but in the rotating system its path will look like Fig. 4.25*b*. This acceleration is called the *Coriolis acceleration*, and the fictitious force derived from it is called the *Coriolis force* (see Fig. 4.25). The third terms in Eqs. (4.34) are just the centripetal acceleration terms, and the fictitious force derived from them is the *centrifugal force*. As long as v is small compared to ωr, the Coriolis force is small compared with the centrifugal force.

As an illustration of these fictitious forces and a reconciliation of the views from inertial and rotating frames, let us consider a supersonic plane moving eastward at the equator at a ground speed of 2000 mi/h (or 8.8×10^4 cm/s). Figure 4.26 illustrates the situation. We consider the path of the plane to be "level," so that it curves with the earth's surface; and since its altitude is small with respect to the size of the earth, we consider it to be at distance r, the earth's radius, from the center.

First we shall view the situation from the standpoint of the inertial frame. In this view the plane moves in a circle of radius r with a speed $\omega r + V$. This speed is the combination of the motion of the earth's surface at the equator plus the speed of the plane relative to the ground. There must be a centripetal force to provide the centripetal acceleration related to this circular path, and that centripetal force is provided by the combination of the force of gravity and the aerodynamic lift force. We therefore write

$$-\frac{GMm}{r^2} + f = -\frac{m(\omega r + V)^2}{r} = -m\omega^2 r - 2m\omega V - \frac{mV^2}{r}$$

where f represents the lift force, m is the mass of the plane, and the negative signs denote force or acceleration directed toward the center. By solving for f we obtain

$$f = \frac{GMm}{r^2} - m\omega^2 r - 2m\omega V - \frac{mV^2}{r}$$

or

$$f = mg - 2m\omega V - \frac{mV^2}{r}$$

In the second expression we have simply recognized that $GMm/r^2 - m\omega^2 r$ is the local effective "gravity" force mg at the equator, as described on page 105. Our conclusion is that because of the combined effects of the plane's ground speed

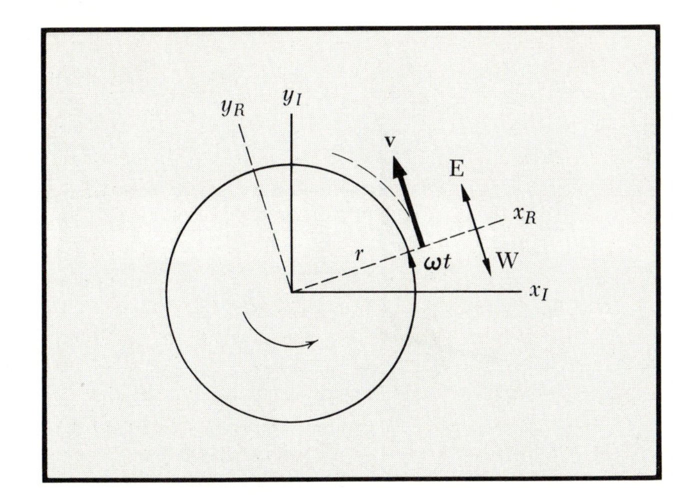

FIG. 4.26 Inertial frame and rotating frame for the earth as viewed from space above the North Pole. x_R and y_R are in the equatorial plane. The vector **v** is the velocity of a supersonic plane relative to the ground, flying a "level" path eastward as indicated by the curved dashed line. The local east and west directions are indicated.

and the earth's rotation, the required aerodynamic lift force is a little less than mg by the last two subtractive terms. If we use the value of V given above, with $\omega = 7.3 \times 10^{-5}$ rad/s and $r = 6.4 \times 10^8$ cm, we find $2\omega V = 12.85$ cm/s^2 and $V^2/r = 12.10$ cm/s^2. These are to be subtracted from $g = 978$ cm/s^2, the acceleration of a falling object at the equator. The required lift force is reduced by about 2.6 percent.

Now let us view the same situation from the standpoint of the rotating frame. We shall utilize the first equation of (4.34) and choose our time to be at the instant $t = 0$ when the rotating axes coincide with the inertial axes. The first equation then becomes

$$\ddot{x}_I = \ddot{x}_R - 2\omega\dot{y}_R - \omega^2 x_R$$

Of course, $m\ddot{x}_I$ is equal to the true force F, and by following the pattern of Eq. (4.7) we can write our present equation as

$$F + 2m\omega\dot{y}_R + m\omega^2 x_R = m\ddot{x}_R$$

The second and third terms on the left side constitute the fictitious forces that must be present if $m\ddot{x}_R$ is to be equated to forces.

The conditions of our problem give the following values to our variables:

$$x_R = r \qquad \dot{x}_R = 0 \qquad \ddot{x}_R = -\frac{V^2}{r}$$

$$y_R = 0 \qquad \dot{y}_R = V \qquad \ddot{y}_R = 0$$

because in the rotating frame the plane moves in a curved path with speed V. [Incidentally, these conditions make all terms zero in the second equation of (4.34).] As before, our true force is

$$F = -\frac{GMm}{r^2} + f$$

where f is the aerodynamic lift force.

The substitution of these values and expressions in our equation yields

$$-\frac{GMm}{r^2} + f + 2m\omega V + m\omega^2 r = -m\frac{V^2}{r}$$

Again solving for f and writing $GMm/r^2 - m\omega^2 r = mg$ as before, we obtain

$$f = mg - 2m\omega V - m\frac{V^2}{r}$$

This is the same result obtained from the standpoint of the inertial frame. The term $2m\omega V$ is a Coriolis force; the last term $m\,V^2/r$ is a centrifugal force due to the plane's speed in its curved path. The centrifugal force due to the earth's rotation has been absorbed into the local force ascribed to gravity. We shall not pursue this example further.

As a further example (mentioned on page 128), we know that in an inertial frame a body projected from the center of rotation will travel in a straight line outward:

$$x_I = v_0 t \qquad y_I = 0 \qquad z_I = 0$$

This gives in the rotating frame,

$$v_0 t = x_R \cos\omega t - y_R \sin\omega t \tag{4.38}$$

$$0 = x_R \sin\omega t + y_R \cos\omega t \tag{4.39}$$

Let us check that this satisfies Eqs. (4.34) with $\ddot{x}_I = 0$, $\ddot{y}_I = 0$. Multiplying the first of those equations by $\cos\omega t$, the second by $\sin\omega t$, and adding, we obtain

$$\ddot{x}_R - 2\omega\dot{y}_R - \omega^2 x_R = 0 \tag{4.40}$$

From Eqs. (4.38) and (4.39),

$$x_R = v_0 t\cos\omega t \qquad y_R = -v_0 t \sin\omega t$$

and when we employ these in Eq. (4.40), we find it is satisfied.

When we consider motion in three dimensions and an arbitrary direction for the coordinate angular velocity vector $\boldsymbol{\omega}$, we obtain

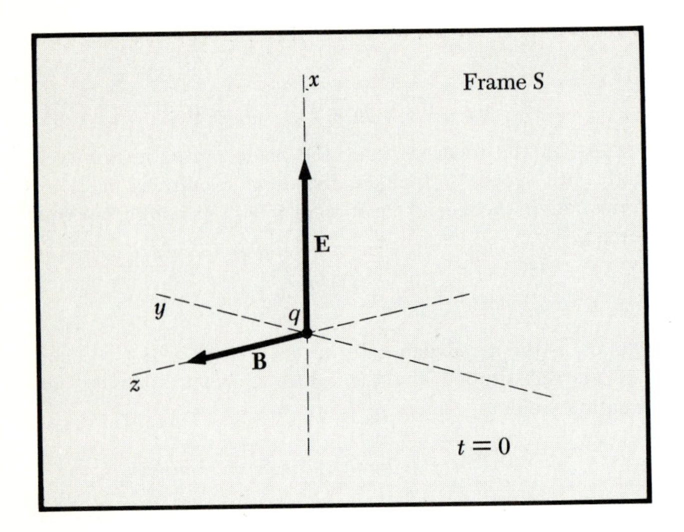

FIG. 4.27 (*a*) Consider a positive charge q at rest at the origin in crossed $\mathbf{E}$ and $\mathbf{B}$ fields.

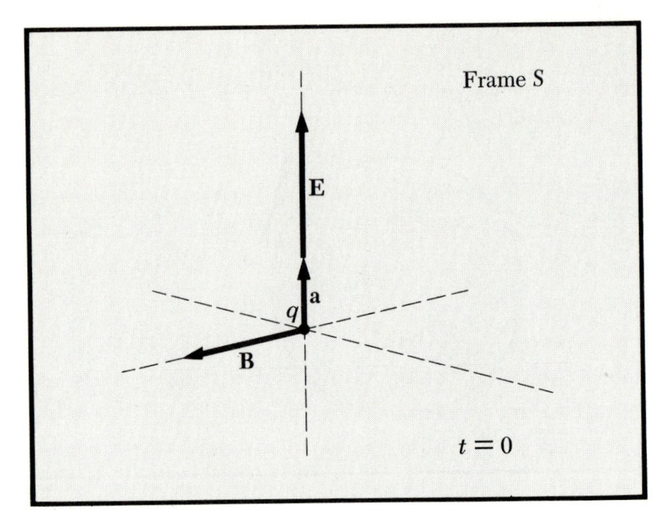

(*b*) The initial acceleration of q is $\mathbf{a} = q\mathbf{E}/M$.

$$\mathbf{a}_I = \mathbf{a}_R + 2\boldsymbol{\omega} \times \mathbf{v}_R + \boldsymbol{\omega} \times (\boldsymbol{\omega} \times \mathbf{r}) = \mathbf{F}/M$$

and so

$$M\mathbf{a}_R = \mathbf{F} - 2M\boldsymbol{\omega} \times \mathbf{v}_R - M\boldsymbol{\omega} \times (\boldsymbol{\omega} \times \mathbf{r})$$

where $\mathbf{F}$ is the true force. Then

$$-2M\boldsymbol{\omega} \times \mathbf{v}_R \quad \text{is the Coriolis force} \tag{4.41}$$

and

$$-m\boldsymbol{\omega} \times (\boldsymbol{\omega} \times \mathbf{r}) \quad \text{is the centrifugal force} \tag{4.42}$$

Motion of a Proton in Crossed Electric and Magnetic Fields

This important example can be solved rather easily, and its interpretation is made simple by a coordinate transformation to a moving frame. Let $\mathbf{B} = B\hat{\mathbf{z}}$ and $\mathbf{E} = E\hat{\mathbf{x}}$, as illustrated in Fig. 4.27*a*. From the definitions of the Lorentz force [given by Eq. (3.19)] and of the gyrofrequency ω_c [given by Eq. (3.27)] we have for the equations of motion of a charged particle

$$\dot{v}_x = \frac{e}{M}E + \omega_c v_y \qquad \dot{v}_y = -\omega_c v_x \qquad \dot{v}_z = 0 \tag{4.43}$$

There is a *special solution* to these equations, describing motion with no acceleration, found by setting $\dot{v}_x = \dot{v}_y = 0$. Then

$$v_x = 0 \qquad v_y = -\frac{eE}{M\omega_c} = -c\frac{E}{B} \tag{4.44}$$

For a charged particle moving with this velocity there is no net force because the electric and magnetic forces cancel each other. Crossed fields are used in this manner as a *velocity selector* in atomic and nuclear research.

If we transform the general problem of the motion to a coordinate frame S' moving with the velocity given by Eq. (4.44), we find that in this new frame the particle moves uniformly in a circle. The velocity component transformations are

$$v_x = v'_x \qquad v_y = -c\frac{E}{B} + v'_y \tag{4.45}$$

Substitution into Eq. (4.43) yields (recall $\omega_c = eB/Mc$)

$$\dot{v}'_x = \omega_c v'_y \qquad \dot{v}'_y = -\omega_c v'_x \tag{4.46}$$

These are the same as Eqs. (3.24), describing uniform circular motion. Thus the behavior of the particle may be simply described as a uniform circular motion with angular velocity $\omega_c = Be/Mc$ in the $x'y'$ plane of the S' frame superimposed upon a steady motion of this frame with velocity $v_y = -c\,E/B$ relative to the laboratory frame. In S' the particle "feels" only the magnetic field; the electric field is zero in this frame.

If we choose initial conditions having the charged particle momentarily at rest at the laboratory frame origin at $t = 0$, the subsequent motion is a cycloid. It is as if the particle were

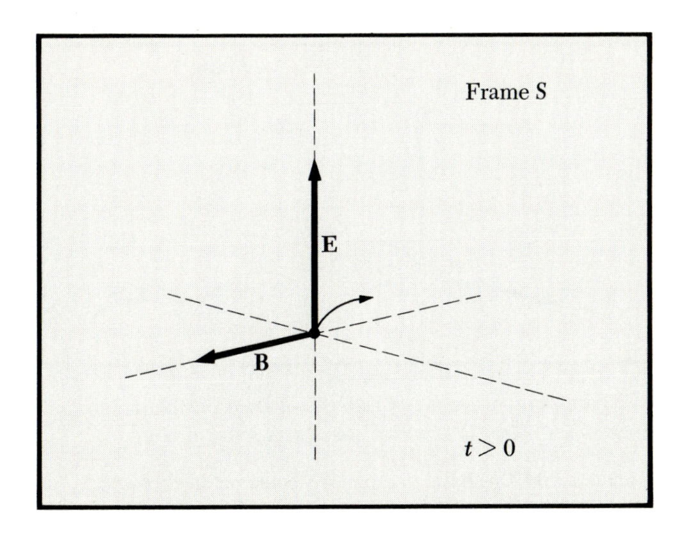

(c) As soon as q acquires velocity in the direction of $\mathbf{E}$, it experiences a force $\mathbf{F} = (q/c)\mathbf{v} \times \mathbf{B}$. The orbit then curves in the $-y$ direction.

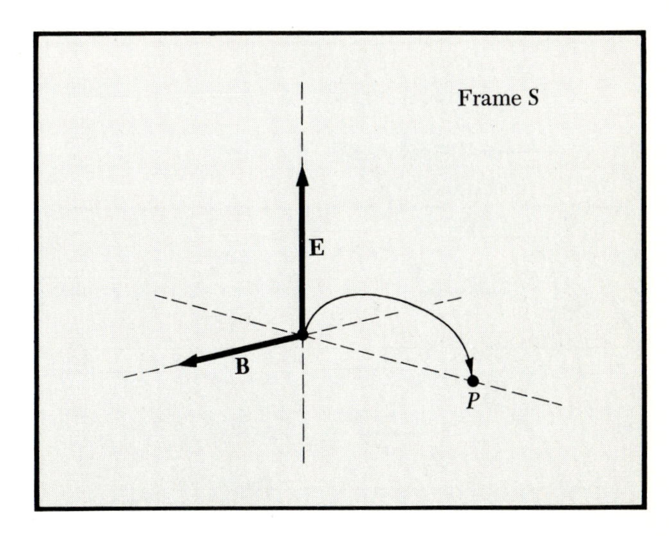

(d) q ultimately comes to rest at P, a point on the y axis. It then begins a new cycle of motion.

a point on the periphery of a wheel rolling with uniform speed $c\,E/B$ along the negative y axis.[1] We shall now demonstrate this fact, which is illustrated in Fig. 4.27a to d.

In S', the initial conditions of velocity corresponding to zero velocity in the laboratory frame are $v_x' = 0$ and $v_y' = c\,E/B$. Solutions of Eqs. (4.46) satisfying this initial condition are

$$v_x' = c\frac{E}{B}\sin\omega_c t \qquad v_y' = c\frac{E}{B}\cos\omega_c t \tag{4.47}$$

These represent uniform circular motion, which is clockwise when viewed from a position on the positive z' axis, with angular velocity ω_c and radius

$$r = \frac{cE}{\omega_c B} \tag{4.48}$$

When transformed into the laboratory frame, Eqs. (4.47) give, by recalling Eq. (4.45),

$$v_x = \frac{dx}{dt} = c\frac{E}{B}\sin\omega_c t \qquad v_y = \frac{dy}{dt} = c\frac{E}{B}(-1 + \cos\omega_c t) \tag{4.49}$$

[1] If the particle has an initial velocity, its motion will be that of a point inside or outside the periphery of the wheel.

Integration of Eqs. (4.49), with initial conditions that $x = y = 0$ at $t = 0$ and recognition of Eq. (4.48), gives

$$x = r(1 - \cos\omega_c t) \qquad y = r(-\omega_c t + \sin\omega_c t)$$

These are just the equations for motion of a point on the rim of a disk of radius r rolling in the negative y direction (see Fig. 4.28a and b). Using SI units, the velocity of the moving frame is $-E/B$, and, of course, $\omega_c = eB/M$. Equation (4.48) becomes

$$r = \frac{E}{\omega_c B} = \frac{ME}{eB^2}$$

The description of the motion is the same.

MATHEMATICAL NOTE

Differentiation of Products of Vectors In Chap. 2 we considered the differentiation of vectors; recall in particular that if

$$\mathbf{r} = x\hat{\mathbf{x}} + y\hat{\mathbf{y}} + z\hat{\mathbf{z}}$$

then

$$\dot{\mathbf{r}} = \dot{x}\hat{\mathbf{x}} + \dot{y}\hat{\mathbf{y}} + \dot{z}\hat{\mathbf{z}}$$

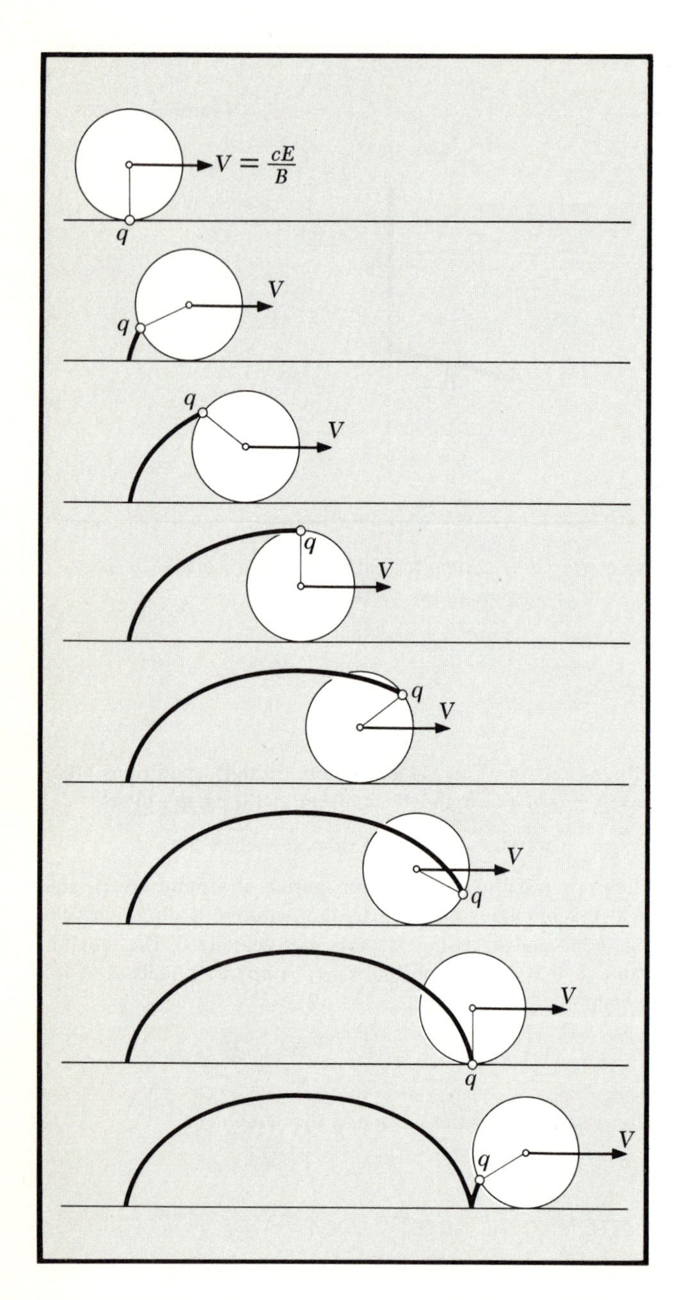

FIG. 4.28 (*a*) The orbit is a common cycloid (if the particle starts from rest), and q has average velocity to the right: $V = cE/B$. Note that the direction of the average velocity cE/B is the direction of $\mathbf{E} \times \mathbf{B}$. In the second Advanced Topic, $\mathbf{E} \times \mathbf{B} = E\hat{\mathbf{x}} \times B\hat{\mathbf{z}} = -EB\hat{\mathbf{y}}$, which is the result of Eqs. (4.44) and (4.45).

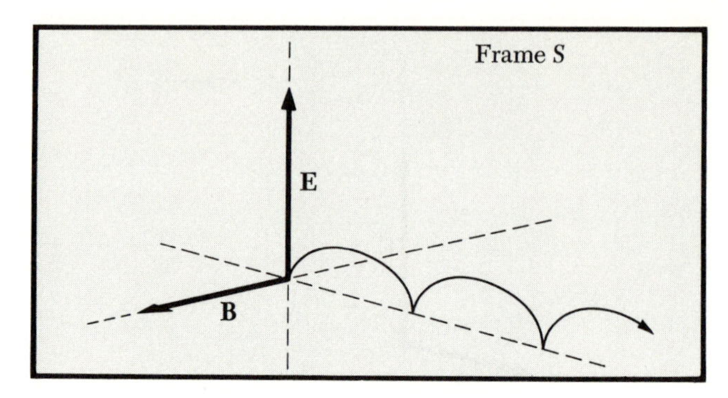

(*b*) The common cycloid is traced out by q on the circumference of a circle which rolls on a straight line.

provided that the basis vectors are constant in direction.

We now derive the relation

$$\frac{d}{dt}(\mathbf{A} \times \mathbf{B}) = \dot{\mathbf{A}} \times \mathbf{B} + \mathbf{A} \times \dot{\mathbf{B}}$$

Let $\mathbf{P}(t)$ denote $\mathbf{A}(t) \times \mathbf{B}(t)$ and consider the expression

$$\begin{aligned}\mathbf{P}(t + \Delta t) - \mathbf{P}(t) &= \mathbf{A}(t + \Delta t) \times \mathbf{B}(t + \Delta t) - \mathbf{A}(t) \times \mathbf{B}(t)\\ &\approx \left[\mathbf{A}(t) + \frac{d\mathbf{A}}{dt}\Delta t\right] \times \left[\mathbf{B}(t) + \frac{d\mathbf{B}}{dt}\Delta t\right] - \mathbf{A}(t) \times \mathbf{B}(t)\\ &= \Delta t\left[\frac{d\mathbf{A}}{dt} \times \mathbf{B} + \mathbf{A} \times \frac{d\mathbf{B}}{dt}\right] + (\Delta t)^2\left[\frac{d\mathbf{A}}{dt} \times \frac{d\mathbf{B}}{dt}\right]\end{aligned}$$

Thus we have

$$\dot{\mathbf{P}} = \lim_{\Delta t \to 0} \frac{\mathbf{P}(t + \Delta t) - \mathbf{P}(t)}{\Delta t} = \dot{\mathbf{A}} \times \mathbf{B} + \mathbf{A} \times \dot{\mathbf{B}}$$

Note that the order of terms in the vector products is important in the result. By a similar argument we have

$$\frac{d}{dt}(\mathbf{A} \cdot \mathbf{B}) = \dot{\mathbf{A}} \cdot \mathbf{B} + \mathbf{A} \cdot \dot{\mathbf{B}}$$

FURTHER READING

PSSC, "Physics," chaps. 20 (secs. 9–11), 22, D. C. Heath and Company, Boston, 1965.

Ernst Mach, "The Science of Mechanics," chap. 2, sec. 6, The Open Court Publishing Company, La Salle, Ill., 1960.

"Collier's Encyclopedia," 1964. An excellent elementary discussion of the Foucault pendulum is given under Foucault.

Mary Hesse, Resource Letter on Philosophical Foundations of Classical Mechanics, *Am. J. Phys.* **32**:905 (1964). This is a quite complete list of works.

CONTENTS

Conservation of Energy

CONSERVATION LAWS IN THE PHYSICAL WORLD

In the physical world there exist a number of *conservation laws,* some exact and some approximate. A conservation law is usually the consequence of some underlying symmetry in the universe. There are conservation laws relating to energy, linear momentum, angular momentum, charge, number of baryons (protons, neutrons, and heavier elementary particles), strangeness, and various other quantities. In Chaps. 3 and 4 we discussed conservation of linear momentum. In this chapter we discuss the conservation of energy. In Chap. 6 we shall generalize the discussion and take up angular momentum. The entire discussion at present will be phrased for the nonrelativistic regime, which means a restriction to galilean transformations, speeds very much less than that of light, and independence of mass and energy. In Chap. 12, after we have introduced the Lorentz transformation and special relativity, we shall give the appropriate forms of the energy and momentum conservation laws in the relativistic regime.

If all the forces in a problem are known, and if we are clever enough and have computers of adequate speed and capacity to solve for the trajectories of all the particles, then the conservation laws give us no additional information. But since we do not have all this information and these abilities and facilities, the conservation laws are very powerful tools. Why are conservation laws powerful tools?

1 Conservation laws are independent of the details of the trajectory and, often, of the details of the particular force. The laws are therefore a way of stating very general and significant consequences of the equations of motion. A conservation law can sometimes assure us that something is impossible. Thus we do not waste time analyzing an alleged perpetual motion device if it is merely a closed system of mechanical and electrical components, or a satellite propulsion scheme that purportedly works by moving internal weights.

2 Conservation laws may be used even when the force is unknown; this applies particularly in the physics of elementary particles.

3 Conservation laws have an intimate connection with invariance. In the exploration of new and not yet understood phenomena the conservation laws are often the most striking physical facts we know. They may suggest appropriate invariance concepts. In Chap. 4 we saw that the con-

servation of linear momentum could be interpreted as a direct consequence of the principle of galilean invariance.

4 Even when the force is known exactly, a conservation law may be a convenient aid in solving for the motion of a particle. Many physicists have a regular routine for solving unknown problems: First they use the relevant conservation laws one by one; only after this, if there is anything left to the problem, will they get down to real work with differential equations, variational and perturbation methods, computers, intuition, and other tools at their disposal. In Chaps. 7 to 9 we exploit the energy and momentum conservation laws in this way.

DEFINITION OF CONCEPTS

The law of conservation of mechanical energy involves the concepts of *kinetic energy, potential energy,* and *work.* These concepts, which can be understood from a simple example, arise very naturally from Newton's Second Law and will be treated in detail later. To start with, we discuss forces and motions in only one dimension, which simplifies the notation. The development is repeated for three dimensions; the student may find the repetition helpful.

To develop the concepts of work and kinetic energy we consider a particle of mass M drifting in intergalactic space and initially free of all external interactions. We observe the particle from an inertial reference frame. A force $\mathbf{F}$ is applied to the particle at time $t = 0$. The force thereafter is kept constant in magnitude and direction; the direction is taken to be the y direction. The particle will accelerate under the action of the applied force. The motion at times $t > 0$ is described by Newton's Second Law:

$$F = M\frac{d^2y}{dt^2} = M\ddot{y}\dagger \tag{5.1}$$

Thus the velocity after time t is

$$\int_{v_0}^{v} dv = \int_0^t \ddot{y}\, dt = \int_0^t \frac{F}{M} dt$$

or

$$v - v_0 = \frac{F}{M}t \tag{5.2}$$

† We use y here rather than x or z purely for convenience in applying the results to the constant gravitational field where we have used y in Chap. 3.

where v_0 is the initial velocity supposed to lie in the y direction.

Notice that Eq. (5.2) may be written as

$$Ft = Mv(t) - Mv_0$$

The right-hand side is the change of momentum of the particle in the time t, and the left-hand side is called the *impulse* of force in the same time. In case F is very large but the time over which it acts is very short, it may be convenient to define

$$\text{Impulse} = \int_0^t F dt = \Delta(Mv) \tag{5.3}$$

Equation (5.3) tells us that the change in the momentum is equal to the impulse.[1]

If the initial position is y_0, on integrating Eq. (5.2) with respect to the time we find

$$y(t) - y_0 = \int_0^t v(t)\, dt = \int_0^t \left(v_0 + \frac{F}{M}t\right) dt = v_0 t + \frac{1}{2}\frac{F}{M}t^2 \tag{5.4}$$

We may solve Eq. (5.2) for the time t:

$$t = \frac{M}{F}(v - v_0) \tag{5.5}$$

Now substitute Eq. (5.5) in Eq. (5.4) to obtain

$$y - y_0 = \frac{M}{F}(vv_0 - v_0{}^2) + \frac{1}{2}\frac{M}{F}(v^2 - 2vv_0 + v_0{}^2)$$

$$= \frac{1}{2}\frac{M}{F}(v^2 - v_0{}^2)$$

thus

$$\tfrac{1}{2}Mv^2 - \tfrac{1}{2}Mv_0{}^2 = F \times (y - y_0) \tag{5.6}$$

If we define $\frac{1}{2}Mv^2$ as the *kinetic energy* of the particle, i.e., the energy it possesses by virtue of its motion, then the left-hand side of Eq. (5.6) is the change in kinetic energy. The change is caused by the force F acting for a distance $(y - y_0)$. It is a useful definition of *work* to call $F \times (y - y_0)$ the *work done on the particle by the applied force*. With these definitions Eq. (5.6) says that the work done by the applied force is equal to the change of kinetic energy of the particle. This is all a mat-

[1] More advanced courses in mechanics often deal with impulse. Problem 16 in this chapter and Prob. 10 in Chap. 8 use the concept.

ter of definition, but the definitions are useful and they follow from Newton's Second Law.

If $M = 20$ g and $v = 100$ cm/s, the kinetic energy

$$K = \tfrac{1}{2}Mv^2 = \tfrac{1}{2}(20)(10^4) = 1 \times 10^5 \text{ g-cm}^2/\text{s}^2$$
$$= 1 \times 10^5 \text{ ergs}$$

The *erg* is the unit of energy in the cgs system of units. If a 100-dyn force is applied through a distance of 10^3 cm,

$$F(y - y_0) = (10^2)(10^3) = 10^5 \text{ dyn-cm} = 10^5 \text{ ergs}$$

One erg is the quantity of work performed by a force of one dyne acting through one centimeter. Work has the dimensions

$$[\text{Work}] \sim [\text{force}][\text{distance}] \sim [\text{mass}][\text{acceleration}][\text{distance}]$$
$$\sim [\text{mass}][\text{velocity}]^2 \sim \left[M\frac{L}{T^2}L\right] \sim [ML^2T^{-2}] \sim [\text{energy}]$$

In the International System the unit of work is the *joule,* which is the work done by a force of one newton acting through one meter. To convert joules to ergs, multiply by 10^7 the value of the work expressed in joules since we have seen in Chap. 3 that $1\text{ N} = 10^5$ dyn and $1\text{ m} = 10^2$ cm. In the above example of kinetic energy, $M = 0.020$ kg, $v = 1$ m/s, and $K = \frac{1}{2}$ $0.02 \times 1.0^2 = 1 \times 10^{-2}$ J.

In talking about work, one must always specify *work done by what.* In the case above, the work is done by the force that accelerates the particle. Such forces are often integral parts of the system that we are investigating; for example, they may be gravitational, electric, or magnetic forces. Later, when we talk about potential energy, we shall call these *forces of the field,* or *forces of the system;* but we shall also consider forces applied by an external agent (perhaps by us), and it will be important to distinguish work done by field forces from that done by the agent. For example, if the agent applies a force always equal and opposite to the field force, then the particle will not be accelerated and no change in kinetic energy will be produced. The work done by the field force is exactly canceled by the work done by the agent, as indeed we should expect since $F_{ag} = -F$. (It is important to note that we are excluding effects of friction forces in the present discussion; we are using ideal situations to establish our definitions and concepts.)

Consider now a body (particle), not in intergalactic space, but released from a height h above the surface of the earth

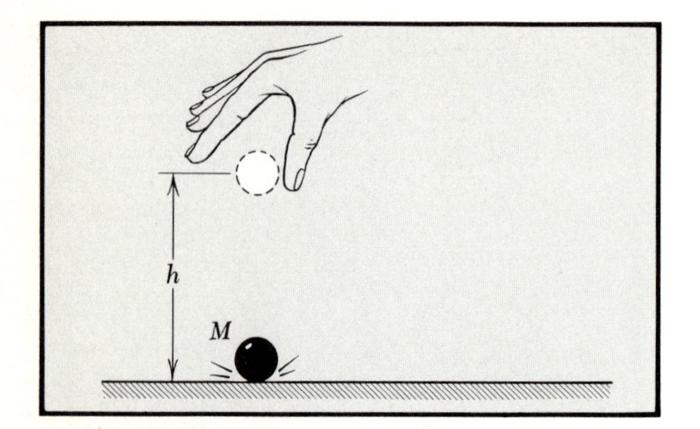

FIG. 5.1 In falling from rest at height h, the force of gravity does work Mgh which is equal to the kinetic energy generated.

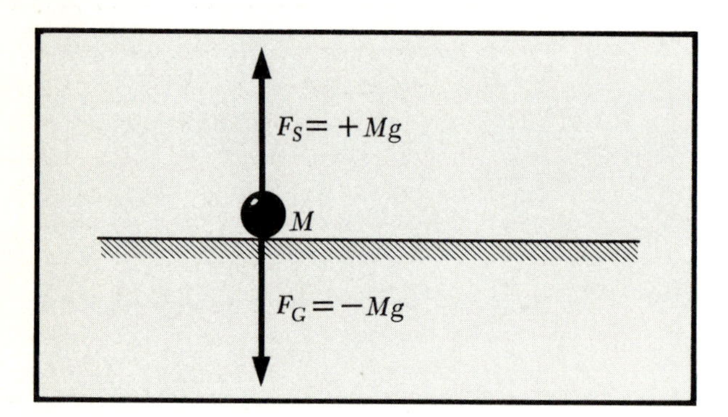

FIG. 5.2 (*a*) A mass M at rest on the earth's surface experiences two equal and opposite forces: F_G, the attractive gravitational force; F_S, the force exerted on M by the supporting surface.

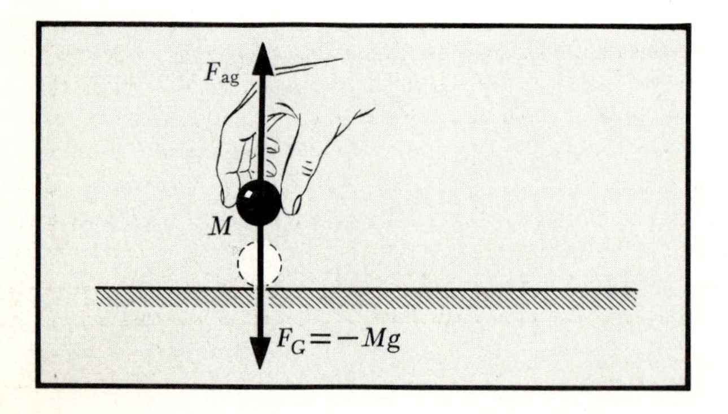

(*b*) To raise M at constant velocity requires an applied force $F_{ag} = +Mg$.

($y_0 = h$; $v_0 = 0$). The gravitational force $F_G = -Mg$ pulls downward on the body. As the body falls toward the surface of the earth, the work done by gravity is equal to the gain in kinetic energy of the body (see Fig. 5.1):

$$W(\text{by gravity}) = F_G \times (y - y_0)$$

or, at the surface ($y = 0$) of the earth,

$$\begin{aligned} W(\text{by gravity}) &= (-Mg)(0 - h) = Mgh \\ &= \tfrac{1}{2}Mv^2 - \tfrac{1}{2}Mv_0^2 = \tfrac{1}{2}Mv^2 \end{aligned} \tag{5.7}$$

where v is the velocity of the body on reaching the surface of the earth. Equation (5.7) suggests that we may say that at height h the body has *potential energy* (capacity to do work or to gain kinetic energy) of Mgh relative to the earth's surface.

What happens to the potential energy when a particle at rest on the earth's surface is raised to a height h? To raise the body, we must apply an upward force $F_{ag}(= -F_G)$ to the body. Now $y_0 = 0$ and $y = h$. We do work

$$W(\text{by us}) = F_{ag} \times (y - y_0) = (Mg)(h) = Mgh \tag{5.8}$$

on the body, thereby giving the body the potential energy Mgh that, as we have said earlier, it has at height h (see Fig. 5.2*a* to *c*). Note that we call the force that *we* exert F_{ag}; in other words, we and the external agent are identical. Of course, it is easy to talk about "we" and "us," and the terms are used below; but the important point to remember is that here an external agent is conceptually brought into the problem only for the purpose of evaluating the potential energy.

In the absence of friction forces a specific definition of the potential energy of a body (particle) at a point of interest can now be formulated: *Potential energy is the work we do in moving the body without acceleration from an initial location, arbitrarily assigned to be a zero of potential energy, to the point of interest.* A few comments may aid our understanding of this definition. We are free to arbitrarily assign the location of zero potential energy according to convenience, and so the value at the point of interest will always be relative to this assignment. Presumably there are field forces acting upon the body, and to move it without acceleration we must exert a force equal and opposite to their resultant force. Under this condition we move the body without acceleration from the *zero position to the point where we wish to evaluate the potential energy.* The work we have done is equal to the potential energy. Since, in the absence of friction, the force we apply is always equal and opposite to the field forces present in the problem, the work we do is equal and opposite to the work done by those forces. Therefore, we can equally well define potential energy as the work done by the forces of the problem, the field forces, in moving the system in the other direction *from the point under consideration to the arbitrary zero.* For example, the work done by gravity [Eq. (5.7)] on the falling body is equal to the work we do [Eq. (5.8)] against gravity in lifting the particle up.

Equally valid is the definition of positive potential energy at a point as the kinetic energy generated by the forces in the free motion of the body to the arbitrary zero, as in Fig. 5.1. This definition, as stated, does not apply to cases in which the potential energy is negative relative to the zero; but an obvious modification of the definition is valid. An example is given on pages 161–162.

Two further points are worth emphasizing. First, the potential energy is purely a function of position, i.e., of the coordinates of the body or system.[1] Second, the zero point must always be specified. It is only the *change* in potential energy that is meaningful; for example, it may be converted into kinetic energy or, conversely, created from it. The absolute value of the potential energy is meaningless. Since this is true, the choice of the location of zero is arbitrary. In many cases a certain zero is particularly convenient, e.g., the surface of the earth, the plane of a table, but *any other* zero will give the same answer to a question of physics.

The dimensions of work and potential energy $[F][L] = [M][L^2]/[T^2]$ are the same as those of kinetic energy. If $F_{ag} = 10^3$ dyn and $h = 10^2$ cm, the potential energy is $10^3 \times 10^2 = 10^5$ dyn-cm $= 10^5$ ergs, or, in SI units, $10^{-2} \times 1.0 = 10^{-2}$ J. We denote the potential energy by U or PE. If in Eq. (5.7) we let v denote, not the velocity after falling a distance h, but the velocity after falling a distance $(h - y)$, then the equation analogous to Eq. (5.7) is

$$\tfrac{1}{2}Mv^2 = Mg(h - y)$$

or

$$\tfrac{1}{2}Mv^2 + Mgy = Mgh = E \tag{5.9}$$

where E is a constant having the value Mgh. This is illustrated in Fig. 5.3. Because E is a constant we have in Eq. (5.9) a statement of the *law of conservation of energy:*

[1] In more complicated problems, the student may encounter useful functions like the potential energy involving other quantities.

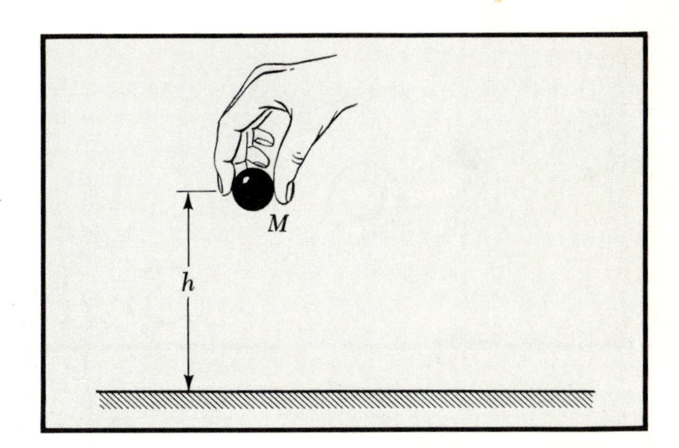

(c) The amount of work done in raising M to height h is $W = F_{ag} \times h = +Mgh$. The potential energy U of the mass M is thereby increased by an amount Mgh.

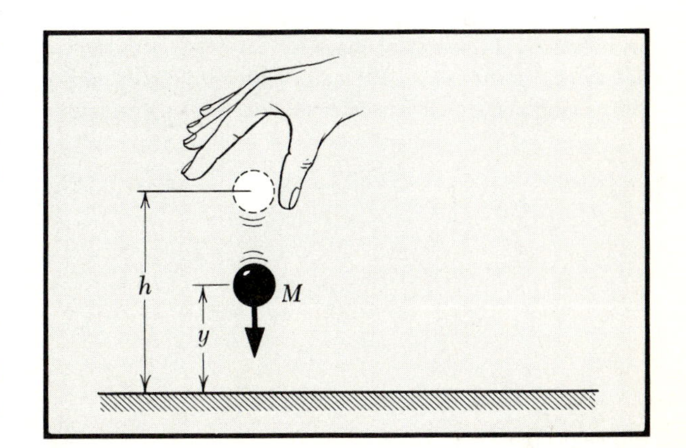

FIG. 5.3 If the mass is released, the potential energy U decreases and the kinetic energy K increases, but their sum remains constant. At height y, $U(y) = Mgy$ and $K(y) = \tfrac{1}{2}Mv(y)^2 = Mg(h - y)$.

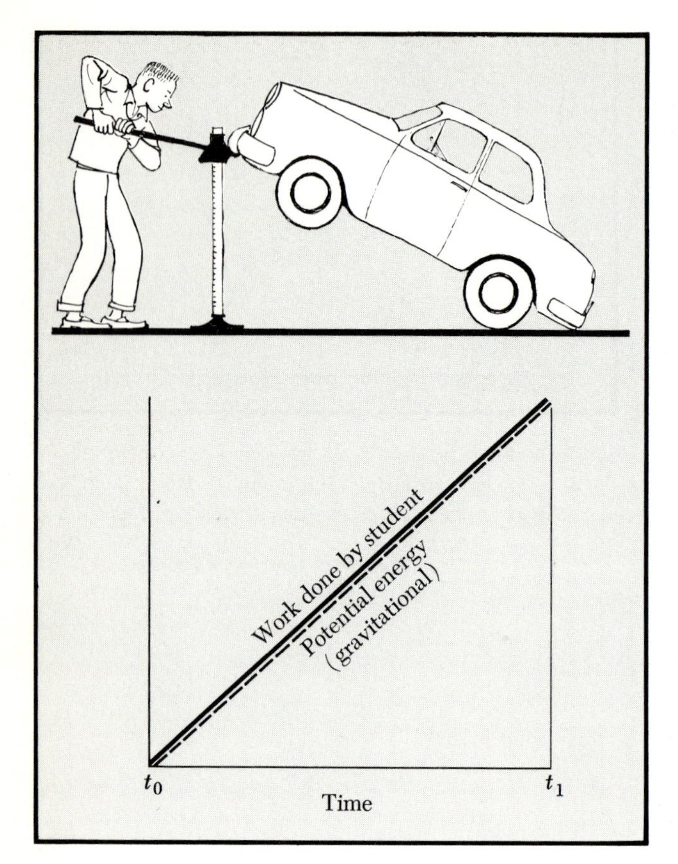

FIG. 5.4 (*a*) Work done by student vs time in jacking a car to change a tire. [The work done on a 1000-kg small car in raising its center of mass by 10 cm will be $Fh = Mgh \approx (10^3 \times 10^3 \text{ g}) \times (10^3 \text{ cm/s}^2)(10 \text{ cm}) = 10^{10}$ ergs.] The work done appears as gravitational potential energy.

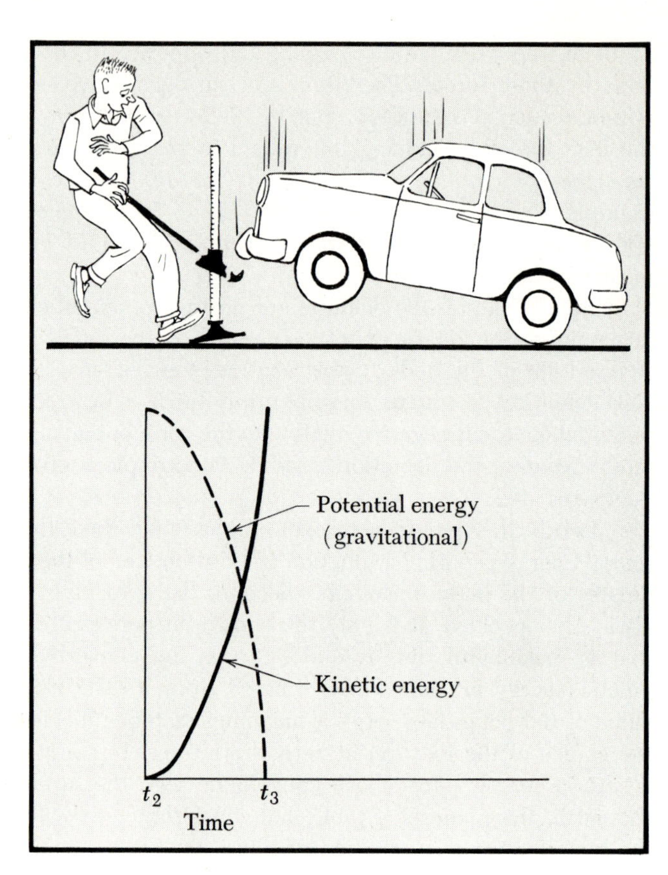

(*b*) Here the jack has slipped and the car falls back. The potential energy is converted into kinetic energy. After the car makes contact with the ground, the kinetic energy is converted to heat in the shock absorbers, springs, and tires.

$$\begin{aligned} E &= K + U \\ &= \text{kinetic energy} + \text{potential energy} = \text{const} \\ &= \text{total energy} \end{aligned}$$

In Eq. (5.9) the term Mgy is the potential energy, where we have chosen $y = 0$ as the zero of potential energy. The symbol E denotes the total energy, which is constant in time for an isolated system. Two illustrations are given in Figs. 5.4*a* and *b* and 5.5*a* and *b*.

Suppose we had chosen the zero of potential energy at $y = -H$. Then we would have

$$E' = K + U = \tfrac{1}{2}Mv^2 + Mg(y + H) = Mg(h + H)$$

which reduces to Eq. (5.9) by subtraction of MgH from each side and exemplifies the fact that the zero of potential energy does not affect the answer to questions of physics.

Sometimes it is convenient to call $E = K + U$, the sum of kinetic and potential energy contributions, the *energy function*. The kinetic energy contribution K is equal to $\tfrac{1}{2}Mv^2$. The potential energy depends on the field force acting, and it has the essential property that $U = -\int F\,dy$, which is an expression for work where the field force F may be a function of position y. Then

$$F = -\frac{dU}{dy} \tag{5.10}$$

where F, the force acting on the particle, results from interactions intrinsic to the problem, such as electrical or gravitational interactions, and is what we have called the *force of the field*, or the *force of the problem*. (In the above example, $U = Mgy$, so that $F = F_G = -Mg$.)

Equation (5.10) illustrates why we call them *forces of the field*. U defines a potential energy field; it is a scalar function of y. The forces are derivable from this field function. Note here that the zero will appear in U as a constant term so when the force is derived in Eq. (5.10), it is the same irrespective of the constant.

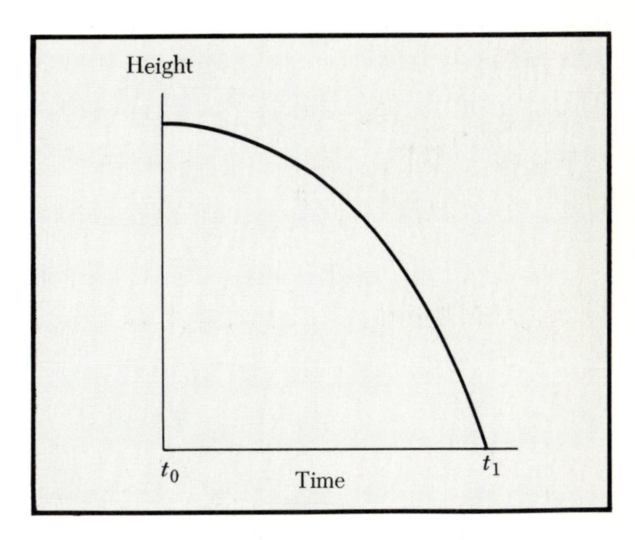

FIG. 5.5 (*a*) Height vs time for a body falling toward the earth, starting at rest.

EXAMPLE

Free Motion of Upward-projected Body If we project a body upward with speed 1000 cm/s, how high will it rise? Assume the level of projection is the position of zero potential energy. Then at the point of projection

$$E = 0 + \tfrac{1}{2}Mv^2 = \tfrac{1}{2}M \times 10^6 \text{ ergs}$$

At maximum height $v = 0$, and so

$$E = Mgh$$

By equating these two expressions for E, we have

$$h = \frac{v^2}{2g} \approx \frac{10^6}{2.10^3} \approx 500 \text{ cm} \qquad \text{or in SI units} \qquad \approx \frac{10^2}{2.10} \approx 5 \text{ m}$$

Try working this out by the method given in Chap. 3, and you will see the advantages of this approach.

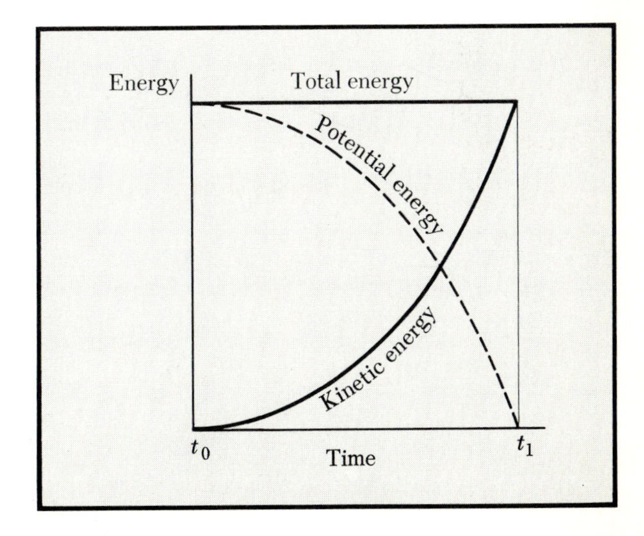

(*b*) Potential energy of the falling body vs time, and kinetic energy of the falling body vs time. The total energy, which is constant, is the sum of the kinetic, plus the potential, energy.

Conservation of Energy We now proceed to develop these ideas in three dimensions so as to be able to use them in full generality. The law of conservation of energy states that for a system of particles with interactions not explicitly[1] dependent on the time, the *total energy of the system is constant.* We accept this result as a very well established experimental fact. More specifically, the law tells us there exists some scalar function [such as the function $\frac{1}{2}Mv^2 + Mgy$ in Eq. (5.9)] of the positions and velocities of the constituent particles that is invariant with respect to a change in time, provided there is no explicit change in the interaction forces during the time interval considered. For example, the mass m or the elementary charge e must not change with time. Besides the energy function, there are other functions that are constant in the conditions specified here. (We treat other functions in Chap. 6, under Conservation of Linear and Angular Momentum.) The energy is a scalar constant of motion. We interpret the phrase *external interaction* to include any change in the laws of physics or in the values of the fundamental physical constants (such as g or e or m) during the relevant time interval as well as any change in external conditions such as gravitational, electric, or magnetic fields. Remember that the law gives us no new information not contained in the equations of motion $\mathbf{F} = M\mathbf{a}$. In our present treatment we do not consider changes of energy from mechanical form (kinetic and potential) into heat. For example, we omit forces of friction; they are not what we define later as conservative forces.

The central problem is to find an expression for the energy function that has the desired time invariance and that is consistent with $\mathbf{F} = M\mathbf{a}$. By consistent we mean that, for example,

$$\frac{d}{dy}E \equiv \frac{d}{dy}(K + U) = \frac{dK}{dy} - F_y = 0$$

is identical with $F_y = Ma_y$. You can check this for Eq. (5.9) to find

$$Mv\frac{dv}{dy} + Mg = M\frac{dy}{dt}\frac{dv}{dy} + Mg = M\frac{dv}{dt} + Mg = 0$$

or

$$M\frac{dv}{dt} = -Mg$$

From advanced points of view, this establishment of the correct energy function is the fundamental problem of classical me-

[1] Consider the system with the particles permanently frozen in place; then a force that depends on time is said to depend *explicitly* on time.

chanics, and its formal solution can be given in many ways, some of which are quite elegant. The hamiltonian formulation of mechanics, in particular, is one way that is very well suited to reinterpretation in the language of quantum mechanics. But here at the beginning of our course we need a simple direct formulation more than we need the generality of the hamiltonian or lagrangian formulations, which are the subject of later courses.[1]

Work We begin by generalizing the definition of work. The work W done by a constant applied force $\mathbf{F}$ in a displacement $\Delta\mathbf{r}$ is

$$W = \mathbf{F} \cdot \Delta\mathbf{r} = F\,\Delta r \cos(\mathbf{F}, \Delta\mathbf{r})$$

in conformity with the definition that follows Eq. (5.6) above. Suppose $\mathbf{F}$ is not constant but is a function of the position $\mathbf{r}$. Then we decompose the path into N line segments within each of which $\mathbf{F}(r)$ is essentially constant and write the following equation:

$$\begin{aligned} W &= \mathbf{F}(\mathbf{r}_1) \cdot \Delta\mathbf{r}_1 + \mathbf{F}(\mathbf{r}_2) \cdot \Delta\mathbf{r}_2 + \cdots \mathbf{F}(r_N) \cdot \Delta\mathbf{r}_N \\ &= \sum_{j=1}^{N} \mathbf{F}(\mathbf{r}_j) \cdot \Delta\mathbf{r}_j \end{aligned} \tag{5.11}$$

where the symbol Σ stands for the sum indicated. Equation (5.11) is strictly valid only in the limit of infinitesimal displacements $d\mathbf{r}$ because in general a curved path cannot be decomposed exactly into a finite number of line segments and $\mathbf{F}$ may not be exactly constant over a segment.

The limit

$$\lim_{\Delta\mathbf{r}_j \to 0} \sum_j \mathbf{F}(\mathbf{r}_j) \cdot \Delta\mathbf{r}_j = \int_{\mathbf{r}_A}^{\mathbf{r}_B} \mathbf{F}(\mathbf{r}) \cdot d\mathbf{r}$$

is the integral of the projection of $\mathbf{F}(\mathbf{r})$ on the displacement vector $d\mathbf{r}$. The integral is called the *line integral* of $\mathbf{F}$ from A to B. The work done in the displacement by the force is defined as

$$\boxed{W(A \to B) \equiv \int_A^B \mathbf{F} \cdot d\mathbf{r}} \tag{5.12}$$

where the limits A and B stand for the positions $\mathbf{r}_A$ and $\mathbf{r}_B$.

[1] The derivation of the lagrangian equations of motion requires several results of the calculus of variations.

Kinetic Energy We now return to the free particle subject to forces. We want to generalize Eq. (5.6), which we here repeat

$$\tfrac{1}{2}Mv^2 - \tfrac{1}{2}Mv_0^2 = \mathbf{F} \times (y - y_0)$$

to include applied forces that vary in direction and magnitude but are known as functions of position throughout the region where the motion occurs. By substituting $\mathbf{F} = M\,d\mathbf{v}/dt$ into Eq. (5.12), where $\mathbf{F}$ is the vector sum of the forces, we find for the work done by these forces

$$W(A \rightarrow B) = M \int_A^B \frac{d\mathbf{v}}{dt} \cdot d\mathbf{r} \tag{5.13}$$

Now

$$d\mathbf{r} = \frac{d\mathbf{r}}{dt}dt = \mathbf{v}\,dt$$

so that

$$W(A \rightarrow B) = M \int_A^B \left(\frac{d\mathbf{v}}{dt} \cdot \mathbf{v}\right) dt \tag{5.14}$$

where the limits A and B now stand for the times t_A and t_B when the particle is at the positions designated by A and B. But we can rearrange the integrand

$$\frac{d}{dt}v^2 = \frac{d}{dt}(\mathbf{v} \cdot \mathbf{v}) = 2\frac{d\mathbf{v}}{dt} \cdot \mathbf{v}$$

so that

$$2\int_A^B \left(\frac{d\mathbf{v}}{dt} \cdot \mathbf{v}\right) dt = \int_A^B \left(\frac{d}{dt}v^2\right) dt = \int_A^B d(v^2) = v_B^2 - v_A^2$$

On substitution in Eq. (5.14) we have an important result:

$$\boxed{W(A \rightarrow B) = \int_A^B \mathbf{F} \cdot d\mathbf{r} = \tfrac{1}{2}Mv_B^2 - \tfrac{1}{2}Mv_A^2} \tag{5.15}$$

for the free particle. This is a generalization of Eq. (5.6).

We recognize

$$K \equiv \tfrac{1}{2}Mv^2 \tag{5.16}$$

as the *kinetic energy* previously defined in Eq. (5.6). We see from Eq. (5.15) that our definitions of work and kinetic energy have the property that *the work done on a free particle by an arbitrary force is equal to the change in the kinetic energy of the particle:*

$$W(A \rightarrow B) = K_B - K_A \tag{5.17}$$

EXAMPLE

Free Fall

(1) We repeat an example given before. If the y direction is normal to the surface of the earth and directed upward, the gravitational force is $\mathbf{F}_G = -Mg\hat{\mathbf{y}}$, where g is the acceleration of gravity and has the approximate value 980 cm/s^2. Calculate the work done by gravity when a mass of 100 g falls through 10 cm. Here we can set

$$\mathbf{r}_A = 0 \qquad \mathbf{r}_B = -10\hat{\mathbf{y}} \qquad \Delta\mathbf{r} = \mathbf{r}_B - \mathbf{r}_A = -10\hat{\mathbf{y}}$$

The work done by gravity is

$$\begin{aligned} W &= \mathbf{F}_G \cdot \Delta\mathbf{r} = (-Mg\hat{\mathbf{y}}) \cdot (-10\hat{\mathbf{y}}) \\ &= (10^2)(980)(10)\hat{\mathbf{y}} \cdot \hat{\mathbf{y}} = 9.8 \times 10^5 \text{ ergs} \end{aligned}$$

Note that W would be independent of any horizontal displacement Δx. Here we have let the gravitational force play the role of the force **F**.

(2) If the particle in (1) initially had speed 1×10^2 cm/s, what would be its kinetic energy and velocity at the end of its 10-cm fall?

The initial value K_A of the kinetic energy is $\frac{1}{2} \times 100(100)^2 = 5 \times 10^5$ ergs; the terminal value K_B according to Eq. (5.17) is equal to the work done by gravity on the particle plus the kinetic energy at A:

$$K_B = W + \tfrac{1}{2}Mv_A{}^2 = 9.8 \times 10^5 + 5 \times 10^5 \approx 15 \times 10^5$$

$$v_B{}^2 \approx \frac{2 \times 15 \times 10^5}{100} \approx 3 \times 10^4$$

$$v_B \approx 1.7 \times 10^2 \text{ cm/s}$$

This result agrees with what we would calculate from $\mathbf{F} = M\mathbf{a}$, but note that we have not specified above the direction of the initial speed 1×10^2 cm/s. If it were in the x direction, it would remain constant and

$$\frac{1}{2}M(v_y{}^2 + v_x{}^2)_B - \frac{1}{2}M(v_x{}^2)_A = 9.8 \times 10^5$$

$$\begin{aligned} v_{yB} &\approx \sqrt{2} \times 10^2 \\ v_B &= \sqrt{v_x{}^2 + v_{yB}{}^2} \\ &= \sqrt{1 \times 10^4 + 2 \times 10^4} \\ &\approx 1.7 \times 10^2 \text{ cm/s} \end{aligned}$$

Or if v_y were downward in the negative y direction, we could call upon the familiar relationships for falling bodies:

$$\begin{aligned} h &= v_0 t + \tfrac{1}{2}gt^2 \\ v - v_0 &= gt \\ h &= v_0 \frac{v - v_0}{g} + \frac{1}{2}g\left(\frac{v - v_0}{g}\right)^2 \\ 2gh &= v^2 - v_0{}^2 \end{aligned}$$

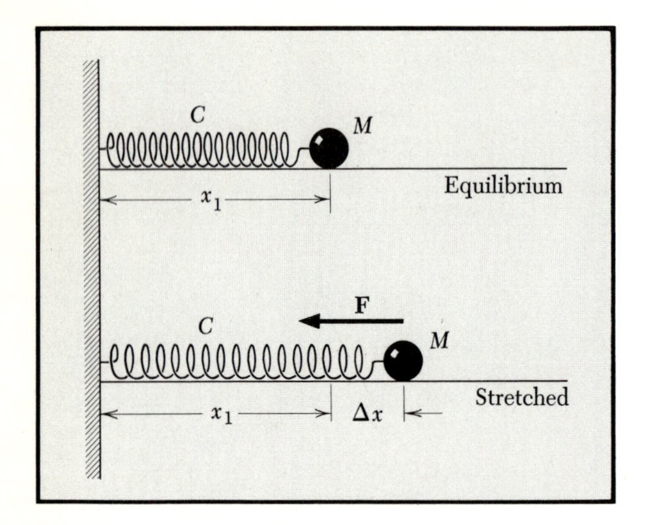

FIG. 5.6 (*a*) A massless spring is connected to mass M. If the spring is stretched a small amount Δx, it exerts a restoring force $F = -C\,\Delta x$ on M in the direction shown. Here C denotes the spring constant of the spring.

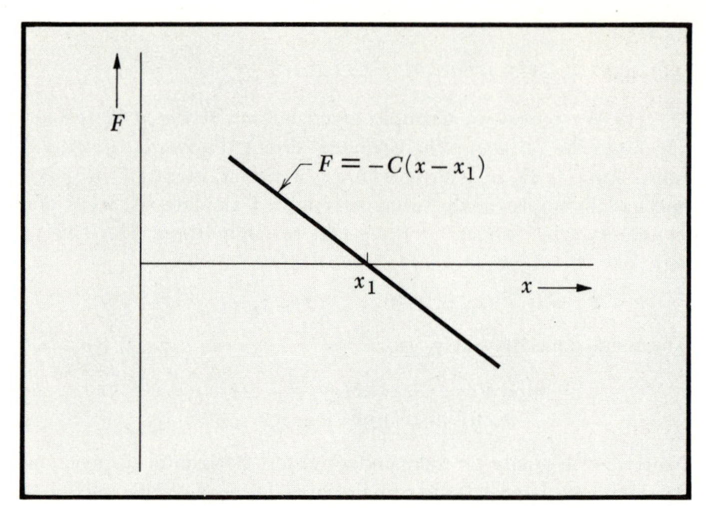

(*c*) The restoring force for small displacements from x_1 is proportional to the displacement.

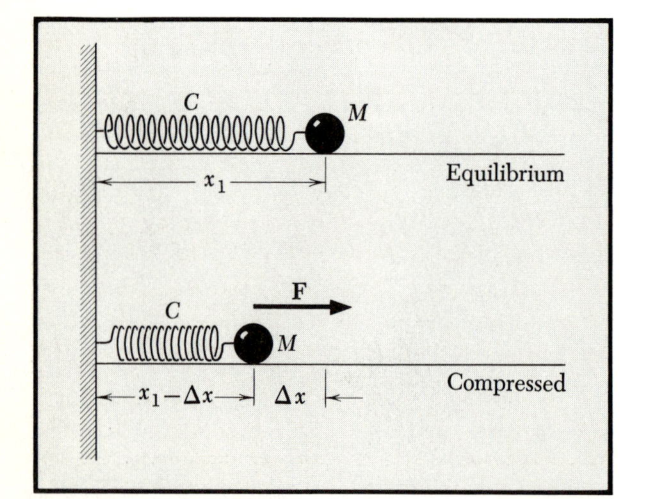

(*b*) If the spring is compressed an amount $-\Delta x$, the spring exerts a restoring force on M equal to $-C(-\Delta x) = C\,\Delta x$, as shown.

From this we again obtain the same work and kinetic energy relationship:

$$Mgh = \tfrac{1}{2}Mv_B^2 - \tfrac{1}{2}Mv_0^2 \qquad \text{since} \qquad v_0 = v_A$$

This is an example of what we mean when we say that the results obtained from the conservation law must be consistent with the equations of motion. Here, using conservation of energy, we obtained the same result as by employing an equation $v^2 - v_0^2 = 2gh$ derived from the equation of motion $\mathbf{F} = M\mathbf{a}$ (cf. Chap. 3).

Potential Energy We have mentioned (page 141) that only *differences* of potential energy are meaningful. Our definition of potential energy indicates that the difference in potential energy at points B and A is the work that *we* must do in moving the system without acceleration from A to B, so that

$$\boxed{U(\mathbf{r}_B) - U(\mathbf{r}_A) = W(A \rightarrow B) = \int_A^B \mathbf{F}_{\text{ag}} \cdot d\mathbf{r}} \tag{5.18}$$

Differences can be positive or negative: That is, if *we expend* work against the field forces, the potential energy is increased $U(\mathbf{r}_B) > U(\mathbf{r}_A)$; if work is expended against us by the field forces (we do a negative amount of work), the potential energy is decreased. We can understand that if the potential energy increases in going from A to B, the kinetic energy of a free particle moving in that direction will decrease (of course, $\mathbf{F}_{ag}$ is not acting), whereas if the potential energy decreases, the kinetic energy will increase. If we now specify $U(A) = 0$ in Eq. (5.18), then the value of $U(B)$ is uniquely defined provided that the forces are conservative (see page 155).

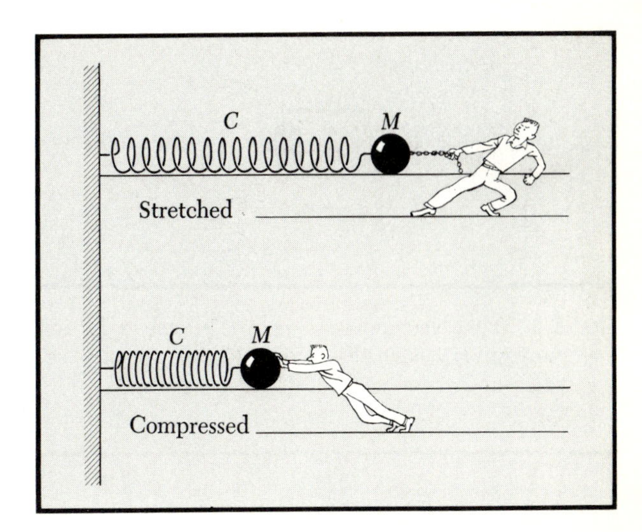

FIG. 5.7 In order to stretch (or compress) the spring, we must exert a force in opposition to the restoring force. In displacing the spring an amount Δx from the equilibrium position x_1, we do work

$$W = \int_{x_1}^{x_1+\Delta x} C(x - x_1)\, dx = \tfrac{1}{2}C(\Delta x)^2$$

EXAMPLE

Linear Restoring Force: Transformations between Kinetic and Potential Energy A particle is subject to a *linear restoring force* in the x direction. A linear restoring force is one that is directly proportional to the displacement measured from some fixed point and in a direction tending to reduce the displacement (see Fig. 5.6a to c). If we take the fixed point as the origin,

$$\mathbf{F} = -Cx\hat{\mathbf{x}} \qquad \text{or} \qquad F_x = -Cx \tag{5.19}$$

where C is some positive constant, the *spring constant*. This is called *Hooke's law*. For sufficiently small displacements such a force may be produced by a stretched or compressed spring. For large elastic displacements we must add terms in higher powers of x to Eq. (5.19). The sign of the force is such that the particle is always attracted toward the origin $x = 0$.

(1) With the particle attached to the spring, we now supply the force that takes the particle from a point x_1 to another point x_2. What is the work we do in this displacement?

Here the force on the particle is a function of position. To calculate the work we do, we use the definition [Eq. (5.12)] and write $\mathbf{F}_{ag} = -\mathbf{F} = Cx\hat{\mathbf{x}}$.

$$W(x_1 \to x_2) = \int_{x_1}^{x_2} \mathbf{F}_{ag} \cdot d\mathbf{r} = C\int_{x_1}^{x_2} x\, dx = \tfrac{1}{2}C(x_2^2 - x_1^2)$$

If we choose $x_1 = 0$, the equilibrium position, as the zero of potential energy, then

$$\boxed{U(x) = \tfrac{1}{2}Cx^2} \tag{5.20}$$

This is the famous result: The potential energy related to a linear restoring force is proportional to the square of the displacement (see Figs. 5.7 and 5.8).

(2) If the particle of mass M is released at rest at the position x_{max}, what is its kinetic energy when it reaches the origin?

We obtain the answer directly from Eqs. (5.15) and (5.20): The work done by the *spring* in going from x_{max} to the origin is

$$W(x_{max} \to 0) = \tfrac{1}{2}Mv_1^2$$

where we have used the fact that $v = 0$ at x_{max}; the particle is assumed to be at rest there. The velocity at the origin is v_1. Thus

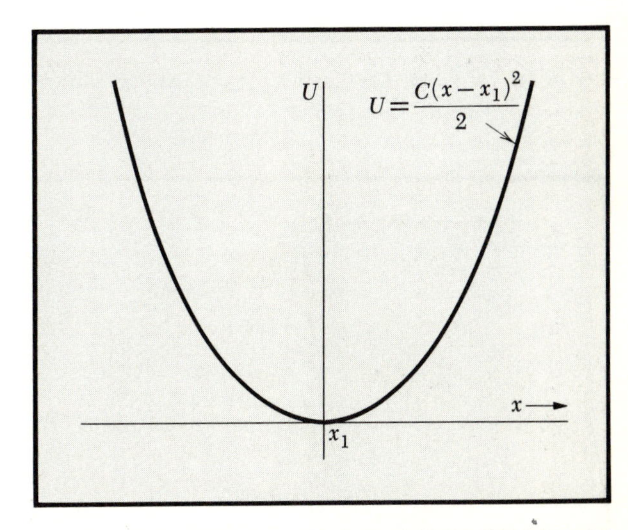

FIG. 5.8 In doing this work you increase the potential energy of the spring-mass system. A spring-mass system displaced $\Delta x = x - x_1$ from equilibrium has potential energy $U = \tfrac{1}{2}C(\Delta x)^2 = \tfrac{1}{2}C(x - x_1)^2$.

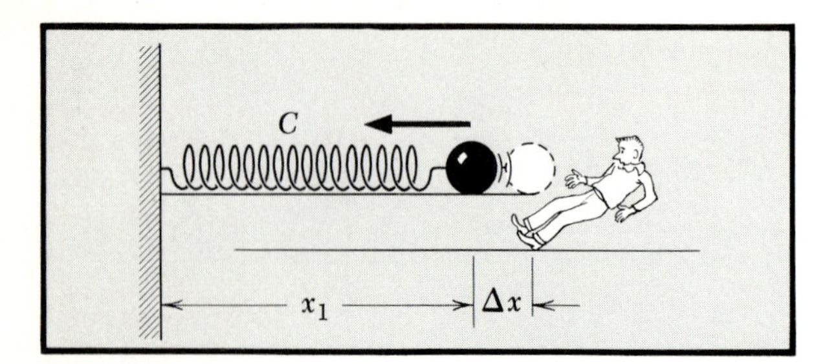

FIG. 5.9 If the spring-mass system is stretched Δx and then released, U will initially decrease and K will increase.

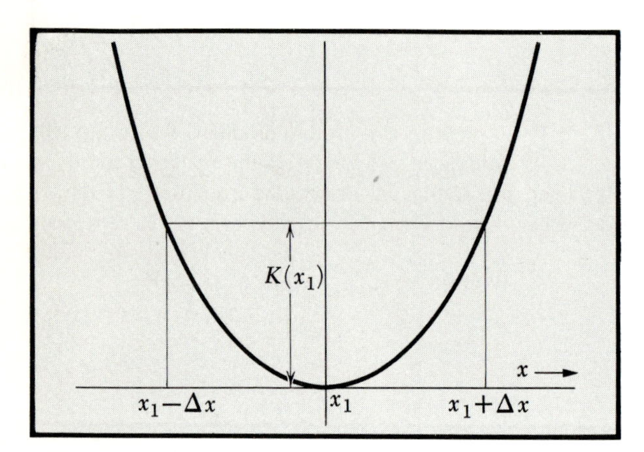

FIG. 5.10 At $x=x_1$, $U=0$ and $K(x_1)=\frac{1}{2}C(\Delta x)^2$ as shown.

$$\tfrac{1}{2}Cx_{max}{}^2 = \tfrac{1}{2}Mv_1{}^2$$

is the kinetic energy at the origin $x = 0$. Alternatively, we can use the conservation of energy. At x_{max}, $K = 0$; so that $U = \frac{1}{2}Cx_{max}{}^2 = E$. Then at $x = 0$,

$$\tfrac{1}{2}mv_1{}^2 = E = \tfrac{1}{2}Cx_{max}{}^2 \tag{5.21}$$

(see Figs. 5.9 and 5.10).

(3) What is the connection between the velocity of the particle at the origin and the maximum displacement x_{max}?

$$v_1{}^2 = \frac{C}{M}x^2_{max}$$

or

$$v_1 = \pm\sqrt{\frac{C}{M}}x_{max} \tag{5.22}$$

EXAMPLE

Energy Conversion in a Waterfall The conversion of energy in one form (potential) into energy of another form (kinetic) is illustrated by the waterfall of Fig. 5.11. The water at the top of the waterfall has gravitational potential energy, which in falling is converted into kinetic energy. A mass M of water in falling from a height h loses potential energy Mgh and gains kinetic energy $\frac{1}{2}M(v^2 - v_0{}^2) = Mgh$. (The velocity v is determined by this equation if the initial velocity v_0 of the water is known.) The kinetic energy of the falling water can be converted in a powerhouse into the rotational kinetic energy of a turbine; otherwise the kinetic energy of the falling water is converted at the foot of the falls into thermal energy or heat. Thermal energy is simply the random energy of molecular motion

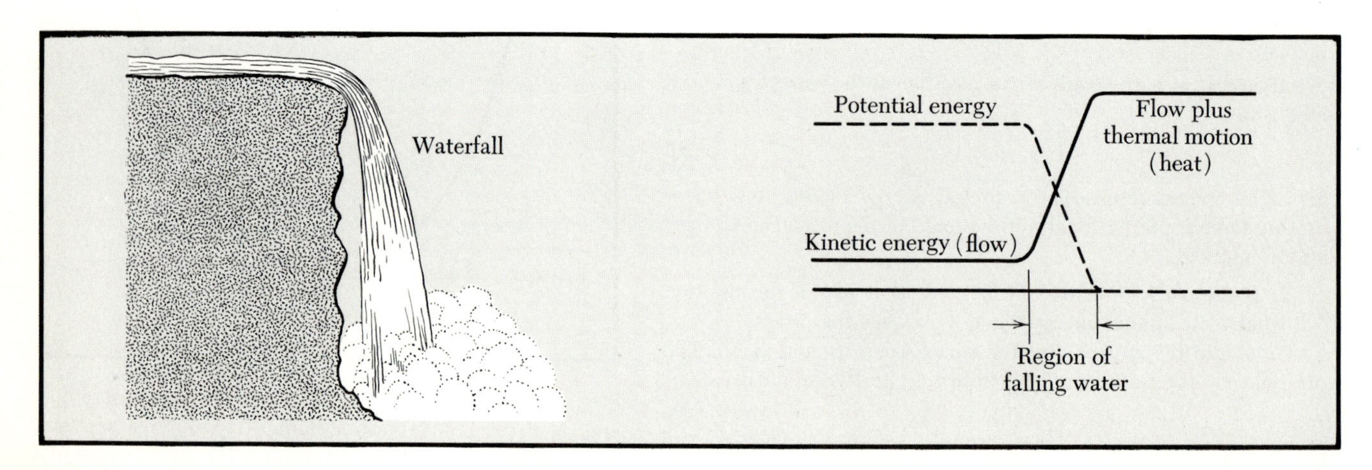

FIG. 5.11 Waterfall as illustration of conversion between forms of energy.

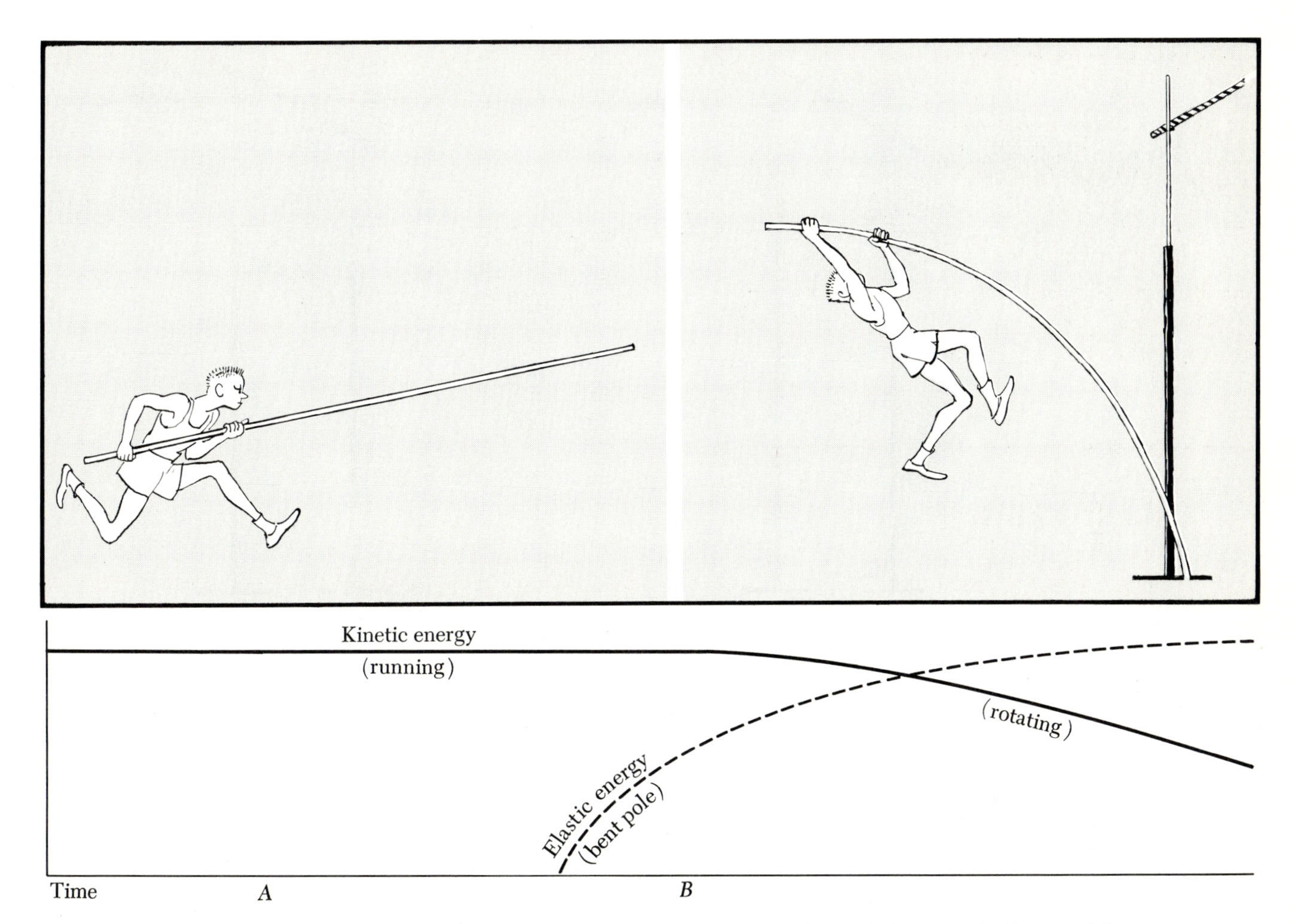

FIG. 5.12 Motion of a pole vaulter. At A his energy is all kinetic, associated with his running velocity. At B he puts the forward end of the pole on the ground and (especially with the new fiberglass poles) stores elastic potential energy in the pole by bending it. At C he is rising in the air. He has considerable kinetic energy left, now associated with

in the water. (At a high temperature the random molecular motion is more vigorous than at a low temperature.)

EXAMPLE

Energy Transformations in the Pole Vault A rather amusing example of the interconversion of energy among various forms—kinetic energy, potential energy of the bent elastic pole, and potential energy due to elevation—is afforded by the sequence of pictures in Fig. 5.12 and the legends accompanying them.

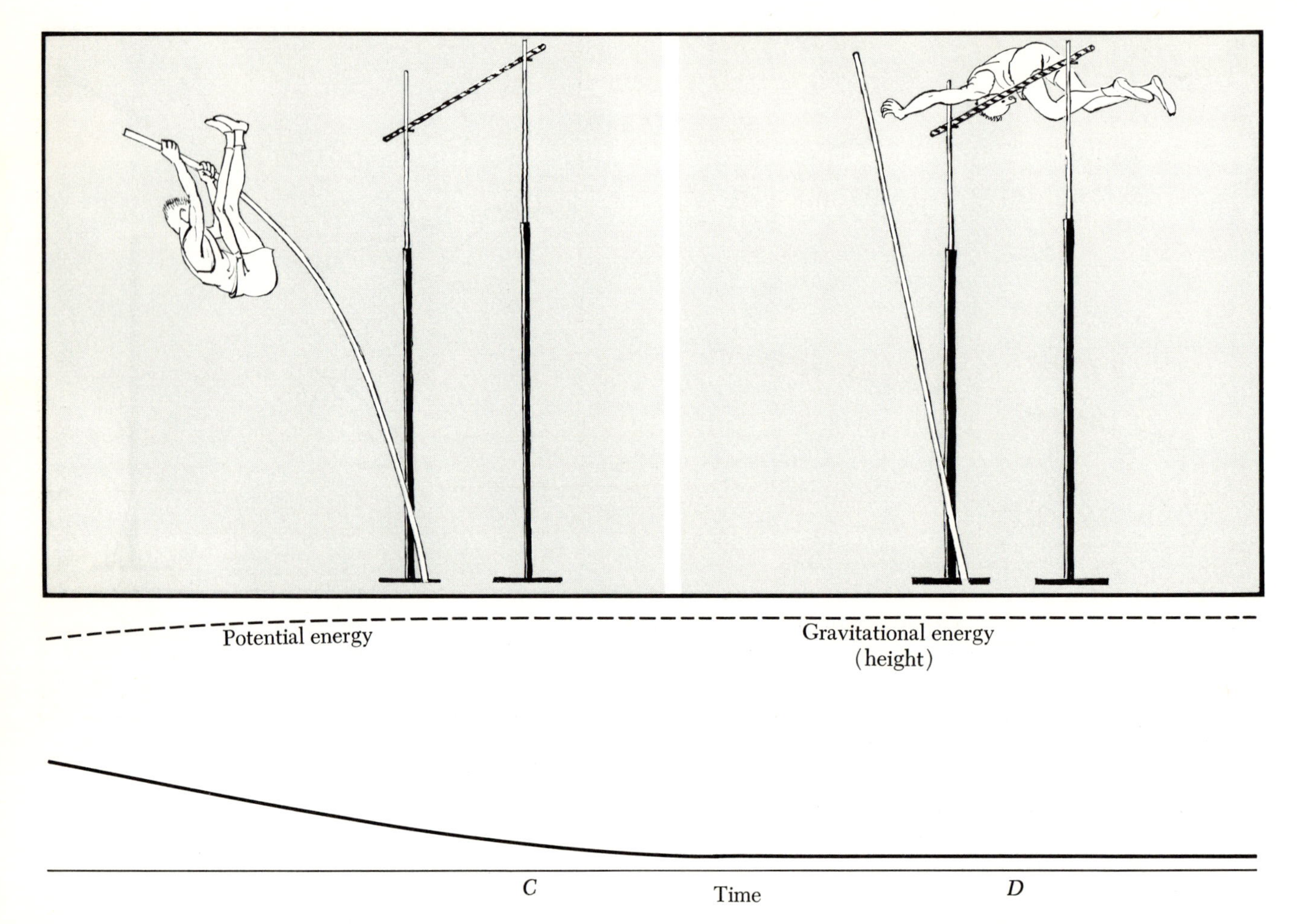

FIG. 5.12 (*cont'd.*) his rotational velocity about the lower end of the pole. He has potential energy both from gravitation and from the remainder of the elastic energy of the pole. At D, going over the bar, his kinetic energy is low because he is moving slowly; his potential energy (gravitational) is high. The total energy is not always constant in the pole vault both because of friction (external and muscular) and because, while bending the pole, the vaulter is doing work. The latter component of work involves "internal" bodily work and energy not accounted for by the man's motion or elevation.

CONSERVATIVE FORCES

A force is *conservative* if the work $W(A \rightarrow B)$ done by the force in moving the particle from A to B is independent of the path by which the particle is moved between A and B. If $W(A \rightarrow B)$ in Eq. (5.15) had a different value in going by one route from that by another route (as might well be the case if friction were present), the importance of Eq. (5.15) would be drastically reduced. Assuming $\int_A^B \mathbf{F} \cdot d\mathbf{r}$ is independent of the path,

$$\int_A^B \mathbf{F} \cdot d\mathbf{r} = -\int_B^A \mathbf{F} \cdot d\mathbf{r}$$

or

$$\int_A^B \mathbf{F} \cdot d\mathbf{r} + \int_B^A \mathbf{F} \cdot d\mathbf{r} = 0 = \oint \mathbf{F} \cdot d\mathbf{r}$$

where $\oint$ means the integral taken around a closed path, for example, starting at A, going to B, and then returning to A possibly by a different route.

We can easily see that a central force is conservative. A *central force* exerted by one particle on another is a force whose magnitude depends only on the separation of the particles and whose direction lies along the line joining the particles. In Fig. 5.13, a central force is directed away from (or toward) the center at the point O. Two paths (labeled 1 and 2) connect points A and B as shown. The dashed curves are sectors of circles centered at O. Consider the quantities $\mathbf{F}_1 \cdot d\mathbf{r}_1$ and $\mathbf{F}_2 \cdot d\mathbf{r}_2$ evaluated on the path segments lying between the dashed circles. (We may regard $\mathbf{F} \cdot d\mathbf{r} = F\,dr \cos\theta$ equally well as the projection of $\mathbf{F}$ on $d\mathbf{r}$ or $d\mathbf{r}$ on $\mathbf{F}$.) Now the magnitudes F_1 and F_2 are equal on the two segments because they lie at equal distances from the point O; the projections $dr\cos\theta$ of the path segments on the respective vectors $\mathbf{F}$ are equal because, as we can see, the separation of the circles measured along the direction of $\mathbf{F}_1$ is equal to the separation measured along $\mathbf{F}_2$. Therefore

$$\mathbf{F}_1 \cdot d\mathbf{r}_1 = \mathbf{F}_2 \cdot d\mathbf{r}_2$$

on the path segments considered. But the identical argument can be employed repeatedly for every comparable path segment, so that

$$\underset{\text{(Path 1)}}{\int_A^B \mathbf{F} \cdot d\mathbf{r}} = \underset{\text{(Path 2)}}{\int_A^B \mathbf{F} \cdot d\mathbf{r}}$$

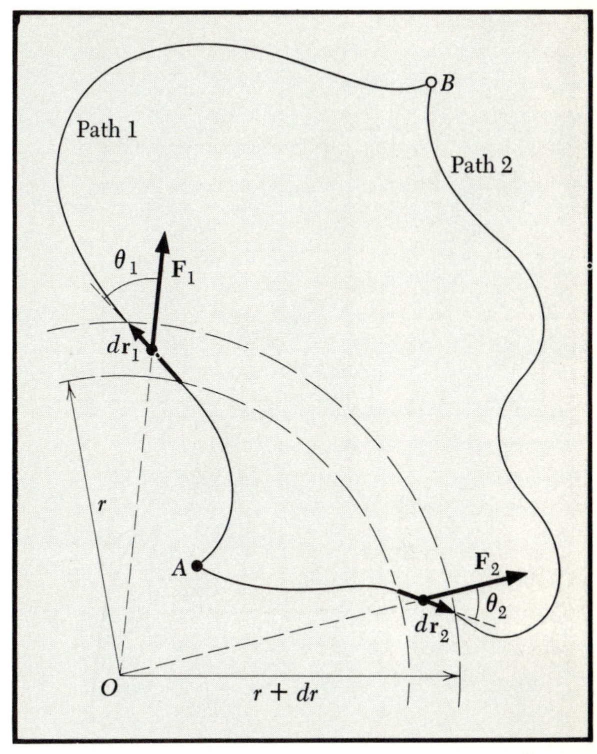

FIG. 5.13 Diagram illustrating the evaluation of $\int_A^B \mathbf{F} \cdot d\mathbf{r}$ for two paths in the case of $\mathbf{F}$ is a central force.

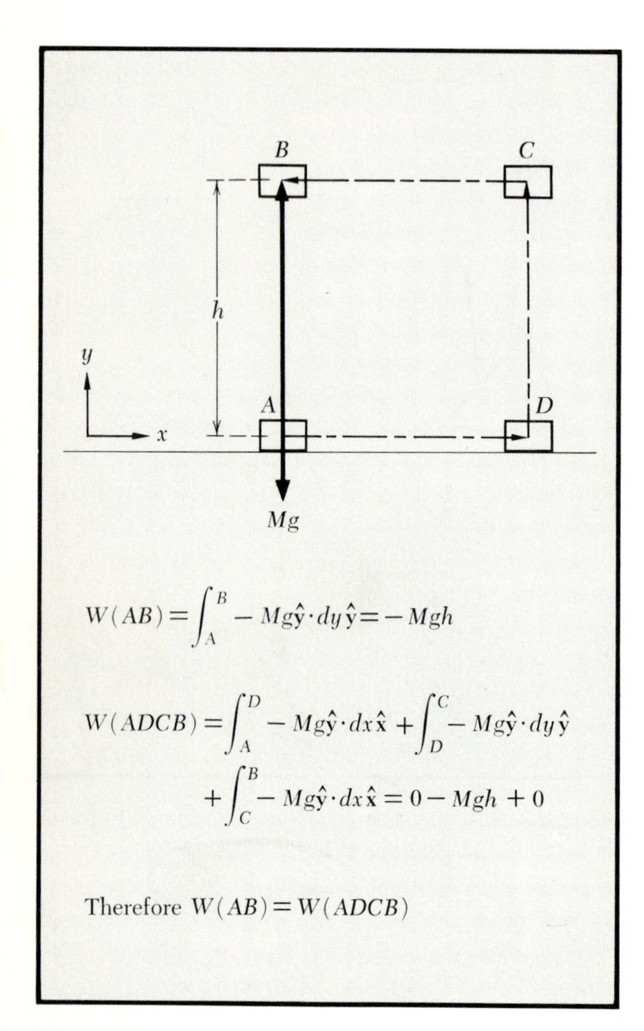

FIG. 5.14 Figure illustrating $\int \mathbf{F} \cdot d\mathbf{r}$ by two different paths for a constant gravitational field.

For the constant gravitational field, the proof is given in Fig. 5.14.

Forces with the property that

$$W(A \to B) = \int_A^B \mathbf{F} \cdot d\mathbf{r} \tag{5.23}$$

is independent of the path are called *conservative* forces. For conservative forces the work done around a closed path is *zero*. Suppose the force depends on the velocity with which the path is traversed. (The force on a charged particle in a magnetic field depends on velocity.) Can such a force be conservative? It turns out that the important fundamental velocity-dependent forces are conservative because their direction is *perpendicular* to the direction of motion of the particle, so that $\mathbf{F} \cdot d\mathbf{r}$ is zero. You can see this for the Lorentz force (Chap. 3), which is proportional to $\mathbf{v} \times \mathbf{B}$. Frictional forces are not really fundamental forces, but they are velocity-dependent and not conservative.

All of our discussion presupposes *two-body forces*. This is an important assumption; it is likely that some of the students in this course will be called upon in their research careers to do battle with many-body forces. A discussion of what is involved in the two-body assumption is given in Volume 2 (Sec. 1.6).

It is known experimentally that $W(A \to B)$ is independent of the paths for gravitational and electrostatic forces. This result for interactions between elementary particles is inferred from scattering experiments; for gravitational forces the result is inferred from the accuracy of the prediction of planetary and lunar motions, as discussed in the Historical Notes at the end of this chapter. We also know that the earth has made about 4×10^9 complete orbits around the sun without any important change in distance to the sun, as judged from geologic evidence on the surface temperature of the earth. The relevant geologic evidence extends back perhaps 10^9 yr and cannot be taken as entirely conclusive because of the numerous factors, including solar output, that affect the temperature, but the observation is suggestive. (Further examples are discussed in the Historical Notes.)

We need to say more about central vs noncentral forces. In consideration of the force between two particles there are two possibilities: the particles have no coordinates other than their positions; one or both particles have a physically distinguished axis. In the first possibility there can *only* be a central force, while in the second the specification that the particle be moved from A to B is incomplete—we have also to specify that the axis be kept in the same direction relative to something. A bar magnet has a physically distinguished axis; if we move the magnet bodily around a closed path in a uniform magnetic field, we may or may not do a net amount of work on the magnet. If the magnet ends up at the same location and in the *same orientation* as it started out, no work is done. If the location is the same but the orientation is different, work will have been done. (The work may have a positive or negative sign.)

It is easy to see that friction is not a conservative force. It is always opposed to the direction of motion, and so the work done by a constant frictional force in a motion from A to B, a distance d, will be $F_{\text{fric}}\,d$; if the motion is from B to A, it will also be $F_{\text{fric}}\,d$. But if friction is a manifestation of fundamental forces and they are conservative, how can friction be nonconservative? This is a matter of the detail of our analysis. If we analyze all motion on the atomic level, that of fundamental forces, we shall find the "motion" conservative; but if we see some of the motion as heat, which is useless in the mechanical sense, we shall consider that friction has acted. The identity of heat and random kinetic energy is treated in Volume 5. In the discussion of conservation of momentum in Chap. 4, we considered an inelastic collision of two particles. Kinetic energy was not conserved; but the sum of the kinetic and internal excitation energy for the two particles was called the *total energy* and was assumed to be conserved, in agreement with all known experiments.

We return now to our discussion of potential energy. The discussion of conservative forces emphasizes the remark (on page 149) that the potential energy at a point can be uniquely and hence usefully defined only in the case of conservative forces. We have seen how to calculate the potential energy from a knowledge of the forces acting in a problem; we choose a zero and then calculate the work we do (or the agent does) in moving the system slowly, without changing the kinetic energy, from the zero to the desired position. Since $\mathbf{F}_{\text{ag}}$ is always exactly equal and opposite to the force $\mathbf{F}$ of the problem, we see that knowing the forces of the problem enables us to calculate the potential energy:

$$\int_{\mathbf{A}}^{\mathbf{r}} \mathbf{F}_{\text{ag}} \cdot d\mathbf{r} = -\int_{\mathbf{A}}^{\mathbf{r}} \mathbf{F} \cdot d\mathbf{r} = U(\mathbf{r}) - U(\mathbf{A}) = U(\mathbf{r}) \qquad (5.24)$$

assuming $U(\mathbf{A}) = 0$.

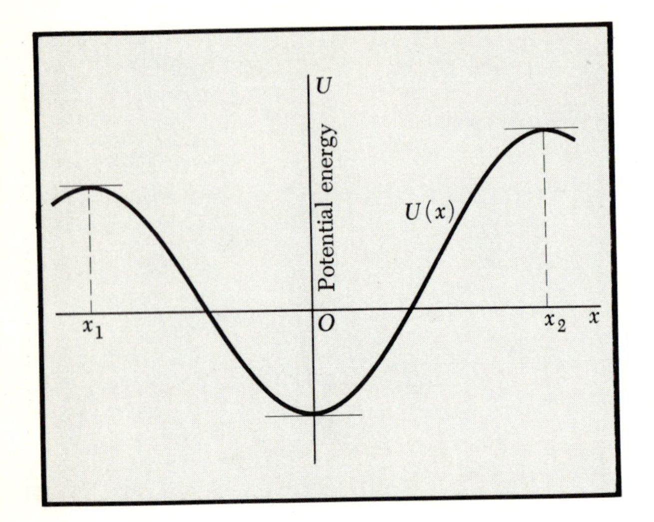

FIG. 5.15 (*a*) A one-dimensional potential energy function $U(x)$ plotted vs x. At points $x = x_1$, O, and x_2 we have $dU/dx = 0$ and thus the force F is zero at these points. These are, therefore, positions of equilibrium, not necessarily stable.

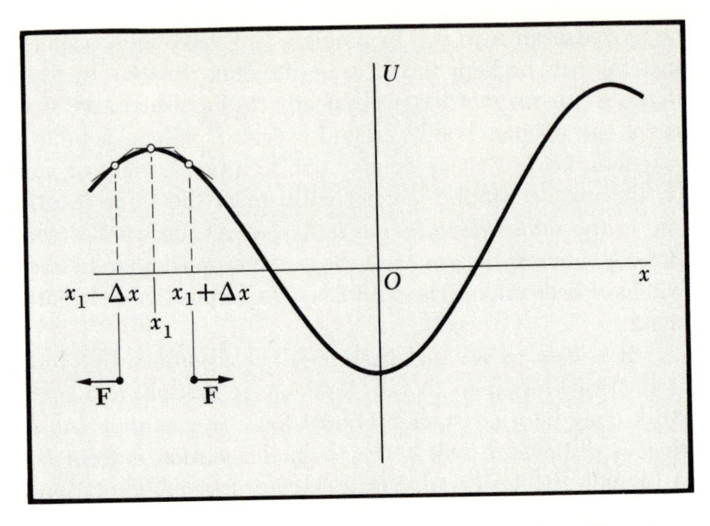

(*b*) At point $x_1 - \Delta x$, $dU/dx > 0$, so that $F < 0$ (to the left). At point $x_1 + \Delta x$, $dU/dx < 0$, so $F > 0$ (to the right). A small displacement from x_1, therefore, results in a force tending to increase the displacement, and so x_1 is a position of *unstable* equilibrium.

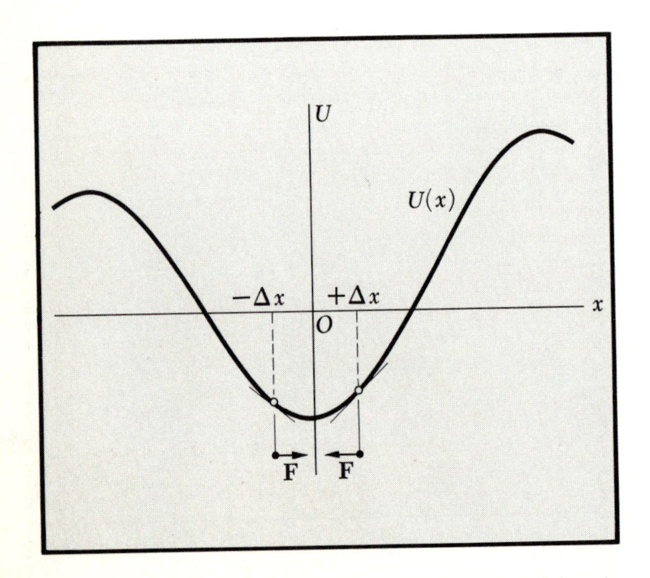

(*c*) At $x = -\Delta x$, $dU/dx < 0$, and F is to the right. At $x = +\Delta x$, $dU/dx > 0$, so that F is to the left. Thus $x = 0$ is a position of *stable* equilibrium. What about x_2?

Does the knowledge of the potential energy enable us to calculate the forces? Yes. In one dimension

$$U(x) - U(A) = -\int_A^x F\,dx \tag{5.25}$$

whence, on differentiation, we have

$$\frac{dU}{dx} = -F \tag{5.26}$$

This result may be checked by substitution of Eq. (5.26) into (5.25):

$$-\int_A^x F\,dx = \int_A^x \frac{dU}{dx}dx = \int_A^x dU = U(x) - U(A) \tag{5.27}$$

Equation (5.27) is an example of the general result that force is the negative of the space rate of change of potential energy. In three dimensions the expression analogous to Eq. (5.26) is[1]

[1]The symbol $\partial/\partial x$ indicates partial differentiation and means that y and z are held constant in the differentiation. The same meaning applies to $\partial/\partial y$ and $\partial/\partial z$. See, for example, page 160.

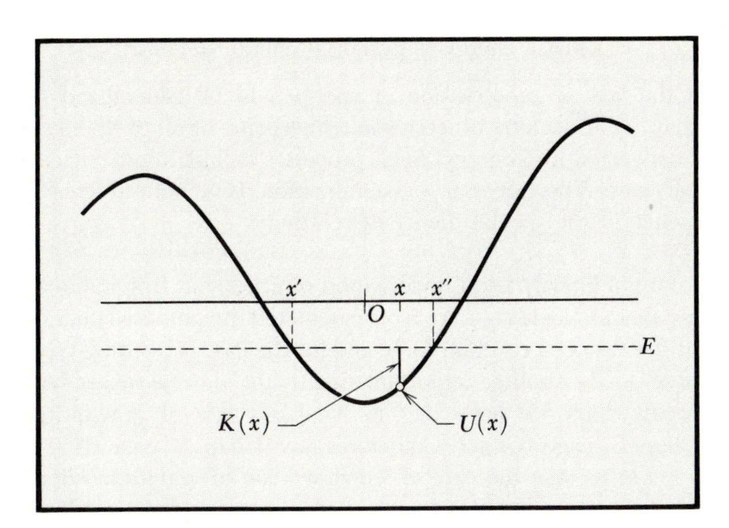

FIG. 5.16 (*a*) The total energy $E = K + U = const$. Thus, given E, the motion can only occur between x' and x'', the "turning points." Between these points $K = Mv^2/2 = E - U \geq 0$.

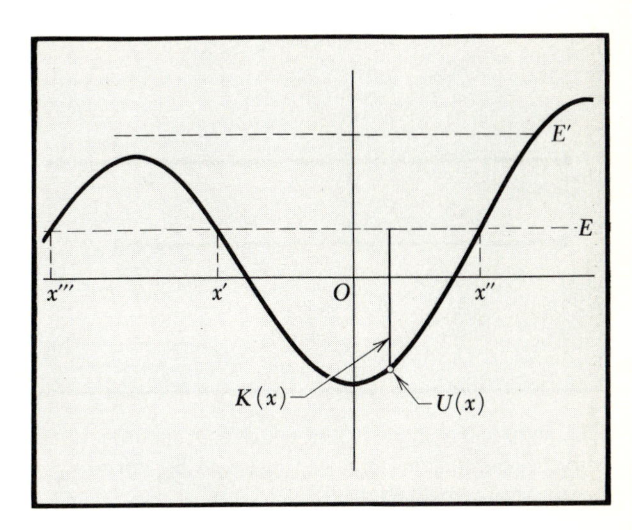

(*b*) If E is increased, the turning points x' and x'' are, in general, changed Now $K(x) = E - U(x)$ is greater. The motion can now also take place to the left of x''', if started at x'''

$$\mathbf{F} = -\hat{\mathbf{x}}\frac{\partial U}{\partial x} - \hat{\mathbf{y}}\frac{\partial U}{\partial y} - \hat{\mathbf{z}}\frac{\partial U}{\partial z} \equiv -\text{grad}\, U \qquad (5.28)$$

where "grad" denotes the gradient operator and is defined by

$$\text{grad} \equiv \hat{\mathbf{x}}\frac{\partial}{\partial x} + \hat{\mathbf{y}}\frac{\partial}{\partial y} + \hat{\mathbf{z}}\frac{\partial}{\partial z} \qquad \text{in cartesian coordinates}$$

$$\text{grad} \equiv \hat{\mathbf{r}}\frac{\partial}{\partial r} + \hat{\boldsymbol{\theta}}\frac{1}{r}\frac{\partial}{\partial \theta} \qquad \text{in plane polar coordinates} \qquad (5.29)$$

The general properties of the gradient operator are considered in Volume 2. It is shown there that the gradient of a scalar is a vector whose direction is that of the maximum spatial rate of increase of the scalar and whose magnitude is equal to the rate of change. The gradient of a scalar U is written variously as grad U or ∇U. The operator ∇ is read as "del," and ∇U is read as "del you."

The application of these ideas to the case $dU/dx = 0$, a position of equilibrium, and to the stability of such equilibrium, is shown in Fig. 5.15*a* to *c*.

Simple graphs of the potential energy U against a coordinate x can often be very informative. Figures 5.16*a* to *c* are examples; they use the fact that the kinetic energy K cannot be negative. What would be the motion if the energy were E' in Fig. 5.16*b*? Our statement

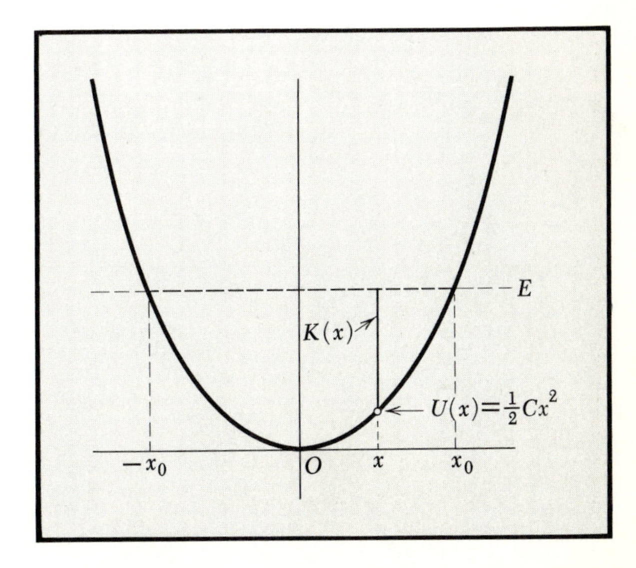

(*c*) The simple harmonic oscillator is in stable equilibrium at $x = 0$. At $x = \pm x_0$, $K = 0$.

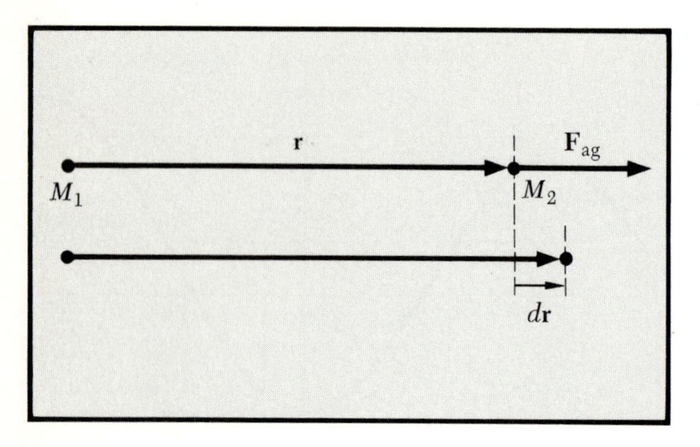

FIG. 5.17 The force $\mathbf{F}_{ag}$ is equal and opposite to the attractive gravitational force and does work $\mathbf{F}_{ag} \cdot d\mathbf{r} = F_{ag}\, dr$ in the displacement $d\mathbf{r}$.

$$\text{Kinetic energy} + \text{potential energy} = \text{const}$$

of the law of conservation of energy will be generalized in Chap. 12 to include processes in which some or all of the mass is converted into energy. Such processes include most nuclear reactions. The necessary generalization is a natural consequence of the special theory of relativity.

Potential Energy and Conservation of Energy in Gravitational and Electric Fields We have calculated the potential energy for the case of a constant force and for the case of a spring-type force $-Cx$. Another important case is the inverse-square-law force, which we have met in Newton's law of gravitation (Chap. 3, page 65) and Coulomb's law (Chap. 3, page 68).

Let us take the case of Newton's law of gravitation first. We have seen that this force is a conservative force, and we shall calculate the work in the easiest way possible. We assume two masses M_1 and M_2 are initially a distance r_A apart and calculate the work *we* do in changing the distance to r. Let M_1 be fixed and let $\mathbf{r}$ be the vector from M_1 to M_2, as in Fig. 5.17. If we move M_2 to a distance $\mathbf{r} + d\mathbf{r}$, as in Fig. 5.17, we must do work

$$dW = \mathbf{F}_{ag} \cdot d\mathbf{r} = \frac{GM_1M_2}{r^3}\mathbf{r} \cdot d\mathbf{r} = \frac{GM_1M_2}{r^2}dr$$

Thus the work done for the entire displacement will be

$$W = \int_{r_A}^{r} \frac{GM_1M_2}{r^2}dr = -\frac{GM_1M_2}{r}\bigg|_{r_A}^{r}$$

$$= -\frac{GM_1M_2}{r} + \frac{GM_1M_2}{r_A} \tag{5.30}$$

We can check that this is of the correct sign, for if $r > r_A$, the work is positive (we put work into the system); if $r < r_A$, the work is negative (we get work done for us).

Let us then apply this to the potential energy. What is a convenient location for zero potential energy? If $U = 0$ at $r = r_A$, the expression Eq. (5.30) will be the value of U:

$$U(r) = -\frac{GM_1M_2}{r} + \frac{GM_1M_2}{r_A}$$

But when we note that $r_A = \infty$ would make the last term vanish so that

$$U(r) = -\frac{GM_1M_2}{r} \tag{5.31}$$

this seems the most convenient choice: $U = 0$ at $r = \infty$. The potential energy then will always be negative since we can always get work from the system by letting the masses come together slowly from infinity.

We can now write the conservation of mechanical energy for a body of mass M_1 moving in the gravitational field of a body of mass M ($M \gg M_1$ so the motion of M can be neglected).

$$E = \frac{1}{2}M_1 v_A^2 - \frac{GM_1M}{r_A} = \frac{1}{2}M_1 v_B^2 - \frac{GM_1M}{r_B} \tag{5.32}$$

where v_A and r_A are the velocity and distance at one time, v_B and r_B at another.

We now return to the electrical case:

$$\mathbf{F} = \frac{q_1 q_2}{r^3}\mathbf{r} = \frac{q_1 q_2}{r^2}\hat{\mathbf{r}}$$

for two charges q_1 and q_2. Let q_1 be fixed. We find the work we do to move q_2 slowly from r_A to r. The force we must exert is

$$F_{\text{ag}} = -\frac{q_1 q_2}{r^3}\mathbf{r}$$

The increment of our work for a displacement $d\mathbf{r}$ is

$$dW = \mathbf{F}_{\text{ag}} \cdot d\mathbf{r} = -\frac{q_1 q_2}{r^3}\mathbf{r} \cdot d\mathbf{r} = -\frac{q_1 q_2}{r^2}dr$$

since $\mathbf{r}$ and $d\mathbf{r}$ are parallel. Then for the total work we obtain

$$W = \int_{r_A}^{r} -\frac{q_1 q_2}{r^2}dr = \frac{q_1 q_2}{r}\bigg|_{r_A}^{r} = \frac{q_1 q_2}{r} - \frac{q_1 q_2}{r_A}$$

Again it is convenient to let $U = 0$ at $r = \infty$, and so

$$\boxed{U = \frac{q_1 q_2}{r}} \tag{5.33}$$

U is positive if q_1 and q_2 are the same sign and negative if they are of opposite signs. We know that this is correct since if they are of the same sign, *we* must do work in pushing them together from infinity. When more than two point charges are present, the total potential energy is the sum of terms like Eq. (5.33) for each particle pair that can be formed in the system of particles.

From the discussion in Chap. 3 (page 68), we can see that U in SI units will be

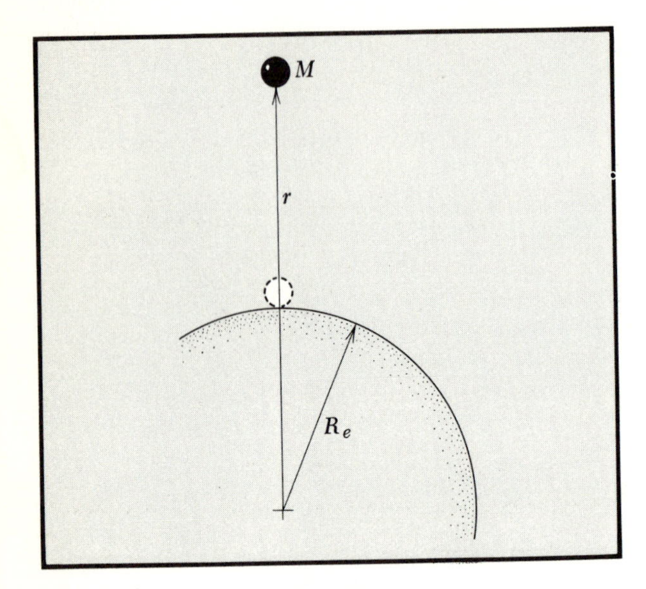

FIG. 5.18 (*a*) Consider the escape velocity required for a mass M to leave the earth's gravitational field starting from the surface.

$$\boxed{U = k\frac{q_1q_2}{r}} \tag{5.33a}$$

We can check that this expression is correct by calculating **F** from it by Eq. (5.29):

$$\mathbf{F} = -\nabla U = -\frac{\partial}{\partial r}\left(\frac{q_1q_2}{r}\right)\hat{\mathbf{r}} = \frac{q_1q_2}{r^2}\hat{\mathbf{r}}$$

and for the gravitational force

$$\mathbf{F} = -\nabla U = -\frac{\partial}{\partial r}\left(-\frac{GM_1M_2}{r}\right)\hat{\mathbf{r}} = -\frac{GM_1M_2}{r^2}\hat{\mathbf{r}}$$

If we write

$$r^2 = x^2 + y^2 + z^2$$

we can calculate F_x, F_y, and F_z:

$$F_x = -\frac{\partial}{\partial x}\left[-\frac{GM_1M_2}{(x^2+y^2+z^2)^{\frac{1}{2}}}\right]$$
$$= -\frac{GM_1M_2x}{(x^2+y^2+z^2)^{\frac{3}{2}}} = -\frac{GM_1M_2x}{r^3}$$

or similarly for the electric field:

$$F_x = -\frac{\partial}{\partial x}\frac{q_1q_2}{(x^2+y^2+z^2)^{\frac{1}{2}}} = \frac{q_1q_2x}{(x^2+y^2+z^2)^{\frac{3}{2}}} = \frac{q_1q_2x}{r^3}$$

The *electrostatic potential* $\Phi(\mathbf{r})$ at $\mathbf{r}$ is defined as the *potential energy per unit positive charge* in the field of force of all the other charges:

$$\Phi(\mathbf{r}) = \frac{U(\mathbf{r})}{q} = \int_{\mathbf{r}}^{\infty} \mathbf{E}(\mathbf{r})\cdot d\mathbf{r} \tag{5.34}$$

This is a very useful quantity. Notice that it is a scalar. It is most important to distinguish Φ from the potential energy U. Beware also of the use in experimental work of the symbol V for both quantities, electrostatic potential and potential energy.

If we know $\mathbf{E}(\mathbf{r})$ everywhere, then we can find the electrostatic potential $\Phi(\mathbf{r})$ everywhere. [This assumes we decide on a zero for $\Phi(\mathbf{r})$.] It is convenient to work with $\Phi(\mathbf{r})$ because it is a scalar, whereas $\mathbf{E}(\mathbf{r})$ is a vector.

The voltage drop or *potential difference* PD between two points $\mathbf{r}_2$ and $\mathbf{r}_1$ is defined as

$$\text{PD} = \Phi(\mathbf{r}_2) - \Phi(\mathbf{r}_1) \tag{5.35}$$

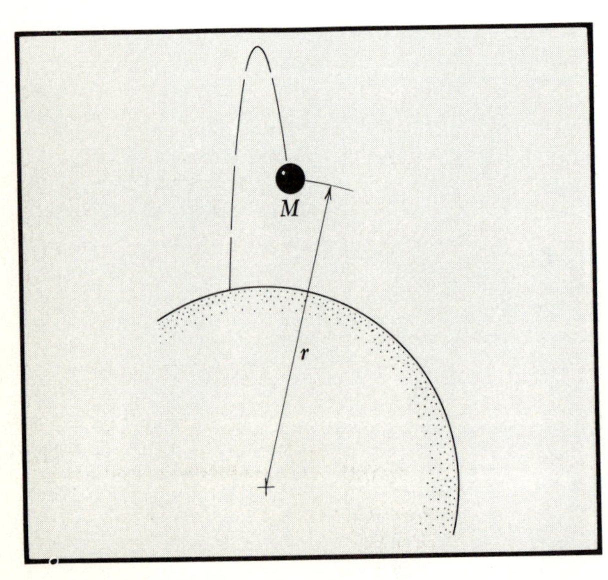

(*b*) Trajectory if the kinetic energy is too low for escape.

This is the change in the electrostatic potential energy of a unit positive charge when taken from $\mathbf{r}_1$ to $\mathbf{r}_2$. Thus, for a charge q taken between these points, the potential energy difference is

$$U(\mathbf{r}_2) - U(\mathbf{r}_1) = q[\Phi(\mathbf{r}_2) - \Phi(\mathbf{r}_1)]$$

The unit of electrostatic potential, or potential difference, in the gaussian cgs system is the *statvolt*. We saw in Chap. 3 that the unit of electric field intensity is called the *statvolt per centimeter* (statvolt/cm); but Φ differs from $\mathbf{E}$ in dimensions by a length, and thus Φ is measured in statvolts. It is also true that Φ has the dimensions of [charge]/[length], so that the statcoulomb/cm is a possible name for the unit of potential.

The *practical* unit of electrostatic potential or potential difference is the *volt* (V). It is also the SI unit. The volt is used in everyday life and widely in the laboratory. The volt is *defined* so that

$$\frac{c}{10^8} \times \text{potential difference in statvolts} = \text{potential difference in volts}$$

where c is the speed of light in cm/s. Or, approximately,

$$300 \times \text{potential difference in statvolts} \approx \text{potential difference in volts}$$

The electric field $\mathbf{E}$ in SI units is measured in volts per meter (V/m); but it is *not* true that Φ can be given in coulombs per meter (C/m). However *joule per coulomb* (J/C) is another name for volt, just as *ergs/statcoulomb* is another name for *statvolt*.

We now give several examples of problems dealing with potential energy and potential, some of which involve the central force interaction of the gravitational and electrical type.

EXAMPLE

Escape Velocity from the Earth and from the Solar System Calculate the initial velocity needed for a particle of mass M to escape (1) from the earth and (2) from the solar system. (Neglect the rotation of the earth.)

Figure 5.18*a* to *f* illustrates the meaning and also the use of potential energy diagrams for this kind of situation. Using Eq. (5.32), we write the total energy E of a particle of mass M at a distance R_e, the radius of the earth, from the center of the earth:

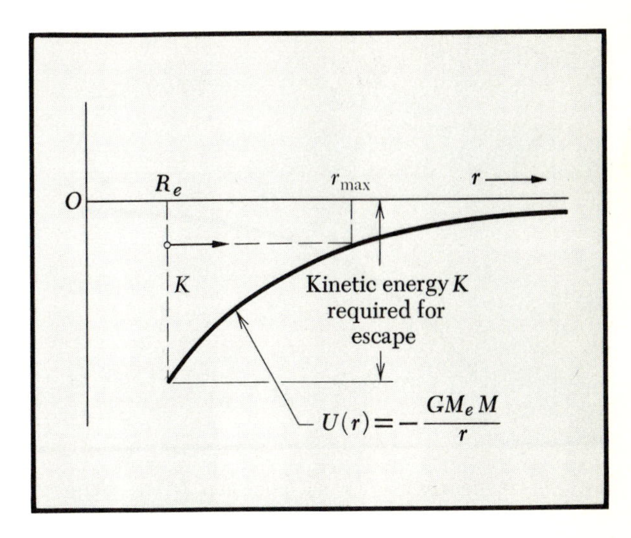

(*c*) Here we represent the initial kinetic energy as being too small for escape. To reach infinity, $K \geq |-U|$.

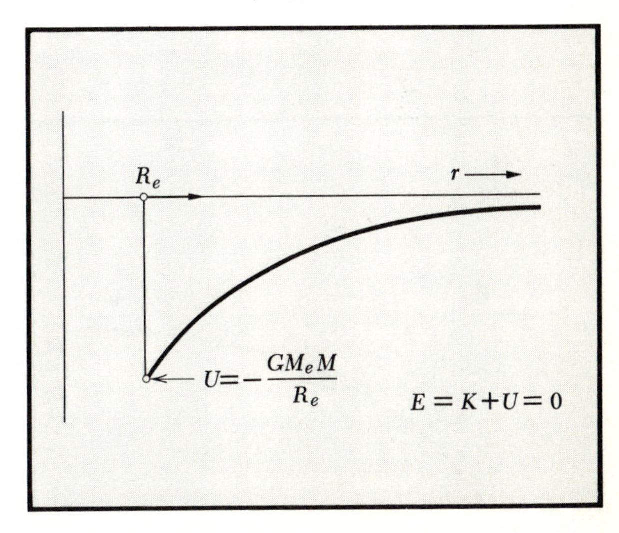

(*d*) Here we see M launched from the earth's surface (radius R_e) with the minimum necessary kinetic energy $K = \frac{1}{2}Mv_e^2 = GM_eM/R_e$. The escape velocity from the earth is denoted by v_e.

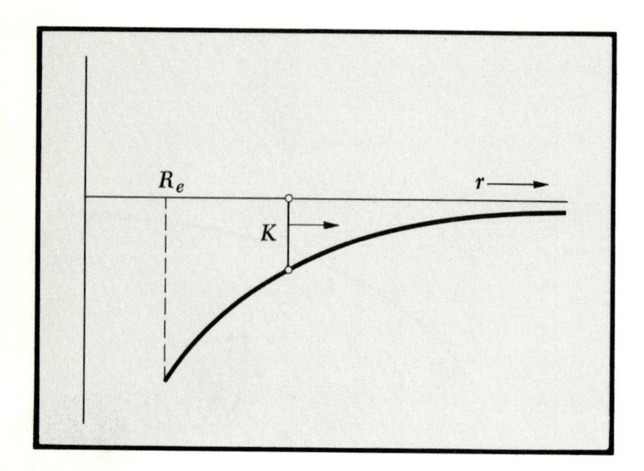

FIG. 5.18 (*cont'd.*) (*e*) Some time later U has increased and K has decreased as M goes further from the center of the earth.

$$E = \tfrac{1}{2}Mv^2 - \frac{GM_eM}{R_e}$$

where

$M_e = 5.98 \times 10^{27}$ g $\qquad R_e = 6.4 \times 10^8$ cm $\qquad G = 6.67 \times 10^{-8}$

or

$M_e = 5.98 \times 10^{24}$ kg $\qquad R_e = 6.4 \times 10^6$ m $\qquad G = 6.67 \times 10^{-11}$

(from Chap. 3, pages 65–67).

To reach an infinite distance from the earth with the least possible (that is, zero) velocity, the total energy must be zero because the kinetic energy is zero and the gravitational potential is also zero. This follows because $U(r) \to 0$ as $r \to \infty$. Thus E must be zero if the total energy of the particle is constant between launching and escape; whence the escape velocity v_e is given by

$$\tfrac{1}{2}Mv_e^{\,2} = \frac{GM_eM}{R_e} \qquad v_e = \sqrt{\frac{2GM_e}{R_e}} \tag{5.36}$$

The acceleration g of gravity at the surface of the earth is $GM_e/R_e^{\,2}$, so that

$$v_e = \sqrt{2gR_e} \approx (2 \times 10^3 \times 6 \times 10^8)^{\frac{1}{2}} \approx 10^6 \text{ cm/s}$$

or in SI units

$$\sqrt{(2 \times 10)(6 \times 10^6)} \approx 10^4 \text{ m/s}$$

To escape from the pull of the sun alone a particle launched from the earth (at a distance R_{es} from the sun) will need an escape velocity

$$v_s = \sqrt{\frac{2GM_s}{R_{es}}} \approx \left[\frac{2 \times (7 \times 10^{-8}) \times (2 \times 10^{33})}{1.5 \times 10^{13}}\right]^{\frac{1}{2}}$$
$$\approx 4 \times 10^6 \text{ cm/s}$$

using the ratio $M_s/M_e = 3.3 \times 10^5$ and the value $R_{es} = 1.5 \times 10^{13}$ cm. For bodies launched from earth, escape from the solar system is more difficult than escape from the earth.

EXAMPLE

Gravitational Potential near the Surface of the Earth The gravitational potential energy of a body of mass M at a distance r from the center of the earth is, for $r > R_e$,

$$U(r) = -\frac{GMM_e}{r}$$

where M_e is the mass of the earth. If R_e is the radius of the earth and y is the height above the surface of the earth, we wish to show that

$$U \approx -MgR_e + Mgy \tag{5.37}$$

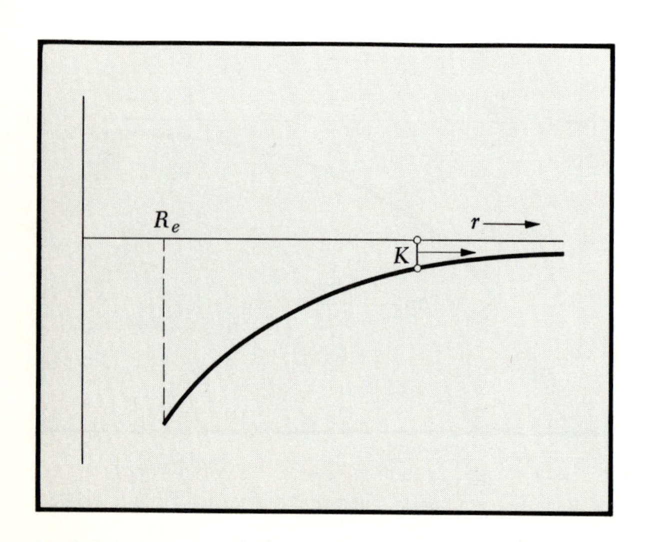

(*f*) Still later, K and $|U|$ have decreased further.

for $y/R_e \ll 1$. Here $g = GM_e/R_e^2 \approx 980\ \text{cm/s}^2$. It is shown as follows. We have

$$U = -GMM_e \frac{1}{(R_e + y)}$$

from above with $r = R_e + y$. Divide numerator and denominator by R_e:

$$U = -\frac{GMM_e}{R_e} \times \frac{1}{(1 + y/R_e)}$$

We can now use the expansion equation (2.49) (Dwight, 9.04) to write ($n = -1$):

$$U = -\frac{GMM_e}{R_e}\left(1 - \frac{y}{R_e} + \frac{y^2}{R_e^2} - \cdots\right)$$

Let $g = GM_e/R_e^2$; then

$$U = -MgR_e\left(1 - \frac{y}{R_e} + \frac{y^2}{R_e^2} - \cdots\right)$$

which reduces to Eq. (5.37) for $y \ll R_e$ and to Eq. (5.8), except for the constant $-MgR_e$.

EXAMPLE

Projectile Motion We give here one more example of two-dimensional motion in a constant gravitational field. Using Newton's Second Law, we have already solved the problem. Let the force be $\mathbf{F}_G = -Mg\hat{\mathbf{y}}$, where g is approximately $980\ \text{cm/s}^2$.

(1) Calculate the work done by gravity when a mass of 100 g moves from the origin to

$$\mathbf{r} = 50\hat{\mathbf{x}} + 50\hat{\mathbf{y}}$$

$$W = \int_{0,0}^{50,50} \mathbf{F}_G \cdot d\mathbf{r} = -Mg\hat{\mathbf{y}} \cdot (50\hat{\mathbf{x}} + 50\hat{\mathbf{y}}) = -100 \times 980 \times 50$$

$$= -4.9 \times 10^6\ \text{ergs}$$

→ düşey boyunca değişim gerekli olduğundan 50'yi kullanıyoruz.

The gravitational force does a negative amount of work; or work is done *against* it by some other means.

(2) What is the change in potential energy in this displacement?

$$F_{\text{ag}} = +Mg\hat{\mathbf{y}}$$

and so

$$\Delta U = -W = +4.9 \times 10^6\ \text{ergs}$$

so that the potential energy increases by 4.9×10^6 ergs; and we see that if $U = 0$ when $x = 0$ and $y = 0$,

$$U = Mgy$$

(3) If a particle of mass M is projected from the origin with

speed v_0 at angle θ with the horizontal, how high will it rise? Here we must use the fact that v_x does not change.

$$E = \tfrac{1}{2}Mv_0^2 = \tfrac{1}{2}M(v_x^2 + v_y^2) = \tfrac{1}{2}M(v_0^2 \cos^2\theta + v_0^2 \sin^2\theta)$$
$$= \tfrac{1}{2}Mv_0^2 \cos^2\theta + Mgy_{\max}$$
$$y_{\max} = \frac{v_0^2 \sin^2\theta}{2g}$$

which can be derived from Eq. (3.9).

EXAMPLES

Electrostatic Field What is the magnitude of the electric field at a distance of 1 Å ($=10^{-8}$ cm) from a proton?

From Coulomb's law

$$E = \frac{e}{r^2} \approx \frac{5 \times 10^{-10}\ \text{statcoulomb}}{(1 \times 10^{-8}\ \text{cm})^2} \approx 5 \times 10^6\ \text{statvolts/cm}$$
$$\approx (300)(5 \times 10^6)\ \text{V/cm} \approx 1.5 \times 10^9\ \text{V/cm}$$

In SI units

$$E = \frac{ke}{r^2} = \frac{(9 \times 10^9)(1.6 \times 10^{-19})}{(1 \times 10^{-10})^2} \approx 1.5 \times 10^{11}\ \text{V/m}$$

The field is directed radially outward from the proton.

Potential What is the potential at this point? From Eqs. (5.33) and (5.34) we have

$$\Phi(r) = \frac{1}{q}\frac{qe}{r} = \frac{e}{r} \approx \frac{5 \times 10^{-10}\ \text{statcoulomb}}{1 \times 10^{-8}\ \text{cm}}$$
$$\approx 5 \times 10^{-2}\ \text{statvolts} \approx 15\ \text{V}$$

from the conversion factor given above. In SI units

$$\Phi(r) = k\frac{e}{r} = (9 \times 10^9)\frac{1.6 \times 10^{-19}\ \text{C}}{1 \times 10^{-10}\ \text{m}} \approx 15\ \text{V}$$

Potential Difference What is the potential difference in volts between positions 1 and 0.2 Å from a proton?

The potential at 1×10^{-8} cm is 15 V; at 0.2×10^{-8} cm it is 75 V. The difference $75 - 15 = 60$ V, or $\frac{60}{300} \approx 0.2$ statvolt.

Charged Particle Energy Derived from Potential Difference A proton is released from rest at a distance of 1 Å from another proton. What is the kinetic energy when the protons have moved infinitely far apart?

By conservation of energy we know that the kinetic energy must equal the original potential energy, which is

$$\frac{e^2}{r} \approx \frac{(4.8 \times 10^{-10}\ \text{statcoulomb})^2}{1 \times 10^{-8}\ \text{cm}} \approx 23 \times 10^{-12}\ \text{erg}$$

If one proton is kept at rest while the other moves, the terminal velocity of the moving proton is given by (using conservation of energy)

$$\tfrac{1}{2}Mv^2 \approx 23 \times 10^{-12}\ \text{erg}$$

$$v^2 \approx \frac{2 \times 23 \times 10^{-12}\ \text{erg}}{1.67 \times 10^{-24}\ \text{g}} \approx 27 \times 10^{12}\ (\text{cm/s})^2$$

or

$$v \approx 5 \times 10^6\ \text{cm/s}$$

Using SI units,

$$U = (9 \times 10^9)\frac{(1.6 \times 10^{-19})^2}{10^{-10}} \approx 23 \times 10^{-19}\ \text{J}$$

$$v^2 \approx \frac{2 \times 23 \times 10^{-19}}{1.67 \times 10^{-27}} \approx 27 \times 10^8\ \text{m}^2/\text{s}^2$$

$$v \approx 5 \times 10^4\ \text{m/s}$$

If both protons are free to move, each proton will have the same kinetic energy when they are widely separated, and so

$$\tfrac{1}{2}Mv_1^2 + \tfrac{1}{2}Mv_2^2 = Mv^2 \approx 23 \times 10^{-12}\ \text{erg}$$

and

$$v \approx \frac{5 \times 10^6\ \text{cm/s}}{\sqrt{2}} \approx 3.5 \times 10^6\ \text{cm/s}$$

Proton Acceleration in a Uniform Electric Field A proton is accelerated from rest by a uniform electric field. The proton moves through a potential drop of 100 V. What is its final kinetic energy? (Note that 100 V $\approx$ 0.33 statvolt.)

The kinetic energy will be equal to the change in potential energy, which is $e\Delta\Phi$, or

$$(4.8 \times 10^{-10}\ \text{statcoulomb})(0.33\ \text{statvolt}) \approx 1.6 \times 10^{-10}\ \text{erg}$$

or in SI units

$$(1.6 \times 10^{-19})(100) \approx 1.6 \times 10^{-17}\ \text{J}$$

EXAMPLE

Electron Volts A convenient unit of energy in atomic and nuclear physics is the *electron volt* (eV), defined as the potential energy difference of a charge e between two points having a potential difference of one volt, or as the kinetic energy gained by a charge e in falling through a potential difference of one volt. Thus

$$\begin{aligned} 1\ \text{eV} &\approx (4.80 \times 10^{-10}\ \text{statcoulomb})(\tfrac{1}{300}\ \text{statvolt}) \\ &= 1.60 \times 10^{-12}\ \text{erg} \\ &= 1.6 \times 10^{-19}\ \text{C} \times 1\ \text{Volt} \\ &= 1.6 \times 10^{-19}\ \text{J} \end{aligned}$$

An alpha particle (He^4 nucleus or doubly ionized helium atom) accelerated from rest through a potential difference of 1000 V has a kinetic energy equal to

$$2e \times 1000 \text{ V} = 2000 \text{ eV}$$

where

$$2000 \text{ eV} = (2 \times 10^3)(1.60 \times 10^{-12}) = 3.2 \times 10^{-9} \text{ erg}$$

We have seen that the difference $K_B - K_A$ in kinetic energy of a particle between two points has the property that

$$K_B - K_A = \int_A^B \mathbf{F} \cdot d\mathbf{r}$$

where $\mathbf{F}$ is the force acting on the particle. But we know from Eq. (5.25) that

$$U_B - U_A = -\int_A^B \mathbf{F} \cdot d\mathbf{r}$$

so that on adding these two equations we have

$$(K_B + U_B) - (K_A + U_A) = 0 \tag{5.38}$$

Thus *the sum of the kinetic and potential energies is a constant, independent of time.* Rewriting Eq. (5.38), we have for a one-particle system the energy function

$$\boxed{E = \tfrac{1}{2}Mv^2(A) + U(A) = \tfrac{1}{2}Mv^2(B) + U(B)} \tag{5.39}$$

where E is a constant called the *energy,* or *total energy,* of the system. Equation (5.32) is just this equation for the case of gravitational potential energy.

Let us write down the generalization of Eq. (5.39) to a two-particle system in the field of an external potential:

$$\begin{aligned} E &= K + U \\ &= \tfrac{1}{2}M_1v_1^2 + \tfrac{1}{2}M_2v_2^2 + U_1(\mathbf{r}_1) + U_2(\mathbf{r}_2) + U(\mathbf{r}_1 - \mathbf{r}_2) \\ &= \text{const} \end{aligned} \tag{5.40}$$

The first term is the kinetic energy of particle 1; the second term is the kinetic energy of particle 2; the third and fourth terms are the potential energies of particles 1 and 2 due to an external potential; the fifth term is the potential energy due to the interaction between particles 1 and 2. Notice that $U(\mathbf{r}_1 - \mathbf{r}_2)$ is put in only once: If two particles interact, the interaction energy is mutual!

If particles 1 and 2 are protons in the earth's gravitational field, the energy E in Eq. (5.40) is

$$E = \tfrac{1}{2}M(v_1^2 + v_2^2) + Mg(y_1 + y_2) - \frac{GM^2}{r_{12}} + \frac{e^2}{r_{12}}$$

where y is measured upward and $r_{12} = |\mathbf{r}_2 - \mathbf{r}_1|$. The last term is the coulomb energy of the two protons; the next-to-last term is their mutual gravitational energy. The ratio of the last two terms is

$$\frac{GM^2}{e^2} \approx \frac{10^{-7} \times 10^{-48}}{10^{-19}} \approx 10^{-36}$$

showing, since the forces depend on the distance in the same way, that the gravitational force between protons is extremely weak in comparison with the electrostatic force. In SI units, we have

$$\frac{GM^2}{ke^2} \approx \frac{10^{-10} \times 10^{-54}}{10^{10} \times 10^{-38}} \approx 10^{-36}$$

POWER

The *power* P is the time rate of transfer of energy. We have defined the work done on the particle in a displacement $\Delta\mathbf{r}$ by an applied force as

$$\Delta W = \mathbf{F} \cdot \Delta\mathbf{r}$$

The rate at which work is done by the force is

$$\frac{\Delta W}{\Delta t} = \mathbf{F} \cdot \frac{\Delta\mathbf{r}}{\Delta t}$$

In the limit $\Delta t \to 0$ we have the power

$$\boxed{P = \frac{dW}{dt} = \mathbf{F} \cdot \frac{d\mathbf{r}}{dt} = \mathbf{F} \cdot \mathbf{v}} \tag{5.41}$$

From the power $P(t)$ as a function of time we can write the work input as

$$W(t_1 \to t_2) = \int_{t_1}^{t_2} P(t)\,dt$$

In the cgs system *the unit of power is one erg per second.* In SI units the unit of power is one joule per second (1 J/s), which is called one *watt* (1 W). To find the power in erg/s, multiply the power expressed in watts by 10^7. To obtain the power in watts from the value expressed in horsepower, multiply by 746, approximately.

PROBLEMS

1. *Potential and kinetic energy—falling body*

(*a*) What is the potential energy of a mass of 1 kg at a height of 1 km above the earth? Express the answer in ergs and in joules, and refer the potential energy to the surface of the earth. *Ans.* 9.8×10^{10} ergs = 9800 J.

(*b*) What is the kinetic energy just as it touches the earth of a mass of 1 kg that is released from a height of 1 km? Neglect friction. *Ans.* 9.8×10^{10} ergs.

(*c*) What is the kinetic energy of the same mass when it has fallen halfway?

(*d*) What is the potential energy when it has fallen halfway? The sum of (*c*) and (*d*) should equal (*a*) or (*b*). Why?

2. *Potential energy above earth*

(*a*) What is the potential energy $U(R_e)$ of a mass of 1 kg on the surface of the earth referred to zero potential energy at infinite distance? [Note that $U(R_e)$ is negative.] *Ans.* -6.25×10^{14} ergs.

(*b*) What is the potential energy of a mass of 1 kg at a distance of 10^5 km from the center of the earth referred to zero potential energy at infinite distance? *Ans.* -3.98×10^{13} ergs.

(*c*) What is the work needed to move the mass from the surface of the earth to a point 10^5 km from the center of the earth?

3. *Electrostatic potential energy*

(*a*) What is the electrostatic potential energy of an electron and a proton at a separation of $1\ \text{Å} \equiv 10^{-8}$ cm referred to zero potential energy at infinite separation? If charge is expressed in esu, the result will be in ergs. *Ans.* -2.3×10^{-11} erg.

(*b*) What is the electrostatic potential energy of two protons at the same separation? (Pay special attention to the sign of the answer.)

4. *Satellite in circular orbit*

(*a*) What is the centrifugal force on a satellite moving in a circular orbit about the earth at a distance r from the center of the earth? The velocity of the satellite relative to the center of the earth is v, and the mass is M.

(*b*) Equate the centrifugal force in (*a*) to the gravitational force (M is in equilibrium in the rotating frame).

(*c*) Express v in terms of r, G, and M_e.

(*d*) What is the ratio of the kinetic energy to the potential energy assuming $U = 0$ at $r = \infty$?

5. *Moon—kinetic energy.* What is the kinetic energy of the moon relative to the earth? The relevant data are given in the table of constants inside the cover of this volume.

6. *Anharmonic spring.* A peculiar spring has the force law $F = -Dx^3$.

(*a*) What is the potential energy at x referred to $U = 0$ at $x = 0$? *Ans.* $\frac{1}{4}Dx^4$.

(*b*) How much work is done on the spring in stretching it slowly from 0 to x?

7. *Gravitational potential energy*

(*a*) What is the potential energy relative to the surface of the earth of a 1.0-kg shell on the edge of a cliff 500 m high?

(*b*) If the shell is projected from the cliff with a speed of 9.0×10^3 cm/s, what will be its speed when it strikes the ground? Does the angle of projection affect the answer?

8. *Atwood's machine.* An Atwood's machine was described in Chap. 3 (page 85).

(*a*) Use the equation of conservation of energy to find the velocities of the two masses when m_2 has descended a distance y after starting from rest.

(*b*) From this expression for the velocity find the acceleration. Compare with the result of Eq. (3.40).

9. *Electron in bound orbit about proton.* Suppose that an electron moves in a circular orbit about a proton at a distance of 2×10^{-8} cm. Consider the proton to be at rest.

(*a*) Solve for the velocity of the electron by equating the centrifugal and electrostatic forces.

(*b*) What is the kinetic energy? Potential energy? Give values both in ergs and in electron volts. *Ans.* $K = 5.8 \times 10^{-12}$ erg = 3.6 eV; $U = 11.5 \times 10^{-12}$ erg = -7.2 eV.

(*c*) How much energy is needed to ionize the system, that is, to remove the electron to infinite distance with no final kinetic energy? (Pay careful attention to the various signs.)

10. *Spring paradox.* What is wrong with the following argument? Consider a mass m held at rest at $y = 0$, the end of an unstretched spring hanging vertically. The mass is now attached to the spring, which will be stretched because of the gravitational force mg on the mass. When the mass has lost gravitational potential energy mgy and the spring has gained the same amount of potential energy so that

$$mgy = \tfrac{1}{2}Cy^2$$

the mass will come to equilibrium. Therefore the position of equilibrium is given by

$$y = \frac{2mg}{C}$$

11. *Escape velocity from the moon.* Using $R_M = 1.7 \times 10^8$ cm and $M_M = 7.3 \times 10^{25}$ g, find:

(*a*) The gravitational acceleration at the surface of the moon

(*b*) The escape velocity from the moon

12. *Potential energy of pair of springs.* Two springs each of natural length a and spring constant C are fixed at points $(-a,0)$ and $(+a,0)$ and connected together at the other ends. In the following assume that either may expand or contract in length without buckling (see Fig. 5.19).

(*a*) Show that the potential energy of the system, for a displacement to (x,y) of the joined ends, is

$$U = \frac{C}{2}\{[(x+a)^2 + y^2]^{\frac{1}{2}} - a\}^2$$

$$+ \frac{C}{2}\{[(a-x)^2 + y^2]^{\frac{1}{2}} - a\}^2$$

(*b*) The potential energy depends on both x and y, and we must therefore use partial differentiation to evaluate the relevant forces. Remember that the partial derivative of a function $f(x,y)$ is taken by the usual rules of differentiation according to

$$\frac{\partial f(x,y)}{\partial x} = \frac{d}{dx} f(x;\ y = \text{const})$$

$$\frac{\partial f(x,y)}{\partial y} = \frac{d}{dy} f(x = \text{const};\ y)$$

Find the force component F_x and show that $F_x = 0$ for $\mathbf{r} = 0$.

(*c*) Find F_y for $x = 0$. Check the signs carefully to make sure the answer makes sense.

(*d*) Sketch a graph of potential energy as a function of $\mathbf{r}$ in the xy plane, and find the equilibrium position.

13. *Loop the loop.* A mass m slides down a frictionless track and from the bottom rises up to travel in a vertical circle of radius R. Find the height from which it must be started from rest in order just to traverse the complete circle without falling off under the force of gravity. *Hint:* What must be the force exerted by the track at the highest point?

14. *Time-of-flight mass spectrometer.* The operation of a time-of-flight mass spectrometer is based on the fact that the angular frequency of helical motion in a uniform magnetic field is independent of the initial velocity of the ion. In practice, the device produces a short pulse of ions and measures electronically the time of flight for one or more revolutions of the ions in the pulse.

(*a*) Show that the time of flight for N revolutions is approximately, for ions of charge e,

$$t \approx 650 \frac{NM}{B}$$

where t is in *microseconds,* M in *atomic mass units,* and B in *gauss.* 1 amu $= 1.66 \times 10^{-24}$g.

(*b*) Show that the gyroradius is approximately

$$R \approx \frac{144\sqrt{VM}}{B} \text{cm}$$

where V is the ion energy in electron volts.

(*c*) Given a magnetic field of 1000 G, calculate the time of flight for 6 revolutions of singly ionized potassium K^{39}.

Ans. 152 μs.

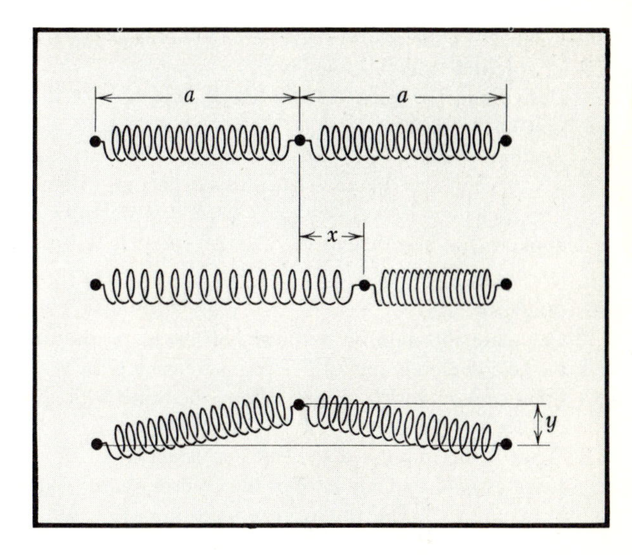

FIG. 5.19

15. *Electron beam in oscilloscope.* Electrons in an oscilloscope tube are accelerated from rest through a potential difference Φ_a and pass between two electrostatic deflection plates. The plates, which have a length l and a separation d, sustain a potential difference Φ_b with respect to each other. The screen of the tube is located at a distance L from the center

of the plates. Use the relation $e\,\Delta\Phi = \frac{1}{2}mv^2$ between the accelerating potential and the velocity v.

(*a*) Derive an expression for the linear deflection D of the spot on the screen.

(*b*) Assume that $\Phi_a = 400$ V; $\Phi_b = 10$ V; $l = 2$ cm; $d = 0.5$ cm; $L = 15$ cm; what is this deflection? (Remember to change volts to statvolts: 300 V = 1 statvolt.) The apparatus is like that in Fig. 3.6, except that the plates are closer together.

16. *Impulse*

(*a*) Calculate the impulse that one ball exerts on the other in the completely inelastic head-on collision of two 500-g balls, each of which is approaching the other with speed 100 cm/s.

(*b*) What is the impulse if the collision is elastic?

(*c*) If we assume that the time of the collision in (*a*) and (*b*) is 1.0×10^{-3} s, find the average force in each case:

$$F_{av} = \frac{\int_0^t F\,dt}{\int_0^t dt}$$

17. *Power.* A moving belt is used to carry sand from one point to another. The sand falls from rest in a hopper onto the belt moving horizontally at speed v. Neglecting friction and what happens at the other end of the belt, find the power necessary to keep the belt running in terms of v and the mass of sand per second $\dot{M} = dM/dt$ that falls on the belt. How much of the power is converted into kinetic energy per second? (Neglect the gravitational energy in the falling sand.)

Ans. $\dot{M}v^2$; $\frac{1}{2}$.

HISTORICAL NOTES

Discovery of Ceres (This discussion illustrates the accuracy of predictions based on classical mechanics.)

The first minor planet to be discovered was Ceres, found visually by Piazzi in Palermo in Sicily on the first day of the nineteenth century, January 1, 1801. Piazzi observed its motions for a few weeks but then became ill and lost track of it. A number of scientists calculated its orbit from the limited number of positions observed by Piazzi, but only the orbit computed by Gauss was accurate enough to predict where it might be the next year. On January 1, 1802, the planet Ceres was rediscovered by Olbers at an angular distance of only 30′ from the predicted location. As more observations accumulated, Gauss and others were able to improve the characteristics of the calculated orbit, and by 1830 the position of the planet was only 8″ from its predicted location. By including the major perturbations of the orbit of Ceres due to Jupiter, Enke found that he could reduce the residual error to an average of 6″/yr. Later computations, taking perturbations into account more accurately, produced predictions that disagreed with observations only by some 30″ after 30 yr.

Accounts of the discovery are given in vol. 12 of the *Philosophical Magazine*, 1802; see the papers by Piazzi (p. 54), Von Zach (p. 62), Tilloch (p. 80), and Lalande (p. 112). It is amusing to find that a society of eminent astronomers of Europe was organized at Lilienthal on September 21, 1800 for "the express purpose of searching out this planet supposed to exist between Mars and Jupiter. . . . The plan of the society was to divide the whole zodiac among the twenty-four members. . . ." Thanks to the postal delays of the Napoleonic war, the invitation to participate in team research was not transmitted to Piazzi until after he had made the discovery. Other accounts of the discovery of Ceres are found in the *Astronomisches Jahrbuch* 1804/5. Calculations by Gauss are in vol. 6 of his *Werke*, pp. 199–211.

The astronomical tradition started in Palermo by the Abbot Piazzi is believed to have reached Lampedusa (hero of the novel "The Leopard," written by his descendant) by way of the Abbot Pirrone, who was Lampedusa's spiritual advisor and astronomical assistant.

Discovery of Neptune During the first half of the nineteenth century, as the precision of observations and of theory improved, it was found that the planet Uranus was not moving according to the law of gravitation and the conservation of energy and angular momentum (see Chaps. 6 and 9). The planet erratically accelerated and decelerated by small but very significant amounts. There was no way of explaining this behavior on the basis of the known properties of the solar system and the laws of physics. Finally, in 1846, Leverrier and Adams independently discovered that the postulation of a hypothetical new planet of a certain mass and of a certain orbit exterior to that of Uranus would completely explain the observed anomalous motion.[1] They solved their equations for the location of this unknown planet and after only a half hour of search, the new planet, named Neptune, was found by Galle

[1] "I proved it was not possible to account for the observations of this planet [Uranus] by the theory of universal gravitation if the planet were subject only to the combined action of the sun and of the known planets. But all the observed anomalies can be explained to the smallest detail by the influence of a new [undiscovered] planet beyond Uranus. . . . We predict [August 31, 1846] the following position for the new planet on 1 January 1847: True heliocentric longitude 326°32′." U. J. Le Verrier, *Compt. Rend.*, **23**:428 (1846).

only 1° from the predicted location.[1] Present predictions of the positions of the major planets agree within a few seconds of arc with observations, even after an extrapolation of many years. The accuracy seems completely dependent upon the completeness of treatment of the various perturbing effects.

It is interesting to note that the planet Pluto, which is still farther from the sun, was discovered in a similar way, and that elements 93 and 94 beyond uranium (element 92) were named neptunium and plutonium.

FURTHER READING

PSSC, "Physics," chaps. 23 and 24, D. C. Heath and Company, Boston, 1960.

HPP, "Project Physics Course," chap. 10 (secs. 1–4), Holt, Rinehart and Winston, Inc., New York, 1970.

E. P. Wigner, Symmetry and Conservation Laws, *Physics Today*, 17 (3): 34 (1964).

Ernst Mach, "The Science of Mechanics," chap. 3, sec. 2, The Open Court Publishing Company, La Salle, Ill., 1960. On the history of the concept of kinetic energy ("vis viva").

[1] "I wrote to M. Galle on the 18th of September to ask his cooperation; this able astronomer saw the planet the same day [September 23, 1846] he received my letter. . . . [Observed] heliocentric longitude 327°24′ reduced to 1 January 1847. . . . Difference [observed and theory] 0°52′." U. J. Le Verrier, *Compt. Rend.*, **23**:657 (1846).

"M. Le Verrier saw the new planet without having need to glance a single time at the sky; he saw it at the *end of his pen*; he determined by the sole power of calculation the position and size of a body situated well beyond the then known limits of our planetary system. . . ." Arago. *Compt. Rend.*, **23**:659 (1846).

For an introduction to a magnificent controversy about the discovery, see pp. 741–754 of the same volume of the *Compt. Rend.* (Paris); see also M. Grosser, "The Discovery of Neptune," (Harvard University Press, Cambridge, Mass., 1962).

Neptune as seen in the Lick Observatory 120-in. reflector. The arrow points to Triton, a satellite of Neptune. (*Lick Observatory photograph*)

CONTENTS

Conservation of Linear and Angular Momentum

In Chap. 4 we considered systems for which galilean invariance was valid, and we showed that the conservation of linear momentum of a system of interacting particles is a necessary consequence of galilean invariance and conservation of energy, provided that no external force acts. The conservation of linear momentum, a law accurately verified by experiment, is an essential part of the "classical package" discussed previously. In this chapter we develop the implications of momentum conservation in the motion of a collection of particles. We define the center of mass of a particle system and learn to view the motion of the system from a reference frame in which the center of mass is at rest. Collision processes among pairs of particles constitute important special cases. We also introduce the important concept of angular momentum, the conservation of angular momentum, and the concept of torque. These are particularly important in the treatment of rigid bodies in Chap. 8 and of central forces in Chap. 9.

INTERNAL FORCES AND MOMENTUM CONSERVATION

In treating the dynamic behavior of a system of particles we find it useful to distinguish between forces of interaction between the particles of the system and other forces due to factors external to the system, such as a gravitational or an electrical field in which the particle system may exist. We refer to the interparticle forces as *internal forces* of the system.

Internal forces cannot affect the total momentum of the collection of particles, where by *total momentum* we mean the vector sum

$$\mathbf{p} = \sum_{i=1}^{N} m_i \mathbf{v}_i \tag{6.1}$$

(A proof has been given in Chap. 3, and the reader is reminded of it in the following discussion.) Considering these interparticle forces to be newtonian, we understand that for the mutual interaction of any two particles the forces obey

$$\mathbf{F}_{ij} = -\mathbf{F}_{ji}$$

where $\mathbf{F}_{ij}$ represents the force exerted upon particle i by particle j and vice versa. Then, by Newton's Second Law, we conclude that in any interval of time the momentum change produced in particle i by the force $\mathbf{F}_{ij}$ is vectorially equal and opposite to that produced in particle j by the force $\mathbf{F}_{ji}$, and so the momentum change due to the mutual interaction of this pair of particles is zero. This argument is valid for any and all particle pairs in the collection, providing that the interaction of any pair is not affected by the presence of other particles. Thus we conclude that internal forces cannot affect the total momentum of a system since the vector sum of all these interparticle forces will be zero.

The foregoing paragraph relates to the total linear momentum of the system. Later in this chapter we shall extend the argument to show that the internal forces also cannot produce a change in the angular momentum of a system of particles. Recognition of these two principles of *momentum conservation* in relation to internal forces greatly simplifies the understanding and analysis of many problems of collective motion.

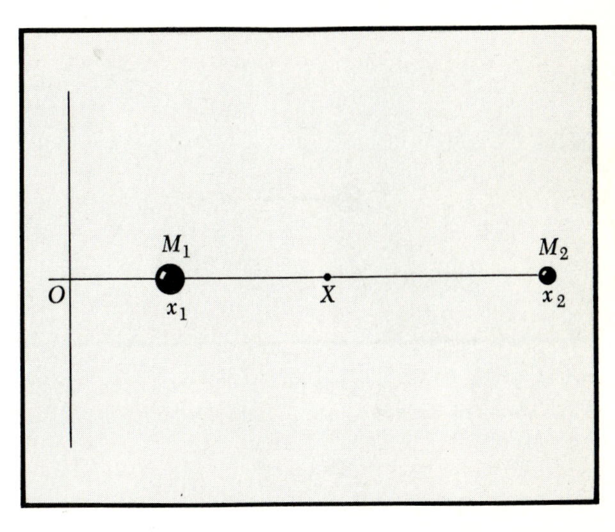

FIG. 6.1 Two masses M_1 and M_2 at positions x_1 and x_2 on the x axis have a center of mass located at $X = (M_1x_1 + M_2x_2)/(M_1 + M_2)$.

CENTER OF MASS

Relative to a fixed origin O, the position $\mathbf{R}_{\text{c.m.}}$ of the center of mass of a system of N particles is defined as

$$\mathbf{R}_{\text{c.m.}} = \frac{\sum_{n=1}^{N} \mathbf{r}_n M_n}{\sum_{n=1}^{N} M_n} \tag{6.2}$$

It is a weighted average position, weighted according to particle mass. For a two-particle system

$$\mathbf{R}_{\text{c.m.}} = \frac{\mathbf{r}_1 M_1 + \mathbf{r}_2 M_2}{M_1 + M_2} \tag{6.3}$$

as shown in Figs. 6.1 and 6.2.

We differentiate Eq. (6.2) with respect to time to obtain the center-of-mass velocity:

$$\dot{\mathbf{R}}_{\text{c.m.}} = \frac{\sum_n \dot{\mathbf{r}}_n M_n}{\sum_n M_n} = \frac{\sum_n \mathbf{v}_n M_n}{\sum_n M_n} \tag{6.4}$$

but $\Sigma_n \mathbf{v}_n M_n$ is just the total momentum of the system. In the absence of external forces the total momentum is constant, so that

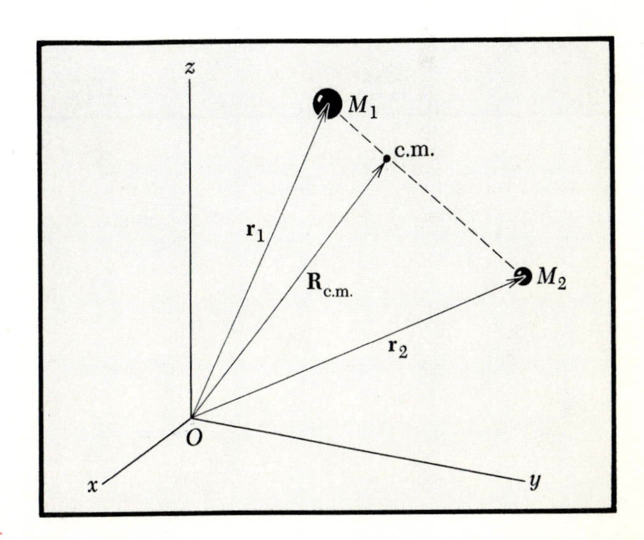

FIG. 6.2 For two masses M_1 and M_2 at arbitrary positions $\mathbf{r}_1$ and $\mathbf{r}_2$, $\mathbf{R}_{\text{c.m.}} = (M_1\mathbf{r}_1 + M_2\mathbf{r}_2)/(M_1 + M_2)$.

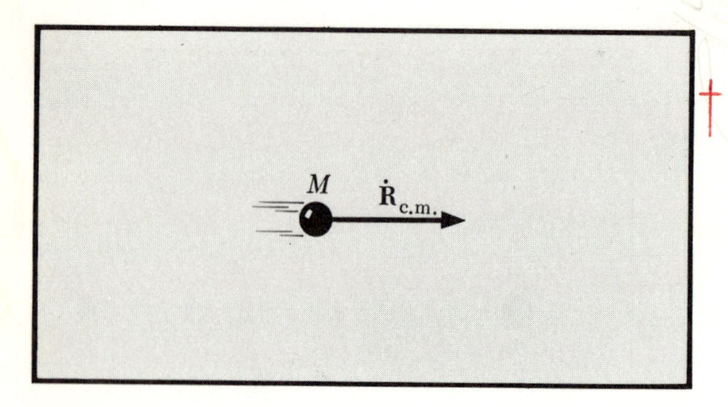

FIG. 6.3 (*a*) In the absence of external forces the velocity of the center of mass is constant. Here a radioactive nucleus with velocity $\dot{\mathbf{R}}_{c.m.}$ is about to decay.

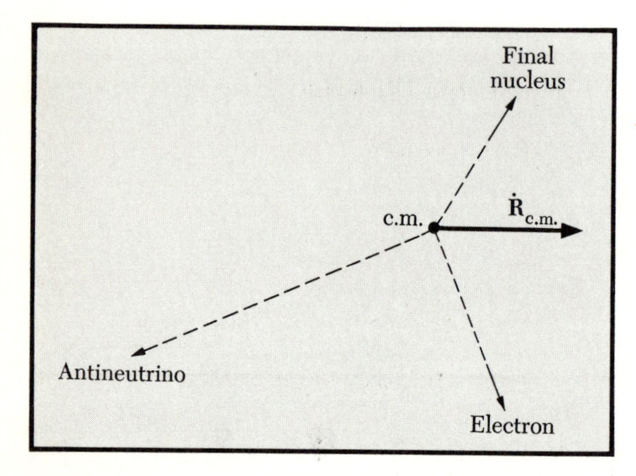

(*b*) The nucleus decays into three particles which go off in separate directions. However, the velocity of the center of mass of these three particles remains unchanged.

$$\dot{\mathbf{R}}_{c.m.} = \text{const} \tag{6.5}$$

This is a remarkable property of the center of mass: The *velocity of the center of mass is constant in the absence of external forces*. This is true, for example, for a radioactive nucleus that decays in flight (see Fig. 6.3*a* and *b*) or for a projectile that explodes into fragments in force-free space.

It is a simple matter to show from Eq. (6.4) that the acceleration of the center of mass is determined by the total external force acting on the system of particles. If $\mathbf{F}_n$ is the force on particle n, then, on differentiating Eq. (6.4) with respect to time, we have

$$\left(\sum_n M_n\right)\ddot{\mathbf{R}}_{c.m.} = \sum_n (M_n\dot{\mathbf{v}}_n) = \sum_n \mathbf{F}_n = \mathbf{F}_{ext} \tag{6.6}$$

where on the right the interparticle internal forces drop out in the sum $\Sigma_n \mathbf{F}_n$ over all particles.

This is another significant result: *In the presence of external forces the vector acceleration of the center of mass is equal to the vector sum of the external forces divided by the total mass of the system*. In other words, we can use the methods we have developed in Chaps. 3 to 5 to treat the motion of the center of mass as if the whole mass of the body were concentrated there and all the external forces acted on it. (This principle is particularly important in the rigid-body problems treated in Chap. 8.) As another example, the center of mass of the earth and moon moves in an approximately circular orbit around the sun. We are now going to illustrate the usefulness of the center of mass by working out some important collision problems. (We have already worked out several problems in Chaps. 3 and 4.)

EXAMPLE

Collision of Particles That Stick Together[1] Consider the collision of two particles of mass M_1 and M_2 that stick together on collision. Let M_2 be at rest on the x axis before the collision, and let $\mathbf{v}_1 = v_1\hat{\mathbf{x}}$ describe the motion of M_1 before the collision.

(1) Describe the motion of $M = M_1 + M_2$ after the collision. Figure 6.4*a* and *b* illustrates the case. Regardless of whether the collision is elastic or inelastic, the total momentum is unchanged in a collision. The collision considered here is inelastic. The initial

[1] The case of equal-mass particles was given in Chap. 4 (page 122). You will find it instructive to calculate the $\Delta\epsilon$ in both frames for this present general case.

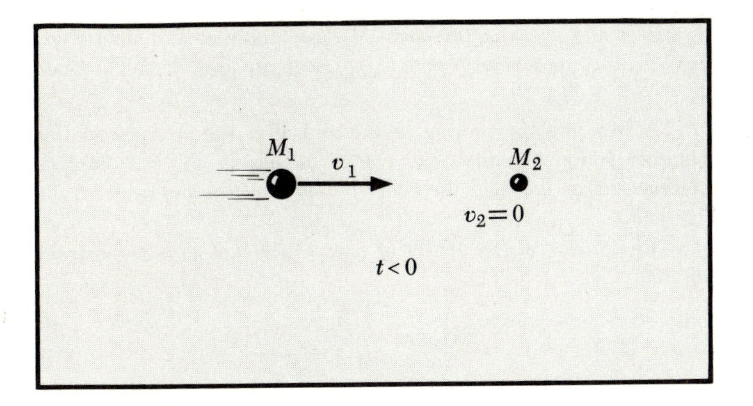

FIG. 6.4 (*a*) Even in an inelastic collision, momentum must be conserved. Consider a collision in which the particles stick together. Before the collision $p_x = M_1v_1$.

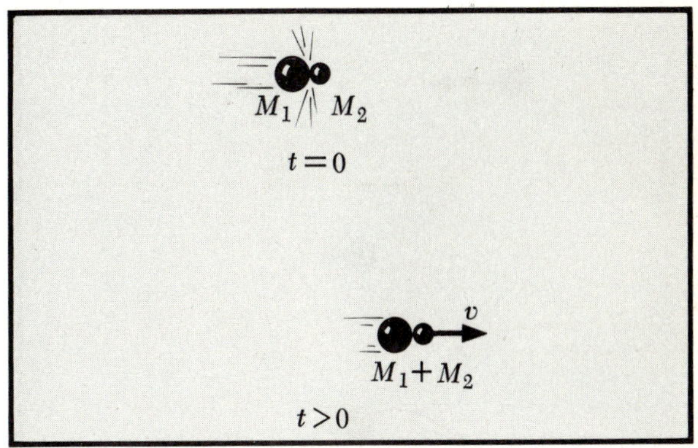

(*b*) After the collision $p_x = M_1v_1 = (M_1 + M_2)v$ so $v = M_1v_1/(M_1 + M_2) < v_1$.

x component of momentum is M_1v_1; the final x component of momentum is $(M_1 + M_2)v$. The other components are zero. By conservation of momentum we have

$$M_1v_1 = (M_1 + M_2)v \tag{6.7}$$

so that the final velocity v is given by

$$v = \frac{M_1}{M_1 + M_2}v_1 \tag{6.8}$$

and, because the particles are stuck together,

$$\mathbf{X}_{\text{c.m.}} = vt\hat{\mathbf{x}} \qquad t > 0$$

describes the motion of the system after the collision. According to Eq. (6.5), this same relation must describe the motion of the center of mass at all times, before or after the collision:

$$\mathbf{X}_{\text{c.m.}} = vt\hat{\mathbf{x}} = \frac{M_1}{M_1 + M_2}v_1t\hat{\mathbf{x}} \tag{6.9}$$

using Eq. (6.7), as shown in Fig. 6.4*c*.

(2) What is the ratio of the kinetic energy after the collision to the initial kinetic energy? The kinetic energy K_f after the collision is

$$K_f = \tfrac{1}{2}(M_1 + M_2)\frac{M_1^2}{(M_1 + M_2)^2}v_1^2 = \frac{M_1^2v_1^2}{2(M_1 + M_2)} \tag{6.10}$$

The initial kinetic energy K_i is equal to $\frac{1}{2}M_1v_1^2$, so that

$$\frac{K_f}{K_i} = \frac{M_1}{M_1 + M_2} \tag{6.11}$$

The remainder of the energy appears in the internal excitations and heat of the composite system after the collision. When a meteorite

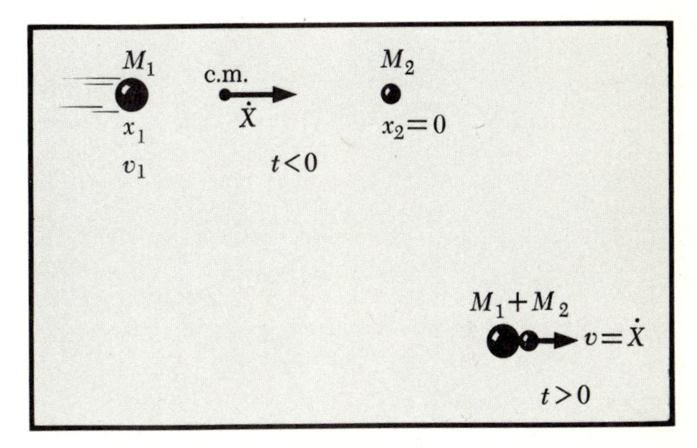

(*c*) $X = (M_1x_1 + M_2x_2)/(M_1 + M_2)$ so $\dot{X} = M_1v_1/(M_1 + M_2)$. $\dot{X}$ is unchanged by the collision.

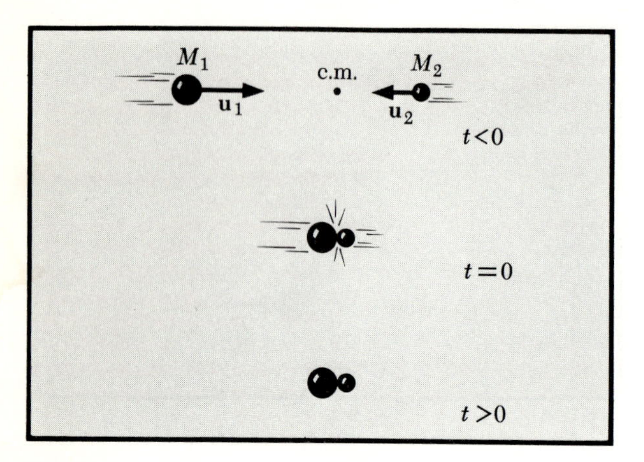

FIG. 6.4 (*cont'd.*) (*d*) In the center-of-mass reference frame the velocities of M_1 and M_2 before collision are $\mathbf{u}_1$, $\mathbf{u}_2$. After collision $(M_1 + M_2)$ is at rest.

M_1 strikes and sticks to the earth M_2, essentially all the kinetic energy of the meteorite appears as heat in the earth because $M_1 \ll M_1 + M_2$.

(3) Describe the motion before and after the collision in the reference frame in which the center of mass is at rest. (Such a reference frame is called the *center-of-mass system* and is shown in Fig. 6.4*d*.)

The position of the center of mass of the system is given from Eq. (6.9):

$$\mathbf{X}_{\text{c.m.}} = \frac{M_1 v_1 t \hat{\mathbf{x}}}{M_1 + M_2}$$

The velocity of the center of mass is given by:

$$\mathbf{V} = \dot{\mathbf{X}}_{\text{c.m.}} = \frac{d}{dt}\mathbf{X}_{\text{c.m.}} = \frac{M_1}{M_1 + M_2} v_1 \hat{\mathbf{x}}$$

In the center-of-mass reference frame the initial velocity $\mathbf{u}_1$ of particle 1 is

$$\mathbf{u}_1 = v_1\hat{\mathbf{x}} - \mathbf{V} = \left(1 - \frac{M_1}{M_1 + M_2}\right) v_1 \hat{\mathbf{x}} = \frac{M_2}{M_1 + M_2} v_1 \hat{\mathbf{x}}$$

In the center-of-mass reference frame the initial velocity $\mathbf{u}_2$ of particle 2 is

$$\mathbf{u}_2 = -\mathbf{V} = -\frac{M_1}{M_1 + M_2} v_1 \hat{\mathbf{x}}$$

Note that

$$M_1\mathbf{u}_1 + M_2\mathbf{u}_2 = \left(\frac{M_1 M_2}{M_1 + M_2} v_1 - \frac{M_2 M_1}{M_1 + M_2} v_1\right)\hat{\mathbf{x}} = 0$$

One can see the advantage of the center-of-mass frame; the momenta are always equal and opposite.

Now, on colliding, the particles stick together; the new combined particle has mass $M_1 + M_2$ and must be at rest in the center-of-mass system. Relative to the laboratory system, then, the new particle has the velocity $\mathbf{V}$ of the center of mass, which is exactly the velocity [Eq. (6.8)] found by the earlier argument.

EXAMPLE

Transverse Momentum Components Two particles of equal mass move initially on paths parallel to the x axis and collide. After the collision one of the particles is observed to have a particular value $v_y(1)$ of the y component of the velocity. What is the y component of the velocity of the other particle after the collision? (Recall that each component x, y, or z of the total linear momentum is conserved separately.)

Before the collision the particles were moving along the x axis, so that the total y component of the momentum is zero. By momentum conservation the total y component of momentum must also be zero after the collision, so that

$$M[v_y(1) + v_y(2)] = 0$$

whence

$$v_y(2) = -v_y(1)$$

We cannot calculate $v_y(1)$ itself without specifying the initial trajectories and the details of the forces during the collision process.

EXAMPLE

Collision of Particles with Internal Excitations Two particles of equal mass and equal but opposite velocities $\pm\mathbf{v}_i$ collide. What are the velocities after the collision?

The center of mass is at rest and must remain at rest, so that the final velocities $\pm\mathbf{v}_f$ are equal but opposite. If the collision is elastic, the conservation of energy demands that the final speed v_f equal the initial speed v_i. If one or both particles are excited internally by the collision, then $v_f < v_i$ by conservation of energy. If one or both particles initially are in excited states of internal motion and on collision they give up their excitation energy into kinetic energy, then v_f can be larger than v_i.

EXAMPLE

General Elastic Collision of Particles of Different Mass This is a famous problem. A particle of mass M_1 collides elastically with a particle of mass M_2 that initially is at rest in the laboratory frame of reference. The trajectory of M_1 is deflected through an angle θ_1 by the collision. The maximum possible value of the scattering angle θ_1 is determined by the laws of conservation of energy and momentum independent of the details of the interaction between the particles. Our problem is to find $(\theta_1)_{\max}$. We shall see that it is convenient at one stage in the calculation to view the collision from the frame of reference in which the center of mass is at rest.

We denote the initial velocities in the laboratory frame (Fig. 6.5) by

$$\mathbf{v}_1 = v_1\hat{\mathbf{x}} \qquad \mathbf{v}_2 = 0$$

and the final velocities (after the collision) by $\mathbf{v}_1'$ and $\mathbf{v}_2'$. The law of conservation of energy requires that in an elastic collision the total kinetic energy before the collision equal the total kinetic energy after the collision. Thus

$$\tfrac{1}{2}M_1v_1^2 = \tfrac{1}{2}M_1v_1'^2 + \tfrac{1}{2}M_2v_2'^2 \tag{6.12}$$

noting the initial condition $v_2 = 0$. The law of conservation of momentum applied to the x component of momentum requires that

$$M_1v_1 = M_1v_1' \cos\theta_1 + M_2v_2' \cos\theta_2 \tag{6.13}$$

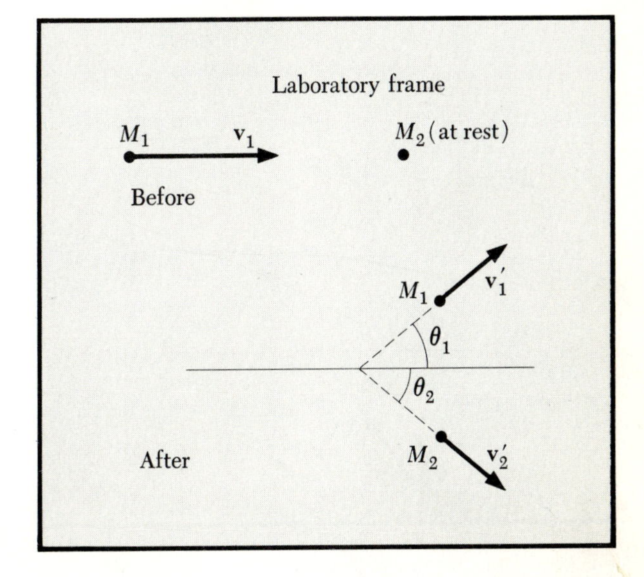

FIG. 6.5 A collision between M_1 and M_2 need not be confined to one dimension. In the laboratory frame M_2 is at rest before the collision.

The entire collision event can be considered to lie in the xy plane, provided that only two particles are involved. The law of conservation of momentum applied to the y component of momentum requires that

$$0 = M_1 v_1' \sin\theta_1 - M_2 v_2' \sin\theta_2 \tag{6.14}$$

because initially the y component of momentum was zero.

It is perfectly possible but a little tedious to solve Eqs. (6.12) to (6.14) simultaneously for whatever quantities interest us. These equations express the entire content of the conservation laws. But it is considerably neater and more informative to view the collision in the center-of-mass reference frame. First we find the velocity $\mathbf{V}$ of the center of mass relative to the laboratory frame. The position of the center of mass is defined by

$$\mathbf{R}_{\text{c.m.}} = \frac{M_1\mathbf{r}_1 + M_2\mathbf{r}_2}{M_1 + M_2}$$

as stated earlier in Eq. (6.3). Differentiating this, we obtain the velocity $\mathbf{V}$ of the center of mass:

$$\dot{\mathbf{R}}_{\text{c.m.}} = \frac{M_1\dot{\mathbf{r}}_1 + M_2\dot{\mathbf{r}}_2}{M_1 + M_2} \qquad \mathbf{V} = \frac{M_1}{M_1 + M_2}\mathbf{v}_1 \tag{6.15}$$

where we have expressed the result in terms of the velocities $\mathbf{v}_1$ and $\mathbf{v}_2$ before the collision, with $\mathbf{v}_2 = 0$. [Note that this is simply Eq. (6.4) writtten for the present conditions.]

We denote by $\mathbf{u}_1$ and $\mathbf{u}_2$ the initial velocities in the reference frame moving with the center of mass, and we denote the final velocities by $\mathbf{u}_1'$, $\mathbf{u}_2'$. Velocity vectors in the laboratory are related to these in the center-of-mass reference system (Fig. 6.6) by

$$\begin{aligned} \mathbf{v}_1 &= \mathbf{u}_1 + \mathbf{V} \qquad & \mathbf{v}_2 &= \mathbf{u}_2 + \mathbf{V} \\ \mathbf{v}_1' &= \mathbf{u}_1' + \mathbf{V} \qquad & \mathbf{v}_2' &= \mathbf{u}_2' + \mathbf{V} \end{aligned} \tag{6.16}$$

The conservation laws allow us to understand immediately certain characteristics of the collision. Momentum conservation in the center-of-mass system requires that the scattering angle of particle 1 equal that of particle 2; that is, the trajectories must be collinear after the collision as well as before. Otherwise the center of mass could not remain at rest in the reference frame we are employing, as it must do in the absence of external forces acting on the particles. Furthermore, if kinetic energy is to be conserved, the velocities in the center-of-mass frame must not be changed; so that $u_1' = u_1$ and $u_2' = u_2$ in an elastic collision. The kinematic description in the center-of-mass frame is seen to be remarkably simple, with no restriction on the scattering angle $\theta_{\text{c.m.}}$ imposed by conservation laws. (The last statement is not generally true for scattering angles θ_1 and θ_2 in the laboratory frame.)

Let us return to the laboratory system. We form, writing θ for $\theta_{\text{c.m.}}$ for convenience,

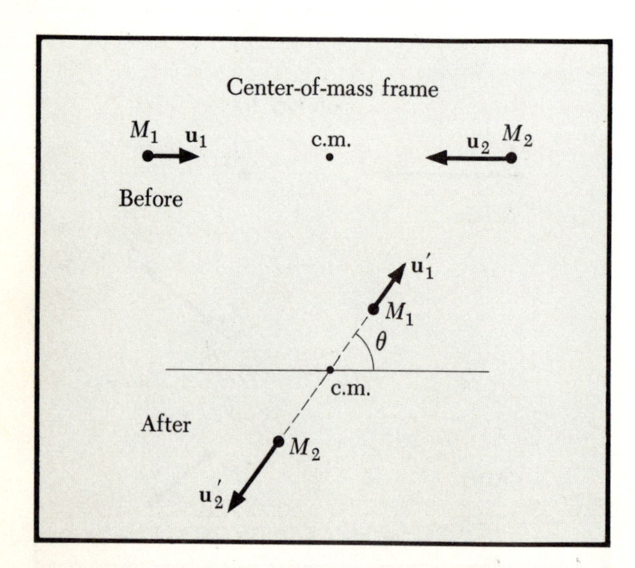

FIG. 6.6 In the center-of mass frame, M_1 and M_2 must go off in opposite directions after collision. All angles $0 \le \theta \le \pi$ are possible, and for elastic collisions $|\mathbf{u}_1'| = |\mathbf{u}_1|$ and $|\mathbf{u}_2'| = |\mathbf{u}_2|$.

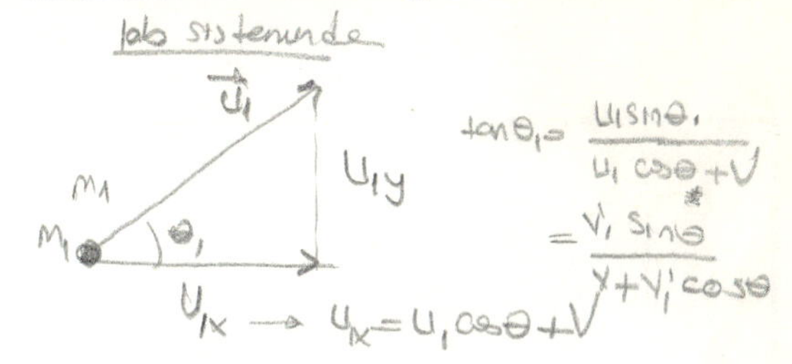

$$\tan\theta_1 = \frac{\sin\theta_1}{\cos\theta_1} = \frac{v_1'\sin\theta_1}{v_1'\cos\theta_1} = \frac{u_1'\sin\theta}{u_1'\cos\theta + V}$$

where we have used the fact that the y component of the final velocity of particle 1 is identical in the two reference frames. Further, $u_1 = u_1'$ by the elastic collision assumption, so that

$$\tan\theta_1 = \frac{\sin\theta}{\cos\theta + V/u_1} \tag{6.17}$$

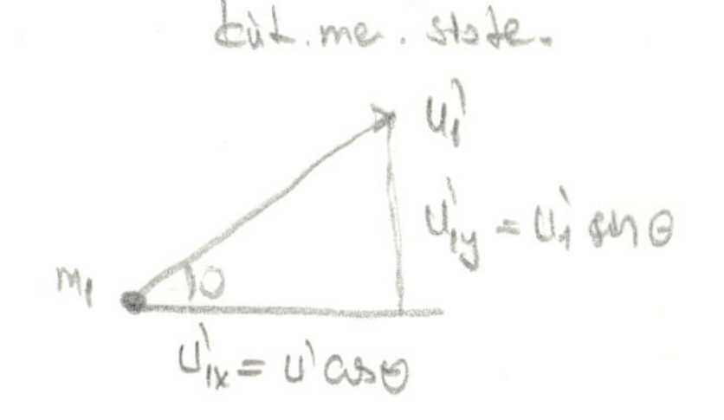

Now Eqs. (6.15) and (6.16) may be combined to give the relations

$$\mathbf{V} = \frac{M_1}{M_1 + M_2}(\mathbf{u}_1 + \mathbf{V}) \qquad \mathbf{V} = \frac{M_1}{M_2}\mathbf{u}_1$$

whence Eq. (6.17) becomes

$$\tan\theta_1 = \frac{\sin\theta}{\cos\theta + M_1/M_2} \tag{6.18}$$

Figure 6.7a and b shows this development.

We want to know the value of $(\theta_1)_{\max}$. If $M_1 > M_2$, this can be found graphically from Eq. (6.18) or by using calculus to determine the maximum of $\tan\theta_1$ as a function of θ. We see by inspection that for $M_1 > M_2$ the denominator can never be zero, and $(\theta_1)_{\max}$

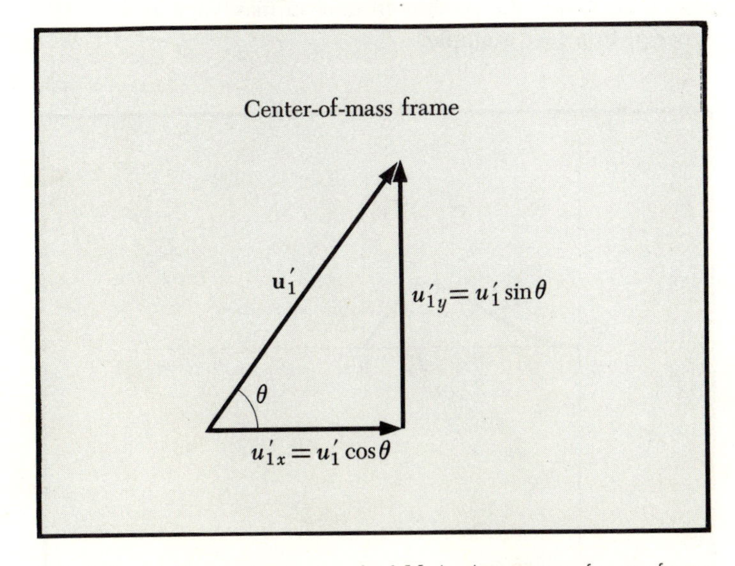

FIG. 6.7 (a) The final velocity $\mathbf{u}_1'$ of M_1 in the center-of-mass frame is resolved into x and y components in the figure.

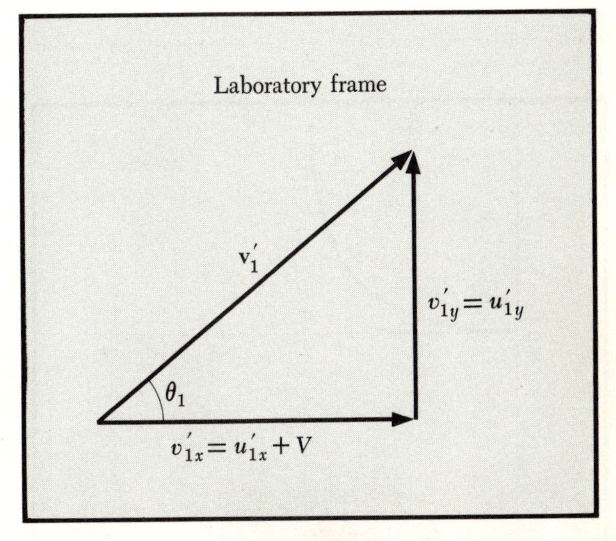

(b) In the laboratory frame the x and y components of $\mathbf{v}_1'$ are as shown. Evidently (recall that $u_1' = u_1$)

$$\tan\theta_1 = \frac{\sin\theta}{\cos\theta + V/u_1} = \frac{\sin\theta}{\cos\theta + M_1/M_2}$$

must then be less than $\frac{1}{2}\pi$. If $M_1 = M_2$, then $(\theta_1)_{\max} = \frac{1}{2}\pi$. If $M_1 < M_2$, any value of θ_1 is allowed. Figure 6.8*a* to *c* shows the relations graphically.

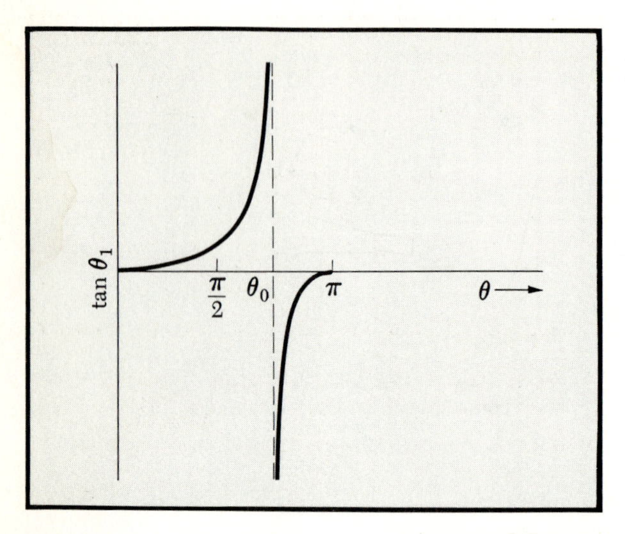

FIG. 6.8 (*a*) For $M_1 < M_2$, $\tan\theta_1 = \sin\theta/(\cos\theta + M_1/M_2)$ goes to infinity at $\theta = \theta_0 = \cos^{-1}(-M_1/M_2)$. All angles $0 \leq \theta_1 \leq \pi$ are possible.

dm/dt V → force

SYSTEMS WITH VARIABLE MASS

In Chap. 3 we expressed Newton's Second Law as

$$\mathbf{F} = \frac{d\mathbf{p}}{dt}$$

where $\mathbf{p}$ is the momentum $M\mathbf{v}$. For objects of constant mass this becomes the familiar $\mathbf{F} = M\mathbf{a}$, but there is a class of mechanics problems in which the mass of a moving object may not be constant and it is necessary to recognize the time dependence of M when we express $d\mathbf{p}/dt$. The Second Law then becomes

$$\boxed{\mathbf{F} = \frac{dM}{dt}\mathbf{v} + M\frac{d\mathbf{v}}{dt}} \qquad (6.19)$$

Numerous interesting and important problems are treated by Eq. (6.19), including rocket motion, the degrading of satellite motion by an atmosphere, and the motions of objects such as chains for which the portion in motion may vary in time. We proceed to a few examples.

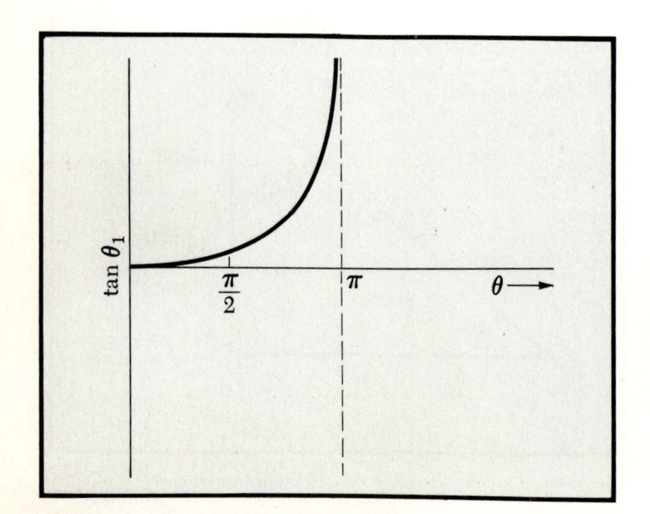

(*b*) For $M_1 = M_2$, $\tan\theta_1$ goes to infinity at $\theta = \pi$. Thus all angles $0 \leq \theta_1 \leq \pi/2$ are possible as roots of the equation for $\tan\theta_1$.

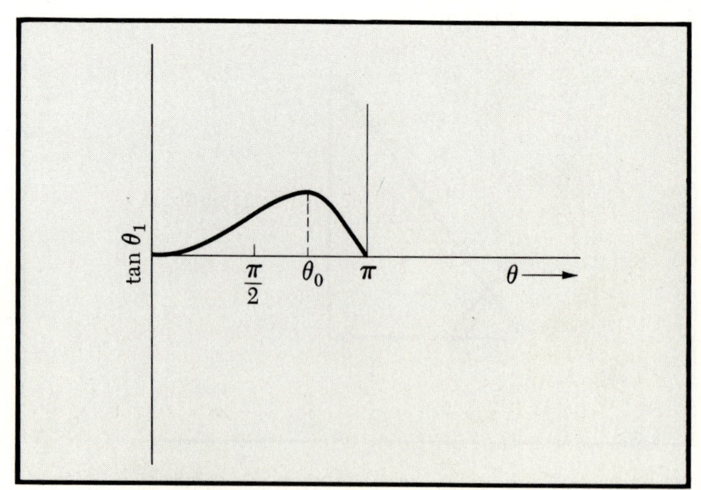

(*c*) For $M_1 > M_2$, $\tan\theta_1$ does not go to infinity. Thus $0 \leq \theta_1 \leq \sin^{-1}(M_2/M_1) < \pi/2$.

EXAMPLE

Satellite in Interplanetary Dust A satellite in force-free space sweeps up stationary interplanetary debris at a rate $dM/dt = cv$, where M is the mass and v the speed of the satellite; c is a constant that depends upon the cross-section area of the volume swept out. What is the deceleration?

We consider this problem from a reference frame in which the interplanetary dust is at rest (see Fig. 6.9). The momentum of the total system consisting of satellite and dust is constant because no external force is present in the problem. Equation (6.19) then requires us to state:

$$\mathbf{F} = \frac{d}{dt}(M\mathbf{v}) = \dot{M}\mathbf{v} + M\dot{\mathbf{v}} = 0$$

or, since the motion is in one dimension with $\dot{M} = cv$, the deceleration is

$$\dot{v} = -\frac{cv^2}{M}$$

(An example is Prob. 16 at the end of this chapter.)

It is possible to view this problem in a different way by considering that a resistive force due to the dust constantly acts to decelerate the satellite. The resistive force will be the opposite (Newton's Third Law) of the force exerted by the satellite upon the dust medium as it moves through it. At any instant the force exerted upon the medium will be the rate of change of momentum imparted to dust particles, which is $v\,dM/dt$ or cv^2. The resistive force on the satellite is then $-cv^2$, and the satellite deceleration will be obtained from Newton's Second Law employed in the usual manner:

$$M\dot{v} = f_{\text{resist}} = -cv^2 \qquad \dot{v} = -\frac{cv^2}{M}$$

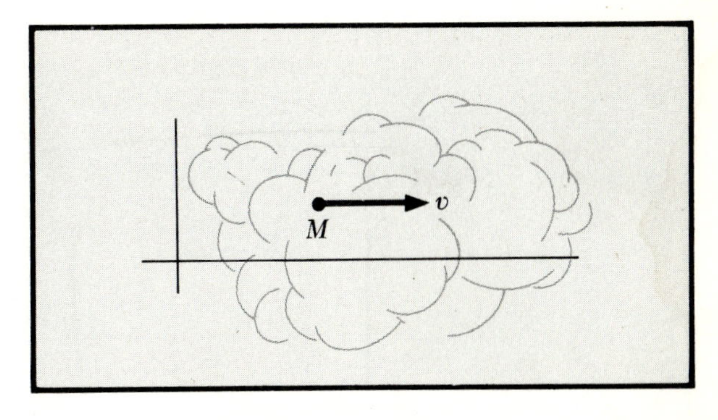

FIG. 6.9 Object moving through dust cloud, referred to reference frame related to cloud.

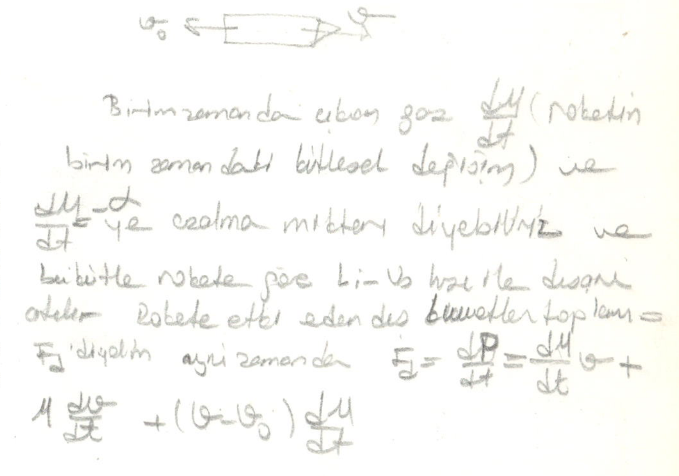

EXAMPLE

Space-vehicle Problem A space vehicle ejects fuel backward at a velocity $\mathbf{V}_0$ relative to the vehicle; the rate of change of mass of the vehicle is $\dot{M} = -\alpha$, a constant. Set up and solve the equation of motion of the space vehicle, neglecting gravity.

Let the velocity of the vehicle at time t be $\mathbf{v}$. The velocity of the fuel viewed in an inertial (*not* in the vehicle) frame of reference is $-\mathbf{V}_0 + \mathbf{v}$. We assume that $\mathbf{V}_0$ and $\mathbf{v}$ are opposite, so that the problem reduces to a one-dimensional problem, as shown in Fig. 6.10.

In the absence of any external force the momentum of the total system consisting of vehicle plus exhausted gases is constant. Thus $F = dp/dt = 0$, and we may write

$$\frac{dp}{dt} = M\dot{v} - v\alpha + (v - V_0)\alpha = 0 \tag{6.20}$$

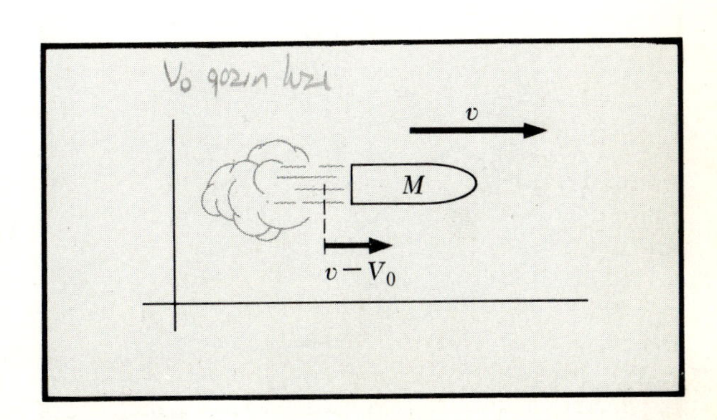

FIG. 6.10 Vehicle moves with speed v in inertial frame; gas ejected with exhaust velocity V_0 moves with speed $v - V_0$.

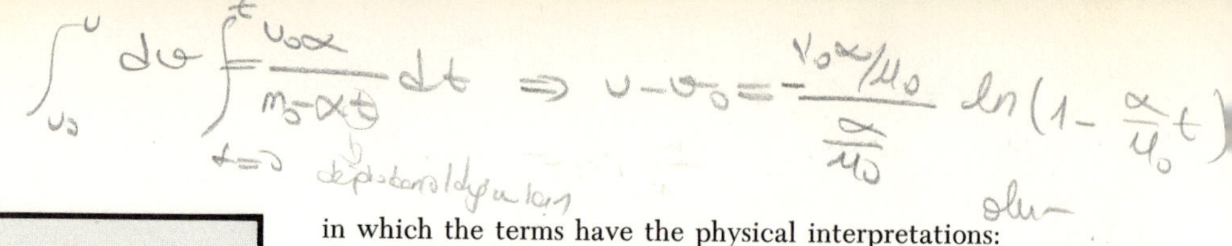

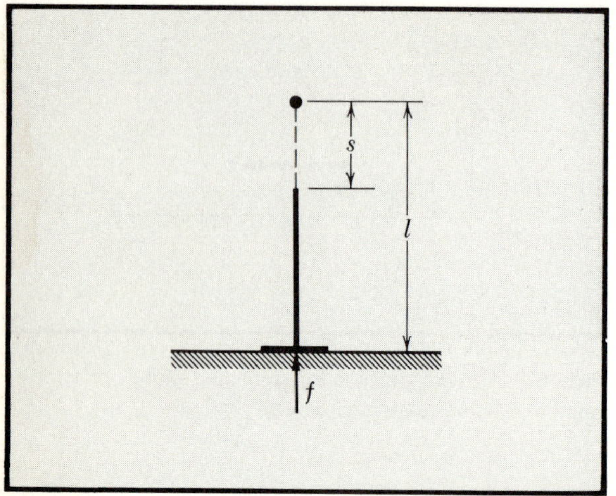

FIG. 6.11 The falling chain, of linear density ρ. In time interval dt a mass increment $\rho\, ds$ arrives moving with speed $\dot{s}$. Thus the rate of change of the chain's momentum due to collapse upon the platform is $\rho(ds/dt) \cdot \dot{s}$ or $\rho \dot{s}^2$ The platform must also support the weight ρgs of what has arrived.

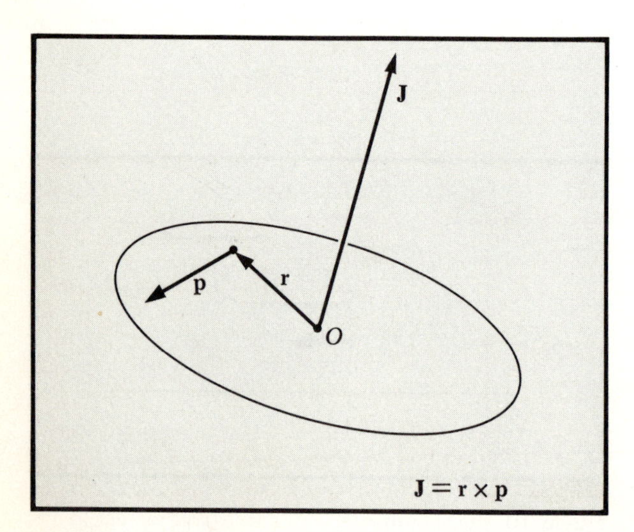

FIG. 6.12 (*a*) The angular momentum **J** *with respect to point O* is defined by the figure.

in which the terms have the physical interpretations:

$$M\dot{v} = \text{vehicle momentum gain rate due to acceleration}$$

$$-v\alpha = \text{vehicle momentum loss rate due to loss of mass}$$

$$(v - V_0)\alpha = \text{rate of increase of momentum possessed by exhausted gas cloud}$$

Equation (6.20) simplifies to

$$M\dot{v} = \alpha V_0 \tag{6.21}$$

Since the rate of mass loss is given as a constant $-\alpha$, the vehicle mass at time t is

$$M = M_0 - \alpha t$$

where M_0 is the initial vehicle mass at time $t = 0$. Equation (6.21) becomes

$$(M_0 - \alpha t)\dot{v} = \alpha V_0$$

In order to obtain the vehicle velocity as a function of time, we express

$$\dot{v} = \frac{\alpha V_0/M_0}{1 - \alpha t/M_0}$$

and integrate, with the assumption that $v = v_0$ at $t = 0$, to obtain

$$v = v_0 + V_0 \ln \frac{M_0}{M_0 - \alpha t} \tag{6.22}$$

The term αt can never become as great as M_0 since the vehicle is not all fuel; but it may be as much as 90 percent fuel. Equation (6.22) exhibits the advantage of high-velocity propellant exhaust. The ultimate in propellant efficiency would be obtained with photons, i.e., light, for which $v_0 = c$. But it is difficult to exhaust much mass in this form!

EXAMPLE

Force due to a Falling Chain A familiar illustration is provided by the force upon a stationary platform while a uniform flexible chain falls upon it from a hanging position. Consider such a chain to be initially suspended by one end with the lower end just contacting the platform. In Fig. 6.11 the chain is shown an instant after release, i.e., when it has fallen a distance s and this length of chain has collapsed upon the platform.

The total upward force required of the platform upon the chain must both support the length of chain that has come to rest and continually reduce to zero the momentum of arriving elements of chain. These two contributions are expressed in[1]

$$f = \rho gs + \rho \dot{s}^2$$

[1] $(\rho\, ds)\, ds/dt$ is the momentum of the mass $\rho\, ds$ which is reduced to 0 in time dt. So the rate of change of momentum $= (\rho\, ds/dt)(ds/dt) = \rho \dot{s}^2$.

where ρ is the linear density (i.e., mass per unit length) of the chain. But since the freely falling part of the chain has acceleration g, we have $\dot{s}^2 = 2gs$. Therefore

$$f = 3\rho g s$$

Thus at any instant the platform exerts a force three times the weight of the amount of chain that has arrived.

(You may find it instructive to view this problem in other ways, such as by considering the acceleration of the center of mass of the chain under the influence of both gravity and the platform force upon it. The foregoing result is readily obtained.)

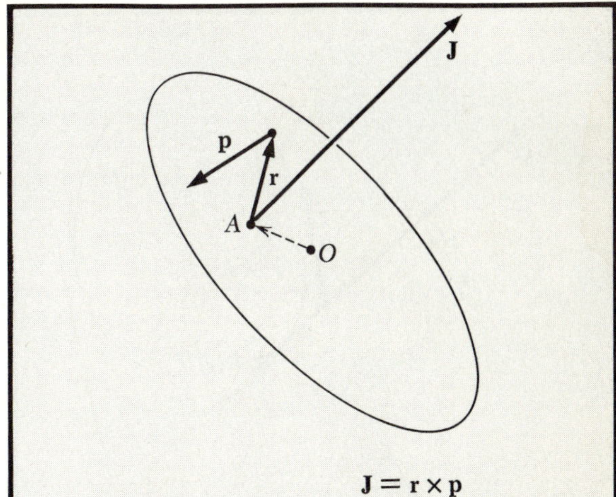

(b) The angular momentum *with respect to another point* A is different even for the *same* particle with the *same* momentum $\mathbf{p}$.

CONSERVATION OF ANGULAR MOMENTUM

We now turn to the important concept of *angular momentum*. The angular momentum $\mathbf{J}$ of a single particle referred to an arbitrary fixed point (fixed in an *inertial* reference frame) as origin is defined as

$$\boxed{\mathbf{J} \equiv \mathbf{r} \times \mathbf{p} \equiv \mathbf{r} \times M\mathbf{v}} \tag{6.23}$$

where $\mathbf{p}$ is the linear momentum (see Fig. 6.12a and b). The units of angular momentum are g-cm^2/s or erg-s. The component of $\mathbf{J}$ along any line (or axis) passing through the fixed reference point is often called the angular momentum of the particle about this axis.

If the force $\mathbf{F}$ acts upon the particle, we define the *torque*, or *moment of force*, about the same fixed point as

$$\boxed{\mathbf{N} = \mathbf{r} \times \mathbf{F}} \tag{6.24}$$

(This, you may remember, appeared in Chap. 2.) The units of torque are dyn-cm. Figure 6.13 illustrates the relation. Now, on differentiating Eq. (6.23), we have

$$\frac{d\mathbf{J}}{dt} = \frac{d}{dt}(\mathbf{r} \times \mathbf{p}) = \frac{d\mathbf{r}}{dt} \times \mathbf{p} + \mathbf{r} \times \frac{d\mathbf{p}}{dt}$$

But

$$\frac{d\mathbf{r}}{dt} \times \mathbf{p} = \mathbf{v} \times M\mathbf{v} = 0$$

and, by Newton's Second Law in an inertial reference frame,

$$\mathbf{r} \times \frac{d\mathbf{p}}{dt} = \mathbf{r} \times \mathbf{F} = \mathbf{N}$$

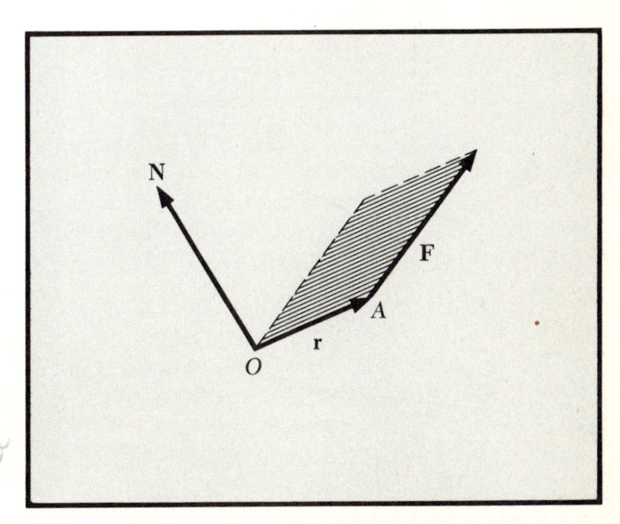

FIG. 6.13 The torque $\mathbf{N}$ effective at point O due to force $\mathbf{F}$ applied at point A at position $\mathbf{r}$ is $\mathbf{r} \times \mathbf{F}$. $\mathbf{N}$ is normal to the plane defined by $\mathbf{r}$ and $\mathbf{F}$.

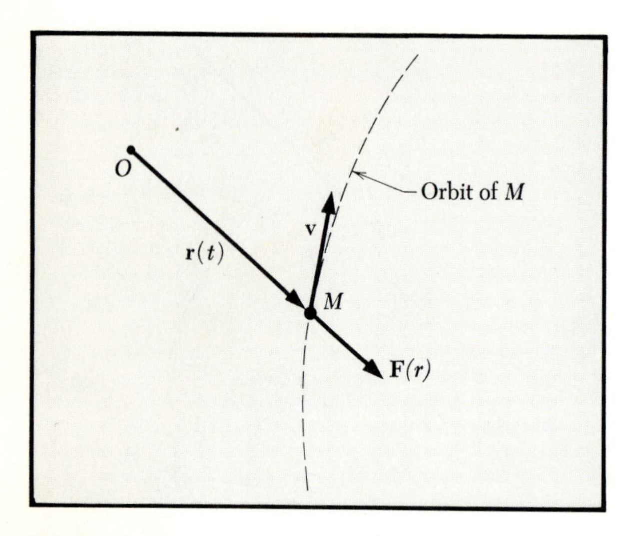

FIG. 6.14 Particle M subjected to repulsive central force $\mathbf{F}(\mathbf{r})$ centered at O. Since the torque $\mathbf{N} = \mathbf{r} \times \mathbf{F} = 0$, we have $\mathbf{J} =$ const. $\mathbf{J}$ is a vector out of the paper.

Thus we have the important result

$$\boxed{\frac{d\mathbf{J}}{dt} = \mathbf{N}} \tag{6.25}$$

The time rate of change of angular momentum is equal to the torque.

If the torque $\mathbf{N} = 0$, then $\mathbf{J} =$ const. *Angular momentum is constant in the absence of torque;* this is a statement of the *law of conservation of angular momentum.* Note that the law of conservation of angular momentum refers not only to particles in closed orbits. It applies as well to open orbits (as in Fig. 6.14) and to collision processes.

Consider a particle subjected to a central force of the form

$$\mathbf{F} = \hat{\mathbf{r}}f(r)$$

A *central* force is one that is everywhere directed exactly toward (or away from) a particular point. The torque is

$$\mathbf{N} = \mathbf{r} \times \mathbf{F} = \mathbf{r} \times \hat{\mathbf{r}}f(r) = 0$$

so that for central forces

$$\frac{d\mathbf{J}}{dt} = 0 \tag{6.26}$$

and the angular momentum is constant. In such a case the particle motion will be confined to a plane normal to the constant $\mathbf{J}$ vector. (In Chap. 9 we make extensive use of this result.) We now consider the extension of the torque equation to a system of N interacting particles.

Torques due to Internal Forces Sum to Zero The interaction forces that may be present among the particles of a system give rise to internal torques. We now show that these sum to zero, so that only torques due to external forces can change the angular momentum of a system of particles.

Including all forces, we write for the total torque

$$\mathbf{N} = \sum_{i=1}^{n} \mathbf{r}_i \times \mathbf{F}_i \tag{6.27}$$

where the index i labels a particle and n is the total number of particles comprising the system. But the force $\mathbf{F}_i$ on particle i is due partly to agencies external to the system and partly to its interactions with other particles; thus

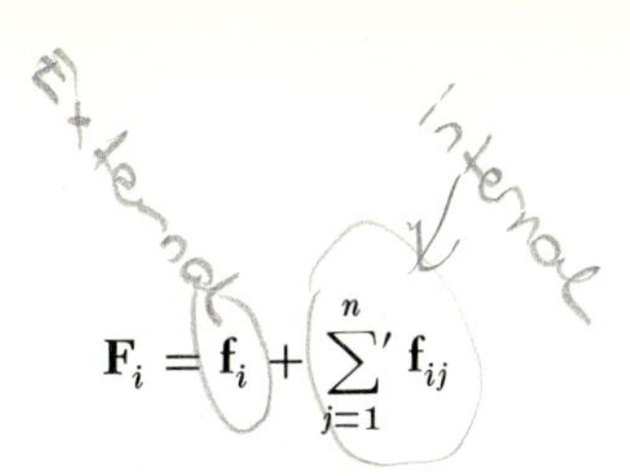

$$\mathbf{F}_i = \mathbf{f}_i + \sum_{j=1}^{n}{}' \mathbf{f}_{ij}$$

where $\mathbf{f}_i$ represents the external force and $\mathbf{f}_{ij}$ is the force exerted upon particle i by particle j. Σ' means that the term $j = i$ is excluded since the force of a particle upon itself has no meaning here. This permits us to write Eq. (6.27) in the form

$$\mathbf{N} = \sum_i \mathbf{r}_i \times \left(\mathbf{f}_i + \sum_j{}' \mathbf{f}_{ij}\right) = \sum_i \mathbf{r}_i \times \mathbf{f}_i + \sum_i \sum_j{}' \mathbf{r}_i \times \mathbf{f}_{ij}$$

where the last term with the double sum is the vector sum of torques due to internal forces of the system, which we shall call $\mathbf{N}_{\text{int}}$.

Detailed examination of this expression for $\mathbf{N}_{\text{int}}$ shows that it can be decomposed into a double sum of pairs of terms:

$$\mathbf{N}_{\text{int}} = \sum_i \left[\sum_j{}' \left(\mathbf{r}_i \times \mathbf{f}_{ij} + \mathbf{r}_j \times \mathbf{f}_{ji}\right)\right] \tag{6.28}$$

where for each value of i taken in turn we sum over all values of j, excluding $j = i$. (This decomposition is readily recognized by treating a system of a small number of particles; see Prob. 5.) Now, according to Newton's Third Law, $\mathbf{f}_{ji} = -\mathbf{f}_{ij}$ and Eq. (6.28) becomes

$$\mathbf{N}_{\text{int}} = \sum_i \left[\sum_j{}' \left(\mathbf{r}_i - \mathbf{r}_j\right) \times \mathbf{f}_{ij}\right]$$

And if the forces between particles are directed along the lines joining interacting pairs, i.e., if they are central forces, this expression will be zero since $\mathbf{f}_{ij}$ will be parallel to $\mathbf{r}_i - \mathbf{r}_j$. Thus

$$\mathbf{N}_{\text{int}} = 0 \tag{6.29}$$

The result is also true for noncentral forces of interaction; however, we shall not prove this fact here.

Torque due to Gravity A question of importance in problems of motion on the surface of the earth is: Can we find a point in an extended body (that is, a body made up of point masses or of a continuous distribution of mass) such that the torque about this point due to all the gravitational forces is zero? For example, if a uniform stick is held at one end, gravity will exert a torque about that end unless the stick is vertical. Where can the stick be held so that there will be no torque?

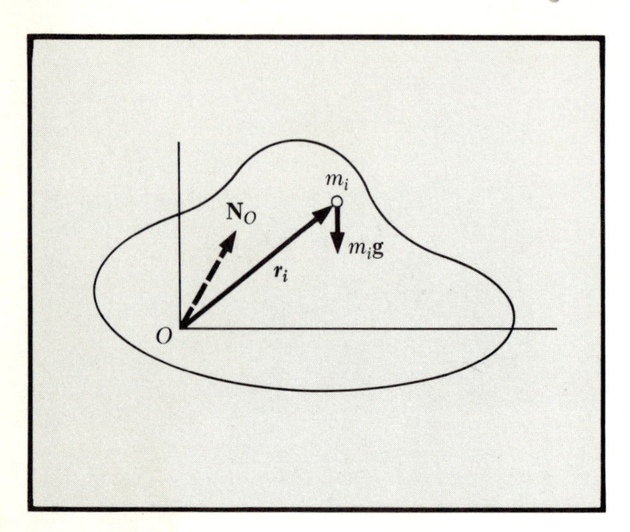

FIG. 6.15 Torque due to force of gravity, $m_i\mathbf{g}$, about O is $\mathbf{r}_i \times m_i\mathbf{g}$.

Obviously, it can be held at the center, but let us treat the problem generally.

Take a point O, as in Fig. 6.15:

$$\mathbf{N}_0 = \sum \mathbf{r}_i \times m_i\mathbf{g}$$

Since $\mathbf{g}$, the acceleration of gravity, is constant, we can rewrite this as

$$\mathbf{N}_0 = \left(\sum m_i\mathbf{r}_i\right) \times \mathbf{g}$$

But

$$\sum m_i\mathbf{r}_i = \mathbf{R}_{\text{c.m.}} \sum m_i$$

Therefore, if O is the center of mass,

$$\sum m_i\mathbf{r}_i = 0 \qquad \text{and} \qquad \mathbf{N}_0 = 0$$

This point is often called the *center of gravity*, which is identical with the center of mass provided $\mathbf{g}$ is constant over the body.

If the point is not the center of mass, what will be the value of $\mathbf{N}$? We know that

$$\sum \mathbf{F}_i = \sum m_i\mathbf{g} = M\mathbf{g} = \mathbf{F}_{\text{grav}} \tag{6.30}$$

when M is the total mass. Is the torque around a point A in Fig. 6.16 simply related to this force $M\mathbf{g}$?

$$\begin{aligned}\mathbf{N}_A &= \sum \mathbf{r}_{Ai} \times m_i\mathbf{g} = \sum (\mathbf{R}_{AO} + \mathbf{r}_{Oi}) \times m_i\mathbf{g} \\ &= \mathbf{R}_{AO} \times \sum m_i\mathbf{g} = \mathbf{R}_{AO} \times M\mathbf{g} = \mathbf{R}_{\text{c.m.}} \times M\mathbf{g}\end{aligned} \tag{6.31}$$

where we have used the fact that $\Sigma m_i\mathbf{r}_{Oi} = 0$ because the point O is the center of mass. Thus we see that the total effect of the forces of gravity can be replaced by that of a single force $M\mathbf{g}$ acting through the center of mass (see Prob. 6).

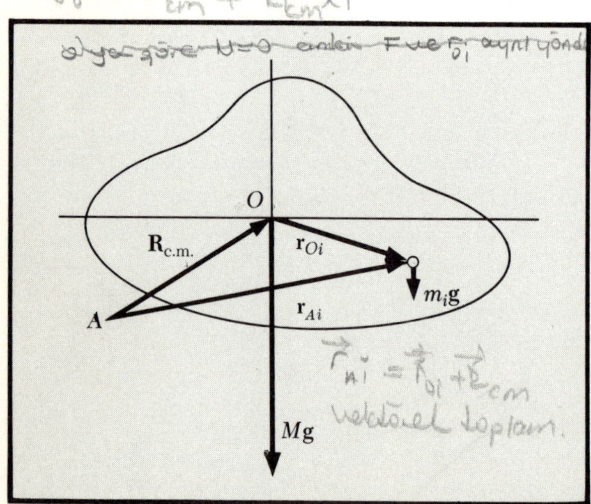

FIG. 6.16 Torque due to forces of gravity about A can be expressed as $\mathbf{R}_{\text{c.m.}} \times M\mathbf{g}$, where O is the center of mass.

Angular Momentum about the Center of Mass The total angular momentum of a system of particles referred to an arbitrary fixed point in an inertial reference frame as origin is, by Eq. (6.23),

$$\mathbf{J} = \sum_{i=1}^{N} M_i\mathbf{r}_i \times \mathbf{v}_i \tag{6.32}$$

Just as for the single particle the value of $\mathbf{J}$ depends on the point we choose for origin O. With $\mathbf{R}_{\text{c.m.}}$ as the vector from the origin to the position of the center of mass, we rewrite

$\mathbf{J}$ in a convenient and important form by subtracting and adding the quantity

$$\sum_i M_i \mathbf{R}_{\text{c.m.}} \times \mathbf{v}_i$$

to expression Eq. (6.32). Thus

$$\mathbf{J} = \sum_{i=1}^{N} M_i(\mathbf{r}_i - \mathbf{R}_{\text{c.m.}}) \times \mathbf{v}_i + \sum_{i=1}^{N} M_i \mathbf{R}_{\text{c.m.}} \times \mathbf{v}_i$$

$$= \mathbf{J}_{\text{c.m.}} + \mathbf{R}_{\text{c.m.}} \times \mathbf{P} \tag{6.33}$$

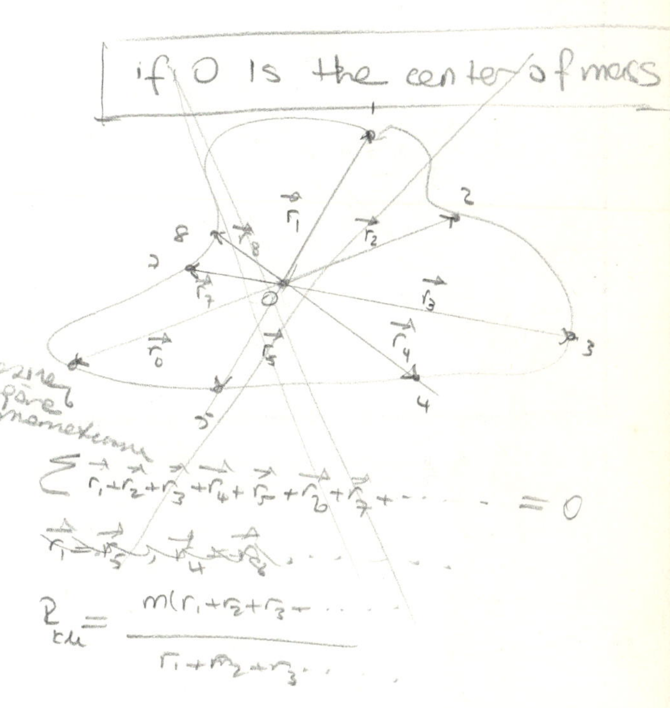

where $\mathbf{J}_{\text{c.m.}}$ is the angular momentum *about* the center of mass and $\mathbf{P} \equiv \Sigma M_i \mathbf{v}_i$ is the total linear momentum. The term $\mathbf{R}_{\text{c.m.}} \times \mathbf{P}$ is the angular momentum due to motion of the center of mass about the origin. This term depends on the choice of origin; the term $\mathbf{J}_{\text{c.m.}}$ does not. In the physics of a molecule, an atom, or a fundamental particle, we find it useful to call $\mathbf{J}_{\text{c.m.}}$ the *spin* angular momentum, or just the spin.

By recognizing that $\mathbf{N}_{\text{int}} = 0$, we have from Eqs. (6.25) and (6.33)

$$\frac{d}{dt}\mathbf{J}_{\text{tot}} = \mathbf{N}_{\text{ext}} \tag{6.34}$$

$$\mathbf{J}_{\text{tot}} = \mathbf{J}_{\text{c.m.}} + \mathbf{R}_{\text{c.m.}} \times \mathbf{P} \tag{6.35}$$

Here $\mathbf{J}_{\text{c.m.}}$ is the angular momentum about the center of mass, whereas $\mathbf{R}_{\text{c.m.}} \times \mathbf{P}$ is the angular momentum of the center of mass about the arbitrary origin. It is usually a very good idea to choose the origin at the center of mass. Then Eq. (6.34) may be written as

$$\frac{d}{dt}\mathbf{J}_{\text{c.m.}} = \mathbf{N}_{\text{ext}}$$

If no external forces act, then $\mathbf{N}_{\text{ext}} = 0$ and $\mathbf{J}_{\text{c.m.}}$ is constant.

We saw that the motion of the center of mass is determined by the total external force acting on the body. We see now that the rotation about the center of mass is determined by the total external torque. (In Chap. 8, we make particular use of this result.)

The geometric meaning of the angular momentum of a particle in an orbit enclosing the origin is suggested by Fig.

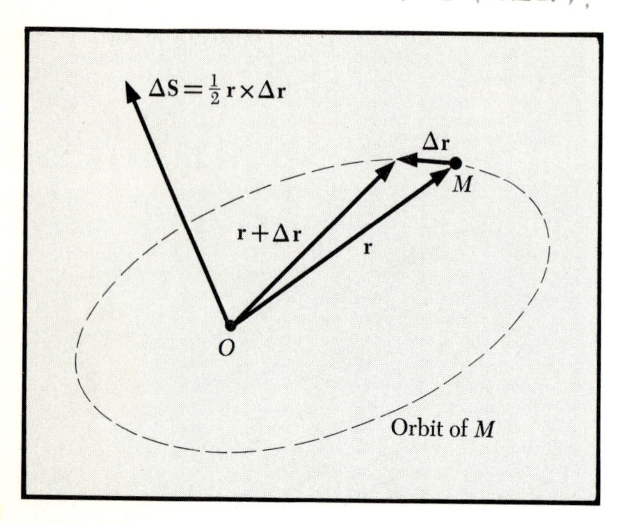

FIG. 6.17 Geometric meaning of angular momentum in terms of area swept out per unit time.

6.17. The vector area $\Delta\mathbf{S}$ of the triangle is given by

$$\Delta\mathbf{S} = \tfrac{1}{2}\mathbf{r} \times \Delta\mathbf{r}$$

Then

$$\frac{d\mathbf{S}}{dt} = \frac{1}{2}\mathbf{r} \times \mathbf{v} = \frac{1}{2M}\mathbf{r} \times \mathbf{p} = \frac{1}{2M}\mathbf{J} \tag{6.36}$$

We have seen that, with suitable choice of the origin, $\mathbf{J}$ = constant for central forces.

If in a planetary problem the origin is taken at the sun, then the angular momentum of a planet is constant, apart from disturbances (perturbations) by other planets. For central forces we see from Eqs. (6.26) and (6.36) that

1. The orbit lies in a plane.
2. The rate of sweeping out of area is a constant—this is one of the three Kepler laws (discussed in Chap. 9).

The first result follows because $\mathbf{r}$ and $\Delta\mathbf{r}$ are in a plane perpendicular to $\mathbf{J}$ and $\mathbf{J}$ is constant in magnitude and direction in a central field.

The planets move in elliptical orbits about the sun at a focus. In order to conserve angular momentum each planet must move faster at the point of closest approach than at the furthest point. This result follows because at these points $\mathbf{r}$ is perpendicular to $\mathbf{v}$ and the angular momentum at these points is Mvr. By the conservation of angular momentum the values of Mvr at these points must be equal, so that the shorter r is associated with the larger v (see Fig. 6.18).

For a particle moving in a circle the velocity $\mathbf{v}$ is perpendicular to $\mathbf{r}$, so that

$$J = Mvr = M\omega r^2 \tag{6.37}$$

For a particle moving along a straight line that misses the origin by distance b, the student should prove for himself that

$$\mathbf{J} = \mathbf{r} \times M\mathbf{v} = Mvb\hat{\mathbf{u}}$$

where $\hat{\mathbf{u}}$ is a unit vector perpendicular to the plane defined by the line of motion and the origin point.

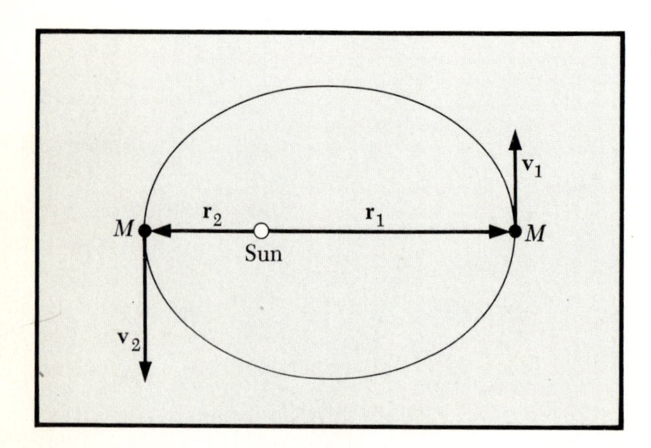

FIG. 6.18 The planet M has constant angular momentum about the sun. Thus $Mr_2v_2 = Mr_1v_1$, where r_1 = greatest distance from sun and r_2 = smallest distance from sun. All the planetary orbits have much less eccentricity than shown here. The figure is exaggerated for clarity.

EXAMPLE

Proton Scattering by a Heavy Nucleus A proton approaches a very massive nucleus of charge Ze. At infinite separation the energy of the proton is $\frac{1}{2}M_pv_0^2$. A *linear extrapolation* to small separations of the trajectory at large separations would pass at a minimum distance b from the heavy nucleus, as in Fig. 6.19. This distance is called the *impact parameter*. What is the distance of closest approach for the actual orbit? (Take the mass of the heavy nucleus to be infinite, so that its recoil energy may be neglected; that is, it may be considered stationary.)

The initial angular momentum of the proton taken about the heavy nucleus is $M_p v_0 b$, where v_0 is the initial velocity of the proton. At the distance of closest approach, denoted by s, the angular momentum is $M_p v_s s$, where v_s is the velocity at this point. The force is central, so that angular momentum is conserved and thus

$$M_p v_0 b = M_p v_s s \qquad v_s = \frac{v_0 b}{s}$$

Note that we have considered the heavy nucleus to remain at rest.

The energy of the proton is also conserved in the collision. The initial energy is all kinetic and is $\frac{1}{2}M_p v_0^2$. The energy at the point of closest approach is

$$\tfrac{1}{2}M_p v_s^2 + \frac{Ze^2}{s}$$

where the first term is the kinetic energy and the second term is the potential energy. Thus the law of conservation of energy tells us that

$$\tfrac{1}{2}M_p v_s^2 + \frac{Ze^2}{s} = \tfrac{1}{2}M_p v_0^2$$

Eliminating v_s gives

$$\frac{Ze^2}{s} = \tfrac{1}{2}M_p v_0^2\left[1 - \left(\frac{b}{s}\right)^2\right]$$

This equation may be solved for s. (Note that the conservation laws have told us quite a lot about the collision process.) If SI units are used, these last three expressions become

$$\tfrac{1}{2}M_p v_s^2 + \frac{kZe^2}{s} \qquad \tfrac{1}{2}M_p v_s^2 + \frac{kZe^2}{s} = \tfrac{1}{2}M_p v_0^2$$

and

$$\frac{kZe^2}{s} = \tfrac{1}{2}M_p v_0^2\left[1 - \left(\frac{b}{s}\right)^2\right]$$

(The methods of Chap. 9 give the complete solution to this problem.)

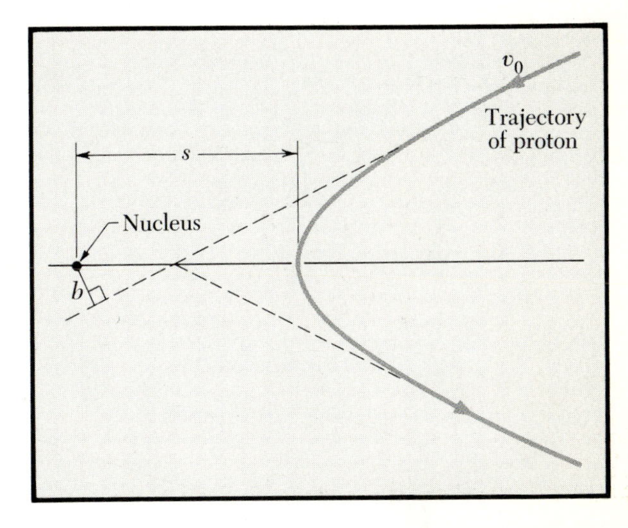

FIG. 6.19 Motion of proton in coulomb field of a heavy nucleus. The trajectory is a hyperbola (Chap. 9). The distance of closest approach is s. The impact parameter b is the normal distance from the nucleus to the linear extrapolation of the initial portion of the trajectory.

Rotational Invariance Just as we found that the conservation of linear momentum was a consequence of galilean invariance and the conservation of energy, so we can deduce that conservation of angular momentum is a consequence of the invariance of the potential energy under rotation of the reference frame (or system). If there is an external torque, we shall, in general, do work against this torque in rotating the system. If we do work, the potential energy must change. If the potential energy U is unchanged by the rotation, there is no external torque. Zero external torque means that the angular momentum is conserved.

The argument can be pursued analytically. Consider the effect of a rotational displacement of a system of particles by which any particle position vector $\mathbf{r}$ is converted into $\mathbf{r}'$. Its length is, of course, unchanged. We assert that the conservation of angular momentum follows from

$$U(\mathbf{r}_1',\mathbf{r}_2', \ldots, \mathbf{r}_N') = U(\mathbf{r}_1,\mathbf{r}_2, \ldots, \mathbf{r}_N)$$

This relationship implies restrictions upon the form of dependence of U upon the $\mathbf{r}$ vectors. A form that can satisfy the relationship is one depending only upon vector differences among the $\mathbf{r}$ vectors, so that for a simple system of only two particles we could write

$$U(\mathbf{r}_1,\mathbf{r}_2) = U(\mathbf{r}_2 - \mathbf{r}_1)$$

The rotation operation changes the direction of $\mathbf{r}_2 - \mathbf{r}_1$ but not the magnitude. U will be invariant if it depends only upon the magnitude, i.e., the distance separating the particles, and not upon the direction of the separation vector; thus

$$U(\mathbf{r}_1,\mathbf{r}_2) = U(|\mathbf{r}_2 - \mathbf{r}_1|)$$

This is equivalent to a statement of the homogeneity and isotropy of space.

For a potential energy function of this form it is ensured that $\mathbf{F}_{12} = -\mathbf{F}_{21}$ and that these forces are directed along the line $\mathbf{r}_2 - \mathbf{r}_1$. Thus, the force is central and the torque vanishes. This ensures that angular momentum will be conserved. For N particles the rotational invariance of the potential is ensured if U depends only on the magnitudes of the separations between the several particles.

The potential seen by an individual electron or ion in a crystal is not rotationally invariant because the electric field due to the other ions in a crystal is highly nonuniform or nonhomogeneous. Therefore, in general, we do not expect to find a conservation law for the angular momentum of the electronic shells of an ion in a crystal, even though the angular momentum is conserved for the same ion considered in free space. The nonconservation of electronic angular momentum of ions in crystals has been observed in studies of paramagnetic ions in crystals, and the effect is called *quenching* of the orbital angular momentum.

The angular momentum $\mathbf{J}$ of the earth is constant with respect to the sun as origin because $\mathbf{r} \times \mathbf{F} = 0$ for each mass point in the earth, where **F** is the gravitational force acting between the sun and the mass point.

EXAMPLE

Angular Acceleration Accompanying Contraction A particle of mass M is attached to a string (Fig. 6.20) and constrained to move in a horizontal plane (the plane of the dashed line). The particle rotates with velocity v_0 when the length of the string is r_0. How much work is done in shortening the string to r?

The force on the particle due to the string is radial, so that *the torque is zero* as the string is shortened. Therefore the angular momentum must remain constant as the string is shortened:

$$Mv_0r_0 = Mvr \tag{6.38}$$

The kinetic energy at r_0 is $\frac{1}{2}Mv_0^2$; at r it has been increased to

$$\frac{1}{2}Mv^2 = \frac{1}{2}Mv_0^2\left(\frac{r_0}{r}\right)^2 \tag{6.39}$$

because $v = v_0r_0/r$ from above. It follows that the work W done from outside in shortening the string from r_0 to r is

$$W = \frac{1}{2}Mv_0^2\left[\left(\frac{r_0}{r}\right)^2 - 1\right]$$

This can also be calculated directly as

$$\int_{r_0}^{r} \mathbf{F}_{\text{centrip}} \cdot d\mathbf{r} = -\int_{r_0}^{r} F_{\text{centrip}}\, dr$$

We see that the angular momentum acts on the radial motion as an effective repulsive force: We have to do *extra* work on the particle on bringing it from large distances to small distances *if* we require that the angular momentum be conserved in the process.

Compare this behavior with that of a particle rotating on a string that is freely winding up on a smooth fixed peg of finite diameter. Why is the kinetic energy now constant as the string winds up? (See Prob. 12 at the end of the chapter.)

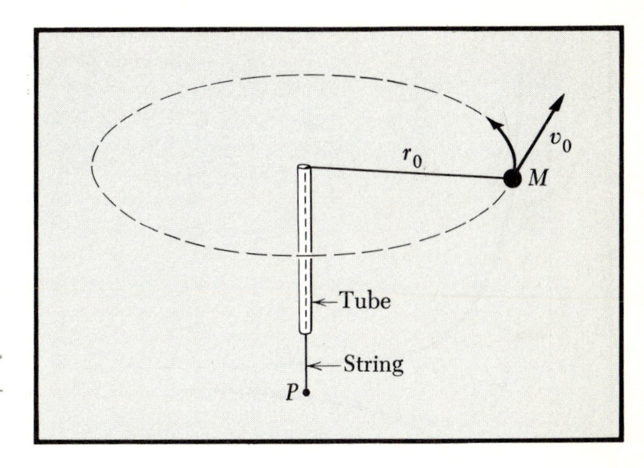

FIG. 6.20 Mass M describes circular motion of radius r_0 and velocity v_0. It is connected to a string which passes through a tube. The distance r_0 can be shortened by pulling on the string at P.

EXAMPLE

Shape of the Galaxy The result of the preceding example has a probable bearing on the shape of the galaxy. Consider a very large mass M of gas endowed initially with some angular momentum,[1] as in Fig. 6.21*a*. The gas contracts under its gravitational interaction. As the volume occupied by the gas gets smaller, the conservation

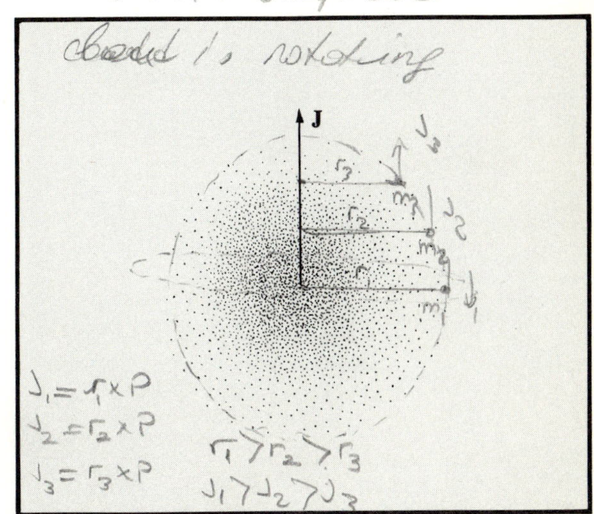

FIG. 6.21 (*a*) Originally a diffuse cloud of gas, possessing some angular momentum.

[1] It is not possible in the present state of knowledge to say where the gas came from in the first place or why a given mass of gas should have an angular momentum. Masses without angular momentum will condense as spheres.

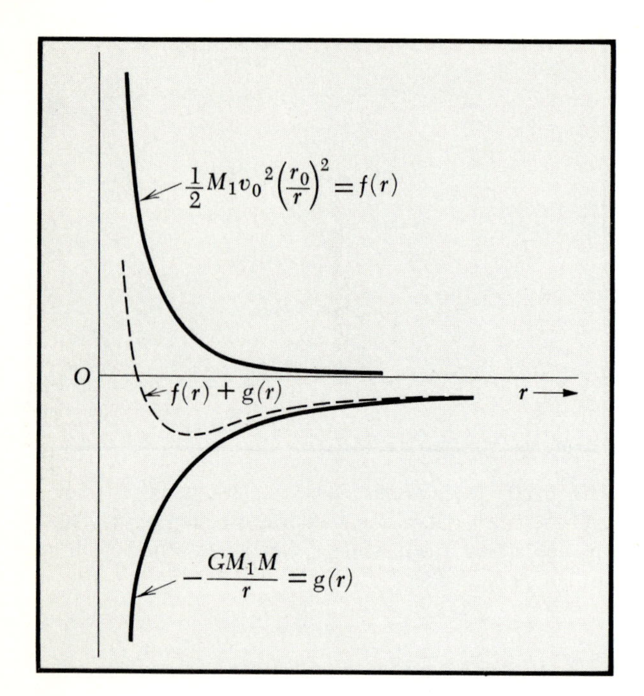

FIG. 6.21 (*cont'd.*) (*b*) Contraction of the galaxy in the plane normal to **J** is limited because the centrifugal "potential energy" $f(r)$ increases so rapidly as $r \to 0$. Thus $f(r) + g(r)$ has a minimum at finite value of r, as shown.

of angular momentum requires an increase in the angular velocity and its kinetic energy increases. But we have just seen that work is needed to produce the increase in angular velocity. Where does the kinetic energy come from? It can only come from the gravitational energy of the gas.

To work out problems such as this, physicists often use energy considerations. A particle of mass M_1 in the outer regions of the galaxy will have a gravitational potential energy due to its interaction with the galaxy of the order of magnitude of

$$-\frac{GM_1M}{r} \tag{6.40}$$

where r is the distance from the center of the galaxy and M is the mass of the galaxy. As r gets smaller and smaller, this gravitational potential energy gets more and more negative but the kinetic energy of Eq. (6.39) gets more and more positive; in fact, it gets positive faster than Eq. (6.40) gets negative. We treat this radius-dependent kinetic energy as if it were a contribution to the potential energy and term it the *centrifugal potential energy* (see Prob. 13). The equilibrium condition is the minimum in the sum of the two, as shown in Fig. 6.21*b*. From Chap. 5, we remember that the derivative of a potential energy is a force, and so the minimum in the sum of these two energies corresponds to setting the sum of two forces equal to zero. This condition is

$$\frac{d}{dr}\left[-\frac{GM_1M}{r} + \frac{1}{2}M_1v_0^2\left(\frac{r_0}{r}\right)^2\right] = 0$$

$$\frac{GM_1M}{r^2} - M_1v_0^2\frac{r_0^2}{r^3} = 0$$

which is equivalent to

$$\frac{GM_1M}{r^2} = M_1\frac{v^2}{r} \tag{6.41}$$

when we replace v_0r_0/r by v from Eq. (6.38). We recognize Eq. (6.41) as just the condition that the centripetal force due to gravity equals the mass times centripetal acceleration.

But the cloud of gas is able to collapse in the direction parallel to the axis of the total angular momentum without changing the value of the angular momentum. The contraction is driven by gravitational attraction; the energy gain in contraction must be dissipated in some way, and this is believed to occur by radiation. The cloud is therefore able to collapse rather completely in the direction parallel to **J**, but the contraction in the equatorial plane is restricted (see Fig. 6.21*c* and *d*). This model of galactic evolution is oversimplified; as yet there is no generally accepted model.

The diameter of our galaxy is of the order of 3×10^4 parsecs, or 10^{23} cm. (1 parsec = 3.084×10^{18} cm = 3.084×10^{16} m.) The thickness of the galaxy in the neighborhood of the sun depends somewhat on how the thickness is defined, but the vast majority of stars cluster about the median plane in a thickness of several hundred

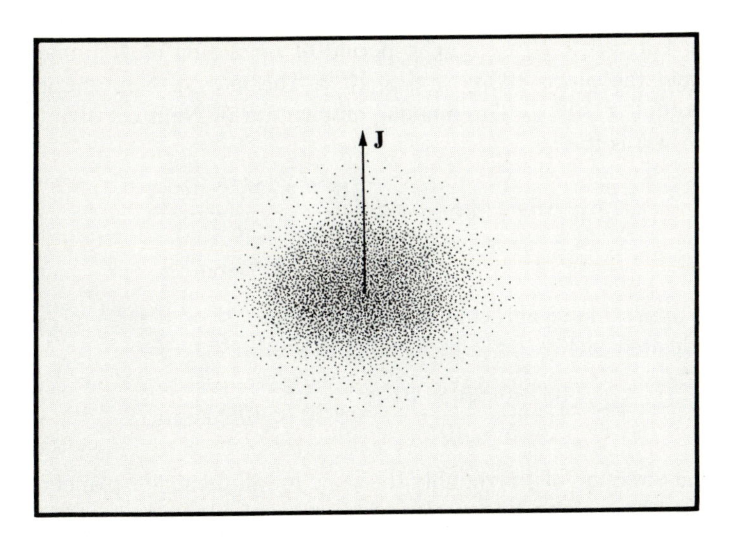

(*c*) The galaxy begins to flatten as it contracts and rotates faster.

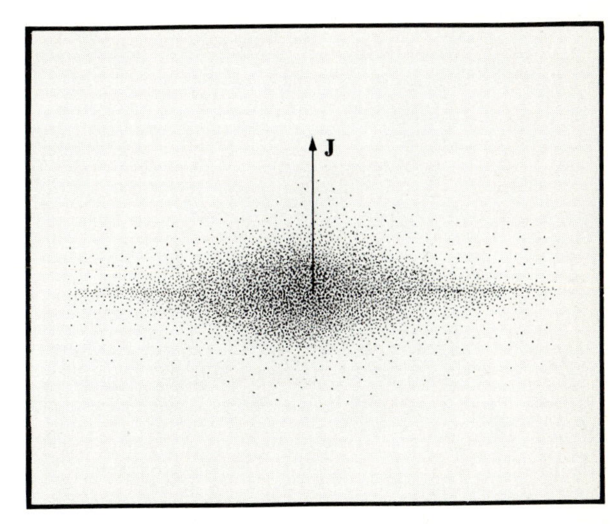

(*d*) Finally it assumes a pancake shape with a more or less spherical central core.

parsecs. Thus the galaxy is greatly flattened. The mass of the galaxy is put at about 2×10^{11} times the mass of the sun, or

$$(2 \times 10^{11})(2 \times 10^{33}) \approx 4 \times 10^{44}\ \text{g} \qquad \text{or} \qquad 4 \times 10^{41}\ \text{kg}$$

An estimate of the mass may be made from Eq. (6.41) by substituting the known values of v and r for the sun. The sun lies toward the outer edge of the galaxy at about 10^4 parsecs $\approx 3 \times 10^{22}$ cm $\approx 3 \times 10^{20}$ m from the axis of the galaxy. The orbital velocity of the sun about the center of the galaxy is approximately 3×10^7 cm/s $\approx 3 \times 10^5$ m/s; from Eq. (6.41) we derive as an estimate of the mass of the galaxy:

$$M = \frac{v^2 r}{G} \approx \frac{(10^{15})(3 \times 10^{22})}{7 \times 10^{-8}} \approx 4 \times 10^{44}\ \text{g}$$

or, in SI units,

$$M = \frac{v^2 r}{G} \approx \frac{(10^{11})(3 \times 10^{20})}{7 \times 10^{-11}} \approx 4 \times 10^{41}\ \text{kg}$$

We have neglected the effect of the mass that lies further from the center of the galaxy than the sun.

Angular Momentum of the Solar System Figure 6.22 shows the angular momenta of the several components of the solar system. Let us make an estimate for ourselves of one of the values given, as a check. Consider the planet Neptune, whose orbit is very closely circular. The mean distance of Neptune from the sun is given in reference works as 2.8×10^9 mi $\approx 5 \times$

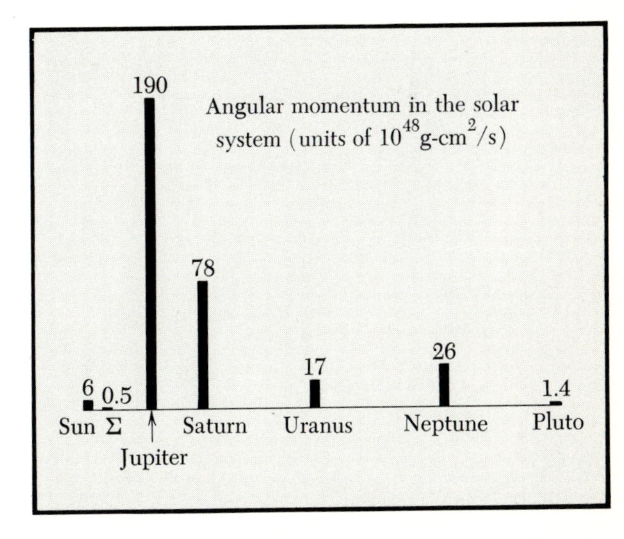

FIG. 6.22 Distribution of angular momentum in the solar system, about the center of the sun. The symbol Σ denotes the sum of the four planets Mercury, Venus, Earth, and Mars. Note the relatively small contribution of the rotation of the sun about its own axis.

10^9 km $\approx 5 \times 10^{14}$ cm. The period of revolution of Neptune about the sun is 165 yr $\approx 5 \times 10^9$ s. The mass of Neptune is about 1×10^{29} g. The angular momentum of Neptune about the sun is then

$$J = Mvr = M\frac{2\pi r}{T}r \approx \frac{(10^{29})(6)(25 \times 10^{28})}{5 \times 10^9}$$
$$\approx 30 \times 10^{48} \text{ g-cm}^2/\text{s}$$

in approximate agreement with the value 26×10^{48} g-cm^2/s indicated in Fig. 6.22. In SI units this value is

$$J \approx \frac{(10^{26})(6)(25 \times 10^{24})}{5 \times 10^9} \approx 30 \times 10^{41} \text{ kg-m}^2/\text{s}$$

The direction of **J** is roughly the same for all the major planets.

The angular momentum of Neptune *about its own center of mass* is much smaller. The angular momentum of a rotating uniform sphere is of the order of MvR, where v is the surface velocity from the rotation and R is the radius. Actually, because the mass of a sphere is not concentrated at a distance R from the axis but is distributed, this result must be reduced for uniform distribution by a numerical factor, which, in Chap. 8, is found to be $\frac{2}{5}$. Thus

$$J_{\text{c.m.}} = \frac{2}{5}\frac{2\pi MR^2}{T}$$

where $T \equiv 2\pi R/v$ denotes the period of rotation of the planet about its own axis. For Neptune $T \approx 16$ h $\approx 6 \times 10^4$ s and $r \approx 1.5 \times 10^4$ mi $\approx 2.4 \times 10^9$ cm, whence

$$J_{\text{c.m.}} \approx \frac{(0.4)(6)(10^{29})(6 \times 10^{18})}{6 \times 10^4} \approx 2 \times 10^{43} \text{ g-cm}^2/\text{s}$$

or

$$J_{\text{c.m.}} \approx \frac{(0.4)(6)(10^{26})(6 \times 10^{14})}{6 \times 10^4} \approx 2 \times 10^{36} \text{ kg-m}^2/\text{s}$$

which is negligible compared with the orbital angular momentum about the sun.

A similar estimate of $J_{\text{c.m.}}$ for the sun gives 6×10^{48} g-cm^2/s (6×10^{41} kg-m^2/s). The rotation of the sun about an axis through its center accounts for only about 2 percent of the total angular momentum in the solar system. A typical hotter star may carry about 100 times as much angular momentum as the sun. It thus appears that the formation of a planetary system is an effective mechanism for carrying off angular momentum from a cooling star. If every star forms a planetary system in passing through the stage of its history similar to the sun, then there may be over 10^{10} stars with planets in our galaxy.

PROBLEMS

1. *Angular momentum of a satellite*
(*a*) What is the angular momentum (referred to the center of the orbit) of a satellite of mass M_s that moves in a circular orbit of radius r? The result is to be expressed in terms only of r, G, M_s, M_e (the mass of the earth).
Ans. $J = (GM_eM_s^2r)^{\frac{1}{2}}$.
(*b*) For $M_s = 100$ kg, what is the numerical value (in cgs units) of the angular momentum of an orbit for which the radius is twice the radius of the earth?

2. *Frictional effects on satellite motion*
(*a*) What is the effect of atmospheric friction on the motion of a satellite in a circular (or nearly circular) orbit? Why does friction increase the satellite velocity?
(*b*) Does friction increase or decrease the angular momentum of the satellite measured with respect to the center of the earth? Why?

3. *Energy—angular momentum relation for a satellite.* Express in terms of the angular momentum J the kinetic, potential, and total energy of a satellite of mass M in a circular orbit of radius r.
Ans. $K = J^2/2Mr^2 \qquad U = -J^2/Mr^2 \qquad E = -J^2/2Mr^2$.

4. *Electron bound to a proton.* An electron moves about a proton in a circular orbit of radius $0.5\ \text{Å} \equiv 0.5 \times 10^{-8}$ cm.
(*a*) What is the orbital angular momentum of the electron about the proton? *Ans.* 1×10^{-27} erg-s.
(*b*) What is the total energy (expressed in ergs and in electron volts)?
(*c*) What is the ionization energy, that is, the energy that must be given to the electron to separate it from the proton?

5. *Internal torques sum to zero.* Consider the isolated system of three particles 1, 2, and 3 (shown in Fig. 6.23) interacting with *central* forces $F_{12} = 1$ dyn, $F_{13} = 0.6$ dyn, and $F_{23} = 0.75$ dyn, where F_{ij} denotes the force on particle i when it interacts with particle j. Pick two different points and show that for each the sum of the torques about that point is zero.

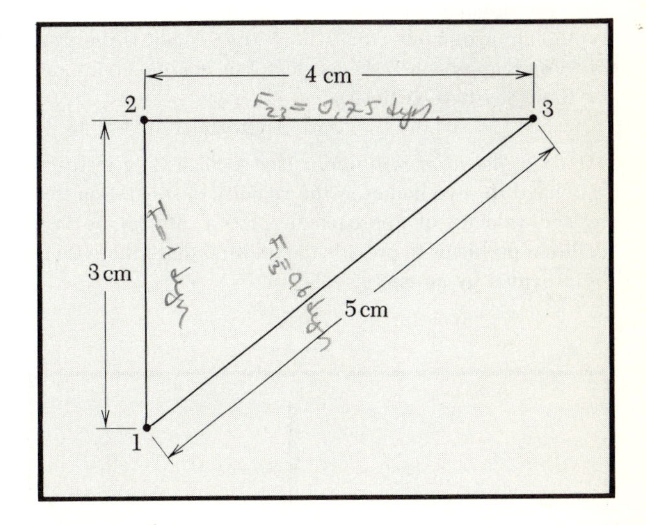

FIG. 6.23

6. *Forces on ladder.* A ladder of mass 20 kg and length 10 m rests against a slippery vertical wall at an angle of 30° with the vertical. The ladder, of uniform construction, is prevented from slipping by friction with the ground. What is the magnitude in dynes of the force exerted by the ladder on the wall? (*Hint:* Use the fact that the torques must sum to zero for a ladder at rest.) *Ans.* 5.7×10^6 dyn.

7. *Kinetic energy of center of mass.* In a collision of a particle of mass m_1 moving with speed v_1 with a stationary particle of mass m_2 not all the original kinetic energy can be converted into heat or internal energy. What fraction can be so converted? Show that this energy is just that equal to the kinetic energy in the center-of-mass system.

8. *Falling chain.* A chain of mass M and length l is coiled up on the edge of a table. A very small length at one end

is pushed off the edge and starts to fall under the force of gravity, pulling more and more of the chain off the table. Assume that the velocity of each element remains zero until it is jerked into motion with the velocity of the falling section. Find the velocity when a length x has fallen off.

Ans. $v^2 = \frac{2}{3}gx$.

When the entire length l is just off the table, what fraction of the original potential energy has been converted into the kinetic energy of translation of the chain?

9. *Angular momentum in near collision of two particles.* A neutron of energy 1 MeV passes a proton at such a distance that the angular momentum of the neutron relative to the proton approximately equals 10^{-26} erg-s. What is the distance of closest approach? (We neglect the energy of interaction between the two particles.)

Ans. Approximately 4×10^{-12} cm.

10. *Coefficient of restitution.* The coefficient of restitution r is defined for two bodies as the velocity of separation divided by the velocity of approach $0 \leq r \leq 1$. It can be used in collision problems to provide the solution that otherwise could be provided by an energy relation.

FIG. 6.24

(*a*) Show that if the coefficient of restitution between a ball and a massive flat horizontal plate is r, the height to which the ball will rise after n bounces is $h_0 r^{2n}$, where h_0 is the height from which it was dropped.

(*b*) Show that for a head-on collision of two bodies of masses m_1 and m_2 with coefficient r, the loss in kinetic energy is $(1 - r^2)$ times the kinetic energy in the center-of-mass system.

11. *Particle-dumbbell collision.* Two equal masses M are connected by a rigid rod of negligible mass and of length a. The center of mass of this dumbbell-like system is stationary in gravity-free space, and the system rotates about the center of mass with angular velocity $\boldsymbol{\omega}$. One of the rotating masses strikes head-on a third stationary mass M, and the two stick together.

(*a*) Locate the center of mass of the three-particle system at the instant prior to collision. What is the velocity of the center of mass? (*Note:* This is not the velocity of the point on the rigid rod that instantaneously coincides with the center of mass.)

(*b*) What is the angular momentum of the three-mass system about the center of mass at the instant prior to collision? At the instant following collision?

(*c*) What is the angular velocity of the system about the center of mass after the collision?

(*d*) What are the initial and final kinetic energies?

12. *Angular momentum of tetherball.* The object of the game tetherball (Fig. 6.24) is to hit the ball hard enough and fast enough to wind its tether cord in one direction about the vertical post to which it is tied before the opposing player can wind it in the opposite direction. The game is exciting, and the dynamics of the ball's motion are complicated. Let us examine a simple type of motion in which the ball moves in a horizontal plane in a spiral of decreasing radius as the cord winds round the post after a single blow that gives the ball an initial speed v_0. The length of the cord is l and the radius of the post is $a \ll l$.

(*a*) What is the instantaneous center of revolution?

(*b*) Is there a torque about the axis through the center of the post? Is angular momentum conserved?

(*c*) Assume that kinetic energy is conserved and calculate the speed as a function of time.

(*d*) What is the angular velocity after the ball has made five complete revolutions?

Ans. $\omega = (l - 10\pi a)v_0/[a^2 + (l - 10\pi a)^2]$.

13. *Effective centrifugal potential energy.* It is convenient to use plane polar coordinates r, θ for motion in a plane perpendicular to an axis of rotation (see Fig. 2.21).

(*a*) Show that the velocity in such a coordinate system may be written

$$\mathbf{v} = v_r\hat{\mathbf{r}} + v_\theta\hat{\boldsymbol{\theta}}$$

where v_r is just dr/dt, the rate of change of the length r, and

$$v_\theta = r\frac{d\theta}{dt}$$

(*b*) Show that the kinetic energy of a particle in this coordinate system is

$$K = \tfrac{1}{2}M(\dot{r}^2 + \omega^2 r^2)$$

where $\omega = d\theta/dt$.

(*c*) Show that the total energy is

$$E = U(r) + \tfrac{1}{2}M\dot{r}^2 + \frac{J^2}{2Mr^2}$$

where J is the angular momentum of the particle about the fixed axis normal to the plane of the motion. *Hint:* Recall Eq. (6.37).

(*d*) Because the force is central there is no torque on the particle and J is a constant of the motion. The term $J^2/2Mr^2$ is sometimes called the centrifugal potential energy. Show that the centrifugal potential energy represents an outward radial force J^2/Mr^3.

(*e*) If $U(r) = \frac{1}{2}Cr^2$, show that $U(r)$ represents an inward radial force $-Cr$.

(*f*) Show from (*d*) and (*e*) that the balance of these forces is equivalent to the condition $\omega^2 = C/M$.

14. *Rocket in earth's field.* A rocket of initial mass M_0 burns an adjustable amount of fuel β g/s. This fuel is ejected straight downward with a velocity V_0.

(*a*) Find β as a function of time in order that the rocket can remain stationary in space a short distance off the ground.

(*b*) Assuming that the amount of fuel used per second remains constant at a value α but is greater than in (*a*), find the velocity of the rocket upward as a function of the time.
Ans. With $M = M_0 - \alpha t$ and $\alpha = dM/dt$, then $v = -gt + V_0 \ln[M_0/(M_0 - \alpha t)]$.

(*c*) Compare this velocity for the case of $M = \frac{3}{4}M_0$ with that given in Eq. (6.22). Calculate these two velocities if $V_0 = 1.65 \times 10^5$ cm/s (five times the speed of sound).

15. *Ice skaters revolving on end of a rope.* Two ice skaters, each weighing 70 kg, are traveling in opposite directions with speed 650 cm/s but separated by a distance of 1000 cm perpendicular to their velocities. When they are just opposite each other, each grabs one end of a 1000-cm-long rope.

(*a*) What is their angular momentum about the center of the rope before they grab the ends? After?

(*b*) Each now pulls in on his end of the rope until the length of the rope is 500 cm. What is the speed of each?

(*c*) If the rope breaks just as they get to 500 cm apart, what mass would it hold up against the force of gravity?

(*d*) Calculate the work done by each skater in decreasing their separation and show that this is equal to his change in kinetic energy.

16. *Slowing down of space vehicle.* A space vehicle has mass 200 kg and cross-sectional area 2×10^4 cm². It travels in a region without appreciable gravitational field through a rarefied atmosphere whose mass density is 2×10^{-15} g/cc with initial speed 7.6×10^5 cm/s. (These would be approximate values for a satellite 500 km above the surface of the earth.) Assume that the conditions of the example on page 183 apply, that all the gas the space vehicle encounters sticks to it.

(*a*) Work out the value of c. Consider the mass of gas picked up in 1 s. *Ans.* $c = 4 \times 10^{-11}$ g/cm.

(*b*) Using the conservation of momentum $Mv = M_0v_0$, find the differential equation for v in terms of t and constants.
Ans. $dv/dt = -cv^3/M_0v_0$.

(*c*) Solve for v and find the time required for the satellite to slow down to 0.9 of its initial speed.
Ans. $t \approx 24$ yr.

FURTHER READING

PSSC, "Physics," chap. 22, D. C. Heath and Company, Boston, 1960.

HPP, "Project Physics Course," chap. 9, Holt, Rinehart and Winston, Inc., New York, 1970.

CONTENTS

Harmonic Oscillator: Properties and Examples

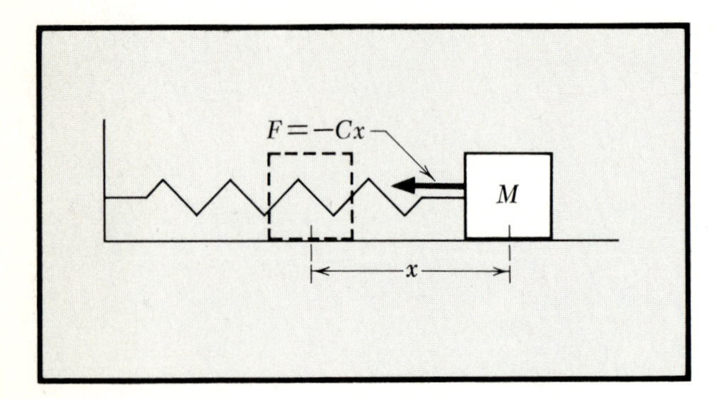

FIG. 7.1 Stretched spring acting on mass M. In the dashed position the spring is unstretched.

The harmonic oscillator is an exceptionally important example of periodic motion because it serves as an exact or approximate model for many problems in classical and quantum physics. The classical systems that are realizations of the harmonic oscillator include any stable system when slightly displaced from equilibrium, such as:

1 A mass on a spring in the limit of small amplitude of oscillation
2 A simple pendulum in the limit of small angle of oscillation
3 An electric circuit composed of an inductance and a capacitance for currents low enough for the circuit elements to be linear

An electric or mechanical circuit element is said to be linear if the response is directly proportional to the driving force. Most phenomena (but not all the interesting ones) in physics are linear if the range is taken to be small enough, just as most curves you encounter may be considered to be straight lines for a sufficiently small range of values.

The most important properties of the harmonic oscillator are the following:

1 The frequency of the motion is independent of the amplitude of the oscillation, within the restrictions of linearity.
2 The effects of several driving forces may be superposed linearly.

In this chapter we shall treat the properties of the harmonic oscillator. We shall consider both free and forced motion with and without damping, although the main elements of forced motion are given as an Advanced Topic at the end of this chapter. We also treat as an Advanced Topic the effects of small nonlinear interactions since it is useful to be acquainted with such types of motion.

MASS ON A SPRING

In Chap. 5 we discussed the potential energy in a compressed or stretched spring in which the force is directly proportional to the compression or stretch:

$$\mathbf{F} = -Cx\hat{\mathbf{x}}$$

where x is positive for a stretch and negative for a compression. What is the motion of a mass M under such a force? Ideally we can imagine, as in Fig. 7.1, the mass moving on a frictionless table:

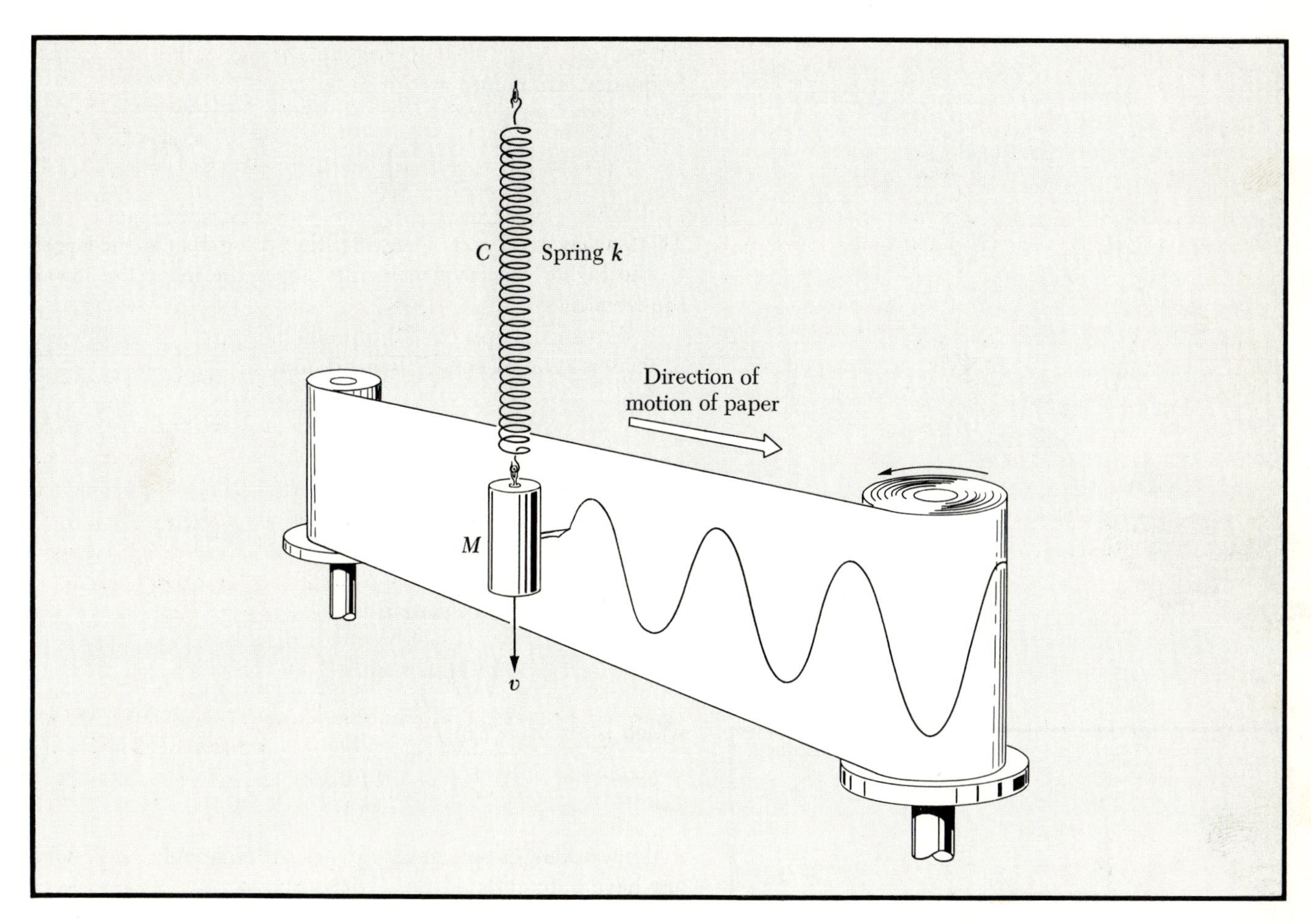

FIG. 7.2 A simple harmonic oscillator consisting of mass M and weightless spring of spring constant C. A pen attached to M will trace out a sine curve on a paper strip driven at constant speed past M.

$$M\frac{d^2x}{dt^2} = -Cx \qquad \frac{d^2x}{dt^2} = -\frac{Cx}{M} \tag{7.1}$$

(This equation is discussed in the Mathematical Notes at the end of this chapter. Those unfamiliar with the solution should study the notes before proceeding.)

The solution can be expressed in the form

$$\boxed{x = A\sin(\omega_0 t + \phi)} \tag{7.2}$$

where

$$\omega_0 = \left(\frac{C}{M}\right)^{\frac{1}{2}} \tag{7.3}$$

At $t = 0$, $x = x_0 = A\sin\phi$, and $dx/dt = v_0 = \omega_0 A\cos\phi$, from which A and ϕ can be determined. Figure 7.2 illustrates the motion. A is called the *amplitude* and ϕ the *phase*. The frequency and period are given by

$$f_0 = \frac{\omega_0}{2\pi} = \frac{1}{2\pi}\left(\frac{C}{M}\right)^{\frac{1}{2}} \qquad T = \frac{1}{f_0} = 2\pi\left(\frac{M}{C}\right)^{\frac{1}{2}} \tag{7.4}$$

This is what we expect: The stiffer the spring, that is, the larger C, the higher the frequency; the bigger the mass, the lower the frequency.

We can also approach this problem from the point of view of conservation of energy [see Eq. (5.21)]:

$$\tfrac{1}{2}Mv^2 + \tfrac{1}{2}Cx^2 = \tfrac{1}{2}M\left(\frac{dx}{dt}\right)^2 + \tfrac{1}{2}Cx^2 = E \tag{7.5}$$

Using A as the value of x when $dx/dt = 0$, $E = \frac{1}{2}CA^2$ and

$$\frac{dx}{dt} = \left(\frac{C}{M}\right)^{\frac{1}{2}}(A^2 - x^2)^{\frac{1}{2}} \tag{7.6}$$

The solution to this equation is

$$\left(\frac{C}{M}\right)^{\frac{1}{2}} t = \sin^{-1}\frac{x}{A} + \text{const}$$

which is just Eq. (7.2)

$$x = A \sin\left[\left(\frac{C}{M}\right)^{\frac{1}{2}} t + \phi\right]$$

if the constant is set equal to $-\phi$. Alternatively, we could just have differentiated Eq. (7.5) to obtain

$$M\frac{dx}{dt}\frac{d^2x}{dt^2} + Cx\frac{dx}{dt} = 0$$

which reduces to Eq. (7.1). (Problems 2 to 4 at the end of this chapter are examples of the use of these ideas.)

SIMPLE PENDULUM

The simple pendulum consists of a point mass M at the lower end of a massless rod or string of length L pivoted freely at its upper end and moving in a vertical plane, as shown in Fig. 7.3. We know from observation that the motion is similar to that of the mass on a spring. One question we might ask is: What is the frequency? The most straightforward way of setting up this problem is to write the appropriate form of $\mathbf{F} = M\mathbf{a}$. Referring to Fig. 7.3, we see that the distance s along the arc, the velocity, and the acceleration of M are

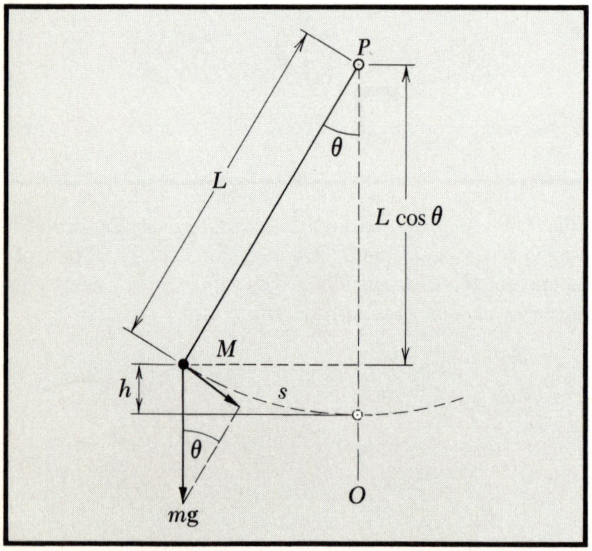

FIG. 7.3 The simple pendulum consists of a point mass M on the end of a massless rod L. The pendulum rotates about an axis through P which is perpendicular to the paper. Line OP is vertical. s is the arc length between O and the position of M.

$$s = L\theta \qquad v = \frac{ds}{dt} = L\frac{d\theta}{dt} = L\dot{\theta} \qquad a = \frac{d^2s}{dt^2} = L\frac{d^2\theta}{dt^2}$$

There are two forces in the problem: the force of gravity and the force exerted by the rod or string. However, the rod exerts no force component along s, and therefore we need consider only the component of mg along s. From Fig. 7.3 we see this is $mg \sin \theta$ in the direction to decrease θ. Therefore $\mathbf{F} = m\mathbf{a}$ reduces for this dimension to

$$mg \sin \theta = -mL\frac{d^2\theta}{dt^2} \tag{7.7}$$

But the series expansion for $\sin \theta$ is

$$\sin \theta = \theta - \frac{\theta^3}{3!} + \frac{\theta^5}{5!} \cdots \tag{7.8}$$

and so for small θ we write

$$mg \sin \theta = mg\,\theta$$

and Eq. (7.7) becomes

$$\frac{d^2\theta}{dt^2} = -\frac{g}{L}\theta \tag{7.9}$$

This is identical with Eq. (7.1), with g/L in place of C/M, and θ in place of x. Therefore

$$\boxed{\theta = \theta_0 \sin (\omega_0 t + \phi)} \tag{7.10}$$

where

$$\boxed{\omega_0 = \left(\frac{g}{L}\right)^{\frac{1}{2}}} \tag{7.11}$$

and the frequency we set out to find is, by Eqs. (7.4) and (7.11), $f_0 = 1/2\pi \sqrt{g/L}$.

The amplitude, or maximum value, of θ is θ_0; $\theta_0 \sin \phi$ is the value of θ at $t = 0$, and $\omega_0\theta_0 \cos \phi$ is the value of $d\theta/dt$. How large can θ_0 be and still have our assumption $\sin \theta = \theta$ valid? Table 7.1 gives some values of the period for various amplitudes. Evidently the amplitude can be over 20° before the actual value of the period departs by as much as 1 percent from the small-amplitude-approximation result.

Let us also look at the conservation-of-energy method of solving the problem. When the rod is deflected through angle θ, the mass M is raised by the distance

$$h = L - L \cos \theta$$

TABLE 7.1

Amplitude, °	*Period* $\div 2\pi \sqrt{l/g}$
0	1.0000
5	1.0005
10	1.0019
15	1.0043
20	1.0077
30	1.0174
45	1.0396
60	1.0719

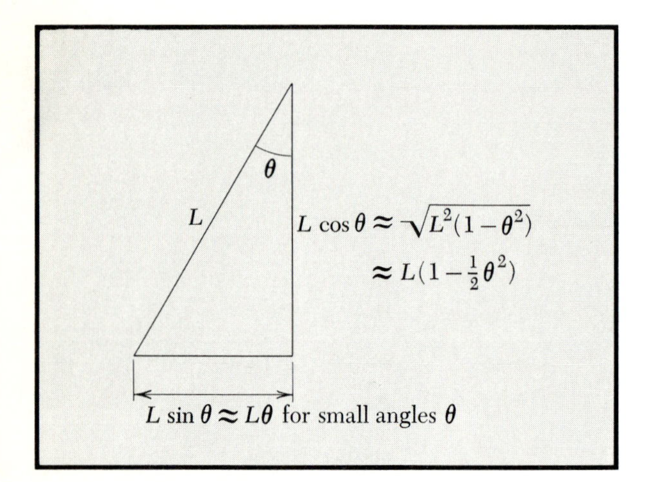

FIG. 7.4 The pythagorean theorem together with the binomial expansion reveals why $\cos\theta \simeq 1 - \frac{1}{2}\theta^2$ for $\theta \ll 1$ rad.

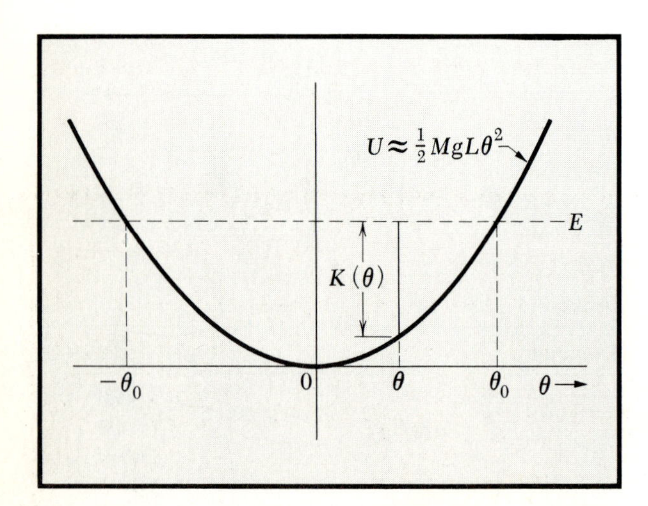

FIG. 7.5 Graph of potential energy vs θ. The pendulum oscillates between the limits θ_0 and $-\theta_0$. At these "turning points," $K = 0$ and $U = E$. At $\theta = 0$, $U = 0$ and $K = E$. For $\theta \ll 1$ rad, $U \simeq \frac{1}{2}MgL\theta^2$.

as is seen by reference to Figs. 7.3 and 7.4. The potential energy of the mass M in the gravitational field of the earth is

$$U(h) = Mgh = MgL(1 - \cos\theta)$$

referred to the undeflected (vertical) position as the zero of potential energy. The kinetic energy of the pendulum is

$$K = \tfrac{1}{2}Mv^2 = \tfrac{1}{2}ML^2\dot{\theta}^2$$

where $v = L\dot{\theta}$ relates the velocity and the rate of change of the angle of deflection. The total energy is

$$E = \text{KE} + \text{PE} = K + U = \tfrac{1}{2}ML^2\dot{\theta}^2 + MgL(1 - \cos\theta) \quad (7.12)$$

By the law of conservation of energy we know that this sum must be constant. We use this fact plus the smallness of the angle θ to obtain a solution for the frequency of motion. Now

$$\cos\theta = 1 - \tfrac{1}{2}\theta^2 + \tfrac{1}{24}\theta^4 \cdots$$

Thus, for $\theta \ll 1$ rad, we may neglect the terms in θ^4 and higher powers and approximate the energy in Eq. (7.12) by

$$E = \tfrac{1}{2}ML^2\dot{\theta}^2 + \tfrac{1}{2}MgL\theta^2 \quad (7.13)$$

Solving Eq. (7.13) for $\dot{\theta}$ we find

$$\frac{d\theta}{dt} = \left(\frac{2E - MgL\theta^2}{ML^2}\right)^{\frac{1}{2}} = \left(\frac{g}{L}\right)^{\frac{1}{2}}\left(\frac{2E}{MgL} - \theta^2\right)^{\frac{1}{2}} \quad (7.14)$$

We denote the turning points of the motion by θ_0 and $-\theta_0$; the amplitude of the oscillation is θ_0. At these points the pendulum is momentarily at rest and the kinetic energy is zero. Figure 7.5 illustrates this. From Eq. (7.13) with $\dot{\theta} = 0$ we have

$$E = \tfrac{1}{2}MgL\theta_0^2 \qquad \theta_0^2 = \frac{2E}{MgL}$$

Thus we may rewrite Eq. (7.14) as

$$\frac{d\theta}{dt} = \left(\frac{g}{L}\right)^{\frac{1}{2}}(\theta_0^2 - \theta^2)^{\frac{1}{2}}$$

or

$$\frac{d\theta}{(\theta_0^2 - \theta^2)^{\frac{1}{2}}} = \left(\frac{g}{L}\right)^{\frac{1}{2}} dt$$

This is identical with Eq. (7.6), and we give below some details of the solution.

If the initial condition or the phase of the motion is such that θ has the value θ_1 at $t = 0$, then

$$\int_{\theta_1}^{\theta} \frac{d\theta}{(\theta_0^2 - \theta^2)^{\frac{1}{2}}} = \left(\frac{g}{L}\right)^{\frac{1}{2}} \int_0^t dt$$

The integral on the left is given in Dwight 320.01, and we obtain

$$\int_{\theta_1}^{\theta} \frac{d\theta}{(\theta_0^2 - \theta^2)^{\frac{1}{2}}} = \left[\sin^{-1}\frac{\theta}{\theta_0}\right]_{\theta_1}^{\theta} = \sin^{-1}\frac{\theta}{\theta_0} - \sin^{-1}\frac{\theta_1}{\theta_0} \qquad (7.15)$$

$$= \left(\frac{g}{L}\right)^{\frac{1}{2}} t$$

We know that $\sin \sin^{-1}(\theta/\theta_0) = \theta/\theta_0$, so that we may rewrite Eq. (7.15) as

$$\frac{\theta}{\theta_0} = \sin\left[\left(\frac{g}{L}\right)^{\frac{1}{2}} t + \sin^{-1}\frac{\theta_1}{\theta_0}\right]$$

or

$$\theta = \theta_0 \sin(\omega_0 t + \phi)$$

where we identify the angular frequency ω_0 and phase ϕ with

$$\omega_0 = \left(\frac{g}{L}\right)^{\frac{1}{2}} \qquad \phi = \sin^{-1}\frac{\theta_1}{\theta_0}$$

These agree with the results given in Eqs. (7.10) and (7.11).

Even though ϕ in Eq. (7.2) or (7.10) has the dimensions of an angle, it is not an angle that you can visualize immediately. It is important to understand the quantities A and ϕ in the case of the mass on the spring and θ_0 and ϕ in the case of the pendulum. Figures 7.6 and 7.7 illustrate the meanings for these two cases. All cases of free oscillation will have the same sorts of constants, although we can often choose the

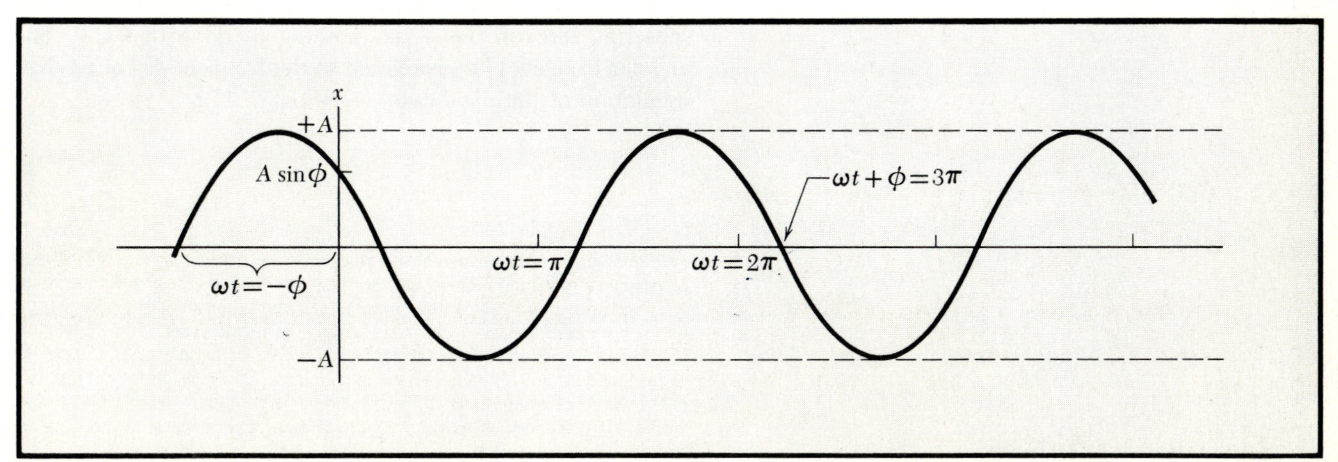

FIG. 7.6 The function $x = A \sin(\omega t + \phi)$ plotted vs ωt $\phi \approx 3\pi/4$. At $t = 0$, $x = A \sin\phi$, which is shown and $dx/dt = \omega A \cos\phi$, which is negative. (Note: For convenience we have dropped the subscript from ω_0.)

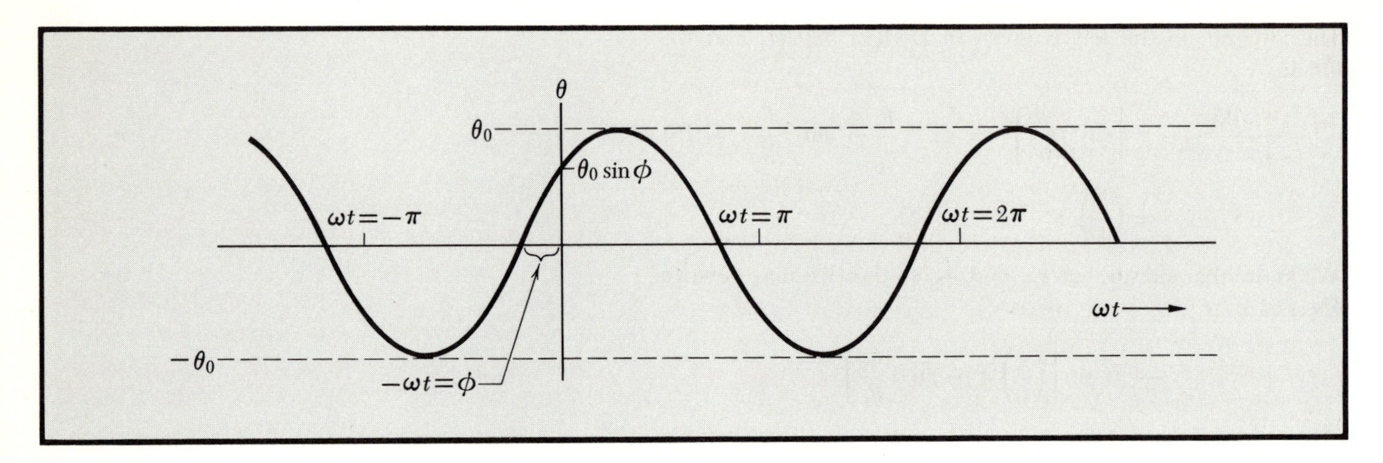

FIG. 7.7 The function $\theta = \theta_0 \sin(\omega t + \phi)$ plotted vs ωt. $\phi \approx \pi/4$. At $t = 0$, $\theta = \theta_0 \sin\phi$, which is shown, and $d\theta/dt = \omega\theta_0 \cos\phi$, which is positive. (The ω_0 of the text is here written as simply ω.)

moment we call $t = 0$ in such a way as to make the value of ϕ be zero or $\pi/2$.

1 A and θ_0 are the maximum amplitudes of oscillation; that is, the motion goes from $+A$ to $-A$ or $+\theta_0$ to $-\theta_0$.

2 In terms of the angle $\omega_0 t$, ϕ is the angle such that at $\omega_0 t = -\phi$, x or θ would be zero and increasing from negative to positive values. This is, of course, just another way of saying that at $t = 0$, $x = A \sin\phi$ or $\theta = \theta_0 \sin\phi$. Note carefully that in Figs. 7.6 and 7.7 the horizontal axis measures $\omega_0 t$, not t.

3 The initial conditions determine A and ϕ or θ_0 and ϕ even though the values of the initial conditions are not directly either of these two quantities.

The symbol ω_0 is used to denote the angular frequency of the natural or free motion of an oscillating system. The subscript zero on the ω has nothing to do with $t = 0$. The angular frequency[1] ω_0 is related to the frequency f_0 of the free oscillation of the pendulum, as in Eq. (7.11):

$$f_0 = \frac{\omega_0}{2\pi} = \frac{(g/L)^{\frac{1}{2}}}{2\pi} \tag{7.16}$$

[1]We shall often refer to the angular frequency simply as the frequency. Many physicists do this, and no particular confusion is caused. The use of the symbol ω rather than f or ν will usually identify a quantity as an angular frequency. As for numerical values, ν and f are usually given in cycles per second (cps) or Hz; ω is given in radians per second (rad/s) or simply as s^{-1}, with the radians understood. A radian is dimensionless. Also it is rather common to differentiate between them by expressing the frequency ν in vibrations/s, cycles/s, revolutions/s or Hz, and the angular frequency ω in rad/s. Both have the dimensions s^{-1}.

If $L = 100$ cm, we have $\omega_0 \approx (\frac{980}{100})^{\frac{1}{2}} \approx 3$ rad/s, or $L = 1.0$ m, $\omega_0 \approx (\frac{9.8}{1.0})^{\frac{1}{2}} \approx 3$ rad/s. The frequency is independent of the mass M and the amplitude θ_0 of the motion, provided that $\theta_0 \ll 1$. Note that there is no way the mass could enter into the right-hand side of Eq. (7.16) and still give a quantity with the dimensions of frequency.

In setting up Eqs. (7.5) and (7.13) we used the law of conservation of energy. Notice that each of these is a first-order differential equation, and we had to carry out only one integration with respect to time to obtain the result. The equation of motion [Eq. (7.1) or (7.9)] is a second-order differential equation. To solve for the displacement or the deflection angle we had to integrate twice with respect to time. It is good to remember that the explicit use of energy conservation can often save mathematical labor by eliminating one integration.

Still a third method of setting up the equation of motion for the pendulum is to use torque = rate of change of angular momentum. Let us take the x axis normal to the plane of the motion, as in Fig. 7.8. The torque N_x due to gravity about O, the point of suspension, is, with $F = Mg$,

$$N_x = (\mathbf{r} \times \mathbf{F})_x = LMg \sin\theta$$

The angular momentum J_x about the same point is, with the linear momentum $p = ML\dot{\theta}$,

$$J_x = (\mathbf{r} \times \mathbf{p})_x = -ML^2\dot{\theta}$$

where $\dot{\theta}$ is positive for θ increasing. Setting the rate of change of angular momentum equal to the torque gives

$$ML^2\ddot{\theta} = -LMg \sin\theta$$

so that the equation of motion of the pendulum is

$$\ddot{\theta} + \frac{g}{L}\sin\theta = 0$$

In the limit $\theta \ll 1$, we again approximate $\sin\theta$ by θ:

$$\ddot{\theta} + \frac{g}{L}\theta = 0$$

which is just Eq. (7.9) again.

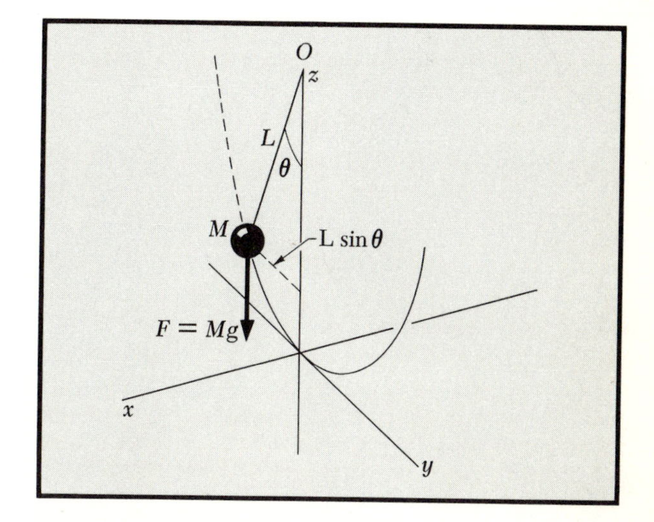

FIG. 7.8 The pendulum oscillates in the yz plane. The force due to gravity on M is $F = Mg$, in the $-z$ direction. The torque N_x arising from this force is $MgL \sin\theta$, in the $+x$ direction, as evaluated at point O.

LC CIRCUIT

Some of the most important examples of oscillating systems occur in electricity. The familiarity of the term *alternating current* (ac), which is an oscillating electric current, shows the importance of this example. Those students who have had some background in electric circuits will readily see the relations to mechanical systems; others may omit this section and perhaps return to it when they have studied Chap. 8 of Volume 2.

The voltage across a capacitance C is

$$V_C = \frac{Q}{C}$$

where Q is the charge on the capacitance. The current in a circuit in series with this capacitance is

$$I = \frac{-dQ}{dt} \qquad \text{or} \qquad Q = -\int I\,dt$$

where we use the minus sign to indicate that the current flows in such a direction as to decrease the charge on the capacitance. The voltage across an inductance L is

$$V_L = -L\frac{dI}{dt}$$

If we consider a circuit in which there is only a capacitance and an inductance, as in the third column of Fig. 7.9, and remember that the sum of the voltages around the circuit is zero, we obtain

$$-L\frac{dI}{dt} + \frac{Q}{C} = 0 = L\frac{d^2Q}{dt^2} + \frac{Q}{C}$$

This is just the equation for the displacement of a spring with

$$Q \leftrightarrow x \qquad L \leftrightarrow M \qquad \frac{1}{C} \leftrightarrow C_{\text{sp}}$$

The solution will be

$$Q = Q_0 \sin(\omega_0 t + \phi) \qquad \omega_0 = \left(\frac{1}{LC}\right)^{\frac{1}{2}} \tag{7.17}$$

and the current can be found from this. Figure 7.9 illustrates these relations and compares them with a pendulum and a mass on a spring.

FIG. 7.9 Three separate harmonic oscillators with the same period: a simple pendulum, a mass-spring system, and an LC circuit. Time increases from Fig. A to Fig. H; the next cycle begins again with Fig. A.

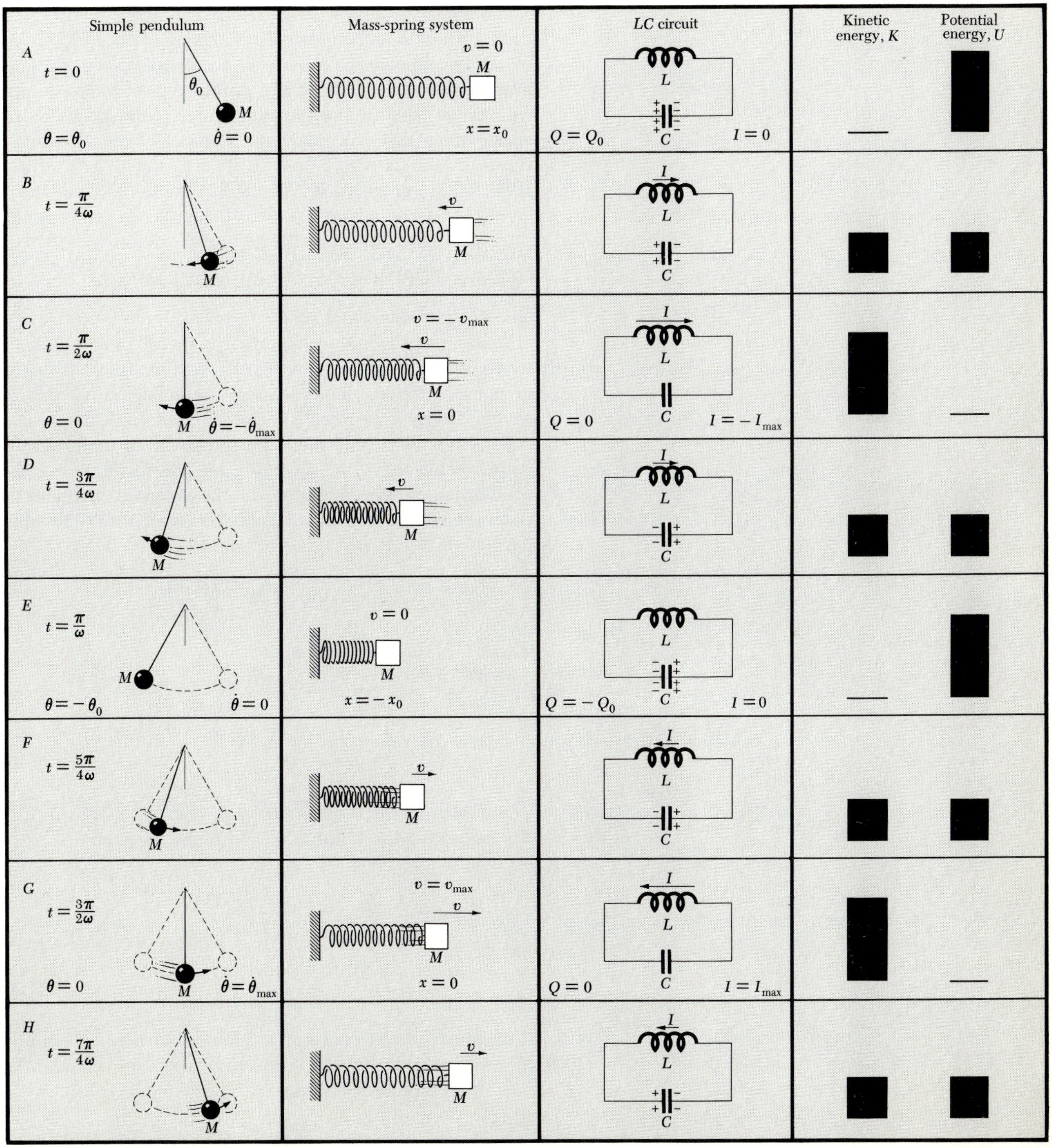

	Simple pendulum	Mass-spring system	*LC* circuit	Kinetic energy, *K*	Potential energy, *U*
A	$t=0$ θ_0 M $\theta=\theta_0$ $\dot{\theta}=0$	$v=0$ M $x=x_0$	L $Q=Q_0$ C $I=0$		
B	$t=\frac{\pi}{4\omega}$ M	v M	I L C		
C	$t=\frac{\pi}{2\omega}$ $\theta=0$ M $\dot{\theta}=-\dot{\theta}_{max}$	$v=-v_{max}$ v M $x=0$	I L $Q=0$ C $I=-I_{max}$		
D	$t=\frac{3\pi}{4\omega}$ M	v M	I L C		
E	$t=\frac{\pi}{\omega}$ M $\theta=-\theta_0$ $\dot{\theta}=0$	$v=0$ M $x=-x_0$	L $Q=-Q_0$ C $I=0$		
F	$t=\frac{5\pi}{4\omega}$ M	v M	I L C		
G	$t=\frac{3\pi}{2\omega}$ $\theta=0$ M $\dot{\theta}=\dot{\theta}_{max}$	$v=v_{max}$ v M $x=0$	I L $Q=0$ C $I=I_{max}$		
H	$t=\frac{7\pi}{4\omega}$ M	v M	I L C		

Students familiar with electricity will notice that the voltage RI where R is the resistance is missing. $RI = R\,dQ/dt$ and by our correspondence above will occupy the place of a force proportional to dx/dt. But this is just the type of frictional force we discuss later in the chapter, and R corresponds to the coefficient b that will relate the frictional force to velocity. Cases of circuits with L and R, with C and R, and with L, R, and C are discussed in Volume 2, Chaps. 4, 7, and 8.

MOTION OF SYSTEMS DISPLACED FROM A POSITION OF STABLE EQUILIBRIUM

One of the reasons that *simple harmonic motion* (as this type of motion is called) is so important is that for small displacements of any system in stable equilibrium, the resulting motion except for frictional forces is simple harmonic. To see this, let us describe by a coordinate α the deviation from the position of stable equilibrium; α might be a distance, an angle, or some more complicated type of coordinate. The condition of stable equilibrium requires that at $\alpha = 0$ the potential energy of the system must be at a minimum, and the force, if α is a distance, the torque, if α is an angle, etc., must be zero. Thus

$$F(\alpha = 0) = 0 = -\left(\frac{dU}{d\alpha}\right)_{\alpha=0} \tag{7.18}$$

where U is the potential energy.

We can always write by a Taylor's series expansion

$$U(\alpha) = U_0 + \left(\frac{dU}{d\alpha}\right)_0 \alpha + \tfrac{1}{2}\left(\frac{d^2U}{d\alpha^2}\right)_0 \alpha^2 + \tfrac{1}{6}\left(\frac{d^3U}{d\alpha^3}\right)_0 \alpha^3 \cdots \tag{7.19}$$

where the subscript 0 refers to $\alpha = 0$. Using Eq. (7.18) in (7.19) and neglecting the terms in α^3 and higher because α is small, we obtain

$$U(\alpha) = U_0 + \frac{1}{2}\left(\frac{d^2U}{d\alpha^2}\right)_0 \alpha^2$$

and

$$F(\alpha) = -\frac{dU}{d\alpha} = -\left(\frac{d^2U}{d\alpha^2}\right)_0 \alpha \tag{7.20}$$

F, of course, is not necessarily a force, but may be a torque or more complicated stress quantity. Now the condition for stability of equilibrium is that

$$\frac{d^2U}{d\alpha^2} > 0$$

In Eq. (7.20) this means that the "force" tends to return the system to $\alpha = 0$. The equation of motion of the system will then be

$$M\frac{d^2\alpha}{dt^2} = -\left(\frac{d^2U}{d\alpha^2}\right)_0 \alpha \tag{7.21}$$

where M is some sort of masslike term. Eq. (7.21) indicates that α will describe simple harmonic motion. In practice, of course, friction is important, but in its absence the above analysis would apply to a bridge, a building, etc.—indeed to any system for which a potential energy function exists that possesses a minimum and is differentiable.

AVERAGE KINETIC AND POTENTIAL ENERGY

We now prove an important characteristic of a harmonic oscillator related to the time average of the kinetic and potential energies. The time average of a quantity K over a time interval T is

$$\boxed{\langle K \rangle = \frac{1}{T}\int_0^T K(t)\, dt} \tag{7.22}$$

Since the motion of an oscillator is repetitive, the time average over one period is the same as over many periods and is unique. Using the period $T = 2\pi/\omega_0$, we write for the time average of kinetic energy of an oscillator whose motion obeys $x = A \sin(\omega_0 t + \phi)$

$$\langle K \rangle = \frac{\int_0^T \frac{1}{2}M\dot{x}^2\, dt}{T} = \frac{1}{2}M\omega_0{}^2A^2 \frac{\int_0^{2\pi/\omega_0} \cos^2(\omega_0 t + \phi)\, dt}{2\pi/\omega_0}$$

Because the integral is extended over a complete period, it does not matter what the value of the phase ϕ is, and we may conveniently set $\phi = 0$. Then if we write $y = \omega_0 t$, we have

$$\frac{\omega_0}{2\pi}\int_0^{2\pi/\omega_0} \cos^2 \omega_0 t\, dt = \frac{1}{2\pi}\int_0^{2\pi} \cos^2 y\, dy = \frac{1}{2}$$

by using the facts that

$$\int_0^{2\pi} \sin^2 y\, dy = \int_0^{2\pi} \cos^2 y\, dy$$

and

$$\int_0^{2\pi} (\sin^2 y + \cos^2 y)\, dy = 2\pi$$

Therefore the time average of the kinetic energy is

$$\langle K \rangle = \tfrac{1}{4} M\omega_0{}^2 A^2 \tag{7.23}$$

The potential energy is (again with $\phi = 0$ since its value will not matter)

$$U = \tfrac{1}{2} Cx^2 = \tfrac{1}{2} CA^2 \sin^2 \omega_0 t$$

Since the average value of the sine squared is the same as that of the cosine squared over one period[1] and since $\omega_0{}^2 = C/M$,

$$\langle U \rangle = \tfrac{1}{4} CA^2 = \tfrac{1}{4} M\omega_0{}^2 A^2 \tag{7.24}$$

Thus $\langle U \rangle = \langle K \rangle$ and the total energy of the harmonic oscillator is

$$E = \langle K \rangle + \langle U \rangle = \tfrac{1}{2} M\omega_0{}^2 A^2$$

Note that $E = \langle E \rangle$ because the total energy is a constant of the motion.

The equality of the time average kinetic energy and potential energy is a special property of the harmonic oscillator and is worth remembering. It is not true for anharmonic oscillators but is a good approximation for weakly damped oscillators, as will be shown later.

FRICTION

Thus far we have neglected frictional effects in the harmonic oscillator. We have discussed friction in Chap. 3, including only the case of a constant frictional force. We now discuss the case of a force that is proportional to the first power of the velocity. For small velocities, this is in many cases a good approximation, and so we shall find that our solution with this kind of a retarding force is realistic. We shall first, however, consider some cases in which only this force acts. We thus have

$$M\frac{d^2x}{dt^2} = F_{\text{fric}} = -b\frac{dx}{dt} = -b\dot{x} \tag{7.25}$$

where b is a positive constant called the *damping coefficient.* The negative sign describes the fact that this is a force always

[1] This is easily seen by drawing the two curves and noting that they are identical when displaced by one quarter period. This type of argument can also be applied to the average value of $\langle x^2 \rangle$ over the surface of a sphere. If $x^2 + y^2 + z^2 = R^2$, then $\langle x^2 \rangle + \langle y^2 \rangle + \langle z^2 \rangle = R^2$. Because the sphere is symmetric with respect to x, y, z, we must have $\langle x^2 \rangle = \langle y^2 \rangle = \langle z^2 \rangle = \tfrac{1}{3}R^2$. This result may be confirmed by direct calculation.

opposite to the velocity. It is sometimes more useful to define a constant, called the *relaxation time*, by the relation

$$\tau = \frac{M}{b}$$

for then Eq. (7.25) becomes

$$M\left(\frac{d^2x}{dt^2} + \frac{1}{\tau}\frac{dx}{dt}\right) = 0$$

We see that τ has the dimensions of time since b possesses dimensions of force divided by velocity or simply mass over time. The reason for the name *relaxation time* will become apparent as we progress.

In terms of the velocity $v = dx/dt = \dot{x}$, this equation becomes

$$\dot{v} + \frac{1}{\tau}v = 0$$

The solution is given at the end of this chapter (page 232). It is

$$v(t) = v_0 e^{-t/\tau} \tag{7.26}$$

where v_0 is the velocity at $t = 0$. The velocity decreases exponentially with time; we say that the velocity is damped with a time constant τ. The behavior is plotted in Fig. 7.10.

The decay of the kinetic energy K of a free particle subject to this friction force is given from Eq. (7.26) by

$$K = \tfrac{1}{2}Mv^2 = \tfrac{1}{2}Mv_0{}^2e^{-2t/\tau} = K_0e^{-2t/\tau} \tag{7.27}$$

On differentiation of Eq. (7.27) we see that

$$\dot{K} = -\frac{2}{\tau}K$$

The effective relaxation time for the kinetic energy is one-half that for the velocity.

What sort of mechanism leads to a damping force of the form $-b\dot{x}$? The case of a sphere moving slowly through a viscous medium was first solved by G. G. Stokes, and the expression for the force

$$F_{\text{fric}} = -6\pi\eta r v \tag{7.28}$$

where r is the radius of the sphere and η is the coefficient of viscosity, is often called *Stokes' law*.

A good representation of the damping force $F_{\text{fric}} = -b\dot{x}$ is also provided by a flat plate moving normal to its plane

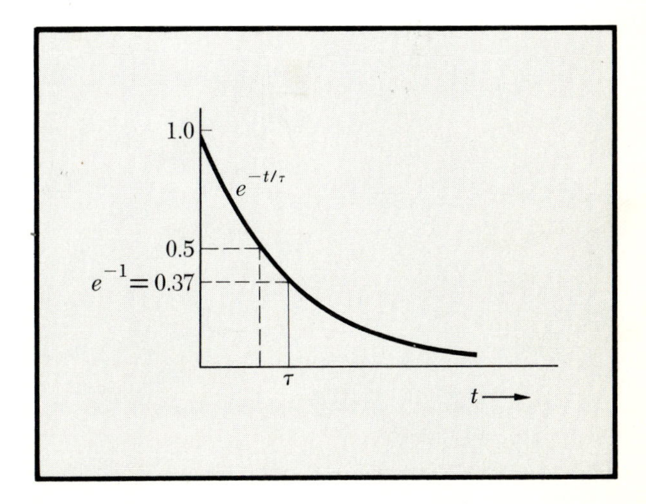

FIG. 7.10 The function $e^{-t/\tau}$ is plotted vs t. Note that since $e^{-0.69} = 0.5$ the function has decreased to half its initial value when $t = 0.69\tau$.

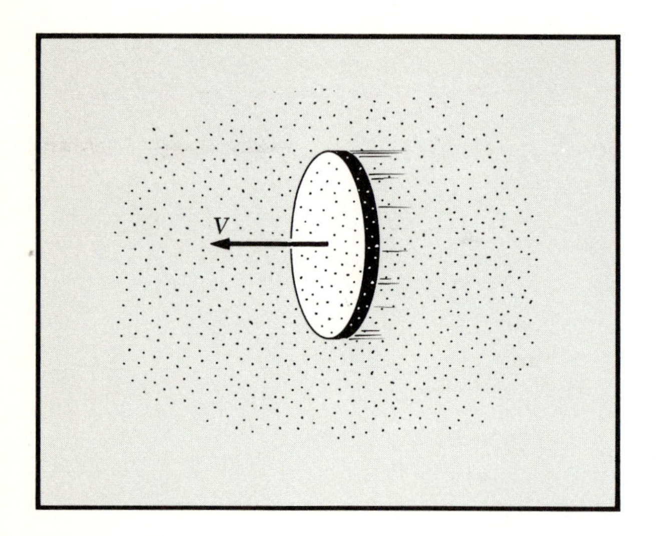

FIG. 7.11 A flat plate moving normal to its plane through a gas at very low pressure is subjected to a retarding force proportional to its velocity V, if V is much less than the average speed of the molecules of the gas.

through a gas at very low pressure, as in Fig. 7.11, provided the speed V of the plate is much slower than the average speed v of the molecules of the gas. The pressure must be sufficiently low that we can neglect the collisions of the molecules with each other. The rate at which molecules strike the plate is proportional to the relative velocity of the incoming molecules and the plate. Suppose that the molecules move only in one dimension. On one side of the plate the relative velocity is $v + V$; on the other side it is $v - V$. The pressure is proportional to the rate at which molecules strike the plate times the average momentum transfer to the plate per molecule. The average momentum transfer is itself proportional to the relative velocity, so that the pressures P_1 and P_2 on the two sides of the plate are

$$P_1 \propto (v + V)^2 \qquad P_2 \propto (v - V)^2$$

The net pressure P is the difference of the pressures on the opposite sides of the plate:

$$P = P_1 - P_2 \propto 4vV$$

so that the drag (the net force on the moving plate) is directly proportional to the speed V of the plate. The direction of the drag can be seen to oppose the motion of the plate.

Terminal Velocity If a constant force such as gravity is applied to a particle under the action of a frictional force such as discussed above, the velocity will increase if the particle starts at rest or at a small velocity, or will decrease if it starts at a very high velocity until the acceleration is zero. This condition is

$$F_{\text{const}} = b\dot{x} \qquad \text{or} \qquad \dot{x} = v = \frac{F_{\text{const}}}{b} \tag{7.29}$$

and the velocity so reached is called the *terminal velocity*. For example, if a particle of mass M falls under the action of gravity and a Stokes' law force, the terminal velocity is

$$v = \frac{Mg}{6\pi\eta r}$$

(See Probs. 10 and 11 at the end of the chapter.)

The concept of terminal velocity is applicable to problems involving frictional forces proportional to other powers of the velocity that often occur at higher speeds. If the frictional force is given by

$$F_{\text{fric}} = -c\dot{x}^n$$

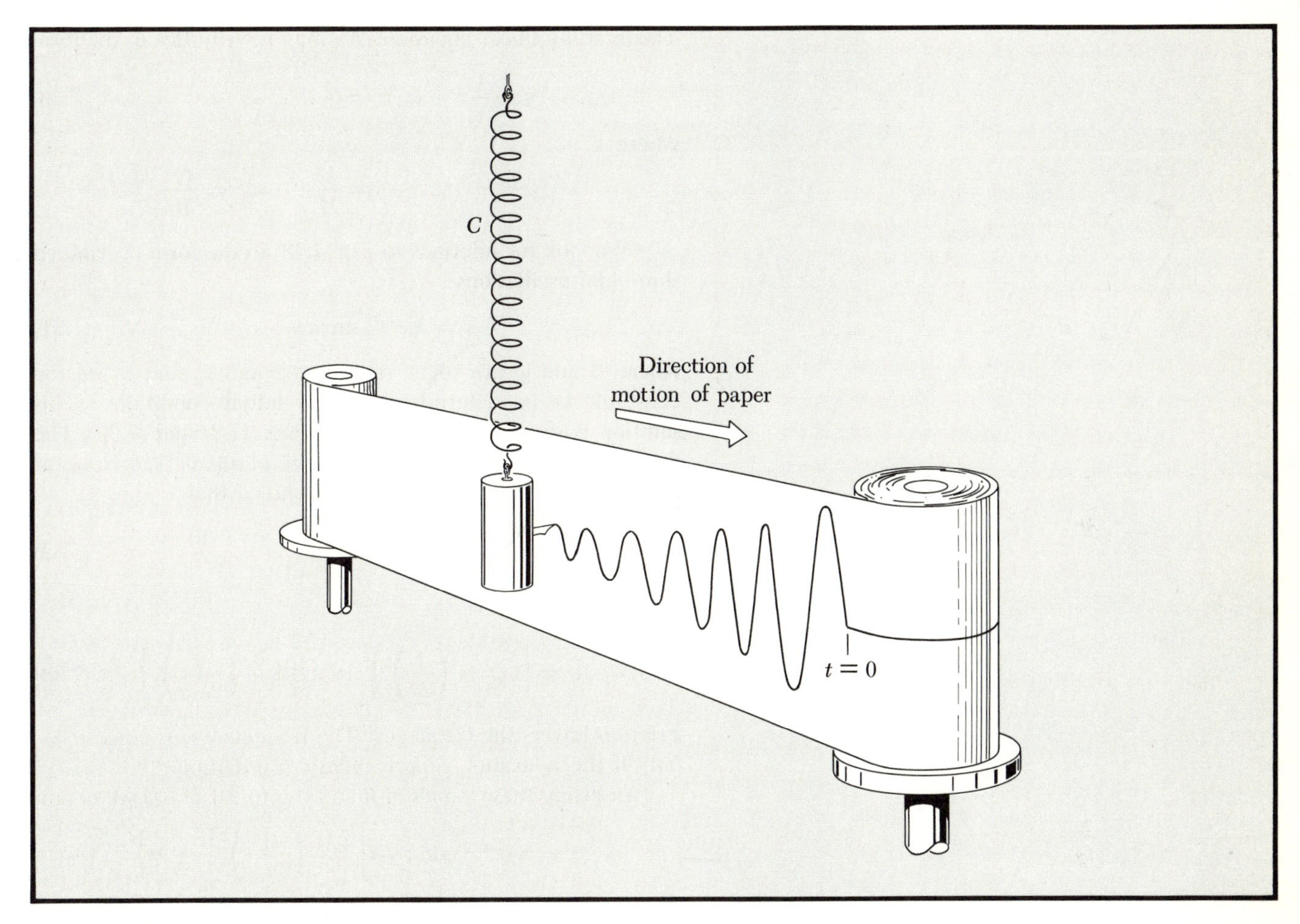

FIG. 7.12 All real harmonic oscillators are subject to damping by frictional forces, such as air resistance. A mass-spring system with light damping would describe such a curve on the paper tape driven at constant speed if the mass-spring system was originally set oscillating at $t = 0$.

where c is some constant and n is positive, the terminal velocity will be

$$v = \left(\frac{F_{\text{const}}}{c}\right)^{1/n}$$

DAMPED HARMONIC OSCILLATOR

We now return to the oscillator and include the damping force $-b\dot{x}$. The type of motion is shown in Fig. 7.12. The equation of motion is

$$M\ddot{x} + b\dot{x} + Cx = 0$$

This is still a linear equation. We may rewrite this in the form

$$\ddot{x} + \frac{1}{\tau}\dot{x} + \omega_0^2 x = 0 \tag{7.30}$$

where

$$\frac{1}{\tau} = \frac{b}{M} \qquad \omega_0^2 = \frac{C}{M}$$

We look for solutions to Eq. (7.30) in the form of damped sinusoidal oscillations:

$$x = x_0 e^{-\beta t} \sin(\omega t + \phi) \tag{7.31}$$

where β and ω are to be determined and x_0 and ϕ are the constants to be determined by the initial conditions.[1] This solution is suggested by combining Eqs. (7.2) and (7.26). The details of the solution are given in the Mathematical Notes at the end of this chapter, where it is shown that

$$\beta = \frac{1}{2\tau} \tag{7.32}$$

and

$$\omega = \left[\omega_0^2 - \left(\frac{1}{2\tau}\right)^2\right]^{\frac{1}{2}} = \omega_0\left[1 - \left(\frac{1}{2\omega_0\tau}\right)^2\right]^{\frac{1}{2}} \tag{7.33}$$

Friction lowers the frequency. The frequency ω is equal to ω_0 only if the relaxation time is infinite (no damping).

Inserting these values of β and ω into Eq. (7.31) we obtain

$$x = x_0 e^{-t/2\tau} \sin\left\{\omega_0 t\left[1 - \left(\frac{1}{2\omega_0\tau}\right)^2\right]^{\frac{1}{2}} + \phi\right\}$$

If $\omega_0\tau \gg 1$, we have the limit of *low damping* in which x may be approximated by

$$x \approx x_0 e^{-t/2\tau} \sin(\omega_0 t + \phi) \tag{7.34}$$

where ω_0 is the natural frequency of the undamped oscillation.

EXAMPLE

Power Dissipation Calculate the rate of energy dissipation by a damped harmonic oscillator, in the weak damping limit with $\omega_0\tau \gg 1$, so that $\omega \approx \omega_0$.

[1]The solution [Eq. (7.31)] is not valid for all values of ω_0 and τ. It will be apparent in the Mathematical Notes that under some conditions the solution is not oscillatory. If $\omega_0^2 < (1/2\tau)^2$, the solution is $e^{-t/2\tau}(Ae^{gt} + Be^{-gt})$ where A and B are the arbitrary constants and $g = [(1/2\tau)^2 - \omega_0^2]^{\frac{1}{2}}$; while if $\omega_0 = 1/2\tau$, the solution is $Ce^{-t/2\tau} + Dte^{-t/2\tau}$ where C and D are the arbitrary constants. This last case is "critically damped." See Eqs. (7.69) and (7.70).

The kinetic energy is $K = \frac{1}{2}M\dot{x}^2$. From the approximate solution [Eq. (7.34)] we have (letting $\phi = 0$)

$$\frac{dx}{dt} = -\frac{1}{2\tau}x_0 e^{-t/2\tau} \sin \omega_0 t + \omega_0 x_0 e^{-t/2\tau} \cos \omega_0 t \tag{7.35}$$

Now

$$\left(\frac{dx}{dt}\right)^2 = \left(\frac{1}{2\tau}\right)^2 x_0{}^2 e^{-t/\tau} \sin^2 \omega_0 t + \omega_0{}^2 x_0{}^2 e^{-t/\tau} \cos^2 \omega_0 t - \left(\frac{\omega_0}{\tau}\right) x_0{}^2 e^{-t/\tau} \sin \omega_0 t \cos \omega_0 t \tag{7.36}$$

The integrals that arise when we take the time average of Eq. (7.36) are given in Dwight, 861.13 to 861.15. But for $\omega_0 \tau \gg 1$, it is a good approximation to take the factor $e^{-t/\tau}$ in Eq. (7.36) outside the angle brackets that denote time average. We can do this to reasonable accuracy if the amplitude of the oscillation $x_0 e^{-t/2\tau}$ does not change much in one cycle of the motion. We are left with the averages:

$$\langle \cos^2 \theta \rangle = \langle \sin^2 \theta \rangle = \tfrac{1}{2} \qquad \langle \cos \theta \sin \theta \rangle = 0$$

The last average, which is new to us, is of considerable importance

$$\langle \cos \theta \sin \theta \rangle = \langle \tfrac{1}{2} \sin 2\theta \rangle = 0$$

because the average of a sine or cosine is zero. Then the kinetic energy averaged over the time of one cycle is

$$\langle K \rangle \approx \frac{1}{2}M\left[\left(\frac{1}{2\tau}\right)^2 \langle \sin^2 \omega_0 t \rangle + \omega_0{}^2 \langle \cos^2 \omega_0 t \rangle - \frac{\omega_0}{\tau}\langle \cos \omega_0 t \sin \omega_0 t \rangle\right] x_0{}^2 e^{-t/\tau}$$

$$\approx \frac{1}{4}M\left[\left(\frac{1}{2\tau}\right)^2 + \omega_0{}^2\right] x_0{}^2 e^{-t/\tau}$$

but $(1/2\tau)^2$ is assumed to be negligible in comparison to $\omega_0{}^2$, so that the average kinetic energy is

$$\langle K \rangle \approx \tfrac{1}{4}M\omega_0{}^2 x_0{}^2 e^{-t/\tau} \tag{7.37}$$

We see that the average kinetic energy decays exponentially. The average potential energy is (see Fig. 7.13, page 220)

$$\langle U \rangle = \tfrac{1}{2}M\omega_0{}^2 x_0{}^2 \langle e^{-t/\tau} \sin^2 \omega_0 t \rangle \approx \tfrac{1}{4}M\omega_0{}^2 x_0{}^2 e^{-t/\tau} \tag{7.38}$$

The average power dissipation P is given by the negative of the rate of change of energy:

$$-\langle P \rangle = \frac{d}{dt}\langle E \rangle \approx \frac{d}{dt}(\langle K \rangle + \langle U \rangle) \approx -\frac{1}{\tau}\left(\frac{1}{2}M\omega_0{}^2 x_0{}^2 e^{-t/\tau}\right)$$

or

$$\langle P(t) \rangle = \frac{\langle E(t) \rangle}{\tau} \tag{7.39}$$

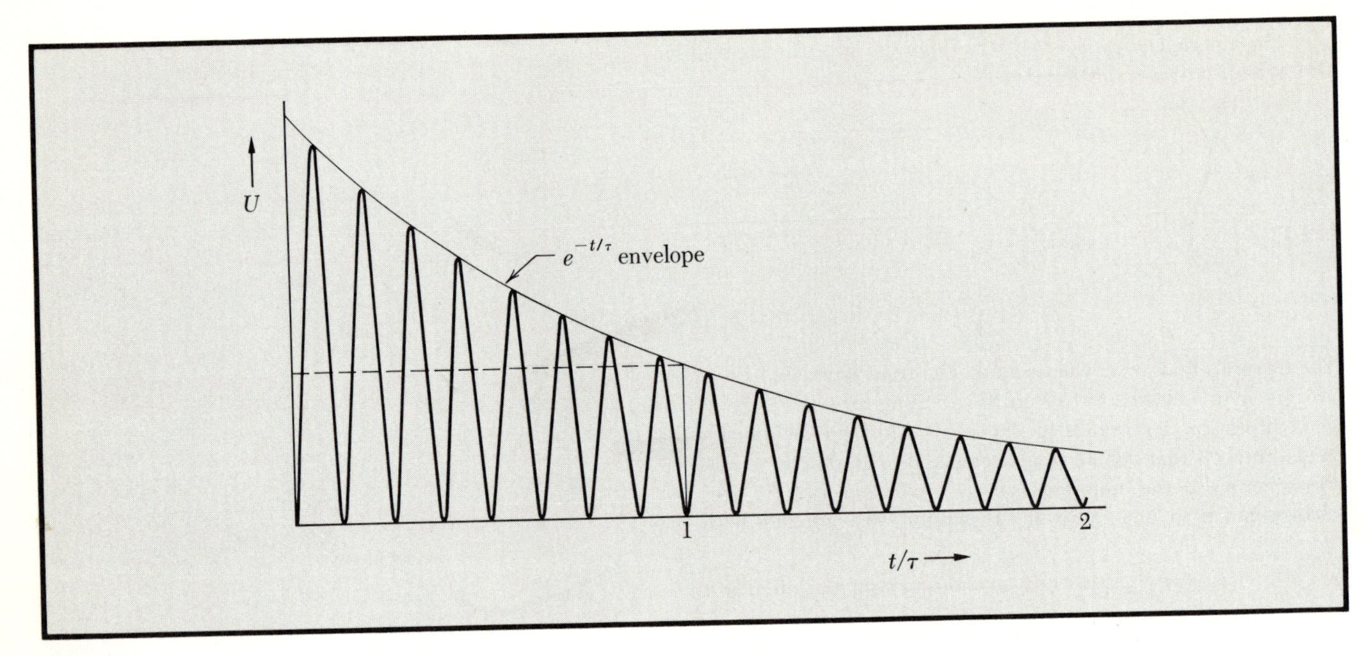

FIG. 7.13 Potential energy of oscillator with $\tau = 8\pi/\omega$ and Q of 8π, plotted vs t/τ. In time τ, in which four oscillations occur, the envelope of the potential energy falls to $1/e$ of its initial value.

We usually omit the $\langle\ \rangle$ on $P(t)$ when the meaning is clear.

The student may be surprised that the averages expressed in Eqs. (7.37) and (7.38) contain the time t, as these are averages over time. We are viewing the motion of a damped oscillator over many cycles, and what we have here is the average (kinetic or potential) energy *over one cycle at about the time* t. Because energy is being dissipated into heat, we expect the average energy (over a cycle) to decrease as more and more cycles are completed.

We expect to find that the power dissipation is equal to the negative of the average rate at which work is done by the frictional force $F_f = -b\dot{x} = -(M/\tau)\dot{x}$ [see Eq. (7.30)]. Using Eq. (7.35) and assuming that $\omega_0\tau \gg 1$ so that $e^{-t/\tau}$ can be taken out of the $\langle\ \rangle$, we obtain for this average rate of doing work

$$\langle F_f v\rangle \approx -\frac{M}{\tau}\omega_0{}^2 x_0{}^2 e^{-t/\tau}\langle\cos^2\omega_0 t\rangle$$

$$\approx -\frac{1}{2\tau}M\omega_0{}^2 x_0{}^2 e^{-t/\tau} \approx -\frac{E(t)}{\tau}$$

in agreement with Eq. (7.39).

Quality Factor Q The Q, or *quality factor*, of an oscillatory system is a very widely used term. It is particularly prevalent in the terminology of alternating current electrical systems but is applicable to all oscillating systems so long as the damping is small. The Q is defined as 2π times the ratio of the energy stored to the average energy loss per period:

$$Q = 2\pi \frac{\text{energy stored}}{\langle\text{energy loss per period}\rangle} = \frac{2\pi E}{P/f} = \frac{E}{P/\omega} \tag{7.40}$$

because the period is $1/f$ and $2\pi f = \omega$. The time of 1 rad of motion is $1/\omega$. The damping must be small enough so that E does not change appreciably in the one period. Note that Q is dimensionless. For the lightly damped harmonic oscillator ($\omega_0\tau \gg 1$) we have

$$Q \approx \frac{E}{E/\omega\tau} \approx \omega_0\tau \tag{7.41}$$

from Eq. (7.39). We see that the value of $\omega_0\tau$ is indeed a good measure of the lack of damping of an oscillator. High $\omega_0\tau$ or high Q means that the oscillator is lightly damped. Note from Eq. (7.38) that the energy of an oscillator decays to e^{-1} of its initial value in the time τ; during this time the oscillator performs $\omega_0\tau/2\pi$ oscillations. Several representative values of Q are given in Table 7.2.

TABLE 7.2 Several Typical Values of Q (Wide variation may be expected)

Earth, for earthquake wave	250–1400
Copper cavity microwave resonator	10^4
Piano or violin string	10^3
Excited atom	10^7
Excited nucleus (Fe^{57})	3×10^{12}

DRIVEN HARMONIC OSCILLATOR

The case of a harmonic oscillator driven by sinusoidally varying force is an extremely important one in many branches of physics. Because of the complexity of the problem, it is treated in the Advanced Topics (at the end of this chapter). However, several results are worth noting here.

1 As might be expected, the steady-state motion (after the friction has damped out any motion corresponding to the natural period of the undriven oscillator, $t \gg \tau$) has the same frequency as the driving force.

2 As also might be expected, particularly for the lightly damped oscillator, the amplitude of the steady-state motion depends strongly on the driving frequency being large when that is close to the natural frequency. Figure 7.14 shows the amplitude as a function of the driving frequency for large and medium Q. Note that the maximum amplitude occurs at

$$\omega = \left(\omega_0^2 - \frac{1}{2\tau^2}\right)^{\frac{1}{2}} = \omega_0\left(1 - \frac{1}{2\omega_0^2\tau^2}\right)^{\frac{1}{2}} = \omega_0\left(1 - \frac{1}{2Q^2}\right)^{\frac{1}{2}}$$

which is very close to ω_0 for a high-Q oscillator. The maxi-

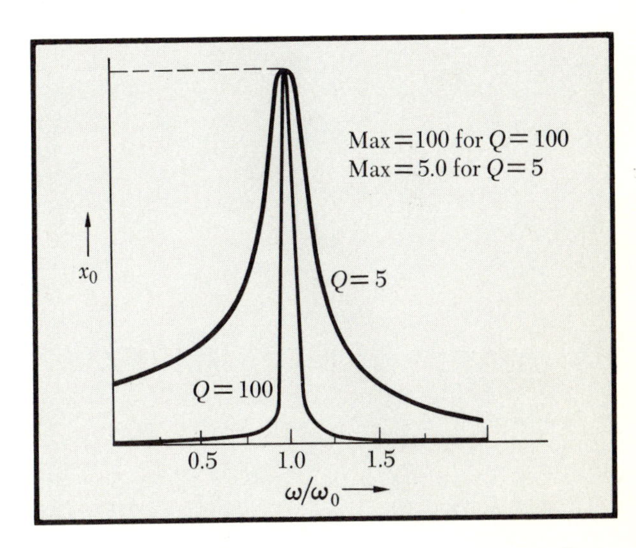

FIG. 7.14 Amplitude of forced harmonic motion as function of driving frequency. Maximum value of x_0 occurs at $\omega = \omega_0\sqrt{1 - 1/(2Q^2)}$, which would fall at slightly below 1.0 on the abscissa, at slightly different points for each curve. The scale of amplitude is purely arbitrary, and for the same driving force and ω_0 the maximum amplitude for $Q = 100$ will be 20 times that for $Q = 5$.

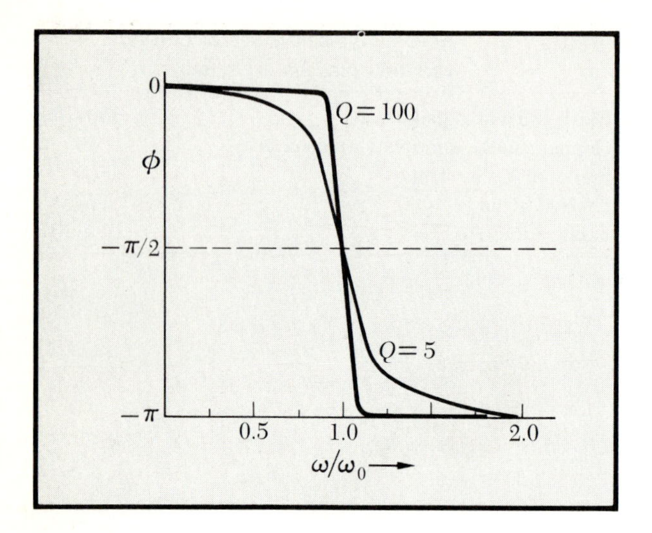

FIG. 7.15 Angle ϕ between displacement and driving force.

mum power absorption occurs at $\omega = \omega_0$. The type of curve shown is often called a *resonance curve.*[1]

3 The shift in time, denoted by the angle ϕ, between the displacement $x = x_0 \sin(\omega t + \phi)$ and the driving force varying as $\sin \omega t$ is also a strong function of the driving frequency, being 0 for low driving frequencies and $-\pi$ for high frequencies. [See Fig. 7.15 and the Advanced Topic (page 227).] The angle is defined here as the angle by which the displacement reaches its maximum before the force described by $F_0 \sin \omega t$. Negative angles therefore mean the displacement oscillation lags the force oscillation. Note that when $\omega = \omega_0$, $\phi = -\pi/2$, the displacement lags the force by one-quarter cycle. See the Advanced Topics (page 226ff.) for more detail.

SUPERPOSITION PRINCIPLE

An important property of the harmonic oscillator is that the solutions are additive: If $x_1(t)$ is the motion under the driving force $F_1(t)$, and $x_2(t)$ is the motion under the driving force $F_2(t)$, then $x_1(t) + x_2(t)$ is the motion under the combined force $F_1(t) + F_2(t)$. That is, if we know the motion x_1 under F_1 alone and the motion x_2 under F_2 alone, then we obtain the motion under the combined forces by adding together x_1 and x_2. This property follows directly from the equation of motion:

$$\left(\frac{d^2}{dt^2} + \frac{1}{\tau}\frac{d}{dt} + \omega_0^2\right)(x_1 + x_2)$$

$$= \left(\frac{d^2}{dt^2} + \frac{1}{\tau}\frac{d}{dt} + \omega_0^2\right)x_1 + \left(\frac{d^2}{dt^2} + \frac{1}{\tau}\frac{d}{dt} + \omega_0^2\right)x_2 \qquad (7.42)$$

$$= F_1 + F_2$$

The validity of the superposition principle for the solutions of the harmonic oscillator equation of motion is a consequence of the linearity of the equation; only the first power of x and its derivatives enters. The picture is quite different when anharmonic terms are included. A term in x^2 in the equation of motion can be shown to mix and to multiply two simultaneous driving frequencies ω_1 and ω_2, thereby producing a full range of harmonic frequencies ($2\omega_1, 3\omega_1, \ldots, 2\omega_2, 3\omega_2, \ldots$) and of combination or "sideband" frequencies ($\omega_1 + \omega_2$; $\omega_1 - \omega_2$; $\omega_1 - 2\omega_2$; etc.).

[1] In some fields the resonance curve is strictly the curve $f(X) = 1/(1 + X^2)$, which is similar in appearance.

PROBLEMS

1. *Simple pendulum.* A pendulum is constructed from a light string of length $L = 100$ cm and a heavy mass $M = 1 \times 10^3$ g. What is the period of the pendulum for small displacements?

Ans. 2.0 s.

2. *Mass on a spring.* Write the equation of motion for a mass M moving in a vertical line under the action of gravity and a spring of spring constant C. What is the effect of gravity on:
(*a*) The period of oscillation?
(*b*) The center of oscillation, the point about which the oscillation occurs?

$Cy_0 = -Mg \quad y_0 = -\frac{Mg}{C}$ center of oscillation

3. *Mass on a spring.* A mass of 1.0×10^3 g is suspended from a linear spring with a spring constant $C = 1.0 \times 10^6$ dyn/cm.
(*a*) What is the period for small oscillations?
(*b*) If at $t = 0$ the displacement from equilibrium is $+0.5$ cm and the velocity is $+15$ cm/s, find the displacement as a function of t.

4. *Mass on a spring—data.* The data in Tables 7.3 and 7.4 were obtained by observing the motion of a mass attached to the end of a spring:
(*a*) Plot the square of the period of oscillation as a function of mass. The tabulated values of mass exclude the mass of the spring. Determine the effective mass of the spring by proper extrapolation of the graph.
(*b*) Determine the spring constant C.
(*c*) Plot the natural logarithm of the amplitude as a function of time and determine the relaxation time.
(*d*) Determine the damping factor b.

TABLE 7.3 Period as a Function of Mass

Mass, g	*Observed periods, s*
50	0.72
100	0.85
150	0.96
200	1.06
250	1.16
300	1.23

TABLE 7.4 Amplitude of Vibration as a Function of Time for Mass 150 g

Time, s	*Amplitude, cm*
0	4.5
30	4.0
80	3.5
125	3.0
180	2.5
235	2.0
340	1.5
455	1.0

5. *Mass in buoyant medium.* A body partly (or completely) submerged in liquid is buoyed up by a force equal to the weight of the liquid displaced (Archimedes' principle). Show that a body of uniform horizontal cross section constrained to move vertically in a liquid of density greater than the density of the body will execute simple harmonic oscillations. What is the period? What is the limit of amplitude of the oscillations?

6. *Pendulum*
(*a*) A pendulum of length 39.2 cm and mass 500 g is set in motion so that at $t = 0$, $\theta = 0.1$, and $\dot{\theta} = -0.02$/s. Find θ as a function of t. Now use the equations of motion to find the force on the mass at $\theta = 0$.
(*b*) The Foucault pendulum was set up by Foucault in 1851 in Paris to show the effect of the rotation of the earth

(see Chap. 4, page 114). Its length is 69 m. Find the period. If the mass is 28 kg and the maximum swing is 10°, find the total energy of motion.

7. *Energy of mass on spring.* A mass of 50 g on the end of a certain spring executes simple harmonic motion according to the equation $x = 2 \sin 10t$ where x is in centimeters and t is in seconds.
(*a*) Find the spring constant C.
(*b*) Find the maximum kinetic energy.
(*c*) What are the maximum potential energy and the total energy?

8. *Two-dimensional oscillator.* A particle is free to move in the x,y plane under the action of a force toward the origin of magnitude $-C(x\hat{\mathbf{x}} + y\hat{\mathbf{y}}) = -C\mathbf{r}$. Assuming the mass is M, find the x and y equations of motion and solve them.
(*a*) What are the conditions for motion in a circle and what is the period?
(*b*) What are the conditions for motion along the line at 45° to the x axis and what is the period?

9. *Mass in spherical bowl.* A mass slides freely in the bottom of a spherical bowl of radius 1.0 m. Find the period for small oscillations. What is the length of the equivalent pendulum?

10. *Viscosity*
(*a*) Consult a reference book to give and explain the definition of viscosity.
(*b*) What are the dimensions of the coefficient of viscosity η?
(*c*) What is the value of the viscosity of water at 20°C?
(*d*) Evaluate (*b*) [see Eq. (7.28)] for a sphere of radius 5 cm in a medium of viscosity 2.0 centipoises. (See Prob. 13.)
(*e*) If in (*d*) the density of the sphere is 2.7 g/cc and the density of the liquid is 1.1 g/cc, find the terminal velocity. Use the net vertical force, as explained in Prob. 5, and the relaxation time.

11. *Motion under a viscous force*
(*a*) A particle of mass M acted on by only the viscous force of the medium $-bv$ is projected from a point with velocity v_0. Write down the velocity as a function of time. Remembering that $v = dx/dt$, find $x(t)$. If $M = 10$ g, $b = 4.0$ dyn-s/cm, and $v_0 = 100$ cm/s, find the distance the mass will travel.
(*b*) In the Millikan oil-drop experiment some drops had radius 2.0×10^{-4} cm. The density of the oil was 0.92 g/cc, and the viscosity of air was 1.8×10^{-4} centipoises. Find the relaxation time and the terminal velocity. Neglect the buoyant correction.

12. *Relaxation time.* An oscillator has $M = 10$ g, $C = 490$ dyn/cm, $b = 1.0$ dyn-s/cm. At $t = 0$, $x = 2.0$ cm, $\dot{x} = 0$:
(*a*) Find x as a function of t.
(*b*) What is the relaxation time for x; for K?
(*c*) What is the Q?

13. *Damped oscillator.* A spherical ball of radius 0.30 cm and mass 0.5 g moves in water under the action of a spring of constant $C = 50$ dyn/cm. η for water is 1.0×10^{-2} dyn-s/cm², or *poises.* Find the number of oscillations that will occur in the time for the amplitude to drop to one-half the initial amplitude. (Note that $e^{-0.693} = \frac{1}{2}$.) What is the Q of the oscillator?

ADVANCED TOPICS

Anharmonic Oscillator Following the discussion on pages 204–206, we consider now a pendulum that is oscillating with an amplitude so large that we may not neglect the θ^3 term in the expansion of $\sin\theta$, as we did in Eq. (7.9). What is the effect on the motion of the pendulum of the term in θ^3? We have seen the effect on the period in Table 7.1. Let us see what we can do analytically.

Anharmonic, or nonlinear, problems are usually difficult to solve exactly (although computers can readily provide any desired precision), but approximate solutions are often adequate to give us a good idea of what is happening. Equation (7.8) gives

$$\sin\theta = \theta - \tfrac{1}{6}\theta^3 + \cdots$$

so that the equation of motion (7.7) becomes, to this order,

$$\frac{d^2\theta}{dt^2} + \omega_0^2\theta - \frac{\omega_0^2}{6}\theta^3 = 0 \tag{7.43}$$

where ω_0^2 denotes the quantity g/L. This is the equation of motion of an *anharmonic oscillator.*

We shall see if we can find an approximate solution to Eq. (7.43) of the form

$$\theta = \theta_0 \sin\omega t + \epsilon\theta_0 \sin 3\omega t \tag{7.44}$$

where ϵ is a dimensionless constant expected to be much less than 1 for $\theta_0 \ll 1$. That is, we shall see if the motion can be represented approximately (or exactly—we don't know yet!) as the superposition of two different motions, one in $\sin \omega t$ and the other in $\sin 3\omega t$. The presence of a term in $\sin 3\omega t$ is suggested by the trigonometric identity [Dwight 403.03]

$$\sin^3 x \equiv \tfrac{3}{4} \sin x - \tfrac{1}{4} \sin 3x \tag{7.45}$$

Thus the θ^3 term in the differential equation (7.43) will generate from the cube of $\sin \omega t$ a term in $\sin 3\omega t$. To satisfy the differential equation we are forced to add to $\sin \omega t$ a term such as $\epsilon \sin 3\omega t$ just to cancel the $\sin 3\omega t$ term generated by θ^3. Going further, the new $\epsilon \sin 3\omega t$ term in the trial solution will generate, on being cubed, a term in $\epsilon^3 \sin 9\omega t$, and so on. There is no apparent reason why the process should stop; but if $\epsilon \ll 1$, the series may be expected to converge rapidly because higher and higher powers of ϵ are involved as factors in the higher-frequency terms. We thus see that Eq. (7.44) can only be an approximate solution at best. It remains for us to determine ϵ and also ω; although ω must reduce to ω_0 at small amplitudes, it may differ at large amplitudes. For simplicity we suppose that $\theta = 0$ at $t = 0$.

An approximate solution of this type is called a *perturbation solution* because one term in the differential equation perturbs the motion that would occur without that term. As you have seen, we arrived at the form of Eq. (7.44) by guided guesswork. It is easy enough to try several guesses and to reject the ones that do not work.

We have from Eq. (7.43)

$$\ddot{\theta} = -\omega^2\theta_0 \sin \omega t - 9\omega^2\epsilon\theta_0 \sin 3\omega t$$
$$\theta^3 = \theta_0^3(\sin^3 \omega t + 3\epsilon \sin^2 \omega t \sin 3\omega t + \cdots)$$

where we have discarded the terms of order ϵ^2 and ϵ^3 because of our assumption that we can find a solution with $\epsilon \ll 1$. Then the terms of Eq. (7.43) become, using the trigonometric identity [Eq. (7.45)],

$$\begin{aligned} \ddot{\theta} &= -\omega^2\theta_0 \sin \omega t - 9\omega^2\epsilon\theta_0 \sin 3\omega t \\ \omega_0^2\theta &= -\omega_0^2\theta_0 \sin \omega t + \omega_0^2\epsilon\theta_0 \sin 3\omega t \\ -\frac{1}{6}\omega_0^2\theta^3 &= -\frac{3\omega_0^2}{24}\theta_0^3 \sin \omega t + \frac{\omega_0^2}{24}\theta_0^3 \sin 3\omega t \\ &\qquad - \frac{\omega_0^2}{2}\theta_0^3\epsilon \sin^2 \omega t \sin 3\omega t \end{aligned} \tag{7.46}$$

Now add vertically the terms in Eq. (7.46) above. The sum on the left-hand side is equal to zero according to Eq. (7.43). If Eq. (7.44) is to be an approximate solution for all time t, it is necessary that the coefficients of $\sin \omega t$ and $\sin 3\omega t$ vanish separately on the right-hand side of Eq. (7.46). For suppose the coefficients did not vanish; then we would have an expression of the form $A \sin \omega t + B \sin 3\omega t = 0$, where A and B are constants. But such an equation *cannot* be satisfied at all times t; hence A and B must each be zero. By cutting off our assumed solution [Eq. (7.44)] at $3\omega t$ we have not included all the terms or frequencies that may occur, but we have included the most important ones.

The requirement that the coefficients of $\sin \omega t$ in Eq. (7.46) should sum to zero is that

$$-\omega^2 + \omega_0^2 - \tfrac{3}{24}\omega_0^2\theta_0^2 = 0$$

or

$$\omega^2 = \omega_0^2(1 - \tfrac{1}{8}\theta_0^2) \qquad \omega \approx \omega_0(1 - \tfrac{1}{16}\theta_0^2) \tag{7.47}$$

using the binomial expansion for the square root. (See Chap. 2, Mathematical Notes, page 53.) Equation (7.47) gives the dependence of ω on θ_0. Here ω_0 is the limit of ω as $\theta_0 \to 0$, that is, the small amplitude limit. For $\theta_0 = 0.3$ rad, the fractional frequency shift is $\Delta\omega/\omega \approx -10^{-2}$, where $\Delta\omega \equiv \omega - \omega_0$. Note that the frequency of the pendulum at large amplitudes does depend on the amplitude.

The solution in Eq. (7.44) also contains a term in $\sin 3\omega t$. The amplitude of this term relative to the amplitude of the term in $\sin \omega t$ is ϵ, which is determined by the condition that the coefficient of the term in $\sin 3\omega t$ in Eq. (7.46) vanish:

$$-9\omega^2\epsilon + \omega_0^2\epsilon + \frac{\omega_0^2}{24}\theta_0^2 = 0$$

If we set $\omega^2 \approx \omega_0^2$, then this equation reduces to

$$\epsilon \approx \frac{\theta_0^2}{192}$$

We think of ϵ as giving the fractional admixture of the $\sin 3\omega t$ term in a solution for θ dominated by the $\sin \omega t$ term. For $\theta_0 = 0.3$ rad, we have $\epsilon \approx 10^{-3}$, which is very small. The coefficient of the term $\sin^2 \omega t \sin 3\omega t$ in Eq. (7.46) is small by $O(\epsilon)$ or by $O(\theta_0^2)$, compared with the terms we have used. We have neglected this term in our approximation.

Why did we not include in Eq. (7.44) a term in $\sin 2\omega t$? Try for yourself a solution of the form

$$\theta = \theta_0 \sin \omega t + \eta\theta_0 \sin 2\omega t$$

and see what happens. You will find $\eta = 0$. The pendulum generates chiefly third harmonics, i.e., terms in $\sin 3\omega t$, and no terms in $\sin 2\omega t$. The situation would be different for a device for which the equation of motion included a term in θ^2. In such a case the solution will have a term in $\sin 2\omega t$, and the same technique can be used. There are many such problems (e.g., the thermal expansion of solids) in which the force is stronger for a positive (negative) value of the displacement than it is for an equal negative (positive) value.

What is the frequency of the pendulum at large amplitudes? There is no single frequency in the motion. We have seen that the most important term (the largest component) is in $\sin \omega t$, and we say that ω is the *fundamental frequency* of the pendulum. To our approximation ω is given by Eq. (7.47). The term in $\sin 3\omega t$ is called the *third harmonic* of the fundamental frequency. Our argument following Eq. (7.44) suggests that an infinite number of harmonics are present in the exact motion but that most of these are very small. The amplitude in Eq. (7.44) of the fundamental component of the motion is θ_0; the amplitude of the third harmonic component is $\epsilon\theta_0$.

Driven Harmonic Oscillator with Damping Force We now consider in detail the driven, or forced, motion of a damped harmonic oscillator. This is a problem of the greatest importance. If, besides friction, there is an external force $F(t)$ applied to the oscillator, the equation of motion is

$$M\ddot{x} + b\dot{x} + Cx = F(t)$$

or, in a more compact notation with $\tau \equiv M/b$ and $\omega_0{}^2 \equiv C/M$,

$$\ddot{x} + \frac{1}{\tau}\dot{x} + \omega_0{}^2 x = \frac{F(t)}{M} \qquad (7.48)$$

Here ω_0 is the *natural frequency* of the system in the absence of friction and in the absence of a driving force. When the system is driven at a different frequency $\omega(\neq\omega_0)$, we shall see that the steady-state response will be at the driving frequency and *not* at the natural frequency. But if the driving frequency is suddenly switched off, the system will revert to a damped oscillation at approximately the natural frequency, provided that the damping is low.

Suppose in Eq. (7.48) that

$$\frac{F(t)}{M} = \frac{F_0 \sin \omega t}{M} \equiv \alpha_0 \sin \omega t \qquad \alpha_0 \equiv \frac{F_0}{M} \qquad (7.49)$$

so that the driving force is sinusoidal at frequency ω. This relation defines the quantity α_0. The *steady-state* (the state of the system after any transient effects have died down) response of the system will be precisely at the driving frequency. Otherwise the relative phase between force and response would change with time. This is an important feature of the result—the steady-state response of a driven harmonic oscillator (even with damping) is at the *driving frequency* and not at the natural frequency ω_0. No frequency other than the driving frequency will satisfy the equation of motion. By response we can mean either the displacement x or the velocity $\dot{x}$. We shall speak here of x as the response.

We look for a solution of Eq. (7.48) of the form

$$x = x_0 \sin(\omega t + \phi) \qquad (7.50)$$

where we have to solve the equation of motion for the values of the amplitude x_0 and the phase angle[1] ϕ. In Eq. (7.50) ω is the frequency of the driving force, *not* the natural frequency of the oscillator; and ϕ gives the phase between the driving force and the displacement of the oscillator. Thus ϕ here has quite a different meaning from what we encountered in the undriven harmonic oscillator, where ϕ was related to the initial conditions. The initial conditions are irrelevant to the driven oscillator if only the steady state is considered.

It is worthwhile to define precisely what we mean by the phase ϕ between the displacement and driving force. Both the driving force and the displacement oscillate with simple harmonic motion. The cycle from maximum to maximum of both the force and displacement takes 360° or 2π rad. *The phase ϕ tells us by what angle the displacement reaches its maximum before the force.* For instance, suppose the force attains its greatest positive value at the instant when the displacement is zero and increasing in the positive direction. Then the displacement will lag the force by $\pi/2$ rad. But ϕ is defined as the angle by which x *leads* F, so that ϕ equals $-\pi/2$ in this instance.

Let us form the derivatives

$$\frac{dx}{dt} = \omega x_0 \cos(\omega t + \phi) \qquad \frac{d^2x}{dt^2} = -\omega^2 x_0 \sin(\omega t + \phi)$$

Then the equation of motion [Eq. (7.48)] is

$$(\omega_0{}^2 - \omega^2)x_0 \sin(\omega t + \phi) + \frac{\omega}{\tau}x_0 \cos(\omega t + \phi) = \alpha_0 \sin \omega t \qquad (7.51)$$

We simplify this by the trigonometric relations

$$\sin(\omega t + \phi) = \sin \omega t \cos \phi + \cos \omega t \sin \phi$$
$$\cos(\omega t + \phi) = \cos \omega t \cos \phi - \sin \omega t \sin \phi$$

Thus Eq. (7.51) becomes

$$\left[(\omega_0{}^2 - \omega^2)\cos\phi - \frac{\omega}{\tau}\sin\phi\right]x_0 \sin \omega t$$
$$+\left[(\omega_0{}^2 - \omega^2)\sin\phi + \frac{\omega}{\tau}\cos\phi\right]x_0 \cos \omega t = \alpha_0 \sin \omega t \qquad (7.52)$$

Equation (7.52) can only be satisfied if the coefficient of $\cos \omega t$ is zero. This condition may be written as

[1] We need to allow the angle ϕ (called the phase angle of x relative to the force F) to be different from zero. No solution can be obtained if ϕ is left out. In speaking of a phase angle, be sure to say the phase of what, relative to what. In electrical problems it is customary to speak of the phase of the current referred to the voltage. Here we speak of the phase of the displacement x referred to the driving force F. The two phases are not equivalent because dx/dt, and not x, is the analog of the current.

$$\tan\phi = \frac{\sin\phi}{\cos\phi} = -\frac{\omega/\tau}{\omega_0^2 - \omega^2} \tag{7.53}$$

It is also necessary that the coefficient of $\sin\omega t$ be equal to α_0:

$$x_0 = \frac{\alpha_0}{(\omega_0^2 - \omega^2)\cos\phi - (\omega/\tau)\sin\phi} \tag{7.54}$$

From Eq. (7.53) it follows that

$$\cos\phi = \frac{\omega_0^2 - \omega^2}{[(\omega_0^2 - \omega^2)^2 + (\omega/\tau)^2]^{\frac{1}{2}}}$$

$$\sin\phi = \frac{-\omega/\tau}{[(\omega_0^2 - \omega^2)^2 + (\omega/\tau)^2]^{\frac{1}{2}}} \tag{7.55}$$

whence Eq. (7.54) becomes

$$x_0 = \frac{\alpha_0}{[(\omega_0^2 - \omega^2)^2 + (\omega/\tau)^2]^{\frac{1}{2}}} \tag{7.56}$$

This is the amplitude of the motion.

Equations (7.55) and (7.56) give us the desired solution. We now know the amplitude x_0 and phase ϕ of the response of the system under the driving force $F = M\alpha_0 \sin\omega t$:

$$x = \frac{\alpha_0}{[(\omega_0^2 - \omega^2)^2 + (\omega/\tau)^2]^{\frac{1}{2}}} \sin\left(\omega t + \tan^{-1}\frac{\omega/\tau}{\omega^2 - \omega_0^2}\right) \tag{7.57}$$

The amplitude in Eq. (7.57) is plotted against ω in Fig. 7.14 and the phase angle ϕ in Fig. 7.15. Note from that graph that the phase angle is always negative. This can be understood from Eq. (7.53) since $\phi = 0$ for $\omega = 0$, $0 > \phi > -\pi/2$ for $\omega < \omega_0$, and $-\pi/2 > \phi > -\pi$ for $\omega > \omega_0$.

We can develop a feeling for this solution by examining limiting cases. In our discussion we assume always that the damping is light, so that $\omega_0\tau \gg 1$.

Low Driving Frequency $\omega \ll \omega_0$ Here we see from Eq. (7.55) that

$$\cos\phi \to 1 \qquad \sin\phi \to -0$$

whence $\phi \to 0$. This response at low frequency is said to be *in phase* with the driving force. From Eq. (7.56)

$$x_0 \to \frac{\alpha_0}{\omega_0^2} = \frac{M\alpha_0}{C} = \frac{F_0}{C} \tag{7.58}$$

The *spring* (and not the mass or the friction) controls the response in this limit; the mass is simply displaced slowly back and forth by the force acting against the restoring force of the spring.

Resonance Response $\omega = \omega_0$ The response may be very large at resonance. We often use resonance response in applications, and we need to treat it carefully. At $\omega = \omega_0$ the driving frequency equals the natural frequency of the system in the absence of friction. We have

$$\cos\phi \to \pm 0 \qquad \sin\phi \to -1 \qquad \phi \to -\frac{\pi}{2}$$

The amplitude at $\omega = \omega_0$ is given by

$$x_0 = \frac{\alpha_0\tau}{\omega_0} \tag{7.59}$$

The lower the damping, the higher are τ and thus x_0. Keeping F_0 constant, the ratio of the response at resonance to the response at zero frequency is given from Eqs. (7.58) and (7.59):

$$\frac{x_0(\omega = \omega_0)}{x_0(\omega = 0)} = \frac{\alpha_0\tau/\omega_0}{\alpha_0/\omega_0^2} = \omega_0\tau = Q$$

with the Q factor as defined by Eq. (7.41). This may be very large—often 10^4 or more! The damping can be said to control the response at resonance.

The maximum response x_0 does not occur at exactly $\omega = \omega_0$. We note that the derivative of Eq. (7.56) has a zero at

$$\frac{d}{d\omega}\left[(\omega_0^2 - \omega^2)^2 + \left(\frac{\omega}{\tau}\right)^2\right] = 2(\omega_0^2 - \omega^2)(-2\omega) + \frac{2\omega}{\tau^2} = 0$$

or

$$\omega^2 = \omega_0^2 - \frac{1}{2\tau^2}$$

This is the position of the maximum response of the curve of x_0 vs ω. If $\omega_0\tau \gg 1$, the maximum is very close to $\omega = \omega_0$.

It might appear odd that the maximum response is obtained with the phase angle close to $-\pi/2$, that is, when the force is 90° out of phase with the displacement. It might seem that resonance would logically occur when $\phi = 0$, not $-\pi/2$. But here is the catch: The power absorbed by the oscillator does not depend directly on the phase between driving force and displacement, but rather on the phase between the driving force and the *velocity*. It takes a moment's reflection to see that we shall obtain the largest deflections when the velocity is exactly in phase with the driving forces. In this manner the mass gets pushed at just the right times and places. When the displacement is zero, the velocity is greatest. If at this point it is moving in the positive direction, we wish the force at this time to attain its greatest value in order to obtain

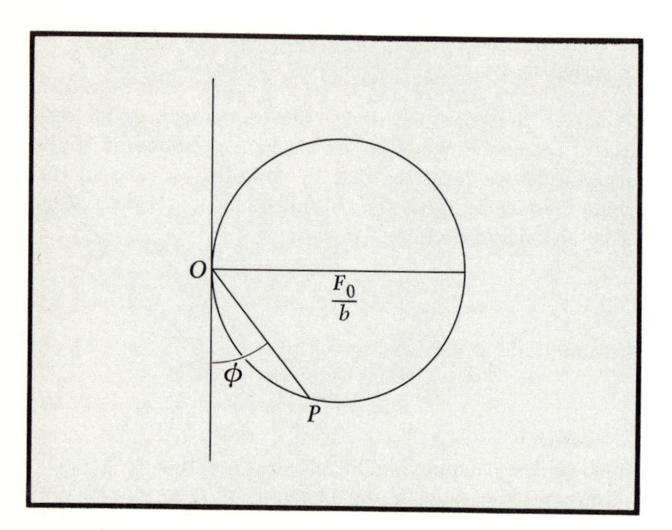

FIG. 7.16 (*a*) The "polar plot" gives a simple graphical representation of the driven harmonic oscillator. A circle with diameter F_0/b is constructed, and a line segment OP making angle ϕ with the ordinate is drawn.

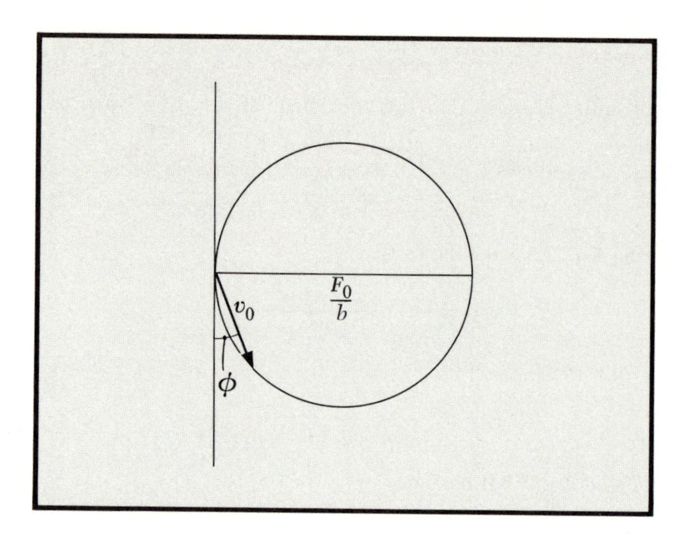

(*c*) For $\omega \ll \omega_0$, $\phi \approx 0$ and $v_0 \ll F_0/b$. The response is very small in this limit.

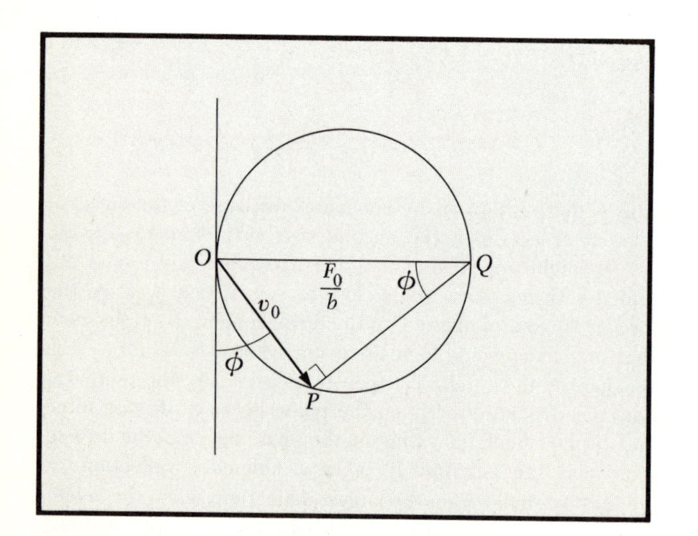

(*b*) For any ϕ the triangle OPQ is a right triangle. Thus line segment $OP = -(F_0/b)\sin\phi$. From Eqs. (7.55) to (7.57) we see that line segment $OP = \omega x_0 = v_0$, the amplitude of the velocity.

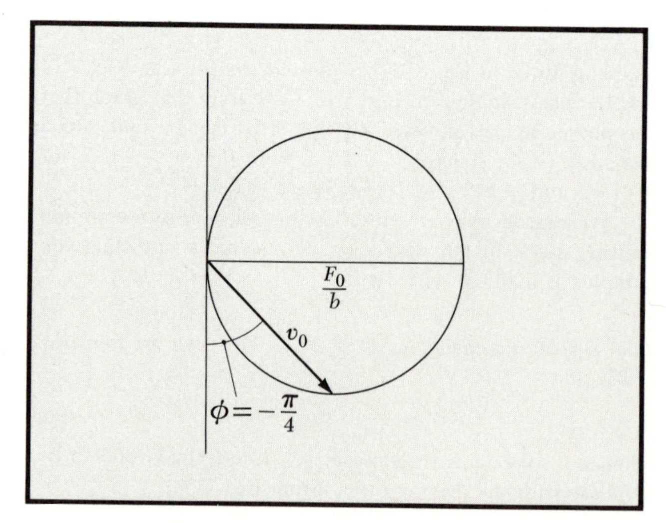

(*d*) As ω grows $|\phi|$ increases and so does v_0. At $\phi = -\pi/4$, $v_0 = F_0/\sqrt{2}b$.

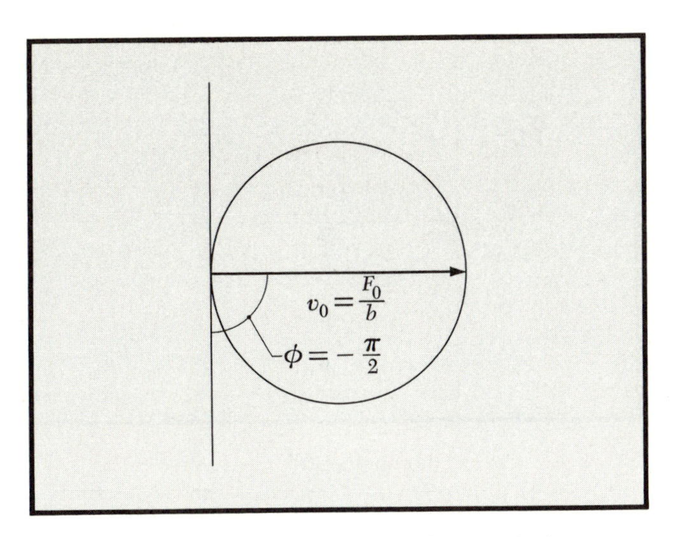

(e) At $\phi = -\pi/2$, $\omega = \omega_0$ and $v_0 = F_0/b$. The velocity amplitude is a *maximum at resonance*.

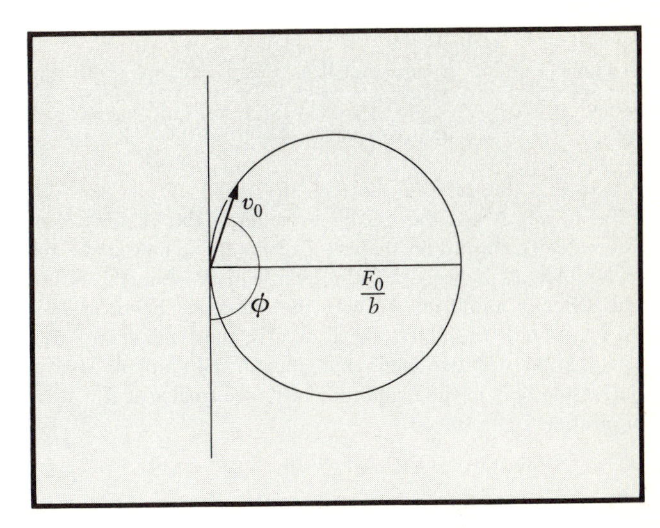

(g) For $\omega \gg \omega_0$, $v_0 \ll F_0/b$ again and $\phi \approx -\pi$.

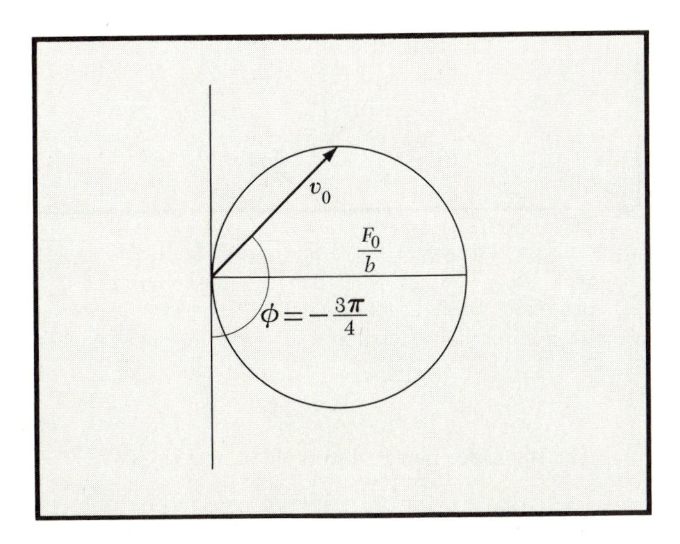

(f) For $\omega > \omega_0$, v_0 decreases again. At $\phi = -3\pi/4$, $v_0 = F_0/\sqrt{2}b$ again.

maximum motion. At the turning points where the velocity changes its direction, if we want resonance, we wish the force to change its direction in the same manner at that moment. Thus resonance is best seen in terms of the phase between velocity and driving force. We know that the velocity of an oscillator leads its displacement by exactly 90°. Thus for resonance, with force and velocity in phase, we must have the force 90° ahead of the displacement, so that $\phi = -\pi/2$. Although, as pointed out above, the maximum amplitude occurs at a frequency slightly lower than ω_0, the maximum power is delivered when $\omega = \omega_0$.

High Driving Frequency $\omega \gg \omega_0$ Here

$$\cos\phi \to -1 \qquad \sin\phi \to 0 \qquad \phi \to -\pi$$

and

$$x_0 \to \frac{\alpha_0}{\omega^2} = \frac{M\alpha_0}{M\omega^2} = \frac{F_0}{M\omega^2}$$

In this limit the response decreases as $1/\omega^2$. The inertia of the mass controls the response in the high-frequency limit, and the mass responds essentially like a free object, being rapidly shaken back and forth by the force. Notice that the phase ϕ of the displacement x with respect to the driving force F starts from zero at low frequencies, passes through $-\frac{1}{2}\pi$ at resonance, and attains $-\pi$ at high frequencies. *The displacement always lags the driving force.*

An interesting geometrical way of understanding these phenomena is given in Fig. 7.16. Instead of plotting the response x_0 or perhaps the power against ω, we use the angle ϕ as our variable. It turns out from Eqs. (7.55) and (7.56) that

$$\omega x_0 = \frac{F_0}{b}\sin(-\phi) = v_0$$

where b is the damping coefficient (M/τ) in Eq. (7.30). The product ωx_0 is just the maximum value of the velocity $\dot{x}$, or the velocity amplitude that we call v_0. If we now make the polar plot as in Fig. 7.16, the length of the line OP is just this velocity amplitude v_0 with the diameter chosen as F_0/b to agree with the above equation. We must remember that ϕ is really a negative angle, but since in the graphing we are interested only in the magnitude of v_0, we treat ϕ as if it were a positive angle since

$$\sin(-\phi) = -\sin\phi \qquad |\sin(-\phi)| = |\sin\phi|$$

The values of ϕ given in the legend to Fig. 7.16 are the correct negative values.

From this polar graph we can see that as ϕ changes, the length of OP starts at a small value (Fig. 7.16*c*), increases to the maximum value of the diameter of the circle when $\phi = -\pi/2$ (Fig. 7.16*e*), and then decreases again as ϕ goes to $-\pi$ (Fig. 7.16*f* and *g*). The power, as will be seen below, is the average over a cycle of $F\dot{x}$ [see Eqs. (7.60) and (7.62)] and has a maximum value at $\phi = -\pi/2$.

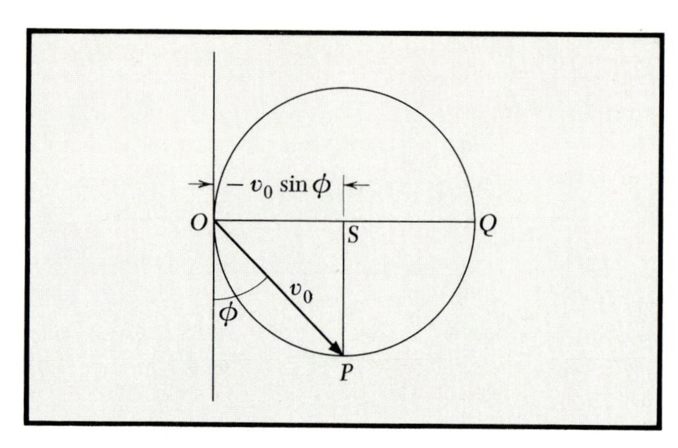

FIG. 7.17 (*a*) The line segment $OS = -v_0 \sin\phi$ is shown. From Eqs. (7.60) to (7.62) we see that the power absorbed is proportional to $-v_0 \sin\phi$, or to line segment OS.

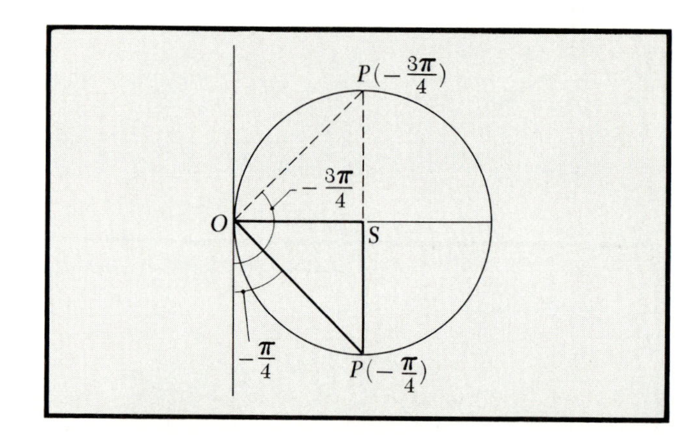

(*b*) At phase angles $\phi = -\pi/4$ and $\phi = -3\pi/4$, line segment $OS = \frac{1}{2}OS_{max}$. Thus *half* the maximum power absorption occurs at these points. Of course *maximum* power absorption occurs at $\phi = -\pi/2$ (resonance).

Power Absorption The time average of the work done per unit time on the oscillating system by the driving force is given, using Eqs. (7.49) and the time derivative of (7.57), by

$$P = \langle F\dot{x}\rangle = \frac{M\alpha_0^2\omega}{[(\omega_0^2-\omega^2)^2 + (\omega/\tau)^2]^{\frac{1}{2}}}\langle \sin\omega t \cos(\omega t + \phi)\rangle \tag{7.60}$$

Using the identity

$$\cos(\omega t + \phi) = \cos\omega t\cos\phi - \sin\omega t \sin\phi$$

we have

$$\begin{aligned}\langle \sin\omega t[\cos\omega t\cos\phi - \sin\omega t\sin\phi]\rangle &= -\sin\phi\langle \sin^2\omega t\rangle \\ &= -\tfrac{1}{2}\sin\phi \end{aligned} \tag{7.61}$$

where we have used the fact that $\langle \sin\omega t \cos\omega t\rangle = 0$. We see that the phase is important here (see Fig. 7.17*a* and *b*). With Eq. (7.55) for $\sin\phi$, we may write Eq. (7.60):

$$\begin{aligned} P &= \frac{1}{2}M\alpha_0^2\frac{\omega^2/\tau}{(\omega_0^2-\omega^2)^2 + (\omega/\tau)^2} \\ &= \frac{1}{2}M\alpha_0^2\tau\left[\left(\frac{\omega_0^2-\omega^2}{\omega/\tau}\right)^2 + 1\right]^{-1} \end{aligned} \tag{7.62}$$

This is an important result illustrated in Fig. 7.17.

The resonance power absorption at $\omega = \omega_0$ is

$$P_{res} = \tfrac{1}{2}M\alpha_0^2\tau$$

The power absorption [Eq. (7.62)] is reduced to one-half of the value at resonance when ω is changed by $\pm(\Delta\omega)_{\frac{1}{2}}$ such that

$$\frac{\omega}{\tau} = \omega_0^2 - \omega^2 \equiv (\omega_0+\omega)(\omega_0-\omega) \approx 2\omega(\Delta\omega)_{\frac{1}{2}} \tag{7.63}$$

Thus the full width $2(\Delta\omega)_{\frac{1}{2}}$ of the resonance at half-maximum power is equal to $1/\tau$. We see, using the expression for Q found in Eq. (7.41),

$$Q = \omega_0\tau = \frac{\omega_0}{2(\Delta\omega)_{\frac{1}{2}}} = \frac{\text{frequency at resonance}}{\text{full width at half-maximum power}}$$

Thus Q measures the sharpness of tuning (see Fig. 7.18).

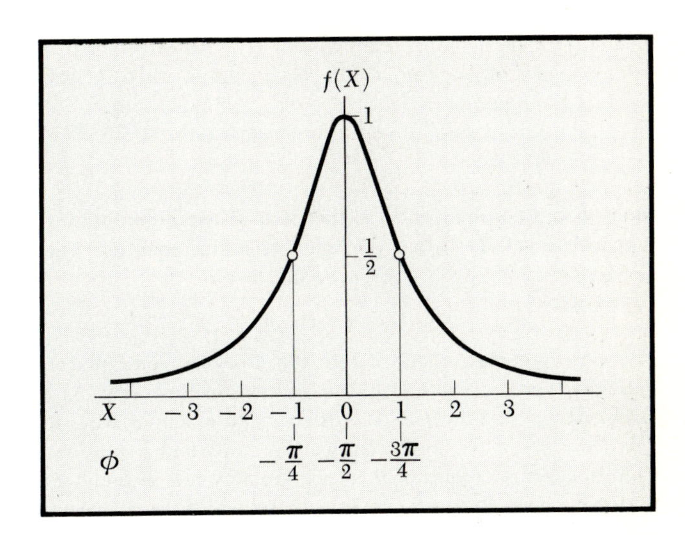

FIG. 7.18 The power absorption is proportional to $f(X) = 1/(1 + X^2)$, where

$$X = \frac{-({\omega_0}^2 - \omega^2)}{\omega/\tau} = \cot\phi$$

according to Eq. (7.62). For $\phi \approx 0$, $X = \cot\phi$ is large and negative. For $\phi = -\pi/2$, $X = \cot\phi = 0$ and the power absorption is a maximum. The half-power points $X = \pm 1$ are also seen. The function $f(X) = 1/(1 + X^2)$ is known as a Lorentz function.

EXAMPLE

Numerical Example of a Harmonic Oscillator Problem Let the mass $M = 1\ \text{g} = 0.001\ \text{kg}$, the spring constant $C = 10^4\ \text{dyn/cm} = 10\ \text{N/m}$, and the relaxation time $\tau = \frac{1}{2}\ \text{s}$. Then from Eq. (7.3)

$$\omega_0 = \left(\frac{C}{M}\right)^{\frac{1}{2}} = \left(\frac{10^4}{1}\right)^{\frac{1}{2}} = \left(\frac{10}{10^{-3}}\right)^{\frac{1}{2}} = 10^2\ \text{s}^{-1}$$

while from Eq. (7.33) the free oscillation frequency is

$$\left[{\omega_0}^2 - \left(\frac{1}{2\tau}\right)^2\right]^{\frac{1}{2}} = [10^4 - 1]^{\frac{1}{2}} \approx 10^2\ \text{s}^{-1}$$

The Q of the system is given by Eq. (7.41):

$$Q \approx \omega_0\tau = (10^2)(\tfrac{1}{2}) = 50$$

The time for the amplitude to damp to e^{-1} of its initial value (for the free system) is

$$2\tau = 1\ \text{s}$$

using Eq. (7.32). The damping constant $b = M/\tau = 1/\frac{1}{2} = 2\text{g/s}$ or

$$\frac{0.001}{\frac{1}{2}} = 2 \times 10^{-3}\ \text{kg/s}$$

Now let the system be driven by the force

$$F = M\alpha_0 \sin\omega t = 10 \sin 90t\ \text{dyn}$$

We see that $\alpha_0 = F_0/M = 10\ \text{dyn/g}$ and the driving frequency is $\omega = 90\ \text{s}^{-1}$. The amplitude is given by Eq. (7.56):

$$x_0 \approx \frac{10}{[4 \times 10^6 + 4 \times 10^4]^{\frac{1}{2}}} \approx 5 \times 10^{-3}\ \text{cm}$$

And the phase is given by Eq. (7.53):

$$\tan\phi \approx -\frac{180}{1.9 \times 10^3} \approx -0.1$$

or $\phi \approx -0.1\ \text{rad} \approx -6°$. Thus in every cycle the maximum of the displacement occurs at $0.1\ \text{rad}/90\ \text{rad/s} \approx 10^{-3}\ \text{s}$ after the maximum of the force.

We may compare the amplitude above with that in the limit $\omega \to 0$ and with that at resonance. From Eq. (7.58) we have $x_0(\omega = 0) = \alpha_0/{\omega_0}^2 = 10/10^4 = 10^{-3}\ \text{cm}$. At resonance we have, from Eqs. (7.59) and (7.41),

$$x_0(\omega = \omega_0) = Qx_0(\omega = 0) = (50)(10^{-3}) = 5 \times 10^{-2}\ \text{cm}$$

The full width of the resonance curve between half-power points is by Eq. (7.63) and the following:

$$2(\Delta\omega)_{\frac{1}{2}} = \frac{\omega_0}{Q} = \frac{100}{50} = 2\ \text{s}^{-1}$$

Note that in this example we have everywhere used *frequency* to mean *angular frequency*. To obtain ordinary frequencies in oscillations per second or cycles per second we must divide by 2π.

MATHEMATICAL NOTES

We shall now investigate some more complicated equations that come up in the study of mechanics. In Chap. 3, (page 93ff.) we have solved the two types of equations corresponding to the *no force* and *constant force* cases. From the point of view of ease in solving the equation, we can next consider the equation:

$$\frac{d^2x}{dt^2} = bt$$

where b is a constant and t is the time. If one differentiates t^3 twice he gets t, so that the solution of this equation will be:

$$x = \tfrac{1}{6}bt^3 + c_0t + d_0$$

where we see that c_0 and d_0 are the arbitrary constants of the same type that we met in the solutions of Eqs. (3.52) and (3.54). At $t = 0$, $x = d_0$ and $dx/dt = c_0$. In a similar way, if $d^2x/dt^2 = bt^2$ or ft^3, and so on, we can easily find a solution. Unfortunately problems of this type are very few in number, so that we must go on to some other cases which we will meet more often in physics.

Resisting Force A fairly common problem in mechanics is the case of a resisting force directly proportional to the velocity. (This equation also arises in the case of the decay of radioactive substances.) In this case Newton's Second Law gives

$$m\frac{d^2x}{dt^2} = -bv = -b\frac{dx}{dt} \tag{7.64}$$

The minus sign indicates that the force tends to reduce or decrease the velocity. We can simplify the equation by remembering that the acceleration d^2x/dt^2 is just equal to dv/dt, so that Eq. (7.64) becomes

$$m\frac{dv}{dt} = -bv$$

or

$$\frac{dv}{v} = -\frac{bdt}{m}$$

We now have an equation which we can integrate by the standard methods.

$$\int\frac{dv}{v} = \log_e v = -\int\frac{b\,dt}{m} = -\frac{bt}{m} + \text{const}$$

$$v = Ae^{-bt/m} \tag{7.65}$$

Here e is the base of natural logarithms and has the value 2.718. . . . The properties of natural logarithms are similar to those of logarithms to the base 10. Values of natural logarithms can be found in most mathematical tables. A number that is worth remembering is $e^{-0.693} = \frac{1}{2} = 0.500$.

What is the constant A in Eq. (7.65)? At $t = 0$, $e^{-0} = 1$, so that $v = A$. Therefore this is an arbitrary constant that we can use to fit the initial conditions. Previously, however, we had two arbitrary constants; here we have only one. But the equation we have solved is a first-order equation and so we have only one constant. We can now go back to $v = dx/dt$ and write:

$$dx/dt = v_0e^{-bt/m} \tag{7.66}$$

where we have written v_0 for the velocity at $t = 0$. Let us try a solution of the form:

$$x = B + Ce^{-bt/m}$$

We put in the exponential because we remember that when we differentiate e^{-t} we get e^{-t} again. Inserting this in Eq. (7.66), we get

$$-b/mCe^{-bt/m} = v_0e^{-bt/m}$$

Therefore the constant $C = -mv_0/b$. What about the constant B? We suspect that it will be the x_0, the initial condition. However, when we insert $t = 0$ in the solution we find that

$$x = B + C = B - \frac{mv_0}{b}$$

Therefore if $x = x_0$ at $t = 0$, $B = x_0 + mv_0/b$. Our final solution is then

$$x = x_0 + \frac{v_0m}{b}(1 - e^{-bt/m})$$

As an example, suppose a particle of mass 25 g is acted on by a force $-5v$ dyn, and starts at $x = -10$ with velocity in the plus x direction of 40. Using the above we find:

$$x = -10 + 40 \times \tfrac{25}{5}(1 - e^{-5/25t}) = -10 + 200(1 - e^{-t/5})$$

$$t = 0 \qquad x = -10$$

$$\frac{dx}{dt} = -200(-\tfrac{1}{5})e^{-t/5} = +40 \text{ at } t = 0$$

Note that as $t \to \infty$, $x \to 190$, and $v \to 0$.

Terminal Velocity We can now proceed to some more complicated equations, such as

$$m\frac{dv}{dt} = F - bv \tag{7.67}$$

where F is a constant.

The final steady-state solution of this is seen to be

$$\frac{dv}{dt} = 0 \qquad F = bv \qquad v = \frac{F}{b}$$

This velocity is often called the terminal velocity. As a solution, try

$$v = \frac{F}{b}(1 - e^{-bt/m})$$

$$\frac{dv}{dt} = \frac{F}{m}e^{-bt/m} = \frac{F}{m}\left(1 - \frac{b}{F}v\right) = \frac{F}{m} - \frac{b}{m}v$$

which agrees with Eq. (7.67). We see from this solution that at $t = 0$, $v = 0$. Therefore, if there is some other initial condition we must arrange to satisfy it. We do note that as $t \to \infty$, $v \to F/b$.

Suppose at $t = 0$, $v = v_0$. Try

$$v = \frac{F}{b}(1 - e^{-bt/m}) + v_0 e^{-bt/m} \tag{7.68}$$

Differentiation shows that this satisfies the original equation.

It is interesting that the general solution Eq. (7.68) above is the sum of the solution of Eq. (7.67) that we tried first plus the solution given by Eq. (7.66) for $m\,dv/dt = -bv$. To look at this another way, consider the solution of $m\,dv/dt + bv = F$. If we can find one solution to this, we can add any solution of $m\,dv/dt + bv = 0$ to it and still have a solution.

Spring Force Now let us solve the equation of motion for a *spring-type* force, one always toward the origin (or toward some point which is conveniently chosen as the origin) and increasing directly as the distance from the point or origin. Mathematically, this is $F = -Cx$; if x is positive, the force is negative; if x is negative, the force is positive. C is a positive constant, often called the *spring constant*. Our equation is now

$$m\frac{d^2x}{dt^2} = -Cx \qquad \frac{d^2x}{dt^2} = -\frac{Cx}{m}$$

As a trial solution let $x = \cos \omega t$. Differentiating we get

$$\frac{dx}{dt} = -\omega \sin \omega t \qquad \frac{d^2x}{dt^2} = -\omega^2 \cos \omega t = -\omega^2 x$$

Comparing this with the original equation, we see that if $\omega^2 = C/m$, our solution is valid. Also it is apparent that $x = \sin \omega t$ works. But where are our constants for the initial conditions? Let us try

$$x = A' \sin \omega t + B' \cos \omega t$$

This is still a solution and we notice that at $t = 0$, $x = B'$. Therefore B' is the x_0 we have used before. When we differentiate and set $t = 0$, we get $dx/dt = \omega A'$. This is the v_0 we have used before. Note that A' is not the initial velocity. An alternative way of writing this solution is

$$x = A \sin (\omega t + \phi)$$

At $t = 0$, $x = A \sin \phi$, and at $t = 0$, $dx/dt = +A\omega \cos \phi$. So the two constants A and ϕ have replaced the two constants A' and B'. More often, this second type of the solution is convenient to use, though you will find it useful to solve some problems with the first type. A few examples may help. Assume that $C/m = 25$ so that $\omega = 5$.

1 The particle starts at the origin at $t = 0$ with a velocity in the negative x direction of 10 cm/s. Assuming $x = A \sin (5t + \phi)$, $x = A \sin \phi$ at $t = 0$ and this must be 0. Therefore ϕ is 0 or π; $dx/dt = +5A \cos \phi = -10$. Therefore we see that ϕ must be π and A must be 2. So our solution is

$$x = 2 \sin (5t + \pi) = -2 \sin 5t$$

2 At $t = 0$ the particle is at $x = +5$ and is at rest. Therefore $+5 = A \sin \phi$ and $dx/dt = \omega A \cos \phi = 0$. Therefore $A = 5$ and $\phi = \pi/2$ and our solution is

$$x = 5 \sin \left(5t + \frac{\pi}{2}\right) = 5 \cos 5t$$

3 At $t = 0$, $x = -5$ and the velocity is -25. This time $-5 = A \sin \phi$ and $-25 = +A\omega \cos \phi = +5A \cos \phi$. Dividing we get $\tan \phi = +1$. Therefore $\phi = \pi/4$ or $5\pi/4$. But $\cos 5\pi/4 = -1/\sqrt{2}$ while $\cos \pi/4 = +1/\sqrt{2}$. Therefore $\phi = 5\pi/4$. Substituting in the first equation gives $A = \sqrt{2} \times 5$. Alternatively, one could choose $A = -5\sqrt{2}$, and $\phi = \pi/4$:

$$x = 5\sqrt{2} \sin \left(5t + \frac{5\pi}{4}\right) \qquad \text{or} \qquad -5\sqrt{2} \sin \left(5t + \frac{\pi}{4}\right)$$

Spring Force and Resisting Force A more difficult equation arises in the case of a damped, harmonic oscillator. The force is now the force $-bv$ plus the force $-Cx$.

$$m\frac{d^2x}{dt^2} = -bv - Cx \text{ where } v = \frac{dx}{dt}$$

$$m\frac{d^2x}{dt^2} + b\frac{dx}{dt} + Cx = 0$$

Try a solution of the form $x = Ae^{-\beta t} \sin (\omega t + \phi)$

$$b\frac{dx}{dt} = -\beta b A e^{-\beta t} \sin (\omega t + \phi) + b\omega A e^{-\beta t} \cos (\omega t + \phi)$$

$$m\frac{d^2x}{dt^2} = -2m\omega\beta A e^{-\beta t} \cos (\omega t + \phi) + \beta^2 m A e^{-\beta t} \sin (\omega t + \phi) - m\omega^2 A e^{-\beta t} \sin (\omega t + \phi)$$

Then

$$\begin{aligned} m\frac{d^2x}{dt^2} + b\frac{dx}{dt} + Cx &= Ae^{-\beta t} \sin (\omega t + \phi)[C - \beta b - m\omega^2 + \beta^2 m] \\ &\quad + Ae^{-\beta t} \cos (\omega t + \phi)[b\omega - 2m\omega\beta] = 0 \end{aligned}$$

The only way this equation can be satisfied for all values of t is to have the coefficients in brackets each equal to zero.

$$b\omega - 2m\omega\beta = 0$$

$$\beta = \frac{b}{2m}$$

$$C - \beta b - m\omega^2 + \beta^2 m = C - \frac{b^2}{2m} - m\omega^2 + \frac{b^2}{4m^2}m = 0$$

$$\omega^2 = \frac{C}{m} - \frac{b^2}{4m^2} = \frac{C}{m} - \beta^2$$

Note that A and ϕ are the arbitrary constants and are not specified by the differential equation. But the frequency ω and the damping constant β are so determined. If $\omega_0 = C/m$, $\omega < \omega_0$; but if β is small, that is, the rate of decrease in the amplitude $Ae^{-\beta t}$ is slow, $\omega \approx \omega_0$.

The form of this solution is shown in Fig. 7.19 for which $\omega/\beta \approx 5$.

It is to be noted that if b is large, ω can be zero or $C = b^2/4m$. What is the solution in this case?

$Ae^{-bt/2m} \sin \phi = A'e^{-bt/2m}$ is a solution as can be seen by trial. But also $Bte^{-bt/2m}$ is a solution. Therefore the solution is

$$x = A'e^{-bt/2m} + Bte^{-bt/2m} \tag{7.69}$$

where A' and B are the arbitrary constants needed to fulfill the initial conditions. This solution is often called the *critically damped solution*. x decreases to zero faster than if

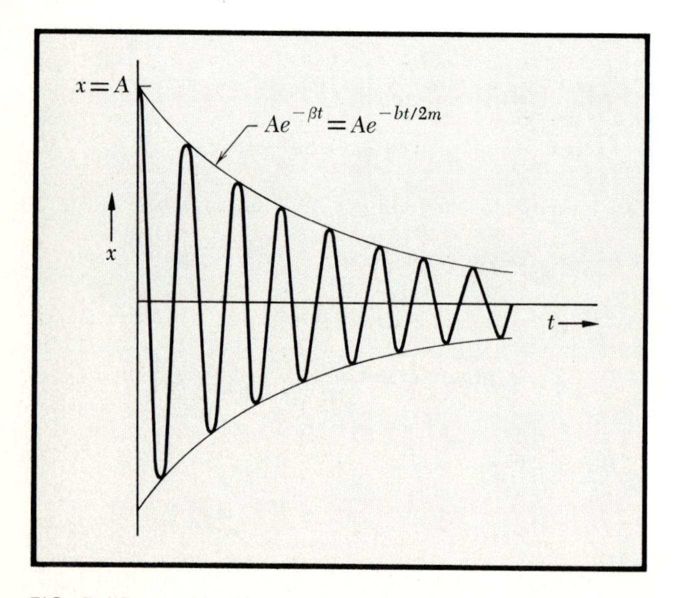

FIG. 7.19 Damped harmonic oscillator.

$$C < \frac{b^2}{4m} \qquad \text{or} \qquad C > \frac{b^2}{4m}$$

If $C < b^2/4m$, the solution is called *overdamped*, and the solution is

$$x = e^{-bt/2m}\left[A \exp\left(\sqrt{\frac{b^2}{4m^2} - \frac{C}{m}}\,t\right) + B \exp -\left(\sqrt{\frac{b^2}{4m^2} - \frac{C}{m}}\,t\right)\right] \tag{7.70}$$

where A and B are the constants needed to fulfill the initial conditions.

Complex Numbers and the Driven Harmonic Oscillator

Students who are familiar with complex numbers may remember De Moivre's theorem, which states

$$e^{i\alpha} = \cos \alpha + i \sin \alpha$$

where $i = \sqrt{-1}$. Such an expression $e^{i\alpha}$ is called a *complex quantity*. $\cos \alpha$ is called the *real part* and $\sin \alpha$ the *imaginary part*. Complex numbers can be visualized by plotting the real part as abscissa and the imaginary part as ordinate. Figure 7.20 shows this representation. The length of the line OA is called the *magnitude of the complex quantity*, or *number*. Its square is obtained by multiplying the number by its complex conjugate; the complex conjugate is obtained by changing the sign of i wherever it appears. The magnitude is of course a real quantity and in this case is

$$e^{i\alpha} \times e^{-i\alpha} = e^0 = 1$$

We see that OA has length unity since $\cos \alpha = OB/OA = OB$.

Addition, subtraction, and multiplication of complex numbers follow the usual rules. For example

$$(a + ib) + (c + id) = a + c + i(b + d)$$
$$(a + ib) - (c + id) = a - c + i(b - d)$$
$$(a + ib) \times (c + id) = ac - bd + i(ad + bc)$$
$$\text{since } i^2 = -1$$

To divide we usually want to manipulate the quotient so that the denominator is a real number and then the real and imaginary parts of the quotient will be easily recognizable:

$$\frac{a + ib}{c + id} = \frac{(a + ib) \times (c - id)}{(c + id) \times (c - id)} = \frac{ac + bd + i(bc - ad)}{c^2 + d^2}$$

Finally it is to be noted that any complex number can be written in the form $\rho e^{i\phi}$. To find ρ and ϕ in terms of the real and imaginary parts we set

$$\rho e^{i\phi} = \rho \cos \phi + i\rho \sin \phi = a + ib$$

$$\rho e^{i\phi} \times \rho e^{-i\phi} = \rho^2 = (a + ib) \times (a - ib) = a^2 + b^2 \tag{7.71}$$

$$\tan \phi = \frac{b}{a} \qquad \phi = \tan^{-1}\frac{b}{a}$$

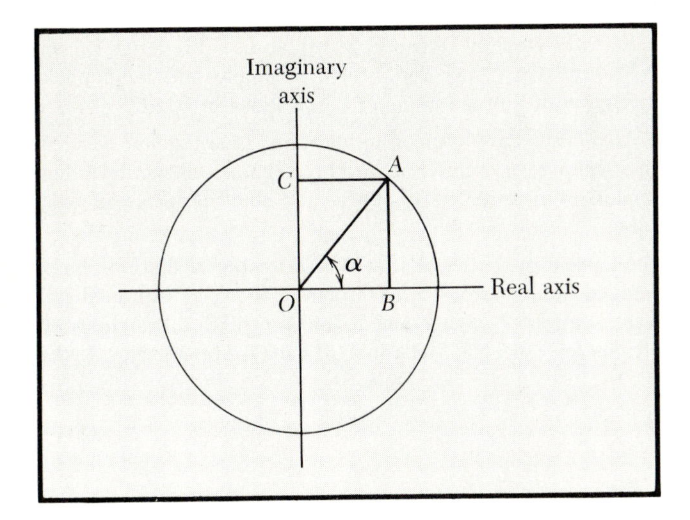

FIG. 7.20 A complex number can be represented by plotting the real part OB along one axis, the imaginary part OC along a perpendicular axis. In this case, OA represents $e^{i\alpha}$, $OB = \cos\alpha$, $OC = \sin\alpha$. OA is the magnitude of the complex quantity.

We now give a very neat solution of the problem of the driven harmonic oscillator, using the complex number scheme.

The equation of motion [Eqs. (7.48) and (7.49)] is (with $\cos\omega t$ written for convenience in place of $\sin\omega t$),

$$\ddot{x} + \frac{1}{\tau}\dot{x} + \omega_0^2 x = \alpha_0 \cos\omega t \tag{7.72}$$

Let us replace the force term by

$$\alpha_0 e^{i\omega t} \equiv \alpha_0(\cos\omega t + i\sin\omega t)$$

At the end of the calculation we may take the response as the real part of x if the driving force is $\alpha_0 \cos\omega t$ (with α_0 real).

We look for a solution to Eq. (7.72)

$$x = X_0 e^{i\omega t} \tag{7.73}$$

where X_0 may be a complex number. On substitution of Eq. (7.73) in (7.72) we have

$$\left(-\omega^2 - \frac{i\omega}{\tau} + \omega_0^2\right)X_0 e^{i\omega t} = \alpha_0 e^{i\omega t}$$

whence

$$X_0 = \frac{\alpha_0}{\omega_0^2 - \omega^2 + i(\omega/\tau)} \tag{7.74}$$

It is useful to consider separately the real and imaginary parts of X_0. We have

$$X_0 = \left[\frac{\alpha_0}{\omega_0^2 - \omega^2 + i(\omega/\tau)}\right]\left[\frac{\omega_0^2 - \omega^2 - i(\omega/\tau)}{\omega_0^2 - \omega^2 - i(\omega/\tau)}\right]$$
$$= \alpha_0 \frac{\omega_0^2 - \omega^2 - i(\omega/\tau)}{(\omega_0^2 - \omega^2)^2 + (\omega/\tau)^2}$$

whence

$$Re\,(X_0) = \frac{(\omega_0^2 - \omega^2)\alpha_0}{(\omega_0^2 - \omega^2)^2 + (\omega/\tau)^2}$$

$$\mathrm{Im}\,(X_0) = \frac{-(\omega/\tau)\alpha_0}{(\omega_0^2 - \omega^2)^2 + (\omega/\tau)^2}$$

In the limit $|\omega_0^2 - \omega^2| \gg \omega/\tau$ we have

$$\mathrm{Re}\,(X_0) \approx \frac{\alpha_0}{\omega_0^2 - \omega^2} \qquad \mathrm{Im}\,(X_0) \approx 0$$

This condition is called *off-resonance*, and here the real part of X_0 is much more important than the imaginary part.

In the limit $|\omega_0^2 - \omega^2| \ll \omega/\tau$ we say we are *near resonance*, and for $\omega_0 = \omega$ we are *on the resonance* or on the center of the resonance. For $\omega_0 = \omega$,

$$\mathrm{Re}\,(X_0) = 0$$

$$\mathrm{Im}\,(X_0) = \alpha_0 \frac{\tau}{\omega}$$

The greater τ is, the weaker the damping and the greater the imaginary part of the response at resonance.

When we remember that off resonance the phase angle ϕ is either close to 0 or to $-\pi$, we can see why the amplitude has a very small imaginary part; whereas when $\omega = \omega_0$, $\phi = -\pi/2$, and the displacement is out of phase with the force, the imaginary part of the amplitude which is correlated with the velocity amplitude will be large and the real part of the amplitude zero.

Let us write X_0 in the form $\rho e^{i\phi}$, as in Eq. (7.71). Then from Eqs. (7.71) and (7.74) we have for the amplitude of the response

$$\rho = (X_0 X_0^*)^{\frac{1}{2}} = \frac{\alpha_0}{[(\omega_0^2 - \omega^2)^2 + (\omega/\tau)^2]^{\frac{1}{2}}}$$

Here X_0^* is the complex conjugate to X_0, so that $X_0 X_0^*$ is real. Also we have for the phase angle of x relative to F,

$$\tan\phi = -\frac{\omega/\tau}{\omega_0^2 - \omega^2}$$

The average power absorption is given by

$$\langle P\rangle = \langle F\dot{x}\rangle = \langle \mathrm{Re}\,(F)\mathrm{Re}\,(\dot{x})\rangle$$
$$= \langle [M\alpha_0 \cos\omega t][-\rho\omega \sin(\omega t + \phi)]\rangle \tag{7.75}$$

We have taken the real part of x to correspond to physical reality if the real part of F is the physically real force. There are other valid forms for the time average—what turns out to be important is to take that part of x which is in phase with F. Using the equivalent of Eq. (7.61) and the relation $\rho \sin \phi = \text{Im}(X_0)$, we have from Eq. (7.75)

$$\langle P \rangle = -M\alpha_0 \rho \omega \langle \cos^2 \omega t \rangle \sin \phi = -\tfrac{1}{2} M\alpha_0 \omega \text{Im}(X_0)$$

$$= \tfrac{1}{2} M\alpha_0{}^2 \frac{\omega^2/\tau}{(\omega_0{}^2 - \omega^2)^2 + (\omega/\tau)^2}$$

This result is identical with the earlier result of Eq. (7.62).

FURTHER READING

PSSC, "Physics," chaps. 20 (sec. 8) and 24 (sec. 12), D. C. Heath and Company, Boston, 1960.

Y. Rocard, "General Dynamics of Vibrations," Frederick Ungar Publishing Co., New York, 1960. A simple and lucid book written with a broad range of applications in mind.

B. L. Walsh, Parametric Amplification, *International Science and Technology*, no. 17, p. 75, May 1963. An elementary discussion of parametric amplifiers and their property of low noise.

For an example of the treatment of harmonic oscillations, free, damped, and driven, in a familiar textbook at intermediate level, see John L. Synge and Byron A. Griffith, "Principles of Mechanics," sec. 6.3, McGraw-Hill Book Company, New York, 1959.

CONTENTS

Elementary Dynamics of Rigid Bodies

The dynamics of rigid bodies is a fascinating and complicated subject, perhaps the highest point of classical mechanics and the most difficult. The prototype of problems in this subject is that of the gyroscope or spinning top whose subtle and intriguing behavior has always challenged understanding. The full development of the description of rigid-body motion arrives at some surprising aspects of simplicity and beauty, but these may not be apparent at the introductory level of treatment required here. Much of the theory of the gyroscope has been covered in a four-volume treatise, *Theorie des Kreisels* by F. Klein and A. Sommerfeld.

By the term *rigid body* we mean an assembly of particles with fixed interparticle distances; we shall exclude consideration of vibrations or deformations attending the motion. The motion of major concern to us will consist of rotation about an axis that may be fixed or changing with time. The understanding we shall develop has applications ranging from the spinning electron and rotating atoms and molecules to rotating machinery, gyroscopes, planets, and inertial guidance.

A certain division is inherent in the subject, based upon whether the axis of rotation maintains a fixed direction in inertial space or changes its direction as time progresses. The treatment of the fixed-axis motion is markedly simpler than the more general case, and many important systems are of that type. Consequently a reduced treatment of rigid-body dynamics at this level is sometimes limited to the fixed-direction-axis case. However, we shall begin our study with the general case; various important special cases will follow, with clear indication of the special conditions or properties of the body and its motion for each special case.

THE EQUATION OF MOTION

We have previously deduced (in Chap. 6) from Newton's Second Law the relationship

$$\boxed{\frac{d\mathbf{J}}{dt} = \mathbf{N}} \tag{8.1}$$

for motion of a system of particles in an inertial reference frame, where $\mathbf{J}$ is the angular momentum vector referred to a chosen origin point, and $\mathbf{N}$ is the moment-of-force, or torque, vector, evaluated at the origin point, due to all external forces effective upon the particles of the system. The internal forces of the system do not produce any resultant torque, as we saw in Chap. 6.

Equation (8.1) contains essentially everything we need to know about the motion: It is the *equation of motion*. Our problem is to correctly apply it to the objects and situations of interest. For this purpose we must see how to express $\mathbf{J}$ for a rigid body. We shall also need to know how to express the kinetic energy associated with its motion, and to these matters we now proceed.

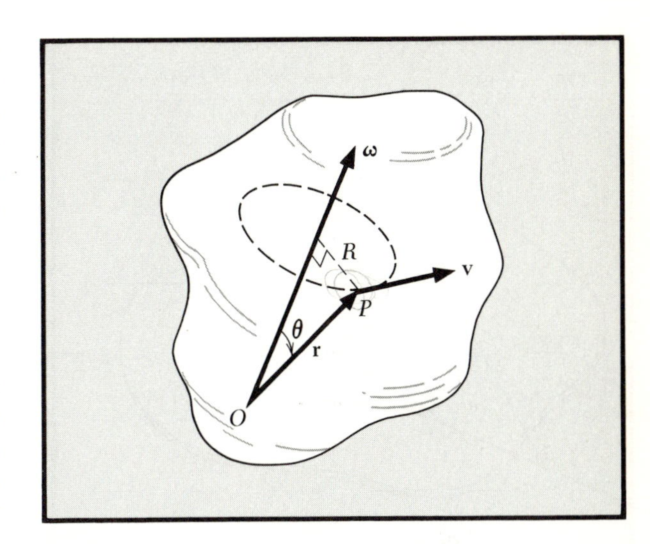

FIG. 8.1 At the moment pictured, rotation of the body causes P to move in a circle of radius $R = r \sin \theta$, in a plane perpendicular to $\boldsymbol{\omega}$. The magnitude of $\mathbf{v}$ is $v = \omega R = \omega r \sin \theta$, and its direction is normal to the plane defined by $\boldsymbol{\omega}$ and $\mathbf{r}$. Thus $\mathbf{v} = \boldsymbol{\omega} \times \mathbf{r}$.

ANGULAR MOMENTUM AND KINETIC ENERGY

Consider a rigid body moving in such a way that some point of it remains constantly at a fixed point in space. The motion at any instant must then be a rotation about some axis through this point. We choose the fixed point, O in Fig. 8.1, as origin for our reference frame, and describe the motion by an angular velocity vector $\boldsymbol{\omega}$ along the instantaneous axis of rotation. Consistent with general practice, the vector $\boldsymbol{\omega}$ will point in the directional sense in which a right-hand-threaded screw would advance along the axis if it were rotated with the body, and the length ω will be numerically equal to its magnitude in rad/s laid out in some chosen length units.

The instantaneous velocity vector $\mathbf{v}$ for a point P at position $\mathbf{r}$ in the body will now be

$$\mathbf{v} = \boldsymbol{\omega} \times \mathbf{r} \tag{8.2}$$

as we may readily recognize from Fig. 8.1 and its caption. If P is the location of a particle of mass m, one of the constituent particles of the body, it will contribute angular momentum $\mathbf{r} \times m\mathbf{v} = \mathbf{r} \times m(\boldsymbol{\omega} \times \mathbf{r})$ to the total angular momentum of the body.

We can now express the total angular momentum as the vector sum of such quantities contributed by all the particles or elements of mass of which the body is composed:

$$\mathbf{J} = \sum \mathbf{r}_i \times m_i(\boldsymbol{\omega} \times \mathbf{r}_i) \tag{8.3}$$

where $\mathbf{r}_i$ is the position vector for mass element m_i, and we sum over all such elements.

The total kinetic energy of the rotating body at the instant pictured is obtained by summing the contributions $\frac{1}{2}mv^2$ for all mass elements. We recall that $v^2 = \mathbf{v} \cdot \mathbf{v}$, so

$$K = \sum \tfrac{1}{2} m_i(\boldsymbol{\omega} \times \mathbf{r}_i) \cdot (\boldsymbol{\omega} \times \mathbf{r}_i) = \tfrac{1}{2} \sum m_i |\boldsymbol{\omega} \times \mathbf{r}_i|^2 \tag{8.4}$$

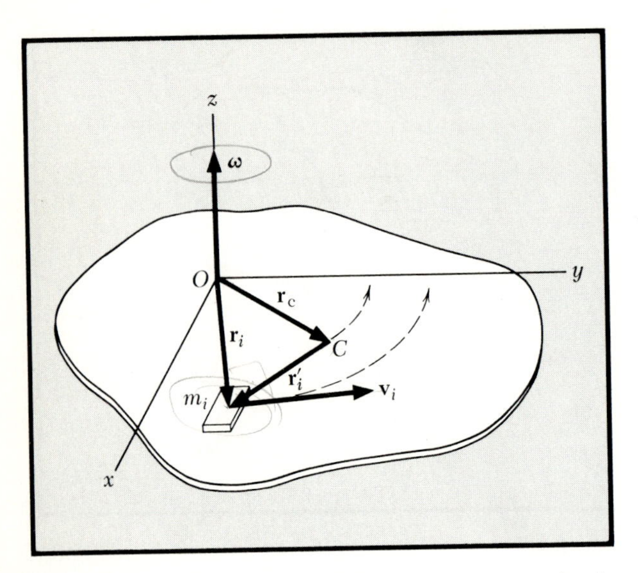

FIG. 8.2 The slab is rotating in its own plane (xy plane) about O. Each point moves in its own circle about O with speed $v_i = \omega r_i$. C is the center of mass.

The foregoing general formulations of the equation of motion, of angular momentum, and of kinetic energy will now be applied to certain important special cases of rigid-body rotation.

MOMENTS OF INERTIA

Consider the thin slab, illustrated in Fig. 8.2, or a plane distribution of matter, to lie in the xy plane and to rotate about the z axis with angular velocity ω. The vector $\boldsymbol{\omega}$ is taken to be constant in its direction. The mass element m_i moves with velocity $\mathbf{v}_i = \boldsymbol{\omega} \times \mathbf{r}_i$, and in this case its speed is simply $v_i = \omega r_i$ since $\boldsymbol{\omega}$ and $\mathbf{r}_i$ are perpendicular. The kinetic energy for the slab is then

$$K = \tfrac{1}{2} \sum m_i v_i^2 = \tfrac{1}{2}\left(\sum m_i r_i^2\right)\omega^2 = \tfrac{1}{2} I_z \omega^2 \qquad (8.5)$$

where we have defined the quantity I_z, called the *moment of inertia*, for the slab with respect to the z axis

$$\boxed{I_z = \sum m_i r_i^2} \qquad (8.6)$$

Parallel Axis Theorem

An important insight is obtained by introducing the center of mass into the picture, as in Fig. 8.2. We substitute

$$\mathbf{r}_i = \mathbf{r}_c + \mathbf{r}'_i$$

and obtain for I_z

$$\begin{aligned} I_z &= \sum m_i r_i^2 \\ &= \sum m_i \mathbf{r}_i \cdot \mathbf{r}_i \\ &= \sum m_i (\mathbf{r}_c + \mathbf{r}'_i) \cdot (\mathbf{r}_c + \mathbf{r}'_i) \\ &= \sum m_i (r_c^2 + 2\mathbf{r}_c \cdot \mathbf{r}'_i + r_i'^2) \end{aligned}$$

But since $\Sigma m_i = M$, that is, the total mass of the slab, we have

$$I_z = M r_c^2 + 2\mathbf{r}_c \cdot \sum m_i \mathbf{r}'_i + \sum m_i r_i'^2$$

We now observe that the middle term will equal zero because $\mathbf{r}'_i$ is the position vector of m_i with respect to the center of mass, and the sum $\Sigma m_i \mathbf{r}'_i = 0$. Furthermore, the last term is simply the moment of inertia with respect to a normal axis through the center of mass at C. Therefore we obtain

$$I_z = I_{cz} + M r_c^2 \qquad (8.7)$$

This result, Eq. (8.7), is the *parallel axis theorem* derived here for the particular system of a plane distribution of matter with normal axis. It can be easily proved as a general theorem for any distribution. Stated in words, it says

> The moment of inertia about any axis is equal to the moment of inertia about a parallel axis through the center of mass, plus the mass of the body times the square of the distance between the two axes. Thus
>
> $$\boxed{I = I_c + Ml^2} \tag{8.8}$$
>
> where l is the separation of the axes.

This is an exceedingly useful result, as we shall see.

In view of Eq. (8.7), the expression (8.5) becomes

$$\boxed{K = \tfrac{1}{2}I_{cz}\omega^2 + \tfrac{1}{2}Mr_c^{\,2}\omega^2} \tag{8.9}$$

which can immediately be interpreted as stating that the total kinetic energy of the rotating slab is composed of the energy due to rotation about its center of mass (first term) plus the energy due to the translational motion of the center of mass about the axis of rotation (second term). This also is a completely general result, although we have proved it for only the case of a slab rotating in its own plane.

The angular momentum for our rotating slab is obtained from Eq. (8.3) applied to this case where $\boldsymbol{\omega}$ and $\mathbf{r}_i$ are perpendicular. The result is easily found to be

$$\mathbf{J} = \sum m_i r_i^{\,2}\omega\hat{\mathbf{z}} = I_z\omega\hat{\mathbf{z}} \tag{8.10}$$

where the moment of inertia again enters. If as before we introduce the center of mass by writing $\mathbf{r}_i = \mathbf{r}_c + \mathbf{r}_i'$, Eq. (8.10) becomes

$$\boxed{\mathbf{J} = \sum m_i r_i'^2\omega\hat{\mathbf{z}} + Mr_c^{\,2}\omega\hat{\mathbf{z}} = I_{cz}\omega\hat{\mathbf{z}} + Mr_c^{\,2}\omega\hat{\mathbf{z}}} \tag{8.11}$$

Again we have a theorem proved in this special case that can be proved in general:

> The angular momentum about any point is equal to the angular momentum about the center of mass plus the angular momentum due to translation of the center of mass with respect to the point.

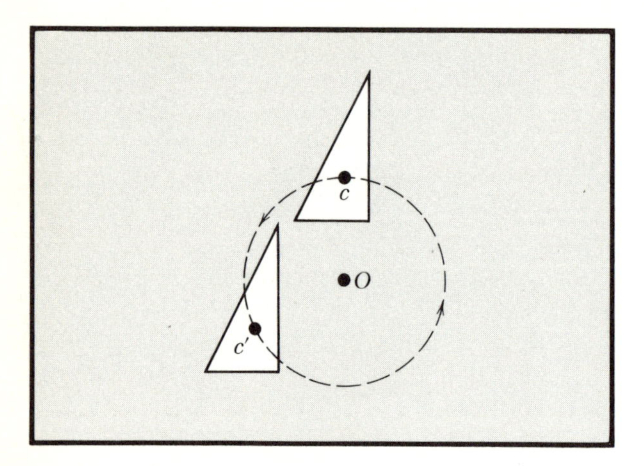

FIG. 8.3 (*a*) The center of mass of a triangular slab is translated in a circle about O, but there is no rotation. Only the second term of Eqs. (8.9) and (8.11) is present.

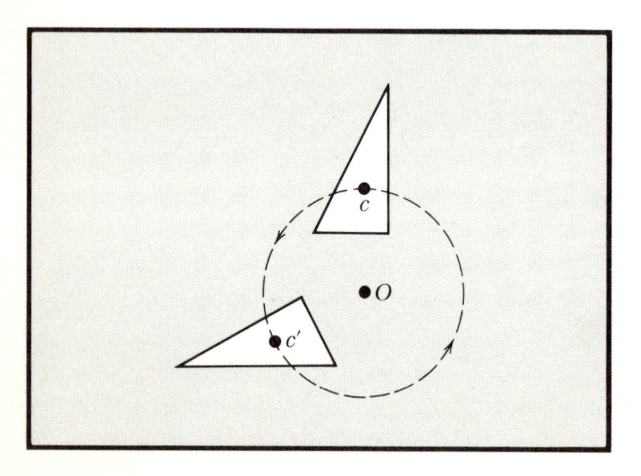

(*b*) Both terms in Eqs. (8.9) and (8.11) contribute.

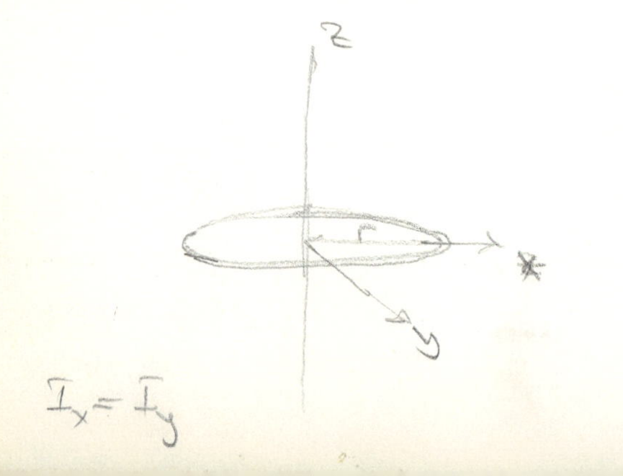

Figure 8.3 illustrates the difference in the two terms present in the theorems.

These theorems that we have established for a slab or plane distribution of matter will also apply to a system formed by stacking many identical thin-slab elements to form a cylindrical or prismatic rigid body constrained to rotate about a fixed axis perpendicular to the slab elements, providing that we interpret r_i to mean distance from the axis of rotation. A number of examples follow and illustrate these matters.

Perpendicular Axis Theorem Before proceeding to treat examples we demonstrate a further useful theorem about moments of inertia for thin, slablike objects. We may refer to Fig. 8.2 illustrating the rotating slab and its representative mass element m_i. The contribution of m_i to the moment of inertia I_z with respect to the z axis is $m_i r_i^2$, that is the mass of the particle times the square of its distance from the axis of rotation. If we should consider rotation about the x axis, the mass element m_i would contribute $m_i y_i^2$ to the moment of inertia with respect to the x axis; similarly it would give $m_i x_i^2$ with respect to the y axis. Thus

$$I_x = \sum m_i y_i^2$$
$$I_y = \sum m_i x_i^2$$

By addition we obtain

$$I_x + I_y = \sum m_i(x_i^2 + y_i^2) = \sum m_i r_i^2 = I_z$$

This proves the *perpendicular axis theorem*, true for plane, thin rigid bodies, which states

> The moment of inertia for a plane rigid sheet with respect to an axis normal to its plane is equal to the sum of the moments of inertia about any two perpendicular axes lying in the plane and intersecting the normal axis.

Some Special Cases

Thin Ring or Rim A ring of mass M, radius R, and of negligible radial breadth will obviously have all its mass elements at the same distance R from a normal axis through its center. The moment of inertia about this axis is thus $I = MR^2$. Clearly the same result applies to a thin-walled circular cylinder to be rotated about its geometric axis. For a parallel axis located at the wall of the cylinder, $I = 2MR^2$, as may be immediately deduced from the parallel axis theorem.

The perpendicular axis theorem applied to a narrow, plane ring tells us that the moment of inertia about an axis lying in the plane of the ring and passing through the center will be $\frac{1}{2}MR^2$.

Uniform Thin Rod We picture in Fig. 8.4 a rod of length L and mass M, whose breadth and thickness are very small compared to L so that the rod may be treated as a weighted line of uniform density. Consider an axis of rotation perpendicular to the rod at one end. An element of length Δx will possess mass $\Delta M = (\Delta x/L)M$, and if its location is at distance x from the axis it will contribute $\Delta I = (\Delta x/L)Mx^2$ to the moment of inertia. Now by considering Δx to be an infinitesimal differential element dx, we may sum the contributions from all such elements by integration. Thus we obtain

$$I = \frac{M}{L}\int_0^L x^2\,dx = \tfrac{1}{3}ML^2$$

for the moment of inertia of a uniform rod about a perpendicular axis of rotation at its end.

By application of the parallel axis theorem we obtain the moment of inertia for an axis at the center, perpendicular to the rod. Thus

$$I_c = I - M\left(\frac{L}{2}\right)^2 = \tfrac{1}{12}ML^2 \tag{8.12}$$

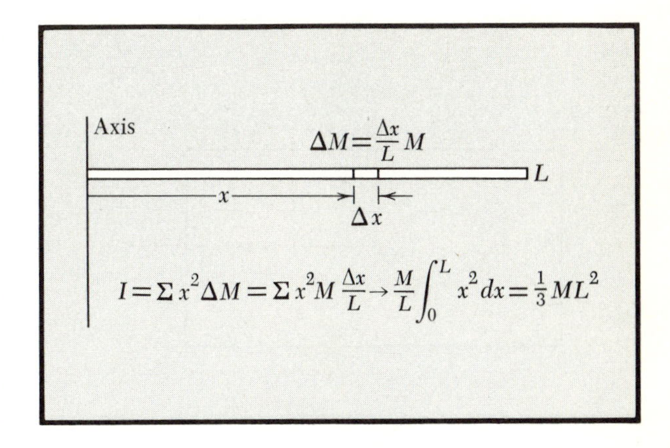

FIG. 8.4 Thin rod, axis at end.

Circular Disk As illustrated in Fig. 8.5, a ring element of breadth Δr at radius r will possess a mass

$$\Delta M = \frac{\pi(r+\Delta r)^2 - \pi r^2}{\pi R^2}M \approx \frac{2r\,\Delta r}{R^2}M$$

where in the final approximation we have neglected the small term involving $(\Delta r)^2$. The contribution made by this ring element to the moment of inertia about a normal axis through the center of the disk will be

$$\Delta I = r^2\,\Delta M \approx \frac{2M}{R^2}r^3\,\Delta r$$

Performing the integration that sums all the ring-element contributions, we obtain

$$I = \frac{2M}{R^2}\int_0^R r^3\,dr = \tfrac{1}{2}MR^2 \tag{8.13}$$

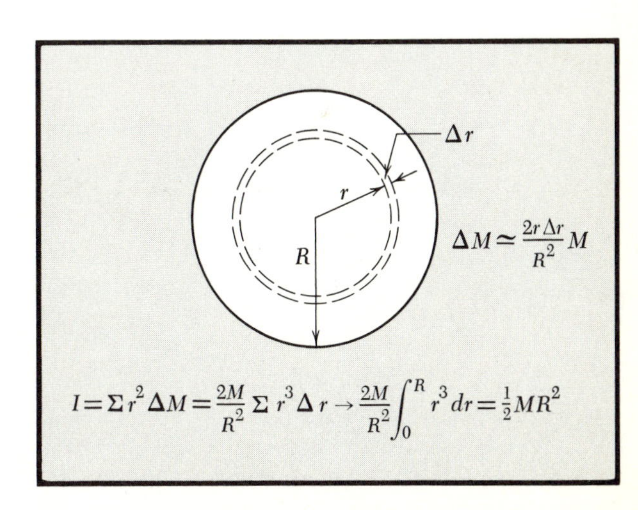

FIG. 8.5 Disk, axis through center normal to plane of disk.

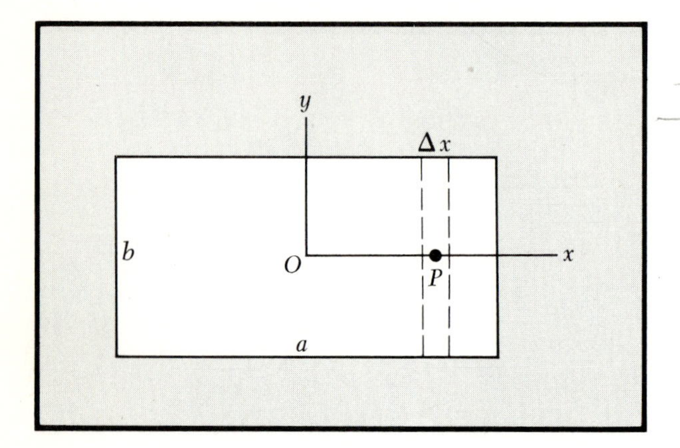

FIG. 8.6 Rectangular plate, normal axis at center.

Since a solid cylinder of uniform density can be regarded as a stack of disks, it is obvious that the moment of inertia of a solid circular cylinder about its axis is $I = \frac{1}{2}MR^2$.

Attending again to the thin disk, we recognize by the perpendicular axis theorem that the moment of inertia about a diametral axis in its plane is

$$I = \tfrac{1}{4}MR^2 \tag{8.14}$$

Rectangular Plate We picture in Fig. 8.6 a plate of length a and breadth b and we desire to calculate the moment of inertia about the z axis normal to the plate at its center of mass. We can regard the plate as composed of narrow strips, one of which is illustrated with breadth Δx located at distance x from the center; and we sum the contributions of the strips to obtain the total moment of inertia about O, treating each strip as a thin rod.

The strip illustrated will possess a mass of $(\Delta x/a)M$ and a moment of inertia about a normal axis at its own center P of $\frac{1}{12}(\Delta x/a)Mb^2$ as shown above in case 2, Eq. (8.12). The contribution by this strip to the moment of inertia about the normal axis at O is then obtained from the parallel axis theorem, Eq. (8.8),

$$\Delta I = \frac{\Delta x}{a}M\left(\frac{b^2}{12} + x^2\right)$$

We now sum all such contributions by integration to obtain

$$I = \frac{M}{a}\int_{-a/2}^{+a/2}\left(\frac{b^2}{12} + x^2\right)dx = \frac{M}{12}(a^2 + b^2) \tag{8.15}$$

Since this calculation is related to rotation about the z axis, we might call its result I_z. The values of I_x and I_y are readily seen to be [again using Eq. (8.12)]

$$I_x = \frac{M}{12}b^2 \qquad \text{and} \qquad I_y = \frac{M}{12}a^2$$

The perpendicular axis theorem $I_z = I_x + I_y$ is obviously fulfilled.

These cases illustrate the manner of calculating moments of inertia. Other cases appear in the problems; among them is the important case of a solid, uniform sphere with the axis through its center. The result is

$$I = \tfrac{2}{5}MR^2 \tag{8.16}$$

Many objects may be treated by superposition, i.e., addition or subtraction, of the moments of inertia of simple forms. For example, the value for a thick-walled hollow cylinder is obtained by taking the difference between the moments of inertia of two solid cylinders of appropriate radii.

ROTATIONS ABOUT FIXED AXES: TIME DEPENDENCE OF THE MOTION

We are now ready to apply Eq. (8.1), the equation of motion $d\mathbf{J}/dt = \mathbf{N}$, to problems dealing with rigid bodies rotating about fixed axes, with the aim of learning the time dependence of the rotation in response to given torques or moments of force.

Since the direction of the rotation axis is constrained to be fixed in space, and also fixed in relation to the body, the inertial properties of the body will be constant in relation to the axis. In deducing the motion we will need to concern ourselves only with the component of angular momentum related to this axis. Likewise, only that component of the applied moment $\mathbf{N}$ need be considered. Consequently the equation of motion can be treated simply as a scalar equation $dJ_a/dt = N_a$, where J_a and N_a refer to the components of $\mathbf{J}$ and $\mathbf{N}$ parallel to the axis. In many important problems only this component exists anyway, the vectors themselves being parallel to the axis, but this is not always the case as we shall see later.

The general formulation of the scalar equation of motion in the axial components is easily obtained by simply projecting the general vector equation upon the axial direction. The axial direction is denoted by the unit vector $\boldsymbol{\omega}/\omega$. The axial component of angular momentum is thus given by [referring to Eq. (8.3)]

$$J_a = \mathbf{J} \cdot \frac{\boldsymbol{\omega}}{\omega} = \frac{1}{\omega}\left[\sum_i \mathbf{r}_i \times m_i(\boldsymbol{\omega} \times \mathbf{r}_i)\right] \cdot \boldsymbol{\omega}$$

The last expression can be simplified as follows:

$$\frac{1}{\omega}\sum_i \left[m_i\mathbf{r}_i \times (\boldsymbol{\omega} \times \mathbf{r}_i)\right] \cdot \boldsymbol{\omega} = \frac{1}{\omega}\sum_i m_i|\boldsymbol{\omega} \times \mathbf{r}_i|^2$$

$$= \frac{1}{\omega}\sum_i m_i(\omega_i r_i \sin\theta_i)^2$$

where in the first step we have used Eq. (2.56) and the fact that

$$r_i^2\omega^2 - (\mathbf{r}_i \cdot \boldsymbol{\omega})^2 = r_i^2\omega^2 - r_i^2\omega^2 \cos^2 \theta_i$$

Thus

$$J_a = \sum_i m_i R_i^2 \omega$$

where, as in Fig. 8.1, $R_i = r_i \sin \theta_i$ is the distance of particle i from the axis of rotation.

But, as in the case of Eq. (8.5), we introduce the moment of inertia of the body with respect to the rotation axis,

$$I_a = \sum_i m_i R_i^2$$

Thus the expression for J_a has been reduced to simply

$$J_a = I_a \omega \tag{8.17}$$

Similarly the axial component of the torque is obtained from

$$N_a = \mathbf{N} \cdot \frac{\boldsymbol{\omega}}{\omega}$$

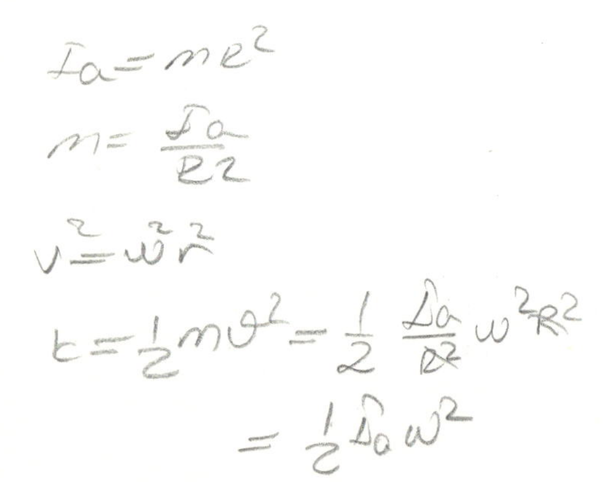

So the equation of motion in axial components becomes

$$\boxed{I_a \frac{d\omega}{dt} = N_a} \tag{8.18}$$

In circumstances where there is no ambiguity the subscripts are omitted.

By such considerations it is apparent that the kinetic energy of rotation about a fixed axis is

$$\boxed{K_a = \tfrac{1}{2} I_a \omega^2} \tag{8.19}$$

Equations (8.17) and (8.19) correspond, respectively, to Eqs. (8.10) and (8.5) for the slab rotating about the fixed z axis, but they apply to a body of any shape rotating about any fixed axis.

EXAMPLE

Angular Acceleration of a Solid Cylinder Subject to Torque A typical simple example is afforded by a solid cylinder, free to rotate about a fixed axis coinciding with its geometric axis, subject to an applied torque. We picture such a case in Fig. 8.7, where the rotation axis is horizontal and the torque is provided by a mass hanging from a string wrapped around the cylinder.

The moment of inertia for the cylinder is, by Eq. (8.13), $I = \frac{1}{2}MR^2$. Its angular momentum, at an instant when its angular velocity is ω, will then be $J = I\omega = \frac{1}{2}MR^2\omega$. The torque is due to the tension T in the string and is thus $N = TR = (mg - ma)R$, where a is the acceleration in the descending motion of mass m. The equation of motion is then, from Eq. (8.18),

$$\frac{MR^2}{2}\frac{d\omega}{dt} = m(g - a)R \tag{8.20}$$

The geometry of the arrangement requires

$$R\omega = v \qquad R\frac{d\omega}{dt} = a$$

and this, together with Eq. (8.20), yields for the angular acceleration of the cylinder,

$$\alpha = \frac{d\omega}{dt} = \frac{m}{M/2 + m}\frac{g}{R}$$

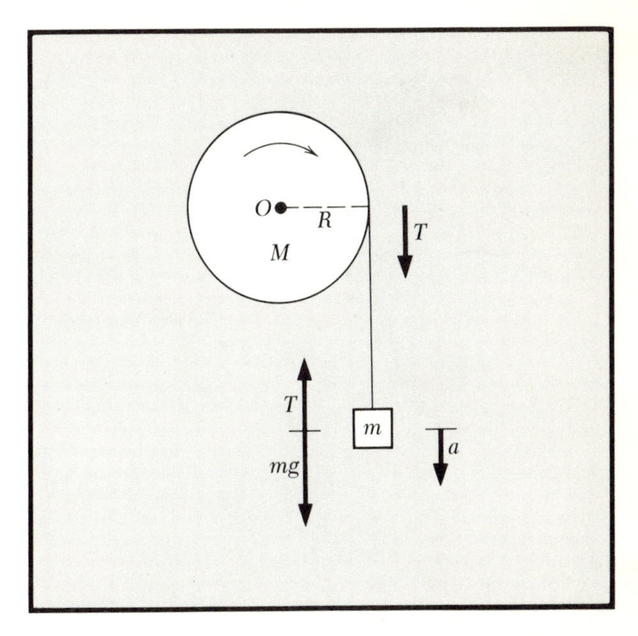

FIG. 8.7 The cylinder rotates freely about the horizontal axis at O. It is subject to the torque TR provided by tension in the string supporting mass m.

Rolling without Slipping We now consider, as in Fig. 8.8, the rolling down an incline of an object with a circular periphery and a mass distribution symmetric about its center. (It might be a solid cylinder, a hollow cylinder, a sphere, etc.) We shall find the translational acceleration of its motion down the plane in three different ways, thus illustrating the consistency of these different points of view in treating the problem.

Rotation about the Instantaneous Point of Contact At any instant the motion consists of rotation about P, the point of contact with the inclined surface. The *direction* of the axis of rotation is constant, although its position advances down the plane. The acceleration in the motion of the rolling object is calculated by recognizing that *instantaneously* the motion is simply a rotation about a point on the periphery of the object. Thus we shall require the moment of force about P to equal the rate of change of angular momentum about P (see Fig. 8.8).

If I represents the moment of inertia for the object about an axis at its center parallel to the rotation axis at P, then we can evaluate the required moment of inertia about P by the parallel axis theorem [Eq. (8.7)]

$$I_p = I + MR^2$$

The angular velocity of instantaneous rotation about P is $\omega = v/r$, where v is the momentary translational speed of the

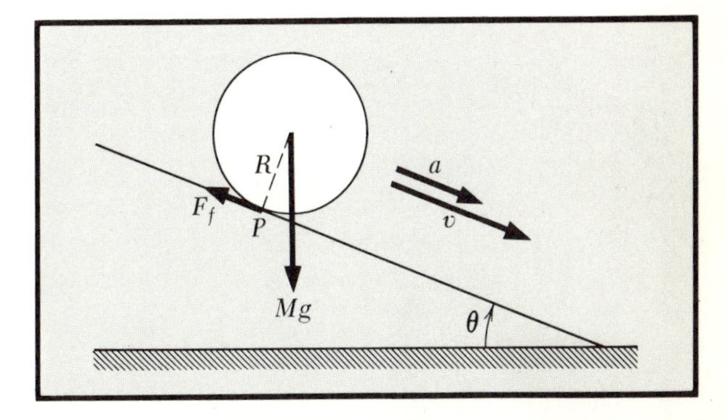

FIG. 8.8 The character of the motion of a rolling body at any instant is rotation about the *instantaneous* point of contact P.

center. Thus the angular momentum about P at any instant is

$$J_P = (I + MR^2)\frac{v}{R} \tag{8.21}$$

The moment of force at P is provided by the gravitational force effective at the center of mass. Thus

$$N_P = MgR \sin\theta \tag{8.22}$$

In view of Eqs. (8.21) and (8.22), the equation of motion $N = dJ/dt$ is

$$MgR \sin\theta = (I + MR^2)\frac{a}{R}$$

where we have written a for dv/dt.

The translational acceleration in the rolling down the plane is then

$$a = \frac{1}{1 + I/MR^2} g \sin\theta \tag{8.23}$$

For a solid cylinder $I = \frac{1}{2}MR^2$; so $a = \frac{2}{3}g \sin\theta$. For a solid sphere, $I = \frac{2}{5}MR^2$; so $a = \frac{5}{7}g \sin\theta$, and so on, for other symmetric objects. Other cases are given as Prob. 6.

Energy Consideration As a second method of determining the translational acceleration we use the conservation of energy. The total energy at any instant will consist of three contributions:

1. KE of translation of center of mass $= \frac{1}{2}Mv^2$
2. KE of rotation about center of mass $= \frac{1}{2}I\omega^2 = \frac{1}{2}Iv^2/R^2$
3. PE due to elevation of center of mass $= Mgh$

where h is the height above a chosen level where PE is assigned the value zero.

To a very good approximation the total energy is conserved. Friction at the contact point produces rolling instead of slipping, but this friction force does no work and does not forbid conservation of the total energy:

$$E = \frac{1}{2}\left(M + \frac{I}{R^2}\right)v^2 + Mgh$$

Since E is constant we may set its time derivative equal to zero:

$$\frac{dE}{dt} = \left(M + \frac{I}{R^2}\right)v\frac{dv}{dt} + Mg\frac{dh}{dt} = 0$$

But $dh/dt = -v \sin\theta$, and $dv/dt = a$; so this becomes

$$\left(M + \frac{I}{R^2}\right)va - Mgv \sin\theta = 0$$

We divide out the common factor v and solve for a to obtain

$$a = \frac{1}{1 + I/MR^2} g \sin\theta$$

in agreement with Eq. (8.23).

Acceleration of the Center of Mass and Angular Acceleration about the Center of Mass As a third method, we consider the acceleration of the center of mass and the angular acceleration about the center of mass.

$$Ma_c = \sum F$$
$$I\frac{d\omega}{dt} = \sum N_c$$

where a_c is the acceleration of the center of mass and ΣF includes all external forces. The second equation assumes that the direction of the axis of rotation is fixed so that $d\mathbf{J}/dt = I\,d\omega/dt$, where ω is the angular velocity about the center of mass.

Referring to Fig. 8.8, we see that there are two forces parallel to the plane: $Mg \sin\theta$ down the plane and F_f, the frictional force up the plane. So

$$Ma_c = M\frac{dv}{dt} = Mg \sin\theta - F_f$$

Now taking moments about the center of mass, we see that only F_f contributes a torque so that

$$N_c = F_f R = I\frac{d\omega}{dt}$$

Because we are considering rolling without slipping,

$$\omega R = v \qquad \text{or} \qquad \frac{d\omega}{dt} = \frac{dv}{dt}\frac{1}{R}$$

and so by combining the three preceding equations it follows that

$$M\frac{dv}{dt} = Mg \sin\theta - \frac{I}{R^2}\frac{dv}{dt}$$

and

$$\frac{dv}{dt} = a_c = \frac{Mg \sin \theta}{M + I/R^2} = \frac{1}{1 + I/MR^2} g \sin \theta$$

which again is the same as Eq. (8.23) since a_c equals the a used there.

This analysis shows clearly what force it is that "slows up" the acceleration. It can also be used with the definition of the coefficient of friction to determine what angle is required for a given coefficient and a given I to cause the body to slip and roll rather than rolling without slipping. (See Prob. 19 at the end of this chapter.)

Torques about a Center of Mass We have not discussed in our previous general discussions what point we should use for calculating angular momentum and taking torques about. In the preceding example of a rolling object we did indeed treat the problem by alternative points of view using two different centers of rotation. But some caution is needed in choosing the center of rotation to be employed in evaluating the torques and calculating the motion. Certainly we can use a point fixed in inertial space. Also in Chap. 6 we derived

$$\sum \mathbf{F}_{\text{ext}} = M\mathbf{a}_{\text{c.m.}}$$

$$\sum \mathbf{r}_i \times \mathbf{F}_{\text{ext}} = \frac{d\mathbf{J}_{\text{c.m.}}}{dt} = \sum \mathbf{N}$$

where the torques are taken about the center of mass. These two points, the fixed point and the center of mass, can *always* be used. Other points, particularly accelerated points, can be used only with great care, and sometimes a "fictitious force" must be introduced. The following example illustrates this, as does Prob. 18.

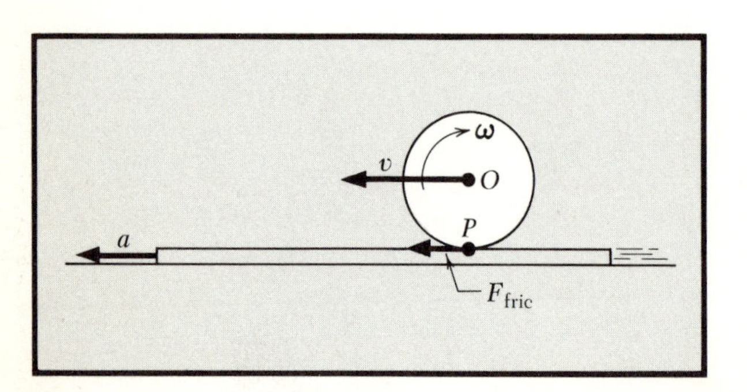

FIG. 8.9 Cylinder being accelerated when the rough surface it rests on is accelerated.

EXAMPLE

Cylinder on an Accelerated Rough Plane Figure 8.9 shows a cylinder resting on a rough horizontal rug that is pulled out from under it with acceleration a perpendicular to the axis of the cylinder. What is the motion of the cylinder, assuming it does not slip?

The only horizontal force on the cylinder is that of friction at P. Therefore let us take moments about P. The forces of gravity and reaction at the surface pass through the point P, as does also the friction force, so the net torque about $P = 0$. Therefore we say

$$\frac{dJ_p}{dt} = 0 \qquad I_p\omega = \text{const} = (MR^2 + I_c)\omega$$

$F_{fric} = mg \rightarrow N = 0$

We recognize this as obviously wrong; the motion certainly is not with constant ω. So let us take the center of mass, point O, for evaluating the moments and angular momentum:

$$M\frac{dv_c}{dt} = F_{\text{fric}}$$

$$F_{\text{fric}}\,R = I_c\frac{d\omega}{dt} = \tfrac{1}{2}MR^2\,\frac{d\omega}{dt}$$

Since we have rolling without slipping, the acceleration of the contact point is

$$\frac{dv_c}{dt} + R\frac{d\omega}{dt} = a$$

so that using

$$M\frac{dv_c}{dt} = \frac{1}{2}MR\frac{d\omega}{dt}$$

we obtain

$$\frac{3}{2}R\frac{d\omega}{dt} = a$$

giving

$$\frac{dv_c}{dt} = \frac{a}{3} \qquad \text{and} \qquad F_{\text{fric}} = M\frac{a}{3}$$

The Compound Pendulum The simple pendulum, which we treated in Chap. 7, is a point mass suspended by a massless thread, swinging in a plane. The compound pendulum is a rigid body, possessing a distribution of mass, free to rotate and oscillate about a fixed horizontal axis rigidly positioned with respect to the body and not passing through its center of mass. Such an object is shown in Fig. 8.10 at an instant in its motion when a reference plane, defined by the axis through P and the center of mass C, is at an angle θ from vertical and is swinging with positive $d\theta/dt$.

Since its motion is constrained to be oscillatory rotation about the fixed axis, we can study the time dependence of θ by considering the component of angular momentum parallel to the axis and the corresponding components of any torques applied to the body.

By the parallel axis theorem [Eq. (8.8)] it is evident that the moment of inertia related to the axis of rotation is

$$I = I_c + Ml^2$$

where l is the distance PC. The angular momentum related to the axis at the moment pictured is then

$$J = I\omega = (I_c + Ml^2)\frac{d\theta}{dt} \tag{8.24}$$

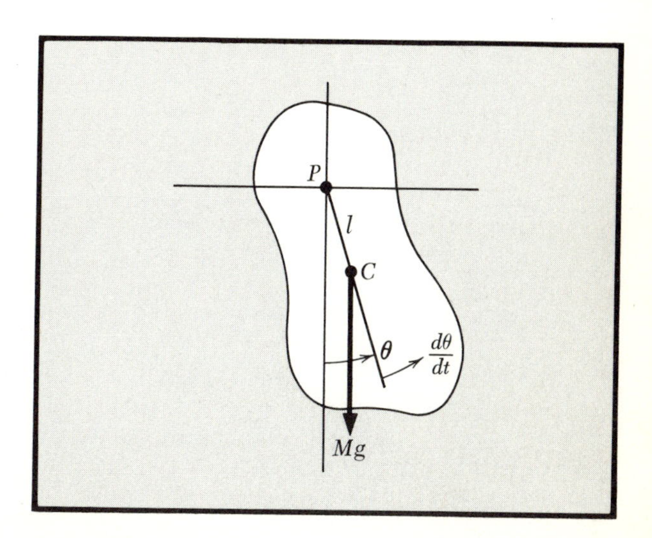

FIG. 8.10 Compound pendulum; C is the center of mass; the axis of rotation is horizontal and passes through P.

The torque about P is provided by the force of gravity Mg, applied at the center of mass C (as shown in Chap. 6, page 188). This torque with respect to the axis is

$$N = -Mgl \sin\theta \tag{8.25}$$

where the negative sign applies because its effect is in the negative sense of the angle θ. The equation of motion $dJ/dt = N$, in view of Eqs. (8.24) and (8.25), is then

$$(I_c + Ml^2)\ddot{\theta} + Mgl \sin\theta = 0$$

We now restrict our consideration to small oscillations and take the familiar approximation for small angles, $\sin\theta \approx \theta$; also we rearrange terms and factors to obtain

$$\ddot{\theta} + \frac{g}{l}\left(\frac{1}{1 + I_c/ML^2}\right)\theta = 0$$

This is of course the differential equation for simple harmonic motion; and if the value of the parenthetic factor were unity it would be the equation for the simple pendulum of length l. This factor carries the effect of the mass distribution of the rigid body, giving a frequency

$$\omega = \sqrt{\frac{g}{l}\left(\frac{1}{1 + I_c/Ml^2}\right)} \tag{8.26}$$

Many interesting and practical applications of these results could be discussed; a few are presented in problems at the end of the chapter. We shall illustrate only one example here, a very simple case. Consider a thin circular rim or hoop of mass M and radius r, suspended from a fixed point like a small nail in the wall. What will be its frequency of small oscillations, and what is the length of a simple pendulum possessing the same frequency?

The parameter l is here equal to r, and the value of I_c is Mr^2. Consequently Eq. (8.26) gives

$$\omega = \sqrt{\frac{g}{2r}}$$

A simple pendulum of length $2r$ has this frequency. So the hoop and a simple pendulum whose length is equal to the hoop diameter will move in unison if small oscillations are initiated in phase.

ROTATION ABOUT FIXED AXES: BEHAVIOR OF THE ANGULAR MOMENTUM VECTOR

In the preceding section our attention was focused upon the *time dependence of rotation* about an axis whose direction was fixed in space and fixed relative to the rigid body. For that purpose we needed only to deal with the component of **J** along the axis and the corresponding component of the applied moment of force **N**. Our equation of motion was simply the scalar relationship $dJ_a/dt = N_a$ in these components. We now must recognize that the angular momentum vector **J** will not be parallel to the axis of rotation unless the latter is related in a particular way to the symmetry properties of the rigid body. When **J** does not coincide with the axis, its time derivative $d\mathbf{J}/dt$ can involve a changing direction of the vector **J** as well as a varying magnitude. Since rotation about a fixed axis implies a circular motion of all features of the body, we anticipate that the changing direction of **J** will be of the nature of a rotation of the **J** vector about the fixed axis.

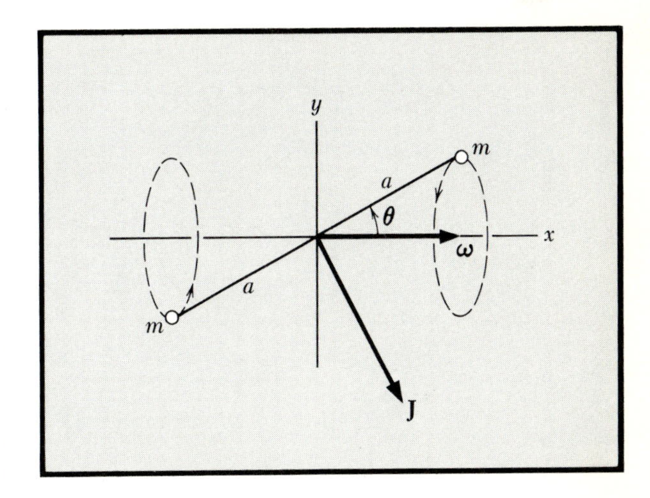

FIG. 8.11 Angular velocity and angular momentum vectors for a light rod with masses at ends.

The simplest illustration of this general problem is provided by a rigid body consisting of two equal mass points joined by a massless rod and made to rotate about a fixed axis through the center of mass and oriented at angle θ from the rod. We picture this system in Fig. 8.11 at an instant in its rotation when the rod coincides with the xy plane. The rod is of length $2a$, and its angular velocity, represented by $\boldsymbol{\omega}$, lies fixed along the x axis.[1] The angular momentum by its general definition [Eq. (8.3)] is

$$\mathbf{J} = \mathbf{r}_1 \times m(\boldsymbol{\omega} \times \mathbf{r}_1) + \mathbf{r}_2 \times m(\boldsymbol{\omega} \times \mathbf{r}_2)$$

We denote the particle in the first quadrant as particle 1, so

$$\begin{aligned} \mathbf{r}_1 &= a\cos\theta\,\hat{\mathbf{x}} + a\sin\theta\,\hat{\mathbf{y}} \\ \mathbf{r}_2 &= -a\cos\theta\,\hat{\mathbf{x}} - a\sin\theta\,\hat{\mathbf{y}} \\ \boldsymbol{\omega} &= \omega\hat{\mathbf{x}} \end{aligned} \tag{8.27}$$

Our expression for **J** then becomes

$$\mathbf{J} = 2m\omega a^2 \sin\theta\,(\hat{\mathbf{x}}\sin\theta - \hat{\mathbf{y}}\cos\theta) \tag{8.28}$$

This vector is perpendicular to the rod, in the orientation shown in the figure. It rotates about the x axis, always keeping the same relationship to the rod and to the axis of rotation.

[1] Note that in this case $\boldsymbol{\omega}$ is along the x axis, while on page 242 it was along the z axis.

Now since **J** thus rotates, its time derivative $d\mathbf{J}/dt$ is not zero. In fact, in this case

$$\boxed{\frac{d\mathbf{J}}{dt} = \boldsymbol{\omega} \times \mathbf{J}} \tag{8.29}$$

for the same reason that we wrote

$$\mathbf{v} = \frac{d\mathbf{r}}{dt} = \boldsymbol{\omega} \times \mathbf{r}$$

in Eq. (8.2) in reference to Fig. 8.1. From Eqs. (8.27) and (8.28), the cross product in Eq. (8.29) yields

$$\frac{d\mathbf{J}}{dt} = -2m\omega^2 a^2 \sin\theta \cos\theta\, \hat{\mathbf{z}}$$

where $\hat{\mathbf{z}}$ points out of the plane of the diagram.

But if $d\mathbf{J}/dt$ is not zero, the general equation of motion, Eq. (8.1), implies that a torque must exist, giving rise to the changing angular momentum vector. In fact

$$\mathbf{N} = \frac{d\mathbf{J}}{dt} = -2m\omega^2 a^2 \sin\theta \cos\theta\, \hat{\mathbf{z}} \tag{8.30}$$

This torque vector (not shown in Fig. 8.11) also rotates with the rod as does **J**. The rotating torque must be supplied by the bearings (not shown) that hold the rod and constrain it to rotate at angle θ about the x axis.

Actually it is easy to see the reason for this torque if we recognize the centripetal forces $m\omega^2 a \sin\theta$ required to hold the two particles in their circular motions about the x axis. These two equal and opposite centripetal forces times the distance $2a \cos\theta$ between them make up the moment of force **N**. They have to be transmitted to the particles via the rigid rod from the bearings.

We note that if the angle θ is 90° then $\boldsymbol{\omega}$ is along the line of symmetry for this simple mass distribution, and **J** coincides with the direction of $\boldsymbol{\omega}$. Then if $\boldsymbol{\omega}$ is constant, so is **J**; and no rotating torque is required to satisfy the motion.

A clear qualitative understanding of these matters is often afforded by contriving an actual model of a system like this one and performing rotations with it.

MOMENTS AND PRODUCTS OF INERTIA: PRINCIPAL AXES AND EULER'S EQUATIONS[1]

Moving now from this simplest illustration of a situation where the direction of **J** is changing, we return to the general definition [Eq. (8.3)] for angular momentum of a rigid body:

$$\mathbf{J} = \sum_i \mathbf{r}_i \times m_i(\boldsymbol{\omega} \times \mathbf{r}_i)$$

and expand the expression for **J** using general expressions for $\mathbf{r}_i$ and $\boldsymbol{\omega}$:

$$\mathbf{r}_i = x_i\hat{\mathbf{x}} + y_i\hat{\mathbf{y}} + z_i\hat{\mathbf{z}}$$
$$\boldsymbol{\omega} = \omega_x\hat{\mathbf{x}} + \omega_y\hat{\mathbf{y}} + \omega_z\hat{\mathbf{z}}$$

and *consider the x, y, and z axes to be fixed in relation to the rigid body*. This operation yields for the x, y, and z components of **J** the results

$$J_x = \sum_i m_i(y_i^2 + z_i^2)\omega_x - \sum_i m_i x_i y_i \omega_y - \sum_i m_i x_i z_i \omega_z$$

$$J_y = -\sum_i m_i y_i x_i \omega_x + \sum_i m_i(z_i^2 + x_i^2)\omega_y - \sum_i m_i y_i z_i \omega_z \tag{8.31}$$

$$J_z = -\sum_i m_i z_i x_i \omega_x - \sum_i m_i z_i y_i \omega_y + \sum_i m_i(x_i^2 + y_i^2)\omega_z$$

For neatness and convenience we rewrite Eq. (8.31) as

$$\begin{aligned} J_x &= I_{xx}\omega_x + I_{xy}\omega_y + I_{xz}\omega_z \\ J_y &= I_{yx}\omega_x + I_{yy}\omega_y + I_{yz}\omega_z \\ J_z &= I_{zx}\omega_x + I_{zy}\omega_y + I_{zz}\omega_z \end{aligned} \tag{8.32}$$

where the I_{xx}, etc., are defined by comparing corresponding terms of Eqs. (8.31) and (8.32).

Examination of the I coefficients reveals that the diagonal members are simply the moments of inertia about the respective axes; thus, for example, I_{zz} is the moment of inertia about the z axis since $x_i^2 + y_i^2$ is simply the square of the distance of particle i from the z axis. The off-diagonal members are called *products of inertia;* they occur in symmetric pairs, for example, $I_{yx} = I_{xy}$.

[1] In an introductory course, the material from here through Eq. (8.42) may well be omitted.

Now it is a remarkable fact, not immediately obvious, that it is always possible to establish a coordinate system fixed in relation to a rigid body in such a way that the products of inertia vanish. For bodies with obvious forms of symmetry (cylinders and rectangular prisms) this is easy to see, but it is true for *any* rigid body and for any point of the body chosen as origin. Sets of coordinate axes for which the products of inertia vanish are termed *principal axes* for the rigid body.

We would like to express the angular momentum components, and the kinetic energy, in reference to principal axes so as to achieve this desirable simplicity; but we must recognize that principal axes are fixed relative to the body, and for this reason they do not in general constitute an inertial reference frame. In fact they rotate with the body, or at least they maintain a relationship to it such that the inertial properties of the body are constant when referred to these axes.

Let us assume that a set of principal axes has been identified in the body, rotating with it, and that we express the angular velocity vector as

$$\boldsymbol{\omega} = \omega_x \hat{\mathbf{x}} + \omega_y \hat{\mathbf{y}} + \omega_z \hat{\mathbf{z}}$$

where the unit coordinate vectors are along the principal axes and the component values of $\boldsymbol{\omega}$ relate to these axes. The angular momentum vector in terms of components pertaining to these axes will be simply

$$\mathbf{J} = I_{xx}\omega_x \hat{\mathbf{x}} + I_{yy}\omega_y \hat{\mathbf{y}} + I_{zz}\omega_z \hat{\mathbf{y}}$$

since the products of inertia pertaining to these axes all vanish. There is no longer a need to carry the double indices on the inertial coefficients since only the moments of inertia about the three principal axes remain. So we write $I_{xx} = I_x$, and so on; and the expression for $\mathbf{J}$ becomes

$$\mathbf{J} = I_x\omega_x \hat{\mathbf{x}} + I_y\omega_y \hat{\mathbf{y}} + I_z\omega_z \hat{\mathbf{z}} \tag{8.33}$$

But now we need to express the time derivative of $\mathbf{J}$ in order to use the equation of motion $d\mathbf{J}/dt = \mathbf{N}$. The moments of inertia are constants, but the angular velocity components ω_x, ω_y, and ω_z may vary, and the unit vectors $\hat{\mathbf{x}}$, $\hat{\mathbf{y}}$, and $\hat{\mathbf{z}}$ are changing with time because they rotate with the body. Therefore, in taking the time derivative of Eq. (8.33), we obtain

$$\frac{d\mathbf{J}}{dt} = I_x \frac{d\omega_x}{dt}\hat{\mathbf{x}} + I_y \frac{d\omega_y}{dt}\hat{\mathbf{y}} + I_z \frac{d\omega_z}{dt}\hat{\mathbf{z}}$$

$$+ I_z\omega_x \frac{d\hat{\mathbf{x}}}{dt} + I_y\omega_y \frac{d\hat{\mathbf{y}}}{dt} + I_z\omega_z \frac{d\hat{\mathbf{z}}}{dt}$$

Now the unit vectors vary only because they rotate with angular velocity $\boldsymbol{\omega}$, so their rates of change are [refer to Eq. (8.29) or (8.2)]

$$\frac{d\hat{\mathbf{x}}}{dt} = \boldsymbol{\omega} \times \hat{\mathbf{x}} \qquad \frac{d\hat{\mathbf{y}}}{dt} = \boldsymbol{\omega} \times \hat{\mathbf{y}} \qquad \frac{d\hat{\mathbf{z}}}{dt} = \boldsymbol{\omega} \times \hat{\mathbf{z}}$$

These relations allow us to write the equation preceding them as

$$\frac{d\mathbf{J}}{dt} = \frac{d'\mathbf{J}}{dt} + \boldsymbol{\omega} \times \mathbf{J} \tag{8.34}$$

where $d'\mathbf{J}/dt$ means the contribution to the variation of $\mathbf{J}$ arising from variation of the angular velocity component values, and $\boldsymbol{\omega} \times \mathbf{J}$ is the contribution due to the rotation of the principal axes to which $\mathbf{J}$ is referred. In the particular case where $\mathbf{J}$ is constant relative to the principal axes, its time derivative with respect to an inertial frame arises solely from the last term, $\boldsymbol{\omega} \times \mathbf{J}$ (as in the problem on page 255, which is treated again below).

By referring the torque vector $\mathbf{N}$ also to the principal axes, so that its components are the moments of force about these axes, we are able to state the equation of motion in the form

$$\frac{d'\mathbf{J}}{dt} + \boldsymbol{\omega} \times \mathbf{J} = \mathbf{N} \tag{8.35}$$

and in principal axis components this reads

$$\begin{aligned} I_x\frac{d\omega_x}{dt} - (I_y - I_z)\omega_y\omega_z &= N_x \\ I_y\frac{d\omega_y}{dt} - (I_z - I_x)\omega_z\omega_x &= N_y \\ I_z\frac{d\omega_z}{dt} - (I_x - I_y)\omega_x\omega_y &= N_z \end{aligned} \tag{8.36}$$

where we have dropped the prime. This set of three equations is known as *Euler's equations for the motion of a rigid body.*

The kinetic energy, fundamentally expressed by Eq. (8.4), takes the form

$$K = \tfrac{1}{2}(I_x\omega_x^2 + I_y\omega_y^2 + I_z\omega_z^2)$$

when formulated with moments of inertia and angular velocity components pertaining to principal axes.

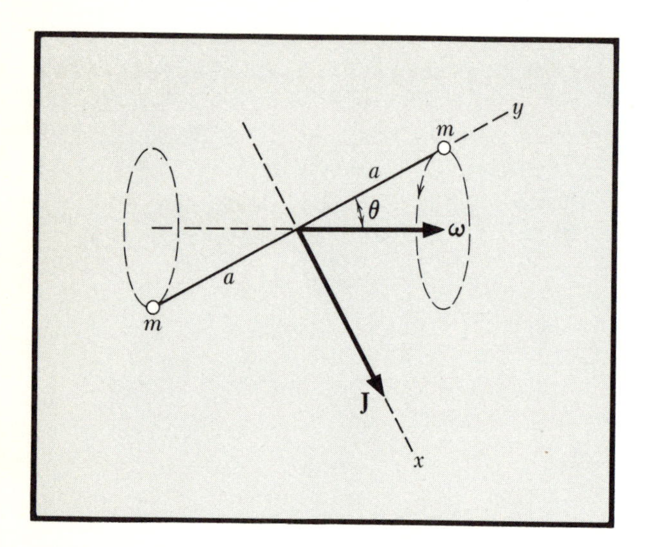

FIG. 8.12 Rigid two-particle rotator: principal axes. Compare with Fig. 8.11.

Some Simple Applications of Euler's Equations

Rigid Two-particle Rotator, Fixed Axis We return to the system of two point masses joined by a massless rod, rotating about a fixed axis through the center of mass but at an arbitrary angle and illustrated in Fig. 8.11. This was treated on pages 255 and 256, but we will now consider the problem using principal axes referring to Fig. 8.12. We choose the y axis to coincide with the rod, with the origin at the center of mass. The x axis is perpendicular to the rod in the plane determined by the rod and $\boldsymbol{\omega}$. The z axis (not shown) then extends outward toward the viewer at the instant pictured. With this choice of axes (note the differences between Figs. 8.11 and 8.12) we have

$$I_x = 2ma^2 \qquad I_y = 0 \qquad I_z = 2ma^2$$
$$\omega_x = \omega \sin\theta \qquad \omega_y = \omega\cos\theta \qquad \omega_z = 0$$

Then, from Eq. (8.33), $\mathbf{J} = 2ma^2\,\omega \sin\theta\,\hat{\mathbf{x}}$. It is thus perpendicular to the rod, as found in Eq. (8.28), and it rotates with the rod along with the x axis. We find the torque necessary to constrain it to rotate about this axis by referring to Euler's equations [Eq. 8.36)], and since ω is considered to be constant the result is

$$N_z = -2ma^2\omega^2 \sin\theta\cos\theta$$

in agreement with Eq. (8.30). It has only a z component, and it rotates with the rod.

Circular Disk, Fixed Axis Inclined from Normal Here we consider a disk, shown in Fig. 8.13, of mass m and radius a, constrained to rotate about a fixed axis at an angle θ from the normal axis. We choose principal axes as indicated, with the z axis normal and the x axis in the plane determined by $\boldsymbol{\omega}$ and $\hat{\mathbf{z}}$. Then applying Eqs. (8.13) and (8.14),

$$I_x = I_y = \frac{ma^2}{4} \qquad I_z = \frac{ma^2}{2}$$
$$\omega_x = -\omega\sin\theta \qquad \omega_y = 0 \qquad \omega_z = \omega\cos\theta$$

So the angular momentum is

$$\mathbf{J} = -\tfrac{1}{4}ma^2\,\omega\sin\theta\,\hat{\mathbf{x}} + \tfrac{1}{2}ma^2\,\omega\cos\theta\,\hat{\mathbf{z}}$$

The angle α between $\boldsymbol{\omega}$ and $\mathbf{J}$ can be found from the scalar product of these vectors, thus

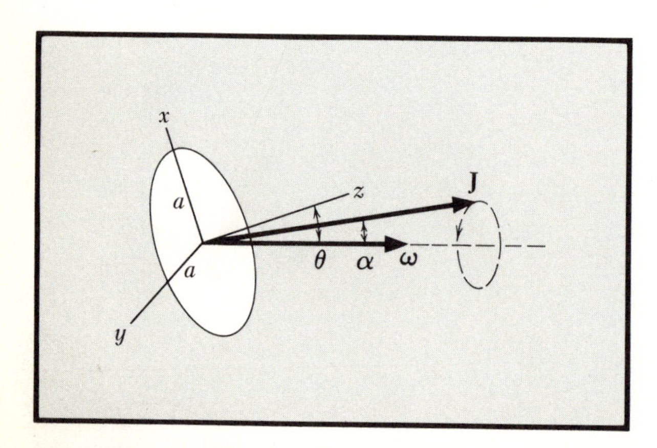

FIG. 8.13 Circular disk rotating about an axis tilted from the normal through the center.

$$\cos \alpha = \frac{\boldsymbol{\omega} \cdot \mathbf{J}}{\omega J} = \frac{1 + \cos^2 \theta}{1 + 3 \cos^2 \theta}$$

As the motion progresses, **J** rotates about $\boldsymbol{\omega}$, generating a cone as pictured.

The rotating torque required to hold the disk in rotation about this tilted direction for $\boldsymbol{\omega}$ is, from Eq. (8.35),

$$\mathbf{N} = \boldsymbol{\omega} \times \mathbf{J} = \tfrac{1}{4} m a^2 \omega^2 \sin \theta \cos \theta \, \hat{\mathbf{y}}$$

The examples in Probs. 13 and 14 belong to a class of rotating systems that are not "dynamically balanced." The fact that **J** does not coincide in direction with $\boldsymbol{\omega}$ means that a rotating torque is required to hold the body in its rotation. The dynamic balancing of crankshafts, wheels, etc., is the operation of adjusting the mass distributions so that the required axis of rotation will be a principal axis, thus causing **J** to lie along $\boldsymbol{\omega}$ and eliminating unwanted rotating torques.

Spinning Top or Gyroscope—Approximate Treatment We picture in Fig. 8.14 a simple form of spinning top consisting of a circular disk of mass M and radius a, with a massless stem. The tip of the stem is at O, and the center of mass of the disk at C is at distance l from the tip. We show an inertial frame XYZ and a rotating frame of principal axes xyz. The principal axes move with the stem of the top, but do *not* spin with the disk about its moving stem. The axis Oz is along the stem, Ox is always in the horizontal XY plane, and Oy inclines below this plane by an angle θ which is the same as the angle by which Oz inclines away from OZ. The projection of the center of mass upon the XY plane falls at C', and OC' is shown at angle φ from the X axis in the horizontal plane. This is the same as the angle between Ox and the negative sense of the Y axis. So the orientation of the stem is specified by the polar angle θ and the azimuthal angle φ, and the motion of the stem is described by variations of these angles. The principal axes follow such motion.

The disk spins about its stem with the rate S rad/s as viewed from the xyz frame, but the total angular velocity of the top will generally involve also variations in φ and θ, so the total angular velocity vector will be the sum

$$\boldsymbol{\omega} = -\dot{\theta}\hat{\mathbf{x}} + \dot{\varphi}\hat{\mathbf{Z}} + S\hat{\mathbf{z}} \tag{8.37}$$

To express this entirely in terms of principal axis components we note that

$$\hat{\mathbf{Z}} = -\sin \theta \, \hat{\mathbf{y}} + \cos \theta \, \hat{\mathbf{z}}$$

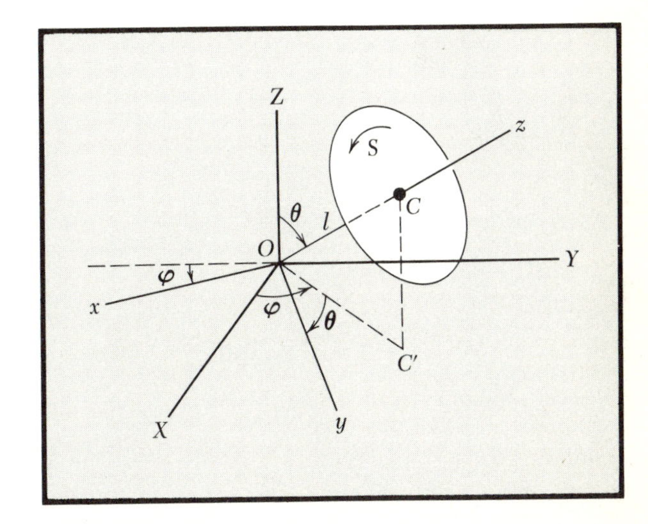

FIG. 8.14 Spinning top. Axes and angles used in the description of the motion. Also shown is the spin angular velocity S.

so Eq. (8.37) becomes

$$\boldsymbol{\omega} = -\dot{\theta}\hat{\mathbf{x}} - \dot{\varphi}\sin\theta\hat{\mathbf{y}} + (\dot{\varphi}\cos\theta + S)\hat{\mathbf{z}}$$

The moments of inertia for the principal axes are [again with reference to Eqs. (8.13) and (8.14)]

$$I_z = \tfrac{1}{2}Ma^2 \qquad I_x = \tfrac{1}{4}Ma^2 + Ml^2 = I_y$$

where in expressing I_x and I_y we have employed the parallel axis theorem. By use of Eq. (8.33) we now write the angular momentum vector about O:

$$\mathbf{J} = (\tfrac{1}{4}Ma^2 + Ml^2)(-\dot{\theta}\hat{\mathbf{x}} - \dot{\varphi}\sin\theta\hat{\mathbf{y}}) + \tfrac{1}{2}Ma^2(\dot{\varphi}\cos\theta + S)\hat{\mathbf{z}} \tag{8.38}$$

At this point we will cease from pursuing the general case of the complex and fascinating motion of the spinning top and limit our attention to the special case of steady precession at angle θ. Thus $\dot{\theta} = 0$; and $\dot{\varphi}$ and S will be constant since there are no torques acting about either OZ or Oz. Moreover, in keeping with familiar actual situations, we make the approximation that $S \gg \dot{\varphi}$, and consequently we neglect terms in $\dot{\varphi}$ in the expression for **J**. Under these conditions, Eq. (8.38) reduces to

$$\mathbf{J} = \tfrac{1}{2}Ma^2S\hat{\mathbf{z}}$$

And the angular velocity of the coordinate axes that do *not* spin with motion $S\hat{\mathbf{z}}$ is simply (with $\dot{\theta} = 0$)

$$\boldsymbol{\omega}' = \dot{\varphi}\hat{\mathbf{Z}}$$

The time derivative of **J**, obtained from Eq. (8.34), is simply $\boldsymbol{\omega}' \times \mathbf{J}$ because of the constancy of $\dot{\varphi}$ and S. Thus from the last two equations we have

$$\begin{aligned}\frac{d\mathbf{J}}{dt} &= \boldsymbol{\omega}' \times \mathbf{J} = \dot{\varphi}\hat{\mathbf{Z}} \times \tfrac{1}{2}Ma^2S\hat{\mathbf{z}} = \tfrac{1}{2}Ma^2\dot{\varphi}S(\hat{\mathbf{Z}} \times \hat{\mathbf{z}}) \\ &= -\tfrac{1}{2}Ma^2S\dot{\varphi}\sin\theta\hat{\mathbf{x}}\end{aligned} \tag{8.39}$$

The torque about O effective upon the top is due to the downward force of gravity acting at C, the only other force being the support force at O. The result is

$$\mathbf{N} = -Mgl\sin\theta\hat{\mathbf{x}} \tag{8.40}$$

The equating of Eqs. (8.39) and (8.40) according to the fundamental equation of motion [Eq. (8.1)] gives us, for the precession rate,

$$\dot{\varphi} = \frac{Mgl}{\frac{1}{2}Ma^2 S} \tag{8.41}$$

We note that it is independent of the inclination θ.

The factor $\frac{1}{2}Ma^2$, the value of I_z for the disk, appears in the denominator of Eq. (8.41). In the special case of steady precession and in the approximation of $S \gg \dot{\varphi}$, we may correctly generalize Eq. (8.41) into

$$\dot{\varphi} = \frac{Mgl}{I_z S} \tag{8.42}$$

so as to include forms of tops and gyros other than simple disks.

This[1] approximate result expressing the rate of steady precession, when the spin angular velocity S is large compared to the precession angular velocity $\dot{\varphi}$, can be obtained directly from application of the fundamental equation of motion, Eq. (8.1). We proceed as follows.

When S is large the angular momentum of the spinning top is almost completely given by

$$\mathbf{J} = I_z S \hat{\mathbf{z}}$$

The rate of change of $\mathbf{J}$ is due to the steady rotation about the vertical $\hat{\mathbf{Z}}$ direction with the precession angular velocity $\dot{\varphi}$. So (refer to Fig. 8.14)

$$\frac{d\mathbf{J}}{dt} = \dot{\varphi}\hat{\mathbf{Z}} \times I_z S \hat{\mathbf{z}} = -I_z S \dot{\varphi} \sin\theta\, \hat{\mathbf{x}}$$

The torque is given by Eq. (8.40); and the equating of that to this present expression for $d\mathbf{J}/dt$ gives again the result in Eq. (8.42):

$$\dot{\varphi} = \frac{Mgl}{I_z S}$$

It should be emphasized that this treatment of the gyroscope problem involves only a simple though important case. More general aspects of the motion can be demonstrated with gyros in the classroom, and they are all subject to analysis by Euler's equations [Eq. (8.36)]. This area of study is central to the technology of inertial navigation and gyroscopic stabilization. It also has applications, with modification, to spinning molecules, atomic nuclei, and elementary particles that are subject to torques in magnetic fields due to their intrinsic magnetic moments.

[1] Those who omitted the discussion of Euler's equations, page 257, can use the simple treatment of the gyroscope beginning here.

PROBLEMS

1. *Parallel axis theorem.* Beginning with the fact that the moment of inertia of a thin disk about a diametral axis is $\frac{1}{4}ma^2$, employ the parallel axis theorem to prove that for a solid circular cylinder of mass M, radius a, and length L, the moment of inertia about a transverse axis through the center of mass is $Ma^2/4 + ML^2/12$.

2. *Additivity of moments of inertia.* Using the principle that moments of inertia are simply additive, calculate the moment of inertia about the central axis of the cylindrical object in Fig. 8.15 if its mass is M, its radius a, the radius of each of the four cylindrical voids is $a/3$, and the axis of each void is at distance $a/2$ from the central axis. *Ans.* $\frac{59}{90}Ma^2$.

3. *Moment of inertia of solid sphere.* Show that the moment of inertia about a diameter of a solid sphere is $\frac{2}{5}Mr^2$. This can be simply done by considering the sphere to be a stack of circular disks of infinitesimal thickness fitting within a spherical bounding surface.

4. *Moments of inertia of triangle.* Three equal mass points at the vertices of an equilateral triangle (see Fig. 8.16) are joined by a rigid triangular sheet of negligible mass.
   (*a*) Find the moment of inertia I_z about the normal axis through the center C.
   (*b*) Evaluate I_y for the y axis as shown.
   (*c*) By invoking the perpendicular axis theorem, evaluate I_x.

5. *Square plate: equality of moments.* Prove that the moment of inertia of a rigid square plate about a diagonal axis in its plane is the same as that about an axis in the plane through the center, parallel to edges of the square. (The perpendicular axis theorem, together with symmetry, allows you to prove this without any calculation.)

6. *Rolling rigid bodies.* A solid cylinder, a thin-walled cylindrical shell, a solid sphere, and a thin-walled spherical shell are all rolled down an inclined plane sloped at angle θ. Each object has the same radius R. Find the acceleration of each.

7. *Rolling of hollow sphere.* A hollow sphere, with inside radius R_1 and outside radius R_2, rolls without slipping down an inclined plane at angle θ from the horizontal.
   (*a*) Find its angular and linear accelerations.

$$\textit{Ans. } \alpha = \frac{a}{R_2} \qquad a = \frac{g \sin\theta}{1 + \frac{2}{5}(1 - R_1^5/R_2^5)/(1 - R_1^3/R_2^3)}$$

   (*b*) At its lower end the plane merges into a curved transition that finally becomes a horizontal plane. With what speed will the object be moving on the final horizontal plane if it started from rest on the inclined plane with its center at height h above the final horizontal plane? (Use conservation of energy.)

$$\textit{Ans. } v^2 = \frac{2g(h - R_2)}{1 + \frac{2}{5}(1 - R_1^5/R_2^5)/(1 - R_1^3/R_2^3)}$$

8. *Frictional torque.* A heavy flywheel in the form of a solid cylinder of radius 50 cm, thickness 20 cm, and mass 1200 kg rotates freely on bearings at an initial rate of 150 rps. It is to be brought to rest by a friction brake, in which a brake

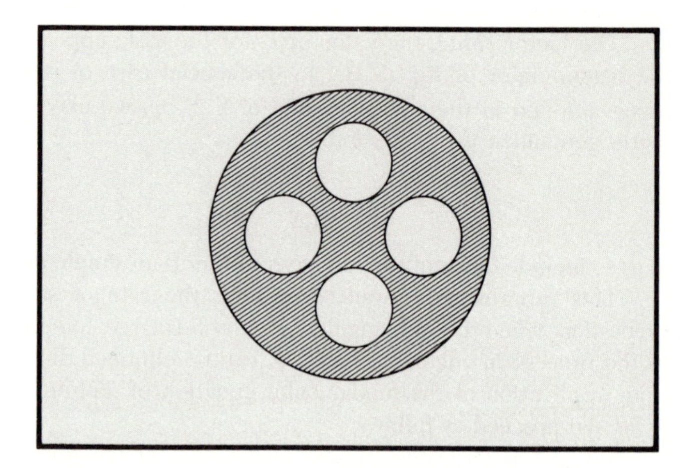

FIG. 8.15

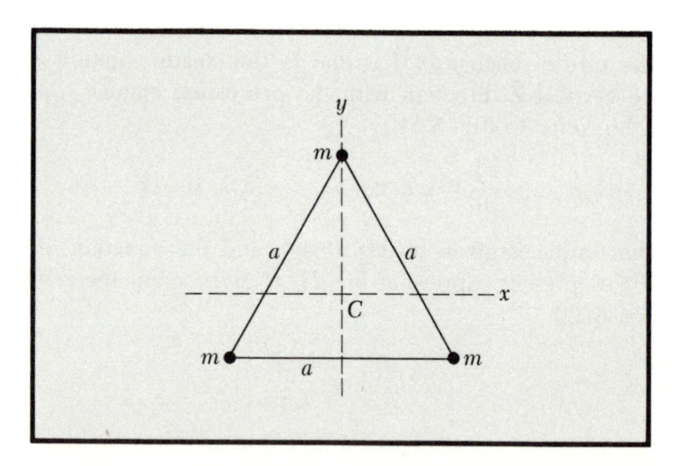

FIG. 8.16

shoe is pressed against the periphery of the flywheel with a force equivalent to a 40-kg weight. The coefficient of friction between the braking surfaces is 0.4 and is assumed to be independent of relative surface speed.

(*a*) Through what angle will the flywheel turn in coming to rest if the brake is steadily applied?

Ans. 8.5×10^5 rad, or 135,000 revolutions (approx.)

(*b*) How long will it be in coming to rest? *Ans.* 1800 s.

9. *Compound pendulum: equivalent length.* Prove that a uniform rod of length L, hanging as a compound pendulum from a pivot at one end, has the same frequency for small oscillations as a simple pendulum whose length is $2L/3$.

10. *Center of percussion.* Consider a rigid rod of length L, suspended from one end by a pivot at P. A force F acting for a brief period (i.e., an impulsive force) is to be applied to set the rod into pendulum motion as shown in Fig. 8.17. The support arrangement at P is very fragile, and it is necessary to apply F at such a distance x that no reaction force occurs at P. Find the value of x to meet this requirement. This position is called the *center of percussion* for the point of suspension P. (*Hint:* The effect of F will be to accelerate the center of mass and to give angular acceleration about P by its moment with respect to P. Compatibility of these accelerations, assuming no reaction force at P, will specify the value of x in terms of L.) *Ans.* $x = 2L/3$.

11. *Unbalanced rigid body.* A thin rim or hoop of mass M and radius R is mounted with massless spokes so as to rotate freely in the vertical plane about a horizontal axis through its center. A particle of mass m is fastened to the rim, causing the system to hang at rest with m at the bottom. Find the frequency of small oscillations. Also find the maximum angular velocity attained if the system is released from a stationary condition with m at the top.

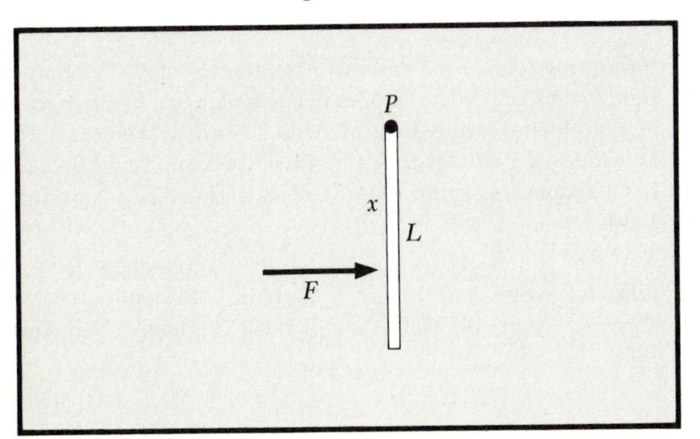

FIG. 8.17

12. *Reversible pendulum.* Prove that for a compound pendulum there are two support distances l_1 and l_2 from the center of mass that will produce the same frequency of small oscillations and that these distances are related by

$$l_1 l_2 = \frac{I_c}{M}$$

Furthermore show that if we have located such a pair of conjugate points and measured their common frequency ω, we may obtain the value of g from

$$g = \omega^2(l_1 + l_2)$$

(This is a technique called the *reversible pendulum* method for measuring g. The support points are on a straight line passing through the center of mass on either side of it; so $l_1 + l_2$ is simply the distance between support points. The position of the center of mass thus is not required.)

13. *Rotating torque.* A rectangular plate of mass M, with sides a and b, is rotated with angular velocity ω about a fixed axis along a diagonal. Evaluate the rotating torque vector that the bearings must apply to the plate to hold it in this mode of rotation. Draw a good diagram showing the angular momentum vector. Express it as a vector. *Partial ans.*

Magnitude of torque $= \frac{1}{12}Mab\omega^2(a^2 - b^2)/(a^2 + b^2)$.

14. *Lack of dynamical balance.* A uniform thin rod of mass M and length L is rotated about a transverse axis through its center. The axis is supposed to be perpendicular to the rod, but through an imperfection it deviates from this by a small angle δ. Find the rotating torque vector required if the rotation is with angular velocity ω. Express the angular momentum vector and show it in a diagram.

15. *Gyroscope.* A certain gyroscope consists of a solid cylinder with radius $a = 4$ cm. It is supported by a massless stem whose tip is pivoted freely at a point 5 cm from the center of mass of the cylinder. It is observed to be moving in steady precession at an angle of inclination from vertical, and the precession occurs at one complete circular excursion every 3 s. Evaluate the angular velocity of spin of the gyroscope about its own axis. *Ans.* 293 rad/s.

16. *Angular acceleration.* A solid cylinder of mass 2.0 kg and radius 4.0 cm is constrained to rotate about its axis, which is horizontal. A string is wrapped around it and one end hanging freely has a mass of 150 g attached (see Fig. 8.7). Find the linear acceleration of the mass, the angular acceleration of the cylinder, the tension in the string, and the vertical force keeping the cylinder up.

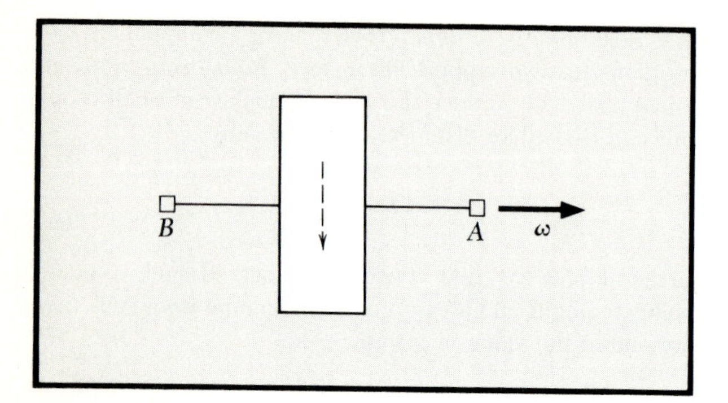

FIG. 8.18

17. *Rotation of gyroscope.* Figure 8.18 represents a gyroscope wheel seen from one side, with its axle mounted in bearings A and B. It is spinning with angular velocity as shown, the near side of the wheel moving downward. Upward support forces exist equally at A and B.

(*a*) It is now desired to reorient the wheel to place A directly over B, without moving the center of mass of the system. Describe the additional forces, besides support, to be applied at A and B.

(*b*) If instead of placing A over B we had wished to bring A out toward the viewer, and B behind A, describe the forces we should apply at A and B.

18. *Torques about center of mass.* A cylinder of mass M_1, radius R_1, with axis horizontal, is constrained to rotate about its axis. A string wrapped around this cylinder is also wrapped around a cylinder of mass M_2, radius R_2, which is free to unwind and fall with its axis horizontal as in Fig. 8.19. Find in the approximation of vertical string the

(*a*) Acceleration of the center of mass of M_2.

Ans. $a = (M_1 + M_2)g/(\frac{3}{2}M_1 + M_2)$

(*b*) Angular acceleration of M_2.

(*c*) Angular acceleration of M_1.

(*d*) Tension in the string.

If one takes moments about point P of the figure, what is the "fictitious force" at the center of mass of the second cylinder?

19. *Minimum coefficient of friction.* For a symmetric body to roll without slipping down an inclined plane, show that

$$\mu \geq \frac{\tan\theta}{MR^2/I_c + 1}$$

where the symbols have the usual meanings.

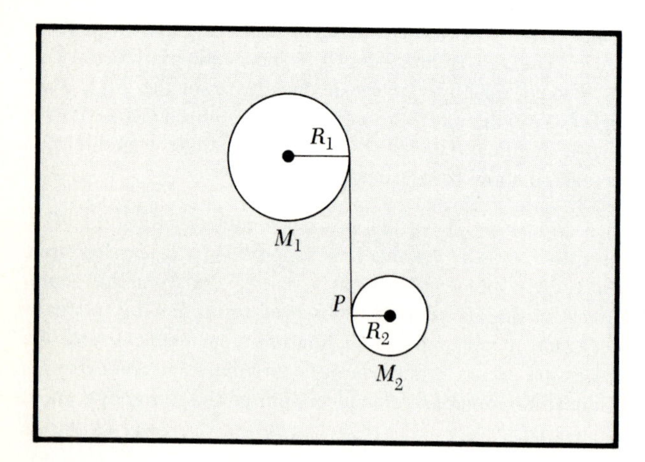

FIG. 8.19

FURTHER READING

Interesting, clear, and extensive treatments of the motion of rigid bodies, and of gyroscopes in particular, are found in some of the older treatises such as Arthur Gordon Webster, "The Dynamics of Particles and of Rigid, Elastic, and Fluid Bodies," B. G. Teubner, Leipzig, 1904 (Stechert-Hafner, Inc., New York, 1920).

A more contemporary treatment at intermediate level is John L. Synge and Byron A. Griffith, "Principles of Mechanics," chap. 14, McGraw-Hill Book Company, New York, 1959.

CONTENTS

Inverse-square-law Force

The magnitudes of the electrostatic and gravitational forces between two point particles at rest are given by

$$F = \frac{C}{r^2}$$

with C a constant. Such forces are called *inverse-square-law central forces*. The word *central* means that the force is directed along the line which connects the two particles. If one particle is at the origin and the second particle is at position $\mathbf{r}$, the force on the second particle is given by

$$\mathbf{F} = \frac{C}{r^2}\hat{\mathbf{r}} \tag{9.1}$$

where for the gravitational force between point masses M_1, M_2:

$$C = -GM_1M_2$$
$$G = 6.67 \times 10^{-8}\ \text{cm}^3/\text{g-s}^2 = 6.67 \times 10^{-11}\ \text{m}^3/\text{kg-s}^2 \tag{9.2}$$

and for the electrostatic force between point charges q_1, q_2:

$$C = q_1q_2 \tag{9.3}$$

provided that the charge is expressed in gaussian cgs units or

$$C = 9.0 \times 10^9 q_1q_2 \tag{9.3a}$$

if the charge, lengths, and forces are expressed in SI units (see Chap. 3 pages 67 to 73). The gravitational force is always attractive. The electrostatic (coulomb) force is attractive if the charges q_1, q_2 have opposite signs, and repulsive if q_1, q_2 have the same sign.

The exponent of r in Eq. (9.1) is known very accurately by experiment to be equal to 2.000 . . . ; for electrostatic forces this is established down to distances as small as of the order of 10^{-13} cm. A wide variety of experimental results would be highly sensitive to small departures from an exact inverse-square law of force. The central experiments are discussed in Volume 2, Chap. 1, with special reference to electrostatic forces. For gravitational forces we appeal for experimental support chiefly to the excellent agreement between prediction and observation of planetary motions in the solar system.

The inverse-square law of force is also expressed as an inverse-first-power law of potential energy. As we saw in Chap. 5, F is equal to $-\partial U/\partial r$. From Eq. (9.1), then

$$F = -\frac{\partial U}{\partial r} = \frac{C}{r^2}$$

and

$$U(r) = \frac{C}{r} + \text{const}$$

If we choose $U(r)$ to be zero when the particles are infinitely far apart, then the constant of integration is zero and we have

$$U(r) = \frac{C}{r}$$

where C is given by Eqs. (9.2), (9.3), or (9.3a) for gravitational or electrostatic forces in the cgs or SI systems. Thus

$$U(r) = -\frac{GM_1M_2}{r} \qquad \text{or} \qquad U(r) = \frac{q_1q_2}{r} \tag{9.4}$$

Alternatively,

$$U = \frac{kq_1q_2}{r} \tag{9.4a}$$

The force law between two protons or two neutrons or a proton and a neutron deviates very strongly from either the gravitation or the coulomb law. The deviation is a very strong attractive force when the particles are very close together, less than about 2×10^{-13} cm, and the force is negligible when they are farther apart. Such forces are treated in books on nuclear physics. The electric force between two electrons is accurately coulombic down to the smallest distances known. Electrons do have magnetic dipole moments in addition to their charge, and the magnetic moments give rise to a noncentral inverse-cube law of force (Volume 2, Chap. 10).

Given an inverse-square law of force, what special characteristics follow? In what vital respects does the universe reflect an inverse-square law? We now turn to these important questions. We shall often discuss the potential energy rather than the force. In solving problems, the student will nearly always find it easier to use the potential energy rather than the force. He can obtain the force components by differentiating the potential and can often use the potential energy in an energy equation. The potential energy is a scalar; the force is a vector.

POTENTIAL ENERGY AND FORCE BETWEEN A POINT MASS AND A SPHERICAL SHELL

One important consequence of the inverse-square law of force is that the force on a point test mass M_1 distant r from the center of a uniform thin spherical shell of radius R is exactly

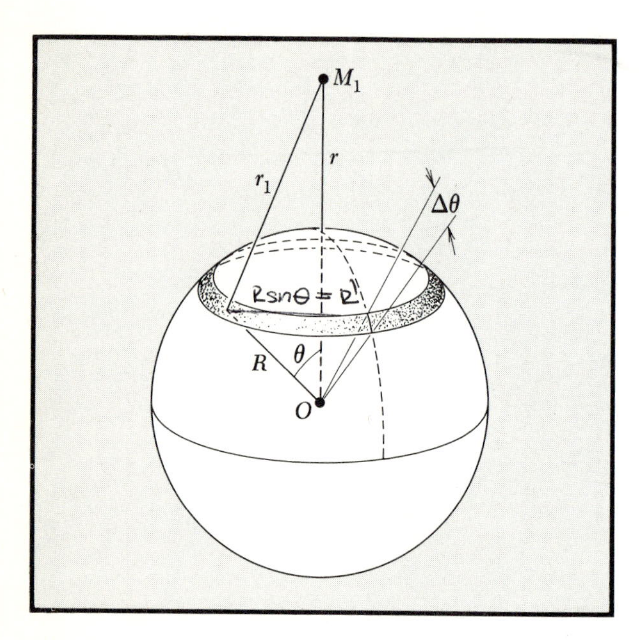

FIG. 9.1 Perspective drawing of spherical shell and point mass M_1, showing how spherical shell is divided into rings. The shell has a mass density σ per unit area.

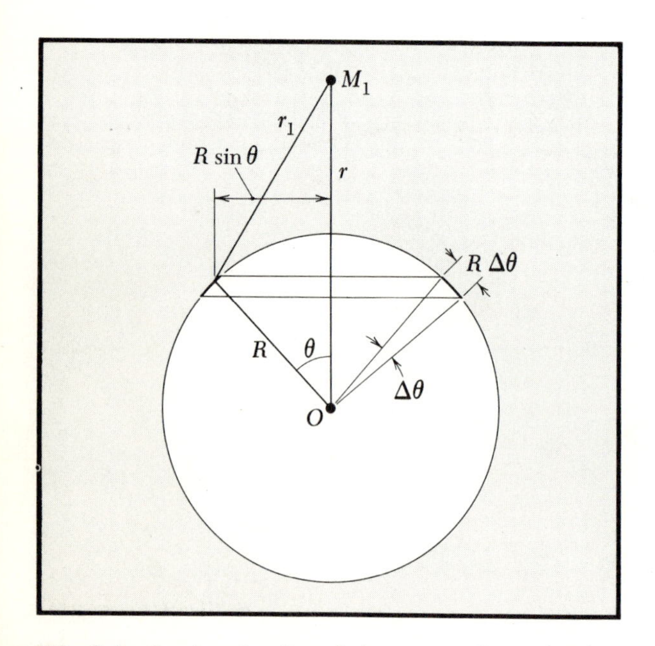

FIG. 9.2 Section drawing of the same sphere showing ring, of total area $2\pi R^2 \sin\theta\,\Delta\theta$.

the same at points $r > R$ outside the shell as if the entire mass of the shell were concentrated at its center. A second consequence is that for points $r < R$ inside the shell the force on the point mass is *zero*. These consequences are so important that we shall give the derivations in full detail. We follow a special method of solution which takes advantage of the geometrical symmetry of the problem.

We first consider a ring on the shell having angular width $\Delta\theta$ or width $R\,\Delta\theta$, as in Fig. 9.1. Let σ be the mass per unit area of the shell. We make this choice because the entire ring is equidistant from the test mass M_1, the distance being r_1. The radius of the ring is $R \sin\theta$ and the circumference is $2\pi R \sin\theta$. The area of the ring is therefore given by (see Fig. 9.2)

$$(2\pi R \sin\theta)(R\,\Delta\theta) = 2\pi R^2 \sin\theta\,\Delta\theta$$

The mass of the ring is given by the product of the area times the mass per unit area σ:

$$M_{\text{ring}} = (2\pi R^2 \sin\theta\,\Delta\theta)\sigma \tag{9.5}$$

By combining Eqs. (9.5) and (9.4), the potential energy U_{ring} of the test mass in the gravitational field of the ring is obtained:

$$U_{\text{ring}} = -\frac{GM_1(2\pi R^2 \sin\theta\,\Delta\theta)\sigma}{r_1} \tag{9.6}$$

Here r_1 is the distance from the test mass to the ring.

By the law of cosines [Eq. (2.8)] applied to the triangle formed by R, r, and r_1, we have

$$r_1^2 = r^2 + R^2 - 2rR\cos\theta \tag{9.7}$$

where R and r are constants. Under these conditions when we evaluate the change in r_1, Δr_1 in terms of the change in θ, $\Delta\theta$, we get

$$2r_1\,\Delta r_1 = -2rR\,\Delta(\cos\theta) = 2rR\sin\theta\,\Delta\theta$$

This useful relation enables us to rewrite Eq. (9.6) to obtain

$$U_{\text{ring}} = -\frac{GM_1(2\pi R\,\Delta r_1)\sigma}{r} \tag{9.8}$$

Notice the denominator is now r, the distance from the test mass to the center of the sphere.

The total potential energy U_{shell} of the test mass in the gravitational field of the spherical shell is given by the sum of U_{ring} over all the rings which make up the shell. When we sum, we have only to sum over Δr_1. When the test mass lies outside the shell, the range of values of r_1 is seen to run from

$r - R$ to $r + R$, so that (see Fig. 9.3)

$$\sum \Delta r_1 = (r + R) - (r - R) = 2R \tag{9.9}$$

It is very fortunate that the problem can be reduced to such a simple summation. Using Eq. (9.9) to sum over Eq. (9.8), we have

$$U_{\text{shell}} = \sum U_{\text{ring}} = -\frac{GM_1 2\pi R\sigma}{r} \sum \Delta r_1 = -\frac{GM_1 4\pi R^2 \sigma}{r} \tag{9.10}$$

But $4\pi R^2$ is the surface area of the spherical shell, and $4\pi R^2 \sigma$ is the mass M_s of the shell. We can therefore rewrite Eq. (9.10) as

$$U_{\text{shell}} = -\frac{GM_1 M_s}{r} \qquad (r > R) \tag{9.11}$$

where r is the distance between the test mass and the center of the spherical shell. We have shown that *the spherical shell acts at points outside as if all its mass M_s were concentrated at the center of the shell.*

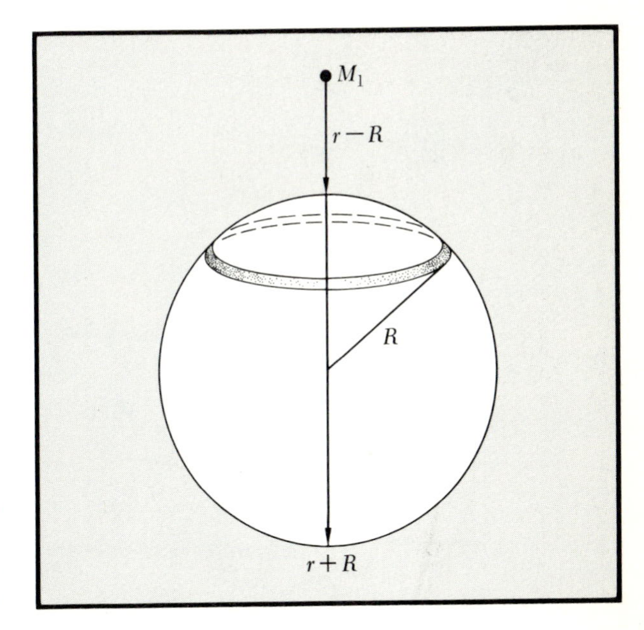

FIG. 9.3 Limits of summation for $r > R$, test mass M_1 outside the spherical shell.

If the test mass lies anywhere within the shell, the derivation is identical except that the range of summation of Δr_1 in $\Sigma\, U_{\text{ring}}$ is from $R - r$ to $R + r$ (see Fig. 9.4), so that now

$$\sum \Delta r_1 = (R + r) - (R - r) = 2r \tag{9.12}$$

Using Eq. (9.12) to sum over Eq. (9.8), we obtain

$$\begin{aligned} U_{\text{shell}} &= \sum U_{\text{ring}} = -\frac{GM_1 2\pi R\sigma}{r} \sum \Delta r_1 \\ &= -GM_1 4\pi R\sigma = -\frac{GM_1 4\pi R^2 \sigma}{R} \\ &= -\frac{GM_1 M_s}{R} \qquad (r < R) \end{aligned} \tag{9.13}$$

The potential [Eq. (9.13)] is constant at all points in the interior of the shell and is equal to Eq. (9.11) evaluated at $r = R$. Figure 9.5*a* shows U both inside and outside the shell.

We have seen before [Eqs. (5.28) and (5.29)] that the magnitude of the force F on the test mass M_1 is equal to $-\partial U/\partial r$ because the force is in the radial direction. From Eqs. (9.11) and (9.13) we have, for the force due to the shell,

$$F = -\frac{\partial U}{\partial r} = \begin{cases} -\dfrac{GM_1 M_s}{r^2} & (r > R) \\ 0 & (r < R) \end{cases} \tag{9.14}$$

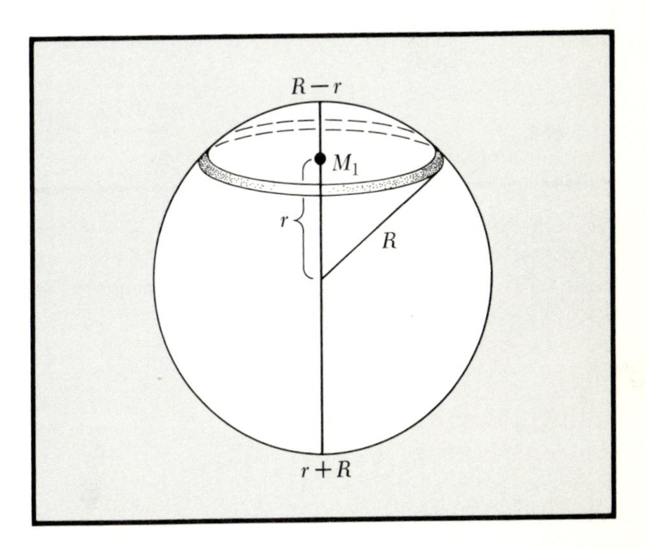

FIG. 9.4 Limits of summation for $r < R$, test mass M_1 inside the spherical shell.

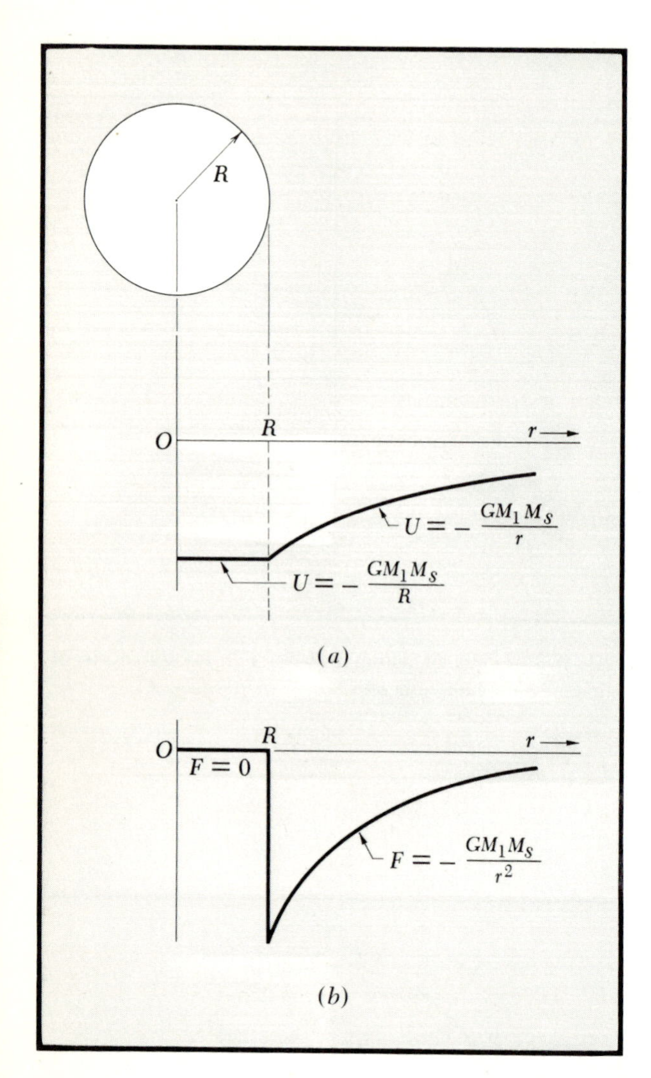

FIG. 9.5 (*a*) Potential energy of point mass M_1 at distance r from center of shell of radius R, mass M_s. (*b*) Force on point mass M_1 (negative sign to indicate attraction). The force is zero for $r < R$.

Thus there is no force on a test mass inside the shell. This is a very special property of an inverse-square-law force. Outside the shell the force varies as $1/r^2$, where r is measured from the center of the shell. Figure 9.5*b* shows the force as a function of r.

POTENTIAL ENERGY AND FORCE BETWEEN A POINT MASS AND A SOLID SPHERE

We may build up a solid sphere of mass M and radius R_0 by adding up a series of concentric shells. For points r outside the sphere we have, using Eq. (9.11), the following result for the potential energy of the test mass M_1 in the gravitational field of the solid sphere:

$$U_{\text{sphere}} = \sum U_{\text{shell}} = -\frac{GM_1}{r}\sum M_s = -\frac{GM_1M}{r}$$

Recall that r is the distance from the test mass to the center of the sphere.

The magnitude of the force on M_1 is, for $r > R_0$,

$$\boxed{F = -\frac{\partial U}{\partial r} = -\frac{GM_1M}{r^2}} \qquad (9.15)$$

This is the central result of our analysis. It could have been obtained also by a direct integration of the force components over the shell, as in Prob. 13, but the mathematics is more concise as we have done the problem. By an easy extension of Eq. (9.15), we see that the force between two uniform spheres of masses M_1, M_2 is equal to the force between two point masses M_1, M_2 at their respective centers. Having replaced one sphere by a point mass, we can then replace the second sphere by a point mass. This happy result simplifies many calculations.

If a point mass is inside a solid sphere, the force will be toward the center of the sphere and will be

kutle kirenin üzerinde

$$\frac{-GM_1M_{\text{ins}}}{r^2}$$

If the sphere is of uniform density ρ then

$$M_{\text{ins}} = \frac{4\pi}{3}r^3\rho \qquad \text{where } M = \frac{4\pi}{3}R_0{}^3\rho$$

and so

$$F = -\frac{GM_1\frac{4}{3}\pi r^3\rho}{r^2} = -\tfrac{4}{3}\pi GM_1\rho r$$

or

$$F = -\frac{GM_1Mr}{R_0^{\ 3}} \tag{9.16}$$

The potential energy for $r < R_0$ is obtained by adding to $-(GM_1M)/R_0$ the energy required to move the mass M_1 from R_0 to r. Using Eq. (9.16) for the force, we get for this energy

$$\int_{R_0}^{r}\frac{GM_1Mr\,dr}{R_0^{\ 3}} = -\frac{GM_1M}{2R_0^{\ 3}}(R_0^{\ 2} - r^2)$$

Addition of the term $-(GM_1M)/R_0$ to this energy gives for the potential energy for $r < R_0$

$$U(r) = -\frac{GM_1M}{R_0} - \frac{GM_1M}{2R_0^{\ 3}}(R_0^{\ 2} - r^2)$$
$$= -\frac{GM_1M}{R_0}\left(\frac{3}{2} - \frac{1}{2}\frac{r^2}{R_0^{\ 2}}\right)$$

When $r = 0$ we have

$$U(0) = -\frac{3}{2}\frac{GM_1M}{R_0}$$

Both $U(r)$ and $F(r)$ are displayed in Fig. 9.6 for $0 \le r \le R_0$ and for $R_0 \le r$.

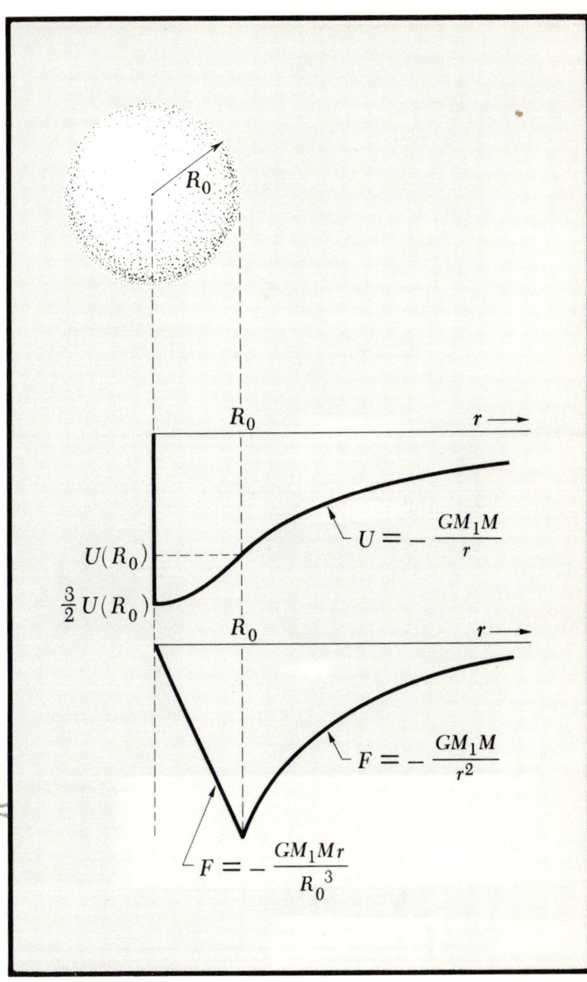

FIG. 9.6 Potential energy of point mass M_1 at distance r from center of solid sphere of radius R_0, mass M. Force on point mass M_1. The force is linear in r for $r < R_0$.

GRAVITATIONAL AND ELECTROSTATIC SELF-ENERGY

The self-energy of a body is defined to be the work done in assembling the body from infinitesimal elements that are initially an infinite distance apart. Let us consider the gravitational self-energy; this will have a negative sign, because the gravitational force is attractive. (We have to do positive work against gravity to separate the atoms of a star, taking each atom to infinity.) We need the gravitational self-energy usually for stellar and galactic problems. The electrostatic self-energy is often calculated for crystals, both insulators and metals, and for nuclei.

The potential energy of N discrete masses due to their mutual gravitational attraction is equal to the sum of the potential energy of all pairs of masses:

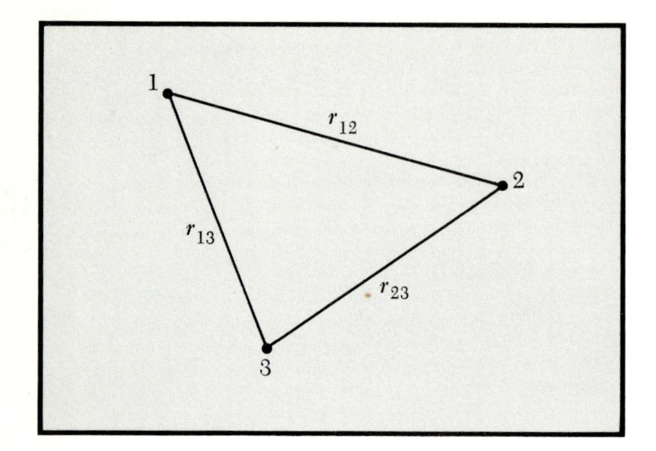

FIG. 9.7 The gravitational potential energy of three atoms of masses M_1, M_2, M_3 is

$$U = -G\left(\frac{M_1M_2}{r_{12}} + \frac{M_1M_3}{r_{13}} + \frac{M_2M_3}{r_{23}}\right)$$

$$U_s = -G \sum_{\substack{\text{All}\\ \text{pairs}\\ i\neq j}} \frac{M_iM_j}{r_{ij}} \tag{9.17}$$

where M_i and M_j are the individual masses and r_{ij} is the distance apart of those individual masses. The case $i = j$ is omitted because this is not a pair at all. The self-energy of the individual mass m_i is also omitted since only the mutual interactions of the masses are considered. A method of evaluating the self-energy of an individual mass is given below.

EXAMPLE

Gravitational Energy of a Galaxy Let us estimate the gravitational energy of the galaxy. If we omit from the calculation the gravitational self-energy of the individual stars, then we need only estimate the value of the expression in Eq. (9.17).

We approximate the gross composition of the galaxy by N stars, each of mass M, and with each pair of stars at a mutual separation of the order of R. Then Eq. (9.17) reduces to

$$U \approx -\frac{1}{2}G(N-1)N\frac{M^2}{R}$$

[In summing over all pairs, we take each of the N stars in turn and sum over the $N - 1$ stars that can be paired with it. In so doing we count each pair twice. (See Fig. 9.7 for $N = 3$.)] If $N \approx 1.6 \times 10^{11}$, $R \approx 10^{23}$ cm, and $M \approx 2 \times 10^{33}$ g (as for the sun), then

$$U \approx -\frac{1}{2}\frac{(7\times 10^{-8})(1.6\times 10^{11})^2(2\times 10^{33})^2}{10^{23}}$$
$$\approx -4 \times 10^{58} \text{ ergs}$$

In SI units

$$U \approx -\frac{1}{2}\frac{(7\times 10^{-11})(1.6\times 10^{11})^2(2\times 10^{30})^2}{10^{21}}$$
$$\approx -4 \times 10^{51} \text{ J}$$

EXAMPLE

Gravitational Energy of a Uniform Sphere It is not difficult to evaluate the self-energy U_s of a uniform sphere of mass M and radius R. We convert the multiple summations implicit in Eq. (9.17) to integrals and then carry out the integrations. But let us first attempt to guess the answer. What would we expect it to be? The answer must involve G, M, and R in the correct dimensions. Why not

$$U_s \approx -\frac{GM^2}{R}$$

This is in fact correct, except for a numerical factor of the order of unity.

In order to calculate exactly this factor we build up the solid sphere in a special way. We first consider (see Fig. 9.8) the energy of interactions between a solid spherical core of radius r and a surrounding spherical shell of thickness dr. If ρ denotes the density, the mass of the core is $(4\pi/3)r^3\rho$ and the mass of the shell is $(4\pi r^2)(dr)\rho$. Thus the gravitational potential energy of the shell due to the presence of the core is, from Eq. (9.11),

$$\frac{-G\left(\frac{4\pi}{3}r^3\rho\right)\left(4\pi r^2\,dr\,\rho\right)}{r} = -\frac{1}{3}G(4\pi\rho)^2r^4\,dr \tag{9.18}$$

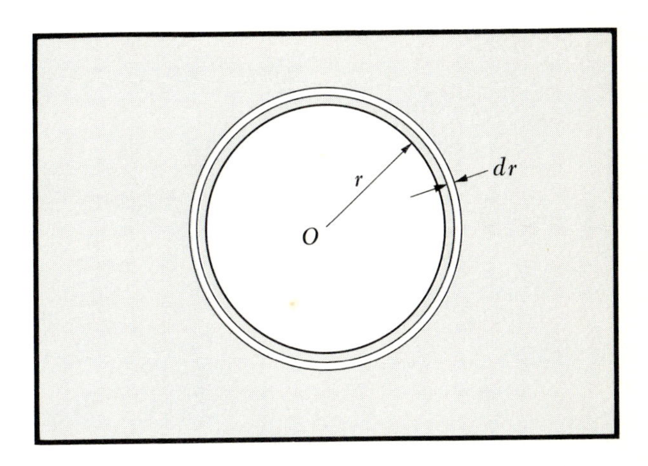

FIG. 9.8 Spherical shell of thickness dr surrounding a solid spherical core of radius r. By adding successive shells we construct a solid sphere of radius R. The area of one surface of the shell is $4\pi r^2$; the thickness is dr, so that the volume of the shell is $4\pi r^2\,dr$.

The self-energy of the solid sphere is given by the integral of Eq. (9.18) between $r = 0$ and $r = R$. The integration corresponds to adding successive shells to the core until the core has radius R. Initially the core has zero radius. The symmetry of the sphere has enabled us to reduce the multiple summations to a single integral. On integrating Eq. (9.18), we obtain the result

$$U_s = -\tfrac{1}{3}G(4\pi\rho)^2\tfrac{1}{5}R^5 = -\frac{3}{5}G\left(\frac{4\pi}{3}\rho R^3\right)^2\frac{1}{R}$$

$$= -\frac{3}{5}\frac{GM^2}{R} \tag{9.19}$$

because the mass is given by

$$M = \frac{4\pi}{3}\rho R^3$$

The gravitational self-energy of the sun is, from Eq. (9.19), $M_s \approx 2 \times 10^{33}$ g and $R_s \approx 7 \times 10^{10}$ cm,

$$U_s \approx -\frac{(3)(7 \times 10^{-8})(2 \times 10^{33})^2}{(5)(7 \times 10^{10})} \approx -2 \times 10^{48}\ \text{ergs}$$

$$\approx -2 \times 10^{41}\ \text{J}$$

This is a lot of energy when one remembers that the rate of energy generation by the sun is 4×10^{33} ergs/s (4×10^{26} J/s) so that it would take the sun $\frac{1}{2} \times 10^{15}$ s or 2×10^7 yr to radiate this much energy.[1] The sun may complete its evolution as a dense white dwarf with a radius about 0.1 of its present radius. It is clear that a large amount of gravitational energy will be released in such a contraction. These considerations are very important in astrophysical studies and may well be involved in the theory of novae stars. The electrostatic self-energy of a uniform spherical distribution of total charge q and radius R is obtained by substituting q^2 for $-GM^2$ (or kq^2 for $-GM^2$) in Eq. (9.19).

[1]The energy now being radiated by the sun comes from nuclear processes, *not* gravitational. At the beginning of the twentieth century physicists had no knowledge of these nuclear processes and estimated the age of the solar system as about 10^6 yr.

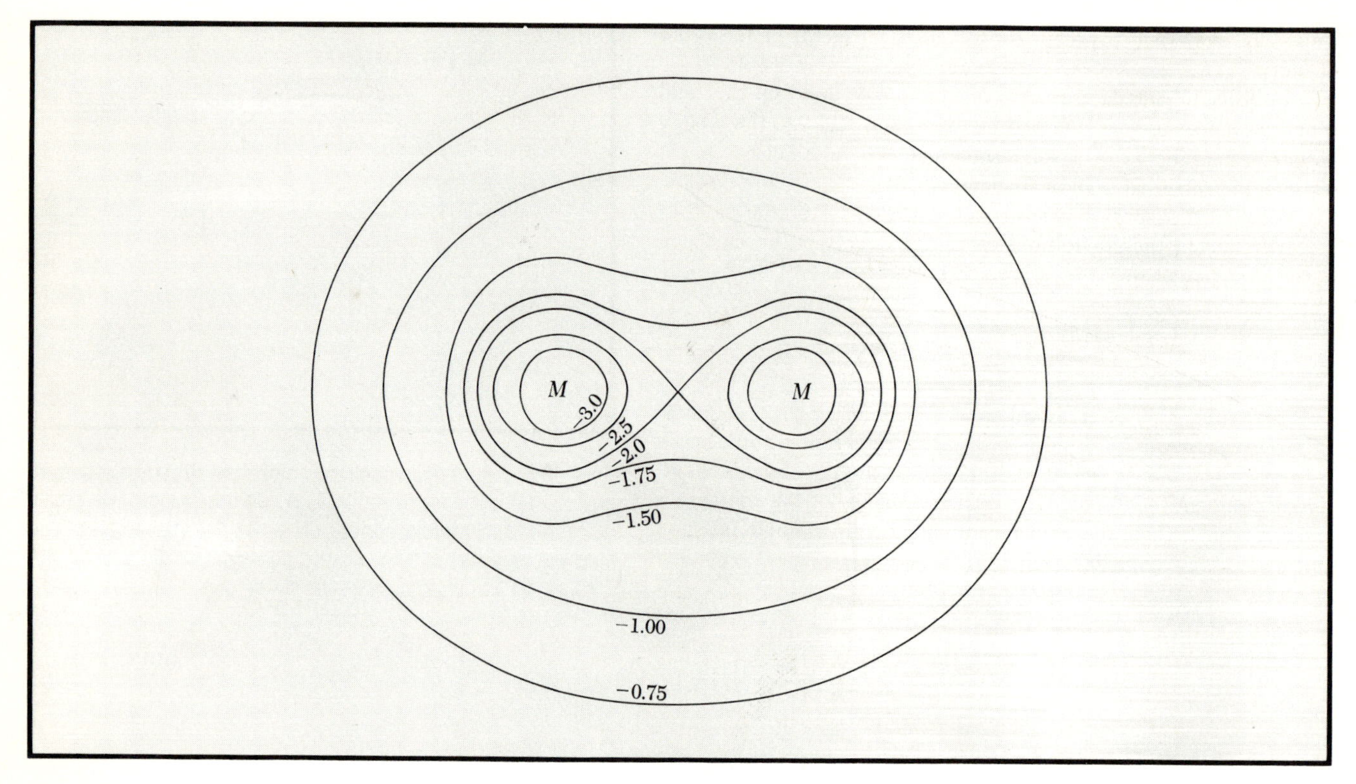

FIG. 9.9 Contour map of equipotential surfaces between two equal masses.

EXAMPLE

Radius of the Electron To estimate the electrostatic self-energy of an electron, we need to know the radius R. Because we do not have a fundamental theory of the electron, all we can do is to proceed backward and estimate a radius from the energy.

There is a famous relation due to Einstein which says that a mass M is always associated with an energy E according to the equation

$$E = Mc^2 \tag{9.20}$$

where c is the velocity of light. (We derive this in Chap. 12.) If the energy of the electron were entirely the electrostatic energy of a uniform charge distribution, then we would have

$$U_s = \frac{3e^2}{5R} = mc^2$$

which would determine the radius of the electron. But we do not know the structure of the electron in detail. The model we have outlined cannot be entirely satisfactory, for what keeps the charge in the electron together? Why doesn't it fly apart under the coulomb repulsion of like-charge elements? At present we have no theory of why there is an electron.

So let us drop the factor $\frac{3}{5}$. It would be pretentious to keep the factor because it suggests a refinement of knowledge about the electron that we do not possess. We define (by universal convention) a length r_0 by the relation

$$\frac{e^2}{r_0} \equiv mc^2 \qquad r_0 \equiv \frac{e^2}{mc^2} = 2.82 \times 10^{-13}\ \text{cm}$$

This length is called the *classical radius of the electron*. It has something to do with the electron, but we don't know exactly what! Nevertheless, it is called a fundamental length and occurs in such expressions as the cross section for x-ray or γ-ray scattering. As a matter of fact we know that the electrical force between electrons is accurately e^2/r^2 down to at least $r = 10^{-15}$ cm.

Inverse-square-law Forces and Static Equilibrium In Volume 2, Chap. 2 (page 62), we show that there can be no stable, *static* equilibrium among a group of masses (or charges) interacting with inverse-square-law forces only. By static we mean with all masses at rest. This result is made plausible by Figs. 9.9 and 9.10. They show lines of equal values of the potential, equipotentials, due to two and four equal masses (marked by M signs) at fixed positions. The position where the equipotentials cross is a position of equilibrium. When displaced from the position of equilibrium, the force is in the direction of the lower equipotential, that is, the more negative equipotential. Note that in the case of the two masses, a test mass displaced up will feel a force back toward the point of equilibrium but when displaced sideways will feel a force away from the point. In order for there to be stable equilibrium, the force must be toward the point of equilibrium no matter what the direction of displacement (see this volume, Chap. 5, page 156). The force is inversely proportional to the distance between equipotentials, and so one might expect that as we go from two to four to eight to a very large number of masses on a sphere the force would go to zero and might even be back toward the center. But as we know, or at least can infer, from Eq. (9.14), the force just goes to zero and we have a state of neutral equilibrium.

ORBITS: EQUATION AND ECCENTRICITY

We have already solved the problem of a mass executing a circular orbit in an attractive inverse-square-law field. For such an orbit there must be a special relation between the velocity and the distance. We derived this relation in Chap. 3 (pages 66–67). Formally $-(Mv^2/r)\hat{\mathbf{r}}$ is the centripetal force, the minus sign indicating it is toward the center of the circle. This

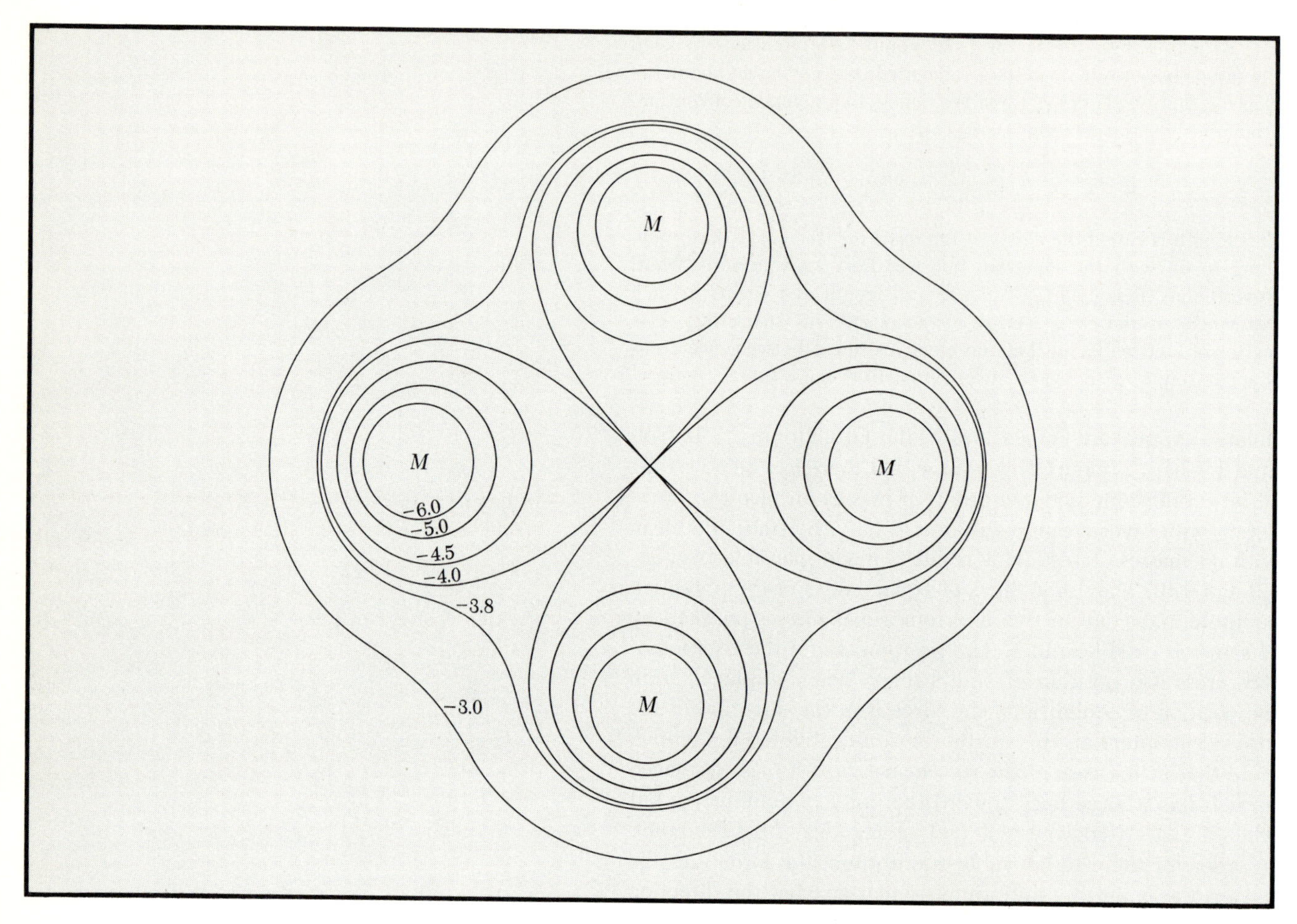

FIG. 9.10 Contour map of equipotential surfaces between four equal masses. The numbers are purely arbitrary.

must be equal to the force $(C/r^2)\hat{\mathbf{r}}$ [Eq. (9.11)], and so we see that unless C is negative, unless the force is attractive, there can be no circular orbit. If $C = -GMM_2$, then $v = (GM_2/r)^{\frac{1}{2}}$.

What is the form of the orbit if this special relation is not met? This problem is often called the *Kepler problem* because Kepler discovered that the orbits of the planets are ellipses in the force field of the sun, which Newton later deduced is an inverse-square-law field. We treat first the case in which the origin of the force at $r = 0$ is fixed. This is conceptually simple, but we shall see in the section on page 289 that the actual problem of any two masses can be reduced to this. Our equation of motion is then

$$M\mathbf{a} = \frac{C}{r^2}\hat{\mathbf{r}}$$

Here we are not assuming simple circular motion, and we leave the kind of inverse-square-law force unspecified. But notice the important fact that if C is negative the force is attractive, and if C is positive the force is repulsive.

What coordinates are most convenient? First, how many coordinates do we need? Three? No, only two since the motion will be in a plane; the plane is that determined by the vector velocity of the mass and the radius vector $\mathbf{r}$, for the velocity component perpendicular to this plane is zero and the force perpendicular is zero. Therefore this component of the velocity must remain zero. It is easy to guess that r and θ, as in Fig. 9.11, are easier to use than x and y. What is the acceleration in terms of these coordinates and the unit vectors $\hat{\mathbf{r}}$ and $\hat{\boldsymbol{\theta}}$? From Eq. (2.30)

$$\mathbf{a} = (\ddot{r} - r\dot{\theta}^2)\hat{\mathbf{r}} + \frac{1}{r}\frac{d}{dt}(r^2\dot{\theta})\hat{\boldsymbol{\theta}}$$

Therefore our equations of motion are

$$M(\ddot{r} - r\dot{\theta}^2) = \frac{C}{r^2} \qquad \frac{d}{dt}(Mr^2\dot{\theta}) = 0$$

The second is easy to integrate once and gives

$$Mr^2\dot{\theta} = J \tag{9.21}$$

where J is the angular momentum defined in Chap. 6. Replacing $\dot{\theta}$ in the first equation by J/Mr^2 gives

$$\ddot{r} - r\frac{J^2}{M^2r^4} = \ddot{r} - \frac{J^2}{M^2r^3} = \frac{C}{Mr^2} \tag{9.22}$$

Unfortunately this differential equation cannot be directly solved, but we remember that we are interested in the form of the orbit with r as a function of θ. So let us eliminate t from Eqs. (9.21) and (9.22):

$$\frac{dr}{dt} = \frac{dr}{d\theta}\frac{d\theta}{dt} = \frac{dr}{d\theta}\frac{J}{Mr^2}$$

$$\frac{d^2r}{dt^2} = \frac{d^2r}{d\theta^2}\left(\frac{J}{Mr^2}\right)^2 - \frac{2J}{Mr^3}\left(\frac{dr}{d\theta}\right)^2\frac{J}{Mr^2}$$

$$= \frac{J^2}{M^2r^4}\left[\frac{d^2r}{d\theta^2} - \frac{2}{r}\left(\frac{dr}{d\theta}\right)^2\right] \tag{9.23}$$

Again this does not provide an equation we are familiar with, and so we try using a function

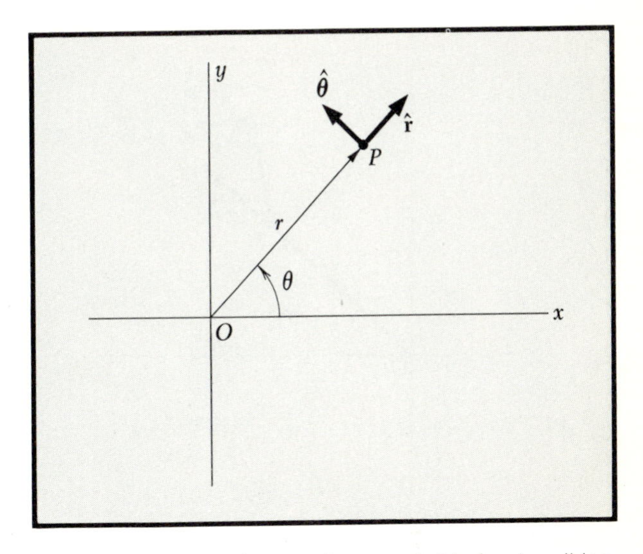

FIG. 9.11 Plane polar coordinates suitable for describing central force motion of mass M, at point P, about fixed force center O; $\hat{\mathbf{r}}$ and $\hat{\boldsymbol{\theta}}$ are unit vectors.

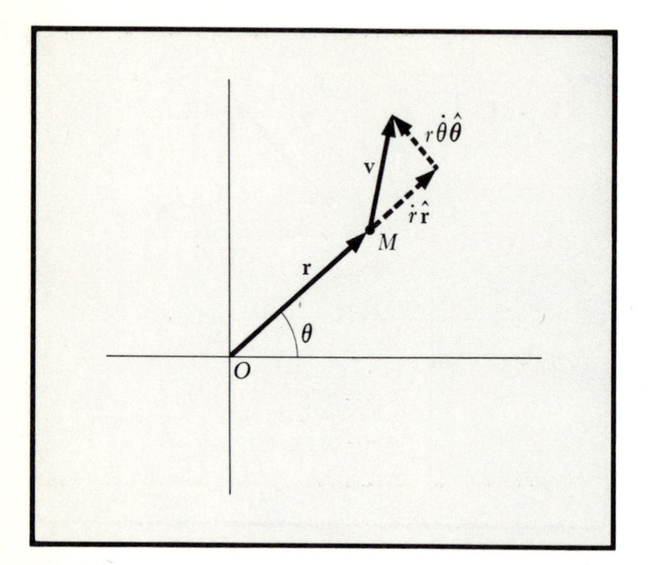

FIG. 9.12 Resolution of the velocity $\mathbf{v}$ of the particle into radial and angular components. The kinetic energy is $K = \frac{1}{2}Mv^2 = \frac{1}{2}M(\dot{r}^2 + r^2\dot{\theta}^2)$. The total energy is $E = K + U = \frac{1}{2}M\dot{r}^2 + \frac{1}{2}Mr^2\dot{\theta}^2 + U$.

$$w(\theta) = \frac{1}{r(\theta)}$$

$$\frac{dw}{d\theta} = -\frac{1}{r^2}\frac{dr}{d\theta}$$

$$\frac{d^2w}{d\theta^2} = -\frac{1}{r^2}\frac{d^2r}{d\theta^2} + \frac{2}{r^3}\left(\frac{dr}{d\theta}\right)^2$$

We then note that the combination of this result with Eq. (9.23) yields

$$\frac{d^2r}{dt^2} = -\frac{J^2}{M^2r^2}\frac{d^2w}{d\theta^2}$$

which gives, when we replace $1/r$ by w and use Eq. (9.22),

$$-\frac{J^2}{M^2}\frac{d^2w}{d\theta^2} - \frac{J^2}{M^2}w = \frac{C}{M} \qquad \text{or} \qquad \frac{d^2w}{d\theta^2} + w = -\frac{CM}{J^2}$$

This equation we have met in Chap. 7, Eq. (7.1), and the solution is

$$w = A\cos(\theta + \phi) - \frac{CM}{J^2}$$

(The cosine rather than the sine is customarily used in this case.)

Since the orientation of the orbit in the $r\theta$ plane is unimportant, we let $\phi = 0$ and get

$$\frac{1}{r} = -\frac{CM}{J^2} + A\cos\theta \tag{9.24}$$

It is convenient to use the equation of energy to determine the constant A since the total energy E is easily interpretable in terms of the types of orbit. The total energy is (see Fig. 9.12)

$$E = \frac{1}{2}Mv^2 + \frac{C}{r} = \frac{1}{2}M(\dot{r}^2 + r^2\dot{\theta}^2) + \frac{C}{r}$$

$$= \frac{1}{2}M\left(\frac{J^2}{M^2r^4}\right)\left[\left(\frac{dr}{d\theta}\right)^2 + r^2\right] + \frac{C}{r} \tag{9.25}$$

where we have employed Eqs. (9.21) and (9.23) to obtain the second line from the first. Note that the potential energy C/r is positive for a repulsive force (C positive), and it is negative for an attractive force (C negative). If we now use Eq. (9.24) and its derivative to substitute for r and $dr/d\theta$ in Eq. (9.25), we obtain an equation involving only A^2 and other constants. Solving for A in terms of the other quantities gives

$$A = \left(\frac{2ME}{J^2} + \frac{C^2M^2}{J^4}\right)^{\frac{1}{2}} = \frac{CM}{J^2}\left(1 + \frac{2EJ^2}{C^2M}\right)^{\frac{1}{2}}$$

so that our final result for the orbit is

$$\frac{1}{r} = -\frac{CM}{J^2}\left[1 - \left(1 + \frac{2EJ^2}{C^2M}\right)^{\frac{1}{2}}\cos\theta\right] \tag{9.26}$$

If the force is attractive C will be negative, as for example in the gravitational case where $C = -GMM_2$. Then Eq. (9.26) is

$$\frac{1}{r} = \frac{GM^2M_2}{J^2}\left[1 - \left(1 + \frac{2EJ^2}{G^2M^3M_2^{\,2}}\right)^{\frac{1}{2}}\cos\theta\right]$$

Now Eq. (9.26) is just what is called the polar form of the equation of a conic section (ellipse, circle, parabola, or hyperbola). You may recall from a text on analytic geometry or from Chap. 2, page 54, that the equation of a general conic section (section of a cone by a plane) may be written as

$$\frac{1}{r} = \frac{1}{se}(1 - e\cos\theta) \tag{9.27}$$

The constant e is known as the *eccentricity*. The constant s determines the scale of the figure. The four types of possible curves described by Eq. (9.27) are

Hyperbola	$e > 1$
Parabola	$e = 1$
Ellipse	$0 < e < 1$
Circle	$e = 0$

It is not difficult to see the main features of the orbit from the values of e. (Also see Fig. 9.13.) If $e = 0$, r is a constant. If $0 < e < 1$, r must remain finite and it varies from $se/(1 - e)$ to $se/(1 + e)$. However, if $e > 1$, there will be two values of $\cos\theta$ at which $1 - e\cos\theta$ goes to zero and r goes to infinity which is of the character of a hyperbola. The parabola with $e = 1$ has r going to infinity at only $\theta = 0$ but it does so both from positive and from negative values of θ. From Eqs. (9.26) and (9.27)

$$e = \left(1 + \frac{2EJ^2}{C^2M}\right)^{\frac{1}{2}} \tag{9.28}$$

From our considerations of the energy E and Eq. (9.25), we see that with a repulsive force $C > 0$, E must be positive, e will always be greater than 1, and the orbit always a hyperbola. On the other hand with an attractive force $C < 0$ (for the gravitational case $C = -GMM_2$ where for the solar system M_2 is the mass of the sun) E will be positive if the kinetic energy

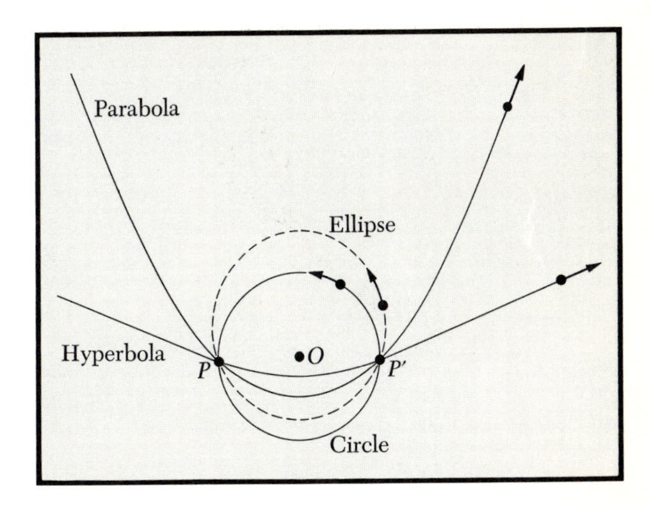

FIG. 9.13 Orbits of particles with the same mass M and angular momentum $\mathbf{J}$, but different energies E, about fixed center of force O. The orbits all cross at P and P'

Orbit	*Eccentricity*
Circle	$e = 0$ } $E < 0$
Ellipse	$e = \frac{1}{3}$ } $E < 0$
Parabola	$e = 1$, $E = 0$
Hyperbola	$e = 3$, $E > 0$

TABLE 9.1

θ	$\cos\theta$	$2(1 - \frac{1}{2}\cos\theta)$	r
0°	1.00	1.00	1.00
20°	0.94	1.06	0.94
40°	0.77	1.23	0.81
60°	0.50	1.50	0.67
80°	0.17	1.83	0.55
90°	0.00	2.00	0.50
100°	−1.17	2.17	0.46
120°	−0.50	2.50	0.40
140°	−0.77	2.77	0.36
160°	−0.94	2.94	0.34
180°	−1.00	3.00	0.33

is greater in magnitude than the potential energy, and the kinetic energy would still be positive at $r = \infty$; E will be negative if the reverse is true, and the particle can therefore never reach infinity. The parabola is the case $E = 0$ where the particle can just get to infinity. It is interesting that the question of whether an orbit is elliptic or hyperbolic in the case of an attractive force is determined only by the sign of E and not by the value of J. Of course the bigger the value of J for a given r, the larger the kinetic energy, and the bigger will be the value of E; but no matter how large the value of J, it is always possible to have orbits such that $E < 0$.

A crude but effective way to satisfy yourself that this equation can give a curve which at least looks like an ellipse is to calculate r for a range of values of θ. The results may be plotted conveniently on polar graph paper, which is readily available. Such graph paper is marked out with lines at constant radius and constant angle. Table 9.1 shows our rough calculations from Eq. (9.27) for the case $s = 1$ and $e = \frac{1}{2}$; you should plot these values of r vs θ on polar graph paper and confirm that the curve looks like an ellipse. Make similar calculations for $s = 1$, $e = 2$; this curve is a hyperbola.

A convenient relation to remember for the eccentricity of an ellipse is obtained by noting that the maximum and minimum values of r are obtained with $\theta = \pi$ and $\theta = 0$, so that

$$e = \frac{r_{\max} - r_{\min}}{r_{\max} + r_{\min}} \tag{9.29}$$

Some additional relations are shown in Fig. 9.14.

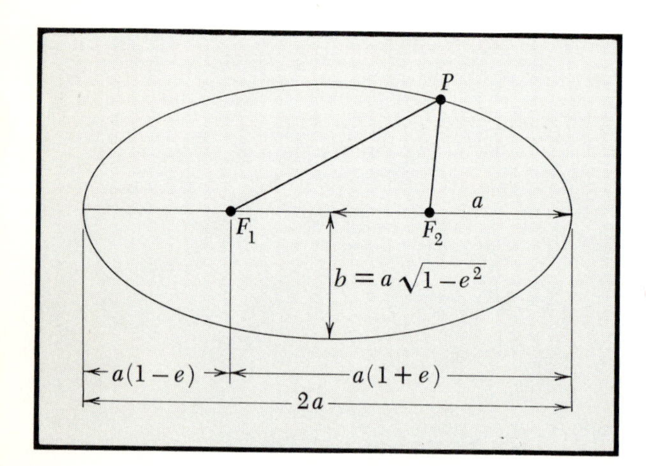

FIG. 9.14 Properties of the ellipse: For any point P the distance $F_1P + F_2P = \text{const} = 2a$. The equation of the ellipse is

$$r = \frac{a(1 - e^2)}{(1 - e\cos\theta)} \qquad 0 < e < 1$$

The semiminor axis is given by $b = a\sqrt{1 - e^2}$. The area of the ellipse is πab.

Circular Orbit We have already worked out the conditions for a circular orbit. Let us check that this leads to $e = 0$. Consider a circular orbit of a planet of mass M about a star of mass M_2. Equating the centripetal acceleration to the gravitational force gives

$$\frac{Mv^2}{r} = \frac{GMM_2}{r^2}$$

The angular momentum is

$$J = Mvr = M\sqrt{\frac{GM_2}{r}}\,r = (GM^2M_2r)^{\frac{1}{2}}$$

The total energy is

$$E = \frac{1}{2}Mv^2 - \frac{GMM_2}{r} = -\frac{1}{2}\frac{GMM_2}{r}$$

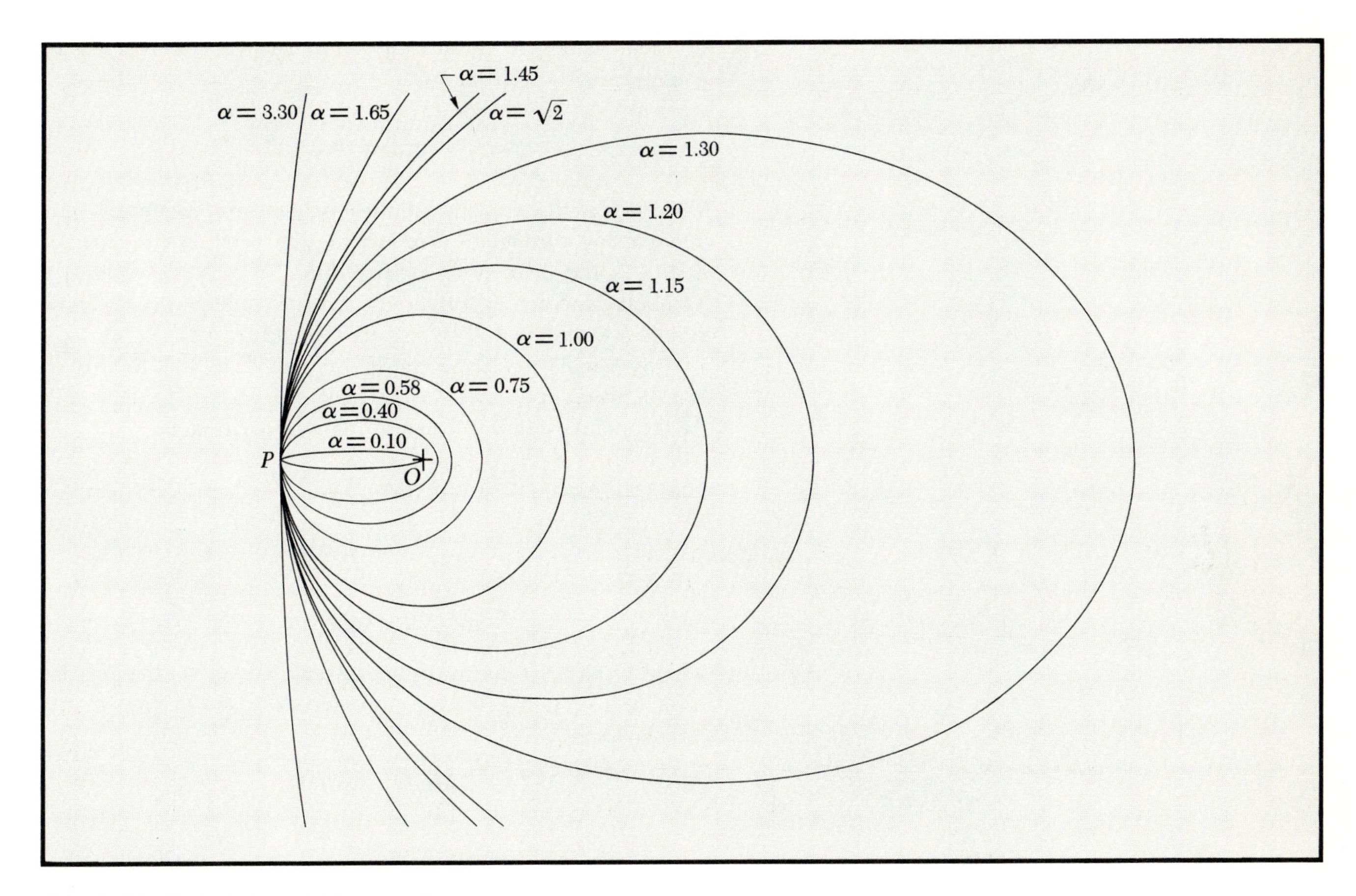

FIG. 9.15 Trajectories which pass through a common point P and normal to the line OP from the force center at O. With $\mathbf{v}_0$ as the velocity of the circular orbit, the parameter α is defined by $v_P(\alpha) = \alpha v_0$. It is shown in Eq. (9.31) that $E(\alpha) = (2 - \alpha^2)E_0$.

and so

$$e = \sqrt{1 + \left(-\frac{GMM_2}{r}\frac{M^2GM_2r}{G^2M^2M_2{}^2M}\right)} = 0$$

We find that some students are inclined to think all closed orbits should be circular. To gain a feeling for elliptical orbits, study Fig. 9.15. In that figure we see a family of trajectories of a particle attracted by an inverse-square force toward the origin at O, as denoted by the cross. The family has been chosen so that all the trajectories pass through a common point P, and at P the velocity is required to be perpendicular to the line between O and P. The different orbits are characterized by different values of the velocity at P. The general velocity v_P is conveniently written as

$$\frac{v_P}{v_0} \equiv \alpha$$

where v_0 is the velocity of the *circular* orbit centered at O and passing through P. For $\alpha = 1$, the orbit is circular; for $\alpha < \sqrt{2}$, the orbit is an ellipse; for $\alpha = \sqrt{2}$, the orbit is a parabola; and for $\alpha > \sqrt{2}$, the orbit is a hyperbola. [These results are illuminated by Eq. (9.31) below.]

By calculation of the energy, we can verify that the transition between open and closed trajectories occurs for $\alpha = \sqrt{2}$. At the point P the total energy may be written as

$$E = \tfrac{1}{2}Mv_P{}^2 - \frac{GMM_2}{r_0} = \tfrac{1}{2}M\alpha^2 v_0{}^2 - \frac{GMM_2}{r_0}$$
$$= \tfrac{1}{2}(\alpha^2 - 1)Mv_0{}^2 + \tfrac{1}{2}Mv_0{}^2 - \frac{GMM_2}{r_0}$$
$$= E_0 + \tfrac{1}{2}(\alpha^2 - 1)Mv_0{}^2 \qquad (9.30)$$

where E_0 and v_0 refer to the energy and velocity of the circular orbit and r_0 is the distance between P and O. Now in the circular orbit

$$\frac{Mv_0{}^2}{r_0} = \frac{GMM_2}{r_0{}^2}$$

where the term on the left is the mass times the centripetal acceleration and the term on the right is the gravitational force. Using this result, we can write the energy in the circular orbit as

$$E_0 = \tfrac{1}{2}Mv_0{}^2 - \frac{GMM_2}{r_0} = \tfrac{1}{2}Mv_0{}^2 - Mv_0{}^2 = -\tfrac{1}{2}Mv_0{}^2$$

and Eq. (9.30) may be written

$$E = E_0 - (\alpha^2 - 1)E_0 = (2 - \alpha^2)E_0 = (\alpha^2 - 2)|E_0| \qquad (9.31)$$

If $\alpha^2 > 2$, the total energy is positive and the orbit is open. If $\alpha^2 < 2$, the total energy is negative and the orbit is closed: the particle cannot escape to an infinite distance. If $\alpha^2 = 2$, the orbit is parabolic.

Kepler's Laws Kepler's determination that the orbits of the planets are ellipses about the sun was one of the great experimental discoveries in the history of science. Together with his formulation of the empirical laws of planetary motion it provided the original experimental evidence for Newton's laws of mechanics and for the theory of gravitational attraction. Kepler stated the three laws essentially as follows:

1 All planets move in elliptical paths, with the sun at one focus.

2 A line drawn from the sun to a planet sweeps out equal areas in equal times.

3 The squares of the periods of revolution of the several planets about the sun are proportional to the cubes of the semimajor axes of the ellipses. (This statement is more general than the original formulation by Kepler.)

In our entire discussion we neglect the effects of the other planets on the one under consideration.

We have demonstrated above that closed orbits are elliptical. Kepler's second law was demonstrated in Chap. 6, Eq. (6.36), where it was shown to be simply a statement of the conservation of angular momentum.

We are now going to derive Kepler's third law. If dS is the area swept out in time dt by the radius vector from the sun to the planet, then we found that

$$\frac{dS}{dt} = \frac{J}{2M} = \text{const} \tag{9.32}$$

where J is the angular momentum and M is the mass. On integrating Eq. (9.32) over one period T of the motion we have

$$S = \frac{JT}{2M} \qquad \text{or} \qquad T = \frac{2SM}{J} = \frac{2\pi abM}{J} \tag{9.33}$$

Here $S = \pi ab$ is the area of an ellipse with a as the semimajor axis and b as the semiminor axis.

Now $2a = r_{\text{max}} + r_{\text{min}}$ is an obvious property of an ellipse; with the use of Eq. (9.27) we have

$$2a = \frac{se}{1+e} + \frac{se}{1-e} = \frac{2se}{1-e^2}$$

With Eqs. (9.26) and (9.27) this becomes

$$2a = \frac{2}{1-e^2} \cdot \frac{J^2}{GM^2M_2} \tag{9.34}$$

On squaring Eq. (9.33) and using Eq. (9.34) for J^2, we have

$$T^2 = \frac{(2\pi abM)^2}{aGMM_2M(1-e^2)} = \frac{4\pi^2ab^2M}{GMM_2(1-e^2)} \tag{9.35}$$

But it is a property of the eccentricity e that (see Fig. 9.14)

$$b^2 = a^2(1 - e^2)$$

whence Eq. (9.35) reduces to

$$T^2 = \frac{4\pi^2 a^3}{GM_2} \tag{9.36}$$

You should verify Eq. (9.36) for a circular orbit.

Table 9.2 gives details of the orbits of the major planets. The inclination as given in the table is the angle between the plane of a planet's orbit and the plane of earth's orbit (the ecliptic). Notice that the orbit of earth is very nearly circular. An *astronomical unit* (AU) of length is defined as one-half the sum of the longest and shortest distances of earth from the sun.

$$1\ \text{AU} = 1.495 \times 10^{13}\ \text{cm}$$

This unit is not to be confused with the parsec. A *parsec* is the distance at which one astronomical unit subtends an angle of one second of arc.

$$1\ \text{parsec} = 3.084 \times 10^{18}\ \text{cm}$$

The distance to the nearest star from the sun is 1.31 parsecs.

Let us test Kepler's third law for the orbit of Uranus compared with that of earth. The cube of the ratio of the lengths of the semimajor axis is

$$\left(\frac{19.22}{1}\right)^3 \approx 71.0 \times 10^2$$

TABLE 9.2

Planet	*Semimajor axis, AU*	*Period, s*	*Eccen-tricity*	*Incli-nation*	*Mass, relative to sun's mass, g*
Mercury	0.387	7.60×10^6	0.2056	7°00	1.671×10^{-7}
Venus	0.723	1.94×10^7	0.0068	3°24′	2.448×10^{-6}
Earth	1.000	3.16×10^7	0.0167	. . .	3.003×10^{-6}
Mars	1.523	5.94×10^7	0.0934	1°51′	3.227×10^{-7}
Jupiter	5.202	3.74×10^8	0.0481*	1°18′	9.548×10^{-4}
Saturn	9.554	9.30×10^8	0.0530*	2°29′	2.858×10^{-4}
Uranus	19.218	2.66×10^9	0.0482*	0°46′	4.361×10^{-5}
Neptune	30.109	5.20×10^9	0.0054*	1°46′	5.192×10^{-5}
Pluto	39.60	7.82×10^9	0.251*	17° 8′	5.519×10^{-7}

*Eccentricity varies with the time because of perturbations of other planets. These are values set for 1972.

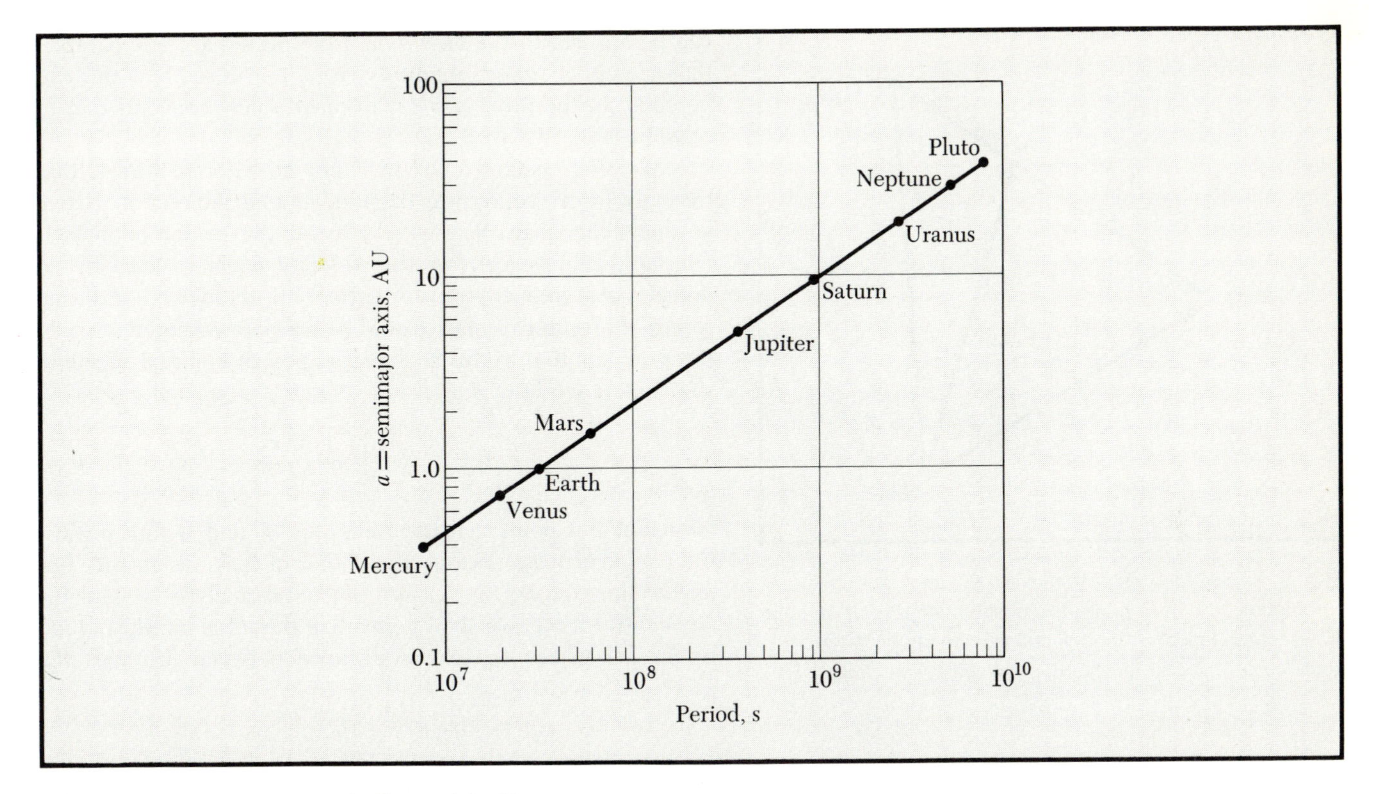

FIG. 9.16 From the slope of this straight line we can easily see that the period T varies as $a^{3/2}$.

The square of ratio of the periods is

$$(84.2)^2 \approx 70.9 \times 10^2$$

which is quite close agreement. (A 10-in. sliderule was used for the calculation; the student should perform the same calculation for the orbit of Mercury compared with that of earth.) In Fig. 9.16 we have used log-log paper to plot the data for the planets. On log-log paper a power law will appear as a straight line; the slope of the line will give the exponent of the power law. (Prove this.)

Newton also tested Kepler's third law against the observed periods of revolution of the four largest moons of Jupiter and found very good agreement.

Two-body Problem: Reduced Mass We have solved the problem of one mass moving in the field of an infinitely large or stationary mass. We have also indicated that this solution

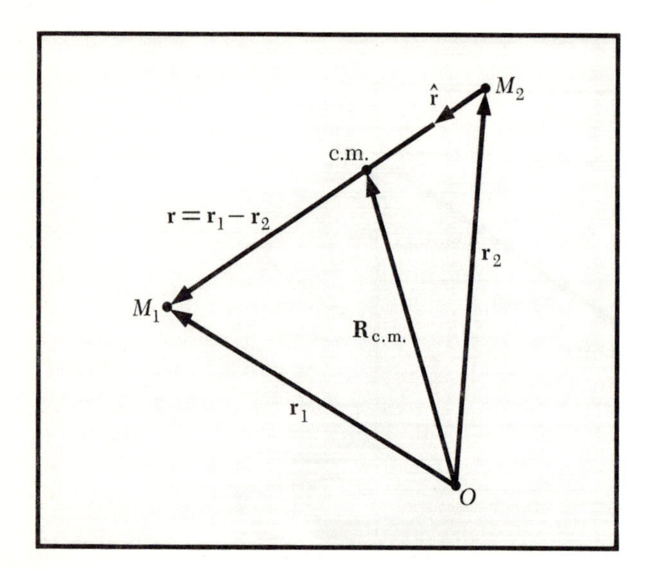

FIG. 9.17 M_1 and M_2 interact via central force collinear with vector $\mathbf{r}$; $\mathbf{r}_1$ and $\mathbf{r}_2$ are the position vectors of M_1 and M_2 referred to some inertial frame with origin O. In the absence of external forces, $\mathbf{R}_{\text{c.m.}} = \text{const}$ (see page 176).

can be applied to the case where the masses are comparable; that is, neither mass is infinitely large. Let us see how this can be done. In the process we shall meet a new concept, that of *reduced mass*.

We shall assume that there are no external forces, that the only forces acting are those of mutual interaction. Then, as shown in Chap. 6, the velocity of the center of mass is constant, and we can set this velocity equal to zero by an appropriate galilean transformation. (If there were external forces, the center of mass would be accelerated and we would refer our solution to this accelerated point.) Figure 9.17 shows the vectors we use:

$$\mathbf{R}_{\text{c.m.}} = \frac{M_1\mathbf{r}_1 + M_2\mathbf{r}_2}{M_1 + M_2}$$

Notice at this point that the forces on M_1 and M_2 are toward the center of mass because the center of mass necessarily lies on the line joining the masses. Since our analysis is valid for any central force, we shall generalize the force on M_1 and M_2 to $F(r_{12})\hat{\mathbf{r}}$, where r_{12} is the distance between M_1 and M_2:

$$M_1\frac{d^2\mathbf{r}_1}{dt^2} = F(r_{12})\hat{\mathbf{r}} \qquad M_2\frac{d^2\mathbf{r}_2}{dt^2} = -F(r_{12})\hat{\mathbf{r}}$$

Instead of adding these two equations, which leads to the constancy of the total momentum, we subtract, after dividing by the mass:

$$\frac{d^2\mathbf{r}_1}{dt^2} - \frac{d^2\mathbf{r}_2}{dt^2} = \frac{d^2(\mathbf{r}_1 - \mathbf{r}_2)}{dt^2} = \left(\frac{1}{M_1} + \frac{1}{M_2}\right)F(r_{12})\hat{\mathbf{r}}$$

From Fig. 9.17 we see that $\mathbf{r}_1 - \mathbf{r}_2 = \mathbf{r}$, the vector position of M_1 relative to M_2, and the unit vector $\hat{\mathbf{r}}$ is along $\mathbf{r}_1 - \mathbf{r}_2$. If we now introduce the *reduced mass* μ,

$$\frac{1}{\mu} = \frac{1}{M_1} + \frac{1}{M_2}$$

$$\boxed{\mu = \frac{M_1M_2}{M_1 + M_2}} \tag{9.37}$$

we get $\mu\, d^2\mathbf{r}/dt^2 = F(r_{12})\hat{\mathbf{r}}$. In the case of gravitational force,

$$\mu\frac{d^2\mathbf{r}}{dt^2} = -\frac{GM_1M_2}{r^2}\hat{\mathbf{r}} \tag{9.38}$$

which we have already solved.

We use Eqs. (9.37) and (9.38) in the following way: Recall that $\mathbf{r}$ is the vector from M_2 to M_1. With Eq. (9.38) we may solve for the motion of M_1 relative to M_2, exactly *as though* M_2 were the fixed origin of an inertial frame, except that we must use μ instead of M_1 as the mass. We have thus reduced the two-body problem to a one-body problem involving the motion of a body of mass μ. But note that the force in Eq. (9.38) is *not* $-G\mu M_2/r^2$! To find the orbit for the two-body problem we need only solve this one-body problem. The reduction of the two-body problem to a one-body problem can be accomplished in the same way for any central force; the reduced mass will always appear.

How are the constants of our one-body solution J and E defined now? We use Fig. 9.18, where the origin is at the center of mass, with the special warning that $\mathbf{r}_1$ and $\mathbf{r}_2$ are different from those of Fig. 9.17. As before $\mathbf{r} = \mathbf{r}_1 - \mathbf{r}_2$, but

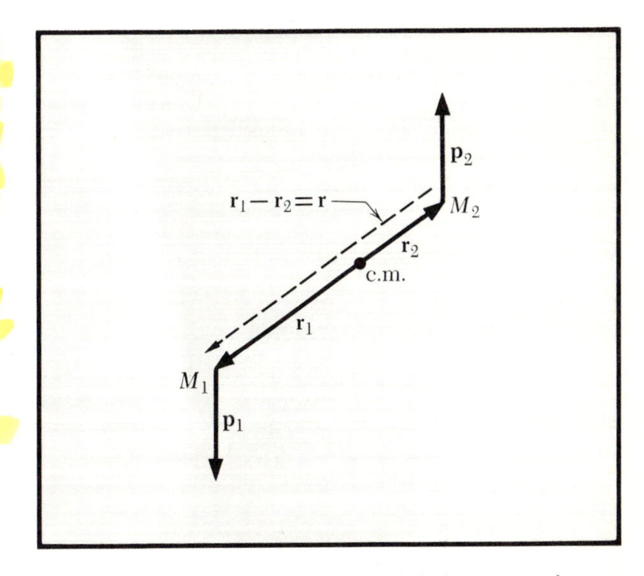

FIG. 9.18 In an inertial frame in which the center of mass is at rest and is at the origin $M_1\mathbf{r}_1 = -M_2\mathbf{r}_2$. The angular momentum of M_1 about the center of mass plus the angular momentum of M_2 about the center of mass is a constant, the total angular momentum $\mathbf{J}$. Note the difference of $\mathbf{r}_1$ and $\mathbf{r}_2$ from those in Fig. 9.17.

$$M_1\mathbf{r}_1 + M_2\mathbf{r}_2 = 0$$

and so

$$M_1\mathbf{r}_1 = -M_2\mathbf{r}_2 \qquad \text{and} \qquad M_1\dot{\mathbf{r}}_1 = -M_2\dot{\mathbf{r}}_2$$

because of the definition of the center of mass. Now $\mathbf{J}$ about the center of mass or origin is, using the last relationship above,

$$\begin{aligned}\mathbf{J} &= \mathbf{r}_1 \times M_1\dot{\mathbf{r}}_1 + \mathbf{r}_2 \times M_2\dot{\mathbf{r}}_2 = (\mathbf{r}_1 - \mathbf{r}_2) \times M_1\dot{\mathbf{r}}_1 \\ &= (\mathbf{r}_1 - \mathbf{r}_2) \times \frac{M_1M_2}{M_1 + M_2}(\dot{\mathbf{r}}_1 - \dot{\mathbf{r}}_2) = \mathbf{r} \times \mu\dot{\mathbf{r}}\end{aligned}$$

where we have utilized the following relationship:

$$\begin{aligned}M_1\dot{\mathbf{r}}_1 &= \frac{M_1(M_1 + M_2)\dot{\mathbf{r}}_1}{M_1 + M_2} \\ &= \frac{M_1}{M_1 + M_2}(-M_2\dot{\mathbf{r}}_2 + M_2\dot{\mathbf{r}}_1) \\ &= \frac{M_1M_2}{M_1 + M_2}(\dot{\mathbf{r}}_1 - \dot{\mathbf{r}}_2)\end{aligned} \tag{9.39}$$

Therefore the angular momentum constant $\mathbf{J}$ is calculated assuming the reduced mass moving around the one mass M_2 as if fixed.

For the energy E we again use the center of mass as fixed:

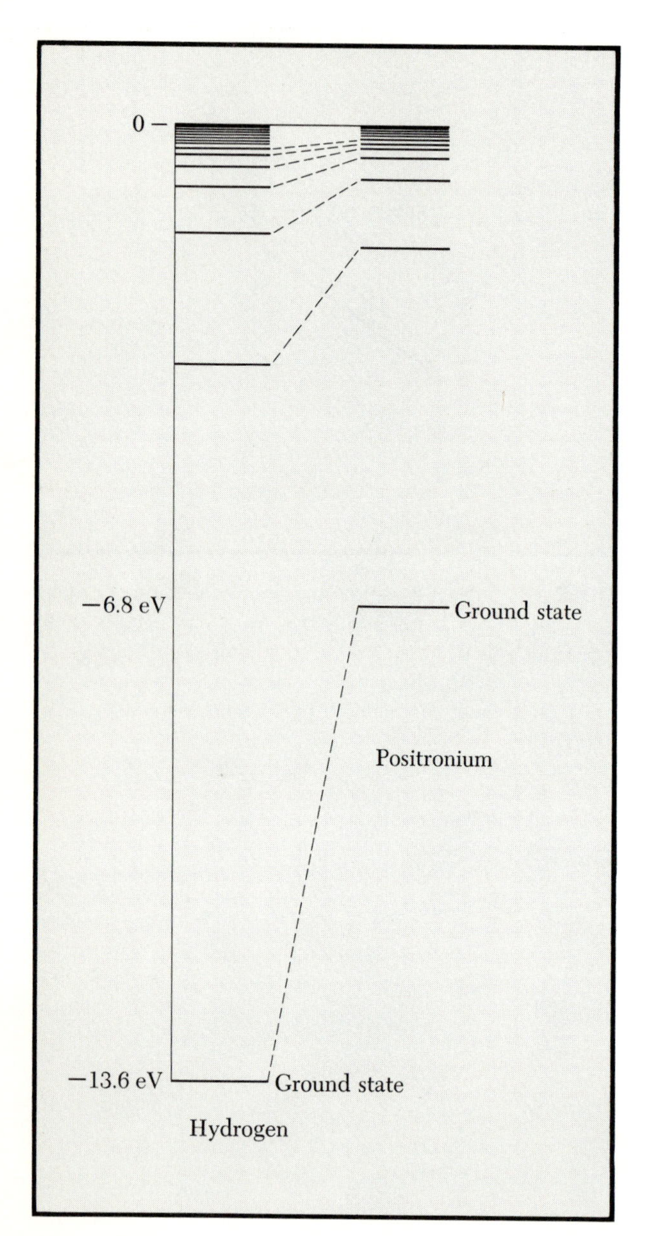

FIG. 9.19 Energy levels of the hydrogen atom and of the positronium atom. The reduced mass for hydrogen is $\mu = m_e/(1 + \frac{1}{1836}) \approx m_e$. The reduced mass for positronium is $\mu = \frac{1}{2}m_e$. This causes a difference of a factor of 2 in the energies.

$$
\begin{aligned}
E &= \tfrac{1}{2}M_1\dot{\mathbf{r}}_1 \cdot \dot{\mathbf{r}}_1 + \tfrac{1}{2}M_2\dot{\mathbf{r}}_2 \cdot \dot{\mathbf{r}}_2 - \frac{GM_1M_2}{r} \\
&= \tfrac{1}{2}\left(M_1 + M_2\frac{M_1^2}{M_2^2}\right)\dot{\mathbf{r}}_1 \cdot \dot{\mathbf{r}}_1 - \frac{GM_1M_2}{r} \\
&= \tfrac{1}{2}M_1\left(\frac{M_2 + M_1}{M_2}\right)\left[\frac{M_2^2(\dot{\mathbf{r}}_1 - \dot{\mathbf{r}}_2)\cdot(\dot{\mathbf{r}}_1 - \dot{\mathbf{r}}_2)}{(M_1 + M_2)^2}\right] - \frac{GM_1M_2}{r} \\
&= \tfrac{1}{2}\mu(\dot{\mathbf{r}}_1 - \dot{\mathbf{r}}_2)\cdot(\dot{\mathbf{r}}_1 - \dot{\mathbf{r}}_2) - \frac{GM_1M_2}{r} \\
&= \tfrac{1}{2}\mu\dot{\mathbf{r}}^2 - \frac{GM_1M_2}{r}
\end{aligned} \tag{9.40}
$$

where we have used Eq. (9.39) to obtain Eq. (9.40) from the first equation for E; so we again consider that the reduced mass moves around M_2 as fixed (see Prob. 11).

The reduced mass must have a value less than either M_1 or M_2. Note that for $M_1 = M_2 = M$,

$$\frac{1}{\mu} = \frac{2}{M} \qquad \mu = \tfrac{1}{2}M \tag{9.41}$$

If $M_1 \ll M_2$, we have from Eq. (9.37)

$$\mu = \frac{M_1M_2}{M_1 + M_2} = M_1\frac{1}{(M_1/M_2) + 1} \approx M_1\left(1 - \frac{M_1}{M_2}\right)$$

We have expanded the fraction by the binomial theorem and retained only the term of lowest order in M_1/M_2. If M_1 is m (the mass of the electron), and M_2 is M_p (the mass of the proton), the reduced mass is

$$\mu \approx m(1 - \tfrac{1}{1836})$$

The lighter of the two masses tends to dominate the value of the reduced mass. The departure of μ from m is easily detectable in the spectrum of atomic hydrogen.

Positronium is a hydrogenlike atom made up of a positron and an electron, with no proton. A positron is a particle which has a mass equal to the electron mass but has positive charge e. The result [Eq. (9.41)] suggests correctly that there may be a similarity between the line spectra of atomic hydrogen and of positronium, provided we make allowance for the fact that the reduced mass of positronium is about one-half that of atomic hydrogen. The coulomb interaction between an electron and a positron has the same form as between an electron and a proton. The energy levels for hydrogen and positronium are shown in Fig. 9.19.

EXAMPLE

Vibration of a Diatomic Molecule Two atoms bound together as a stable molecule will have a potential energy that is a quadratic function of the difference $\mathbf{r} - \mathbf{r}_0$ of their separation $\mathbf{r}$ from the equilibrium separation $\mathbf{r}_0$:

$$U(r) = \tfrac{1}{2}C(r - r_0)^2 \tag{9.42}$$

provided $(r - r_0)/r_0 \ll 1$ (see Fig. 9.20). The force is along the line connecting the atoms and is given by (if the molecule is *not* rotating)

$$F = -\frac{dU}{d(r - r_0)} = -C(r - r_0) \tag{9.43}$$

This describes a harmonic oscillator of force constant C. The masses of the atoms are M_1 and M_2. What is the frequency of vibration?

In a free vibration both atoms will be in motion while the center of mass remains at rest. The equation of motion is naturally given by Eq. (9.38) with Eq. (9.43) substituted for the gravitational force:

$$\mu\frac{d^2\mathbf{r}}{dt^2} = -C(r - r_0)\hat{\mathbf{r}} \tag{9.44}$$

If the molecule is not rotating, the direction of $\hat{\mathbf{r}}$ is fixed and thus

$$\frac{d^2\mathbf{r}}{dt^2} = \frac{d^2r}{dt^2}\hat{\mathbf{r}}$$

(The derivative of $\mathbf{r}$ is not this simple if the direction of $\hat{\mathbf{r}}$ is changing.) Therefore we may rewrite Eq. (9.44) as a scalar equation:

$$\mu\frac{d^2r}{dt^2} = -C(r - r_0)$$

which is the equation of motion of a simple harmonic oscillator of angular frequency

$$\omega_0 = \left(\frac{C}{\mu}\right)^{\frac{1}{2}} \tag{9.45}$$

It is known from spectroscopic measurements that the fundamental vibrational frequencies of the molecules HF and HCl are

$$\omega_0(\text{HF}) = 7.55 \times 10^{14}\ \text{rad/s}$$
$$\omega_0(\text{HCl}) = 5.47 \times 10^{14}\ \text{rad/s}$$

Let us use these data to compare the force constants C_{HF} and C_{HCl}. The reduced mass of HF is, in atomic mass units,

$$\frac{1}{\mu_{\text{HF}}} \approx \frac{1}{1} + \frac{1}{19} = \frac{20}{19} \qquad \mu_{\text{HF}} \approx 0.950$$

$$\frac{1}{\mu_{\text{HCl}}} \approx \frac{1}{1} + \frac{1}{35} = \frac{36}{35} \qquad \mu_{\text{HCl}} \approx 0.973$$

(Here we have used the atomic mass of the most abundant isotope of chlorine, Cl^{35}.) Notice that the reduced masses are quite close

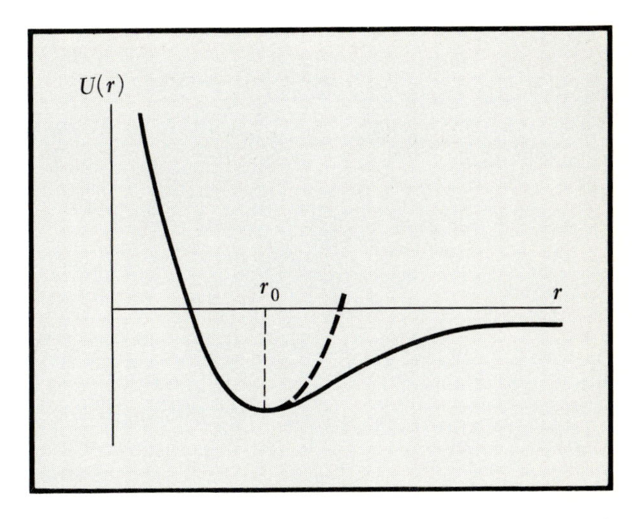

FIG. 9.20 Graph of the potential energy as a function of distance between two atoms combined in a molecule. The position of equilibrium is at r_0. The dashed curve shows the parabola corresponding to a quadratic potential energy function given in Eq. (9.42).

to each other in value. This is because the hydrogen, being lightest, does most of the oscillating.

Now from Eq. (9.45)

$$\frac{C_{HF}}{C_{HCl}} = \frac{(\mu\omega_0^2)_{HF}}{(\mu\omega_0^2)_{HCl}} \approx \frac{54.0 \times 10^{28}}{29.0 \times 10^{28}} \approx 1.86$$

while for an individual force constant

$$C_{HF} \approx (54 \times 10^{28})(1.66 \times 10^{-24}) \approx 9.0 \times 10^5 \text{ dyn/cm}$$

Here we have inserted the factor which converts the mass from atomic mass units to grams.

Is this value of C reasonable? Suppose we stretch the molecule (which is about 1 Å or 1×10^{-8} cm in length) by 0.5 Å. The work needed to do this would probably be nearly enough to break up the molecule into separate atoms of H and F. By Eq. (9.42) the work to stretch 0.5 Å should be of the order of magnitude

$$\tfrac{1}{2}C(r - r_0)^2 \approx \tfrac{1}{2}(9 \times 10^5)(0.5 \times 10^{-8})^2 \approx 1 \times 10^{-11} \text{ erg}$$

or $\approx (1 \times 10^{-11})/(1.6 \times 10^{-12}) \approx 6$ eV. This is not unreasonable for an energy of decomposition into separate atoms. In making this estimate we have used the form [Eq. (9.42)] beyond the region in which it is valid. The actual intermolecular potential energy will be more of the form shown in Fig. 9.20.

PROBLEMS

1. *Gravitational attraction of infinite line.* Show that $2G\rho M_1/R$ is the gravitational force on a mass M_1 located at a distance R from an infinite line of mass ρ per unit length. (Be careful about the direction of the force due to a line element.)

2. *Gravitational attraction of finite line.* You are at a point x on the perpendicular bisector of a line of length $2L$. The mass of the line is M; the origin of the coordinate system is on the line.

(*a*) Find an expression for the potential energy, referred to $U = 0$ at $x = \infty$ of a point mass m.
Ans. $-(GMm/L) \log\{[L + (x^2 + L^2)^{\frac{1}{2}}]/x\}$.

(*b*) Find an expression for the gravitational force exerted by the line on a point mass m at x. In what direction is the force?

(*c*) Show that your result for part (*a*) reduces to $U \approx -GMm/x$ when $x \gg L$.

Consider a thin wire of length 2 m and linear density 2 g/cm.

(*d*) What is the value (in dynes) of the gravitational force exerted by the wire on a point mass $m = 0.5$ g located on the long axis of the wire at a distance of 3 m from the center of the wire? *Ans.* 1.7×10^{-10} dyn.

(*e*) What is the potential energy (in ergs) of the point mass in the force field of the wire, at the position given in part (*d*)? *Ans.* -4.6×10^{-7} erg.

3 *Gravitational potential energy of array of stars.* Find the mutual gravitational potential energy (in ergs) of a system of eight stars, each of mass equal to that of the sun, located at the corners of a cube whose edge is 1 parsec. (Omit the self-energy of each star.) *Ans.* 2×10^{42} erg.

4. *Hole through earth.* Consider a hole drilled through the center of the earth. Show that if the rotation of the earth and friction are neglected the motion of a particle will be simple harmonic. Find the period. Comment on the relation between this and the period of a satellite revolving close to the surface of the earth. (*Note:* The rotation of the earth would not prevent the motion from being simple harmonic, but the period would be slightly changed. Can you show this to be true? How is the period affected?)

5. *Motion in a galaxy.* Consider a uniform spherical distribution of stars in a galaxy of total mass M and radius R_0. A star of mass M_s at some distance $r < R_0$ from the center will move under the action of a central force whose magnitude depends on the mass included within a sphere of radius r.

(*a*) What is the force at r? *Ans.* $F = GM_sMr/R_0^3$.

(*b*) What is the circumferential velocity of the star if it moves about the center in a circular orbit?

Ans. $v = (GMr^2/R_0^3)^{\frac{1}{2}}$.

6. *Meteor orbit.* A meteor has velocity of 7.0×10^6 cm/s at perihelion when its distance from the sun is 5.0×10^{12} cm. Find its distance and velocity at aphelion and the eccentricity of its orbit by using Eqs. (9.21), (9.25), and (9.28).

Ans: Distance $= 5.5 \times 10^{13}$ cm; velocity $= 6.3 \times 10^5$ cm/s; $\epsilon = 0.83$.

7. *Earth satellite.* Imagine that the moon has no mass so that it does not influence the orbit of a satellite. What velocity is necessary for a satellite perpendicular to a radius vector from the center of the earth and 200 mi above the surface of the earth in order that at the other end of its elliptic path it should be at the moon, (240,000 mi from the center of the earth)?

8. *Escape velocity.* Neglecting friction, find the velocity that must be given to a satellite at the surface of the earth in order that it can just reach the point between the earth and moon where the gravitational force is zero. If it just passed through this point, with what velocity would it hit the moon?

Ans. v(escape) $= 1.1 \times 10^6$ cm/s.

9. *Change in sun's mass.* What would happen to the earth's orbit, assumed circular, if the mass of the sun were suddenly reduced to half its value?

10. *Helium orbit.* Assume that the energy level at -13.6 eV in hydrogen results from an electron being in a circular orbit.

(*a*) Calculate the angular momentum.

Ans. 1.1×10^{-27} erg-s.

(*b*) What would be the radius and energy of an electron in circular orbit about a helium nucleus (charge $+2e$) moving with the same angular momentum?

11. *Orbital motion of binary stars.* J. S. Plaskett's star is one of the most massive stars known at present. It is a double, or binary, star[1]; that is, it consists of two stars bound together by gravity. From spectroscopic studies it is known that

(*a*) The period of revolution about their center of mass is 14.4 days (1.2×10^6 s).

(*b*) The velocity of each component is about 220 km/s. Because both components have nearly equal (but opposite) velocities, we may infer that they are nearly equidistant from the center of mass, and hence that their masses are nearly equal.

(*c*) The orbit is nearly circular.

From these data calculate the reduced mass and the separation of the two components.

Ans. $\mu \approx 0.6 \times 10^{35}$ g; separation $\approx 0.8 \times 10^{13}$ cm.

[1]There is a good discussion of binary stars in O. Struve, B. Lynds, and H. Pillans, "Elementary Astronomy," chap. 29, Oxford University Press, New York, 1959. At least half of the 50 stars nearest to the sun are binary or multiple stars.

12. *Form of the sea on a uniform earth.* A uniform spherical earth is covered by water. The surface of the sea assumes the form of an oblate spheroid (a flattened sphere) when the earth spins with angular velocity ω. Find an approximate expression for the difference in the depth of the sea at a pole and at the equator, assuming that the surface of the sea is a surface of constant potential energy. (Why is this a plausible assumption?) Neglect the gravitational attraction of the sea upon itself. *Hint:* We need a potential energy expression which will represent the effect of the earth's rotation. This centrifugal potential has been referred to in Fig. 6.21*b* and in Chap. 6, Prob. 13:

$$U_{\text{centrif}} = -\tfrac{1}{2}\omega^2 r^2$$

where U refers to a unit mass. Since $F = -\partial U/\partial r$ we see

$$F_{\text{centrif}} = \omega^2 r$$

which is the familiar "centrifugal force." Remember we wish to use this expression only for r slightly greater than the radius of the earth R_E.

We wish to set the gravitational potential at the N pole (or S pole) where $U_{\text{centrif}} = 0$ (since the r in this expression is the distance from the axis of rotation) equal to the gravitational potential plus U_{centrif} at the equator. The surface at the pole will be $R_E + D_{\text{pole}}$ from the center of the earth, at the equator $R_E + D_{\text{eq}}$, where D_{pole} and D_{eq} are both much smaller than R_E. *Ans.* $(D_{\text{eq}} - D_{\text{pole}})/R_E \approx \omega^2 R_E/2g \approx \frac{1}{580}$. This is very close to the value of $\frac{1}{298}$ for the actual earth.

13. *Direct calculation of force.* Use direct calculation to show that Eq. (9.14) is true; that is, set up the differential element of force and integrate. (*Hint:* Use the symmetry of the problem to show that the force must lie along the line joining M_1 and the center of the shell so that the integration involves only this component of the force.)

14. *Satellite around moon.* Find the period of a satellite moving around the moon, assuming values on the inside cover of this book.

ADVANCED TOPIC

Substitute for Integrating $1/r$ Equation The labor of solving the differential equation for r may be avoided by the use of another constant of the motion

$$\boldsymbol{\epsilon} = \frac{-1}{MC}\mathbf{J} \times \mathbf{p} + \frac{\mathbf{r}}{r} \tag{9.46}$$

where $\mathbf{p} = M\mathbf{v}$ = the momentum, and C is the constant in the force law $F = C/r^2$. The reader can check that this is a dimensionless quantity. To show that $\boldsymbol{\epsilon}$ is a constant we need to prove

$$\frac{d\boldsymbol{\epsilon}}{dt} = 0$$

We perform the differentiation

$$\frac{d\boldsymbol{\epsilon}}{dt} = \frac{-1}{MC}\left[\left(\frac{d\mathbf{J}}{dt} \times p\right) + \mathbf{J} \times \frac{d\mathbf{p}}{dt}\right] + \frac{\mathbf{v}}{r} - \frac{1}{r^2}\mathbf{r}\frac{dr}{dt}$$

Now

$$\frac{d\mathbf{J}}{dt} = 0 \qquad \frac{d\mathbf{p}}{dt} = \frac{C}{r^3}\mathbf{r} \qquad \mathbf{J} = M\mathbf{r} \times \mathbf{v}$$

and

$$\mathbf{r} \cdot \mathbf{v} = \mathbf{r} \cdot \left(\frac{dr}{dt}\frac{\mathbf{r}}{r} + r\frac{d\theta}{dt}\hat{\boldsymbol{\theta}}\right) = r\frac{dr}{dt}$$

Therefore

$$\begin{aligned}\frac{d\boldsymbol{\epsilon}}{dt} &= -\frac{1}{MC}\mathbf{J} \times \frac{C\mathbf{r}}{r^3} + \frac{\mathbf{v}}{r} - \frac{1}{r^3}(\mathbf{v} \cdot \mathbf{r})\mathbf{r} \\ &= -\frac{(\mathbf{r} \times \mathbf{v}) \times \mathbf{r}}{r^3} + \frac{\mathbf{v}}{r} - \frac{(\mathbf{v} \cdot \mathbf{r})\mathbf{r}}{r^3} \\ &= -\frac{r^2\mathbf{v}}{r^3} + \frac{(\mathbf{r} \cdot \mathbf{v})\mathbf{r}}{r^3} + \frac{\mathbf{v}}{r} - \frac{(\mathbf{r} \cdot \mathbf{v})\mathbf{r}}{r^3} = 0\end{aligned}$$

in which we have made use of the expression for the triple vector product, Eq. (2.55).

From the definition of $\boldsymbol{\epsilon}$, Eq. (9.46),

$$\boldsymbol{\epsilon} \cdot \mathbf{r} = \epsilon r \cos\theta = \frac{-1}{MC}(\mathbf{J} \times \mathbf{p} \cdot \mathbf{r}) + r$$

or

$$r(1 - \epsilon\cos\theta) = -\frac{\mathbf{J} \cdot \mathbf{J}}{MC} = -\frac{J^2}{MC}$$

where we have used the fact that

$$\mathbf{J} \times \mathbf{p} \cdot \mathbf{r} = \mathbf{J} \cdot \mathbf{p} \times \mathbf{r} = -\mathbf{J} \cdot \mathbf{r} \times \mathbf{p} = -\mathbf{J} \cdot \mathbf{J}$$

or

$$\frac{1}{r} = -\frac{MC}{J^2}(1 - \epsilon\cos\theta)$$

which is just Eq. (9.24).

We see that the vector $\boldsymbol{\epsilon}$ is of magnitude equal to the eccentricity and of direction $\theta = 0$, the major axis of the ellipse or axis of the hyperbola; ϵ can of course now be found as previously from the energy equation.

FURTHER READING

PSSC, "Physics," chap. 21, D. C. Heath and Company, Boston, 1960.

HPP, "Project Physics Course," chaps. 5–8, Holt, Rinehart and Winston, Inc., New York, 1970. This gives a good elementary account of the historical aspects of the understanding of planetary motion.

P. van de Kamp, "Elements of Astromechanics," W. H. Freeman and Company, San Francisco, 1964. Paperback; selected elementary topics in celestial mechanics.

O. Struve, B. Lynds, and H. Pillans, "Elementary Astronomy," Oxford University Press, New York, 1959. Emphasizes the main ideas of physics in relation to the universe; excellent.

T. S. Kuhn, "The Copernican Revolution," Vintage Books (paperback), Random House, Inc., New York, 1962.

American Association of Physics Teachers, selected reprints, "Kinematics and Dynamics of Satellite Orbits," American Institute of Physics, New York, 1963.

Arthur Koestler, "The Watershed: A Biography of Johannes Kepler," Anchor Books, Doubleday & Company, Inc., Garden City, N. Y., 1960. A fascinating account of Kepler's intellectual and spiritual pilgrimage in reaching a correct description of planetary orbits.

CONTENTS

The Speed of Light

c AS A FUNDAMENTAL CONSTANT OF NATURE

The speed of light[1] in vacuum c is one of the fundamental constants of physics. Some of its characteristics are

1. It is the speed at which all electromagnetic radiation travels in free space, independent of the frequency of the radiation.
2. No signal can be transmitted by any means whatsoever, in free space or in a material medium, at a speed faster than the speed of light c.
3. The speed of light in free space is independent of the reference frame from which it is observed. If the speed of a light signal is observed to be $c = 2.99793 \times 10^{10}$ cm/s in one galilean frame, it will be observed to be c and not $c + V$ (or $c - V$) in a second galilean frame moving parallel to the signal with a speed V with respect to the first frame.
4. Maxwell's equations in electromagnetic theory and the Lorentz force equation involve the speed of light. This is particularly apparent when they are written in gaussian units.
5. The dimensionless constant (which is called the reciprocal of the fine-structure constant)

$$\frac{\hbar c}{e^2} \approx 137.04$$

   involves the speed of light. Here $2\pi\hbar$ is Planck's constant and e is the charge on the proton. This constant plays an important role in atomic physics and will be discussed in Volume 4. We do not have a theory that predicts the value of this constant.

This chapter is concerned chiefly with experiments and experimental results. We discuss the measurement of the speed of light and experimental evidence for the invariance of the speed of light with respect to the velocity of any inertial frame. We leave for Volume 3 questions about the electromagnetic nature of light and the propagation of light in refractive and dispersive media such as solids and liquids. (A refractive medium is one in which the refractive index, the ratio of the speed

[1] Note that the phrase *speed of light* should always be understood to mean the speed of light in free space (c), unless it is explicitly stated otherwise. Thus the speed of light in a material medium is less than c and may even be less than the speed of a charged particle in the same medium (Cerenkov effect).

of light in vacuum to that in the medium, is not exactly unity. A dispersive medium is one in which the refractive index is a function of the frequency.)

MEASUREMENTS OF c

Many methods have been employed to determine the speed of light.[1] We list and sketch several of the methods here.

Transit Time of Light across the Orbit of the Earth For some centuries before there was experimental proof, it was believed that the speed of light must be finite. The first experimental evidence of the finite speed of light was due to Roemer in 1676. He observed that the motion of Io, the innermost moon of Jupiter, did not follow an entirely regular timetable. There was a slight variation in the periods of the eclipses of Io by Jupiter. When at one time of year (see Fig. 10.1) he predicted the time of eclipse 6 months later (see Fig. 10.2), he was about 22 min in error. He postulated that this was the transit time of light across the orbit of the earth. His best estimate of the average diameter D of the earth's orbit about the sun was 2.83×10^{13} cm, and so he calculated for c

[1] An excellent review in English of measurements of the speed of light is given by E. Bergstrand in "Handbuch der Physik," S. Flügge (ed.), vol. 24, pp. 1–43 (Springer-Verlag OHG, Berlin, 1956). The values of c we quote are those listed by Bergstrand. See also J. F. Mulligan and D. F. McDonald, *Am. J. Phys.*, **25**:180 (1957).

FIG. 10.1 Eclipse of Jupiter's moon M occurs when M disappears behind J as viewed from the earth. The actual time of observation on the earth is L/c later because of the finite speed of light. The period of M is about 42 h.

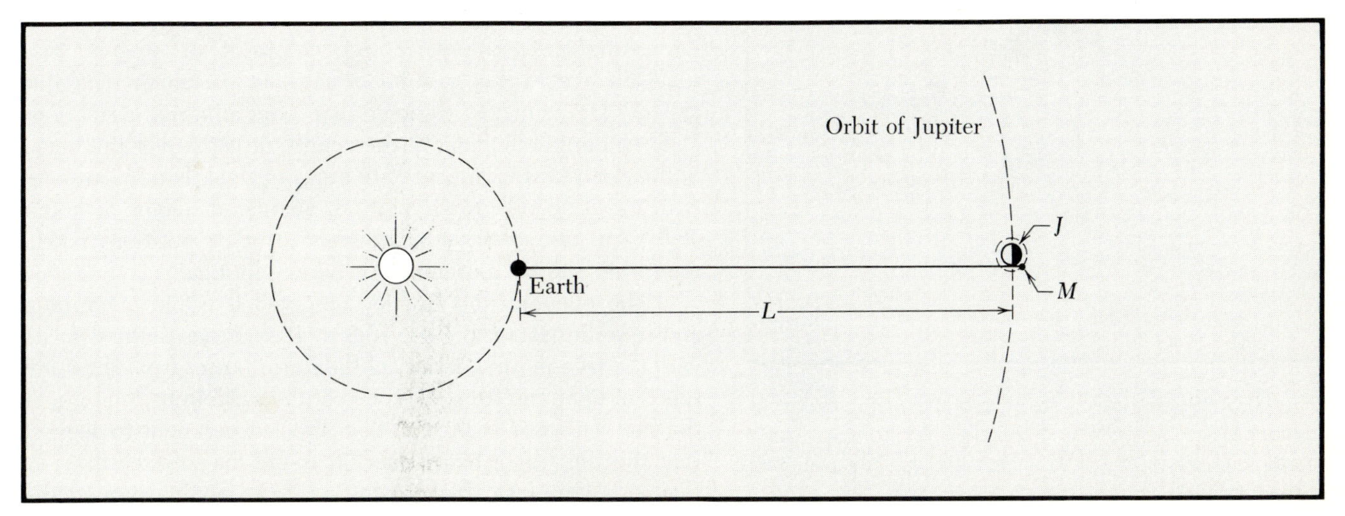

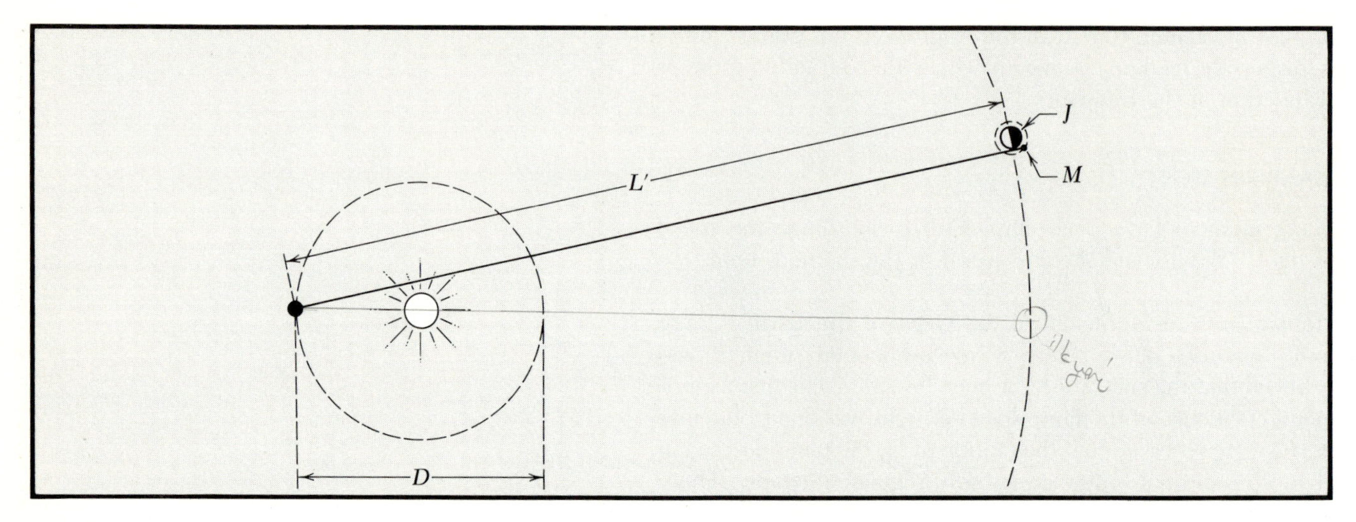

FIG. 10.2 Six months later the earth has completed a semicircle but Jupiter has moved only about 15°. The eclipse is now observed L'/c later where $L' \approx L + D$.

$$c = \frac{2.83 \times 10^{13}}{22 \times 60} = 2.14 \times 10^{10} \text{ cm/s}$$

For the time at which he made the estimate, this value is in good agreement with 3.0×10^{10} cm/s. The angular motion of Jupiter about the sun is slower (12 yr vs 1 yr) than that of the earth; thus it is the diameter of the earth's orbit and not Jupiter's orbit which is chiefly involved in the calculation. The Roemer method is not very accurate, but it did show astronomers that in analyzing planetary observations to find the true motion of a planet or moon, it is necessary to make allowance for the propagation time of the light signal.

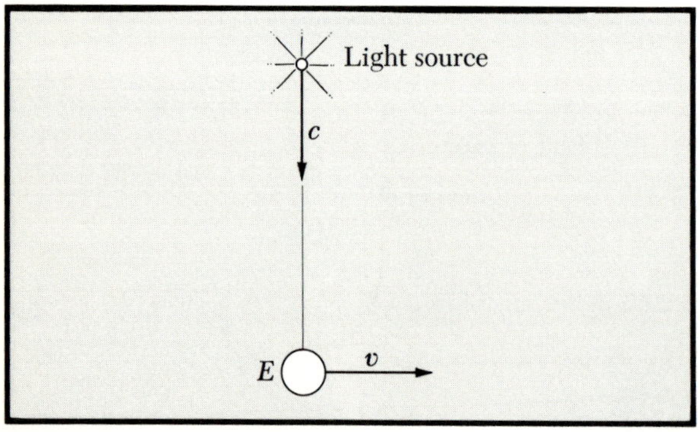

FIG. 10.3 Bradley in 1725 utilized the phenomenon of aberration to determine c. Suppose light from a distant source illuminates object E, which has velocity v normal to the incoming light.

Aberration of Starlight In 1725 James Bradley started an interesting series of precise observations of an apparent seasonal change in the position of stars, in particular of a star called γ Draconis. He observed that (after all other corrections had been applied) a star at the zenith (directly over the plane of the earth's orbit) appeared to move in a nearly circular orbit with a period of a year, with an angular diameter of about 40.5″. He also observed that stars in other positions had a somewhat similar motion—in general, elliptical.

The phenomenon Bradley observed is called *aberration*, and it is illustrated in Figs. 10.3 to 10.5. It has nothing to do with the true motion, if any, of the star; it arises from the finite speed of light and from the speed of the earth in its orbit about the sun. This was really the first direct experiment to suggest that the sun was a better inertial frame than the earth—i.e., that it is better to think of the earth as moving around the

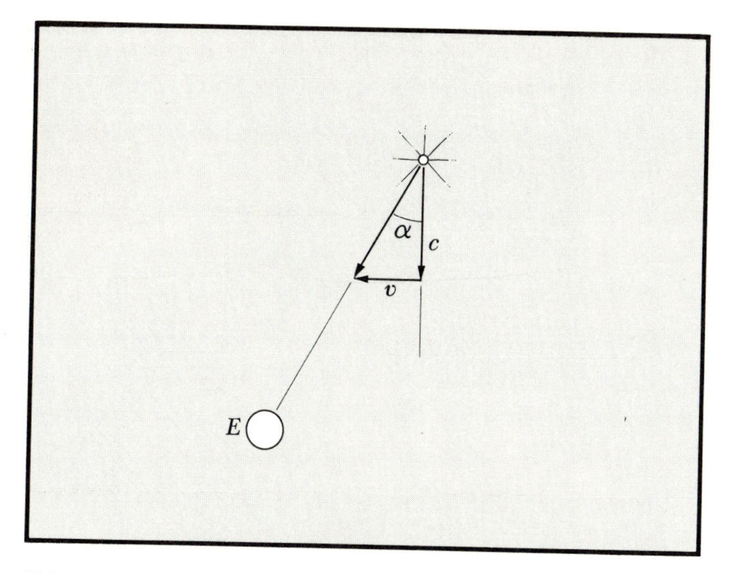

FIG. 10.4 According to an observer on E, the light has horizontal velocity component v as well as vertical component c. Thus the light ray from the source is inclined at angle α, where $\tan \alpha = v/c$.

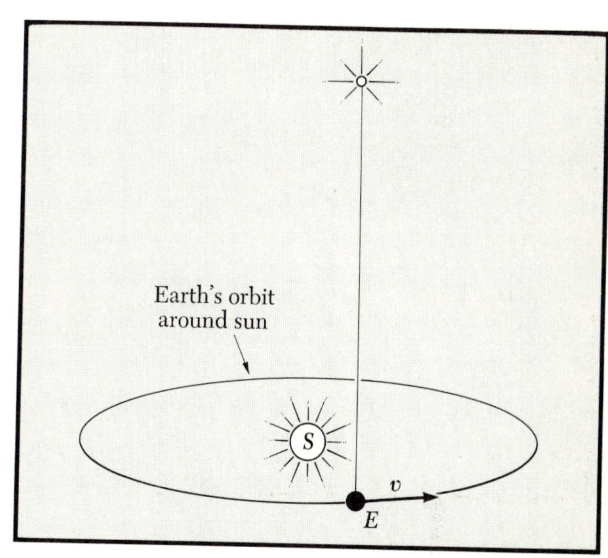

FIG. 10.5 Bradley used light from a distant star at zenith and the known velocity of the earth ($v_e = 30$ km/s) to determine c from measurements of α; $\tan \alpha = v_e/c$.

sun than of the sun as moving around the earth, for this experiment detects directly the annual change in the direction of the velocity of the earth relative to the stars.

The simplest explanation of aberration is the analogy of light propagation to the fall of raindrops (see Fig. 10.6). If no wind is blowing, raindrops fall vertically and a man at rest with an umbrella directly over his head does not get wet. If the man runs, holding the umbrella in the same position, the front of his coat will get wet. Relative to the moving person, the raindrops do not fall exactly vertically.

We quote from an account[1] of how the explanation of his observations came to Bradley:

> At last, when he despaired of being able to account for the phenomena which he had observed, a satisfactory explanation of it occurred to him all at once, when he was not in search of it.[2] He accompanied a pleasure party in a sail upon the river

[1] T. Thomson, "History of the Royal Society," p. 346, London, 1812.

[2] Many inventions and discoveries are made when, after an initial failure, the scientist has taken his thoughts away from the problem. A distinguished mathematician discusses this effect in a fascinating and important little book: J. Hadamard, "An Essay on the Psychology of Invention in the Mathematical Field," Princeton University Press, Princeton, N.J., 1945, reprint Dover Publications, Inc., New York, 1954.

FIG. 10.6 A homely example of aberration: This student is caught in rain coming straight down. If he stands still under his umbrella, he keeps dry. But if he runs for it he gets wet. In his new reference frame the rain has horizontal velocity $-v$, where v is his velocity with respect to the ground.

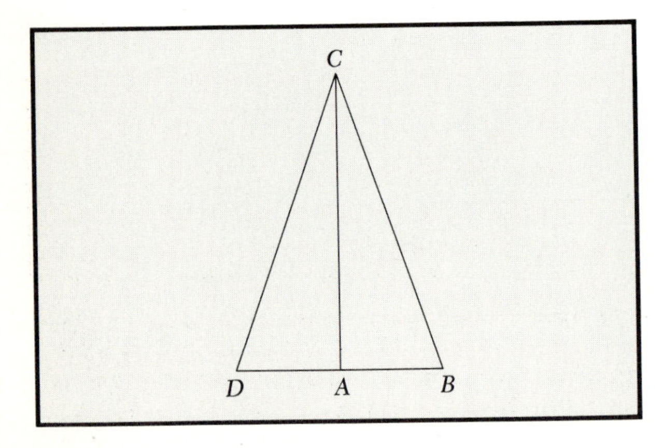

FIG. 10.7 Velocity diagram used by Bradley.

Thames. The boat in which they were was provided with a mast, which had a [weather] vane at the top of it. It blew a moderate wind, and the party sailed up and down the river for a considerable time. Dr. Bradley remarked that, every time the boat put about, the vane at the top of the boat's mast shifted a little, as if there had been a slight change in the direction of the wind. He observed this three or four times without speaking; at last he mentioned it to the sailors, and expressed his surprise that the wind should shift so regularly every time they put about. The sailors told him that the wind had not shifted, but that the apparent change was owing to the change in the direction of the boat, and assured him that the same thing invariably happened in all cases. This accidental observation led him to conclude, that the phenomenon which had puzzled him so much was owing to the combined motion of light and of the earth.

This is the explanation of aberration in Bradley's words:[1]

I considered this matter in the following manner. I imagined *CA* [see Fig. 10.7] to be a ray of light, falling perpendicularly upon the line *BD*; then if the eye is at rest at *A*, the object must appear in the direction *AC*, whether light be propagated in time or in an instant. But if the eye is moving from *B* towards *A*, and light is propagated in time, with a velocity that is to

[1] J. Bradley, *Phil. Trans. Roy. Soc.*, London, **35**:637 (1728).

the velocity of the eye, as CA to BA; then light moving from C to A, whilst the eye moves from B to A, that particle of it by which the object will be discerned when the eye is in motion comes to A, is at C when the eye is at B. Joining the points B, C, I supposed the line CB to be a tube (inclined to the line BD in the angle DBC) of such a diameter as to admit of but one particle of light; then it was easy to conceive that the particle of light at C (by which the object must be seen when the eye, as it moves along, arrives at A) would pass through the tube BC, if it is inclined to BD in the angle DBC, and accompanies the eye in its motion from B to A; and that it could not come to the eye, placed behind such a tube, if it had any other inclination to the line BD.

For a star directly overhead the maximum aberration occurs when the earth's velocity is perpendicular to the line of observation. Then the tilt angle, or aberration, of the telescope is seen from Figs. 10.4 and 10.5 to be given by

$$\tan \alpha = \frac{v_e}{c} \tag{10.1}$$

where v_e is the speed of the earth. The orbital speed of the earth about the sun is 3.0×10^6 cm/s; the speed due to rotation about the earth's own axis, which is about 100 times slower, may be neglected here. The angle α in Eq. (10.1) will be half of Bradley's observed angular diameter of 40.5″. So by using $\alpha = 20''$ and solving for c in Eq. (10.1), we obtain (with $\tan \alpha \approx \alpha$)

$$c = \frac{v_e}{\alpha} = \frac{3 \times 10^6}{\frac{20}{3600} \times 1/57.3} = 3.1 \times 10^{10} \text{ cm/s}$$

This compares well with present values.

Toothed Wheels and Rotating Mirrors The first terrestrial determination of the speed of light was carried out by Fizeau in 1849. He found (see Fig. 10.8*a* to *c*)

$$c = (315{,}300 \pm 500) \text{ km/s}$$

for the speed of light in air.[1] He used a rotating toothed wheel as a light switch to determine the transit time of a light flash over a path length of 2×8633 m.

The toothed-wheel apparatus was soon replaced by a rotating-mirror device, which gives more light and better focusing. The arrangement used by Foucault in 1850 is shown

[1]The speed of light in vacuum is calculated to be about 91 km/s faster than in air.

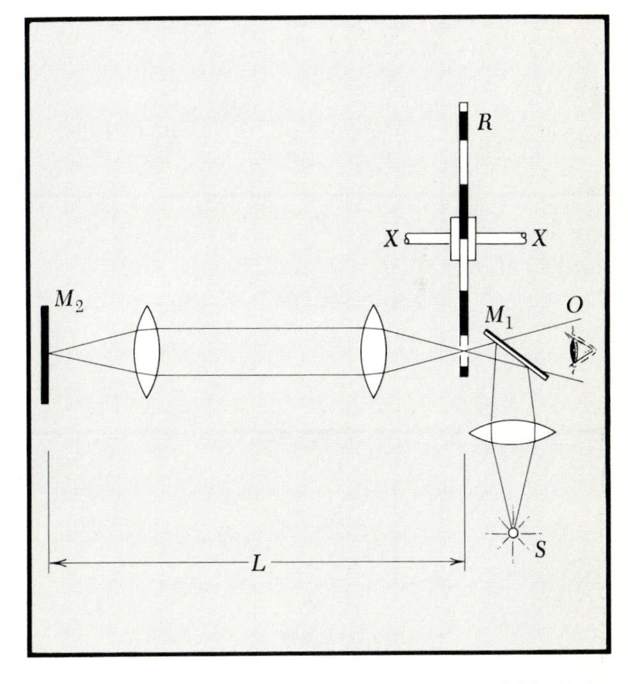

FIG. 10.8 (*a*) Fizeau's cogwheel apparatus, 1849. Light from a point source S is reflected from a half-silvered mirror M_1 past the cogwheel R rotating on axis X-X. Light then goes to mirror M_2 and returns to observer O through R and M_1. A half-silvered mirror reflects half the light incident and transmits half.

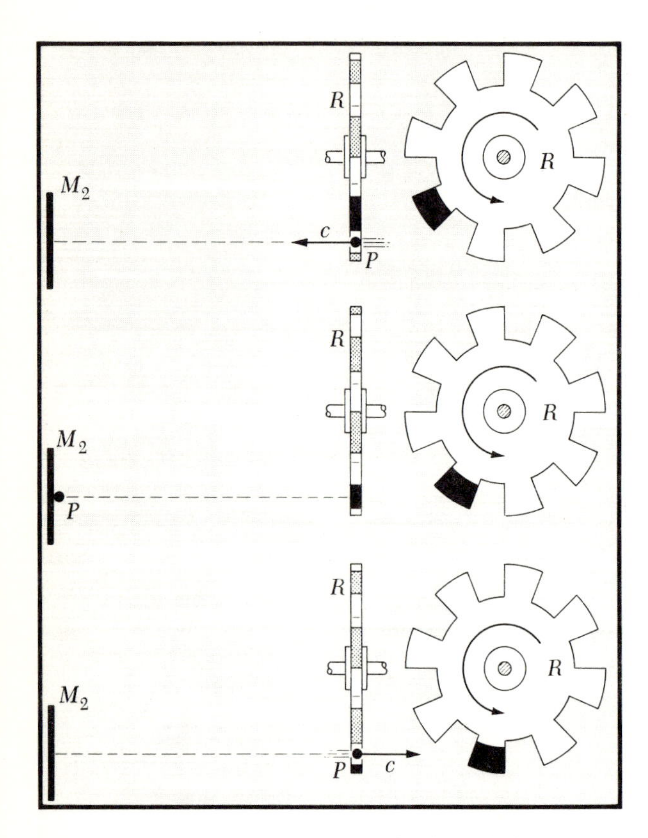

FIG. 10.8 *(cont'd)* (*b*) The pulse P with velocity c must travel to M_2 and return to R (total distance $2L$) in a time during which the cogs move over one space, if the pulse is to be transmitted to O. Fizeau determined c from L and the angular velocity of R.

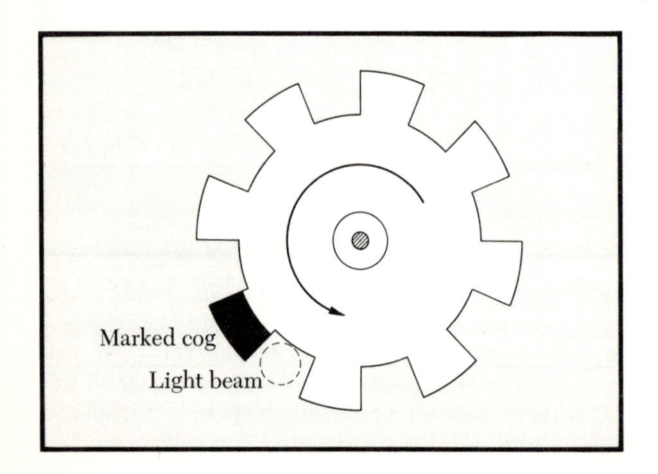

FIG. 10.8 *(cont'd)* (*c*) View of the light beam and cogwheel R seen by observer O. Rotation of R chops the light beam from S, M_1 into short pulses. (Light can only pass from M_1 to M_2 if no cog is in the way.)

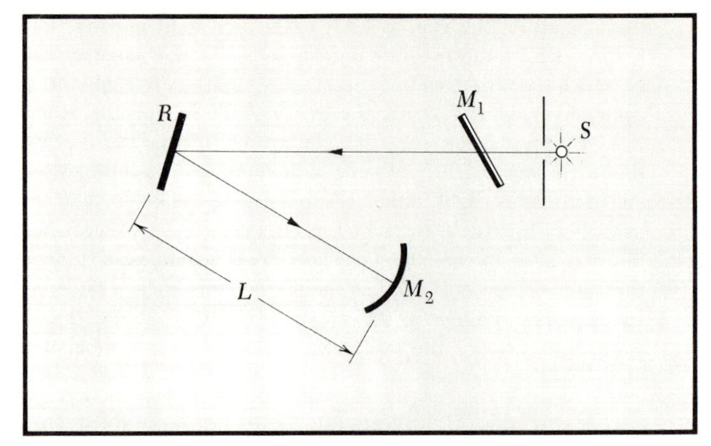

FIG. 10.9 (*a*) Foucault's rotating mirror apparatus, 1850, consisting of a source S, half-silvered mirror M_1, rotating mirror R (axis of rotation normal to the page), and spherical mirror M_2. Beam path from S to M_2 shown.

in Fig. 10.9a to c. His best value (1862) for the speed of light in air is

$$c = (298{,}000 \pm 500) \text{ km/s}$$

A development of the rotating-mirror arrangement was used by Michelson (1927) over a path of 22 mi between Mt. Wilson and Mt. San Antonio in California. His arrangement has the light source at the focal point of a lens, giving parallel light over a long path. He found

$$c = (299{,}796 \pm 4) \text{ km/s}$$

This work greatly exceeded in accuracy all previous work. (Further details are given in Prob. 3.)

Cavity Resonator It is possible to determine very accurately the frequency at which a resonant cavity of known dimensions (a metal box) contains a known number of half wavelengths of electromagnetic radiation. The speed of light is then calculated from the theoretical relation

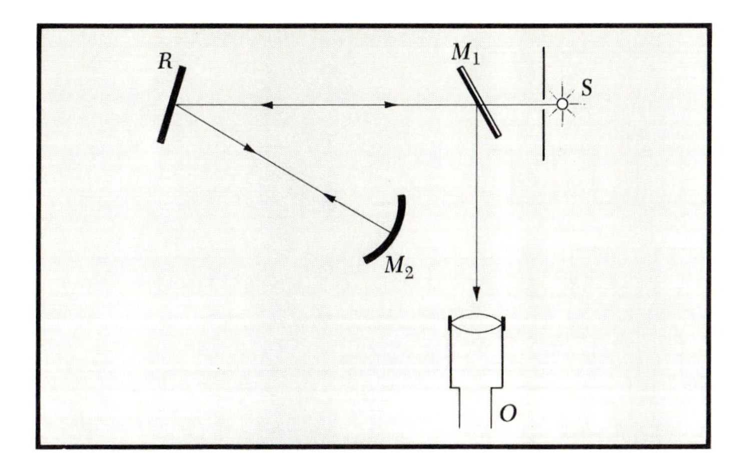

FIG. 10.9 (*cont'd*) (*b*) With R stationary, a light beam from M_1 to R to M_2 is reflected back along same path to M_1, and detected by O.

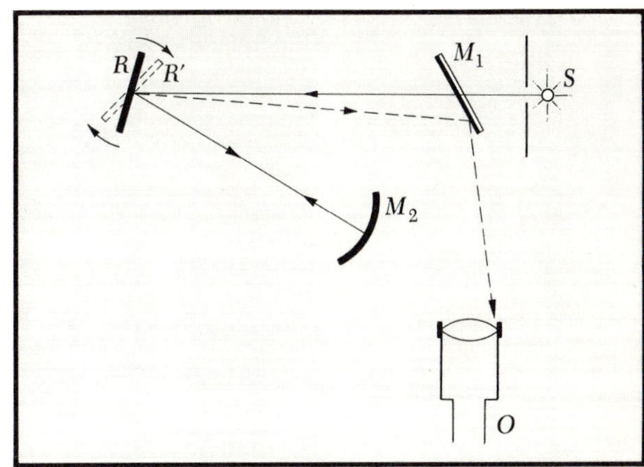

FIG. 10.9 (*cont'd*) (*c*) If mirror R rotates, light from S to R to M_2 returns when rotating mirror is in a new position R'. Thus O observes a displaced image on M_1. Foucault determined c from L, the image displacement, and the mirror angular velocity.

$$c = \lambda\nu \tag{10.2}$$

connecting the wavelength λ and the frequency ν. The cavity is usually evacuated. It is necessary to correct the inside dimensions of the cavity for the small penetration[1] of the electromagnetic field into the surface of the metal. Essen (1950) used frequencies of 5960, 9000, and 9500 Mc/s to find

$$c = (299{,}792.5 \pm 1)\ \text{km/s}$$

Kerr Cell When polarized light passes through a Kerr cell (a liquid in which an electric field can affect the transmission of polarized light) the intensity of light emerging and polarized in the initial direction can be modulated by varying the voltage between the plates producing the electric field. If the same frequency voltage is used to modulate the sensitivity of a photocell that detects the light, then a measurement of the speed of light can be made with the apparatus diagramed in Fig. 10.10. The response of the detector D will be a maximum if light of maximum intensity reaches D at a time of maximum sensitivity. If we assume that maximum intensity and maximum sensitivity occur at the same time, this maximum

[1]The penetration region is known as the *skin depth*. It is of the order of 1 micron (abbreviated μ; $1\ \mu \equiv 10^{-4}$ cm) in thickness in copper at room temperature at 10^{10} cps. There are also other corrections to be applied.

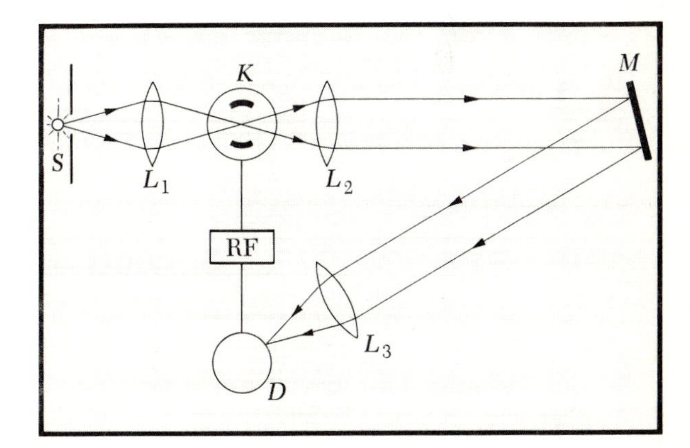

FIG. 10.10 A modern method for determination of c. Light from source S is amplitude-modulated in the Kerr cell K, then passed to mirror M and the photoelectric detector D through lenses $L_{2,3}$. The photodetector sensitivity and Kerr cell are synchronized by a modulated radio-frequency voltage generator RF.

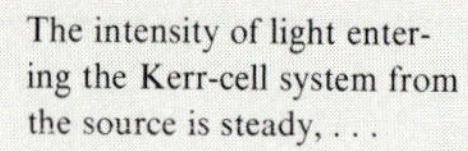

The intensity of light entering the Kerr-cell system from the source is steady, . . .

but the light coming out of the Kerr-cell system is modulated. The time of transit of the light from K to D can be varied by moving M: M can be adjusted so light arrives at D as shown.

If we move M out a little, the light arrives later . . .

for M *further* out, light arrival is *still* later . . .

for M *further* out, light arrival is *still* later . . .

for M further out, light arrival is still later.

Now suppose the sensitivity of the detector is modulated as shown here . . .

The detector responds only when it is sensitive and when light is coming in.

Thus we have this detector response for condition a.

For condition b we have this: The incoming light and the detector sensitivity are in phase.

For c we have this.

For d the arriving light and the detector sensitivity are 180° out of phase, so there is no response.

For e we have this.

As we continuously vary the position of M, we obtain this average detector response.

The distance between two successive maxima of this curve corresponds to a change in the light path $2\Delta L$ caused by displacement of M.

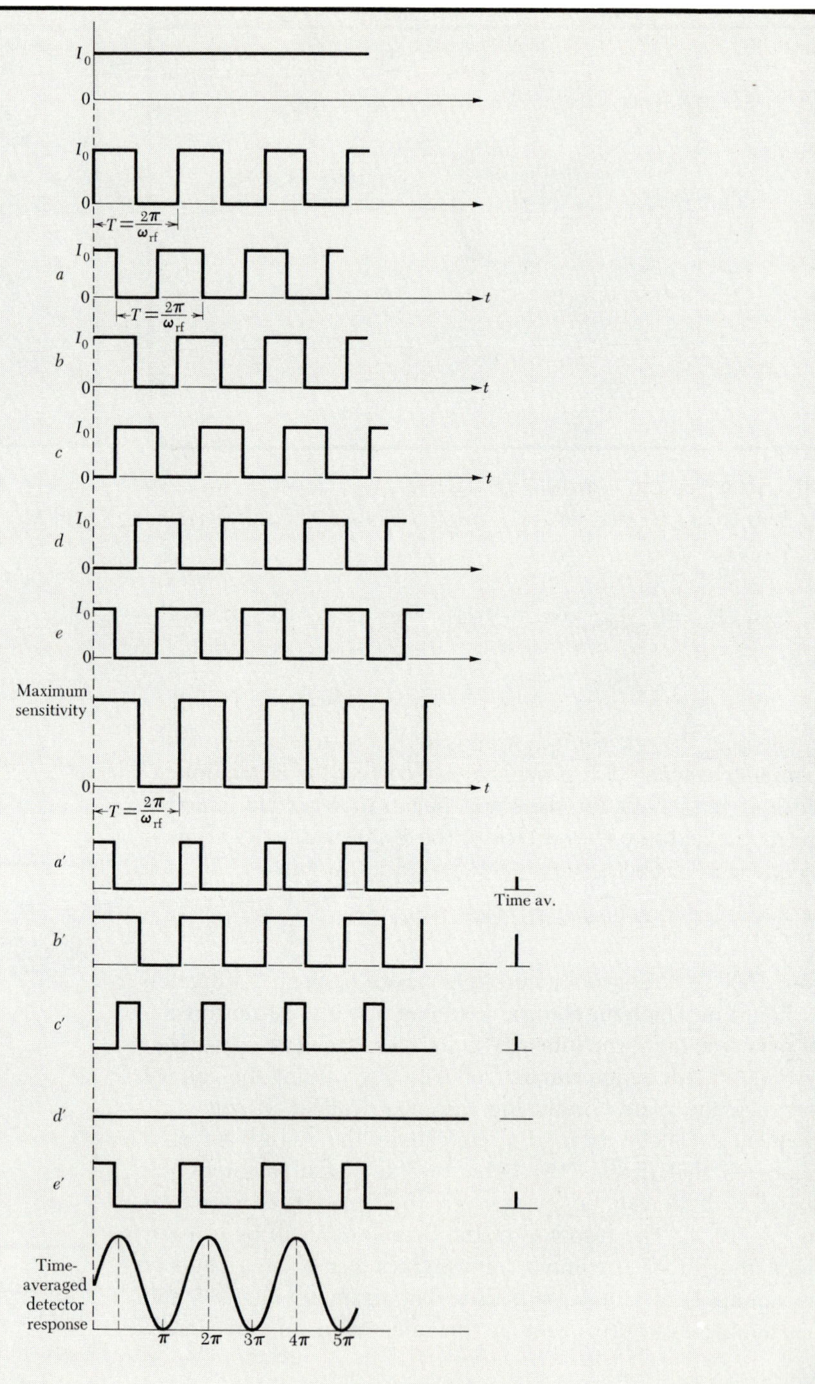

response will occur if the time taken by the light to go from the Kerr cell K to mirror M and back to D is an integral number N of periods of the radio frequency ν. This time is N/ν and so

$$c = \frac{L\nu}{N}$$

where L is the distance from K to D. In the actual experiment it is of the order of 10 km. Some details of the method are given in Fig. 10.11.

Using this method, Bergstrand measured

$$c = (299{,}793.1 \pm 0.3) \text{ km/s}$$

Note that the estimated error is very low. The same device is used (together with a standard value for c) to determine geodetic lengths over distances up to 40 km; in this application it is known as a *geodimeter*.

Hundreds of measurements of c have been made in the past hundred years by these and a dozen or so other methods. The present accepted value is

$$\boxed{c = (2.997\,925 \pm 0.000\,001) \times 10^{10} \text{ cm/s}} \qquad (10.3)$$

This represents a consensus of the most reliable recent measurements by different methods in which electromagnetic waves from 10^8 cps (radio frequency) to 10^{22} cps (γ-rays) have been investigated. The precision at the highest frequency is not as great as at radio or optical frequencies, but there is at present no reason to believe that c varies with the frequency of the radiation.

SPEED OF LIGHT IN INERTIAL FRAMES IN RELATIVE MOTION

An elementary application of the galilean transformation to the problem of a moving receiver requires that the speed of light in the frame of the receiver be different from c. According to common sense we expect the speed of light c_R relative to the moving receiver to be given by

$$c_R = c \pm V \qquad (10.4)$$

FIG. 10.11 Bergstrand's measurement of c is based on the method of "phase-sensitive detection" and is similar to the experiment described here.

where V is the speed of the receiver that is supposed to be moving toward $(+)$ or away $(-)$ from the source. This seems a perfectly reasonable way to add velocities and is illustrated in Fig. 10.12*a* and *b*. The same relation should hold when the source and receiver are at rest and the medium moves with velocity V. The relation [Eq. (10.4)] is apparently obeyed in countless everyday experiences, at least where light is not involved. It holds for sound waves, if the velocity of sound is written for c. But it is *not true,* even approximately, for light waves in free space. It is found experimentally that (as shown in Fig. 10.12*c* and *d*)

$$c_R = c \tag{10.5}$$

for any frame *regardless of its velocity,* and regardless of the velocity relative to an imagined propagation medium. This demonstrated fact lies at the root of the relativistic formulation of physical laws.

We now examine the experimental basis of Eq. (10.5). There are many different types of experiments which support the special theory of relativity; those leading to Eq. (10.5) make a convenient takeoff point. We consider the experiments which show that the velocity of light is independent of the velocity (3×10^6 cm/s) of the earth in its orbit.

First suppose, as did the physicists of the nineteenth century, that light propagates as an oscillation in a medium, just as sound propagates as an oscillation of atoms in a liquid, solid, or gas. The luminiferous medium through which light waves propagate in free space was called the *ether.*

What is the ether? Today we consider ether as only another word for vacuum. But Maxwell and many others could not imagine a field as a self-supporting entity propagating in free space. Maxwell argued:

> But in all these theories the question naturally occurs:—If something is transmitted from one particle to another at a distance, what is its condition after it has left the one particle and before it has reached the other? If this something is the potential energy of the two particles, as in Neumann's theory, how are we to conceive this energy as existing in a point of space, coinciding neither with the one particle nor with the other? In fact, whenever energy is transmitted from one body to another in time, there must be a medium or substance in which the energy exists after it leaves one body and before it reaches the other, for energy, as Torricelli remarked, "is a quintessence of so subtle a nature that it cannot be contained in any vessel except the inmost substance of material things." Hence all these theories lead to the conception of a medium

FIG. 10.12 Velocity addition predicted by the galilean transformation (*a,b*) and as actually observed for light (*c,d*).

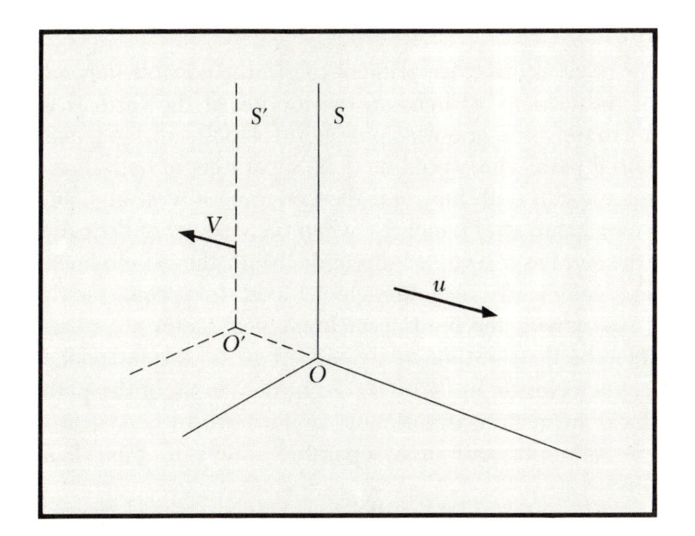

(*a*) If u is an ordinary terrestrial speed as observed in inertial frame S,

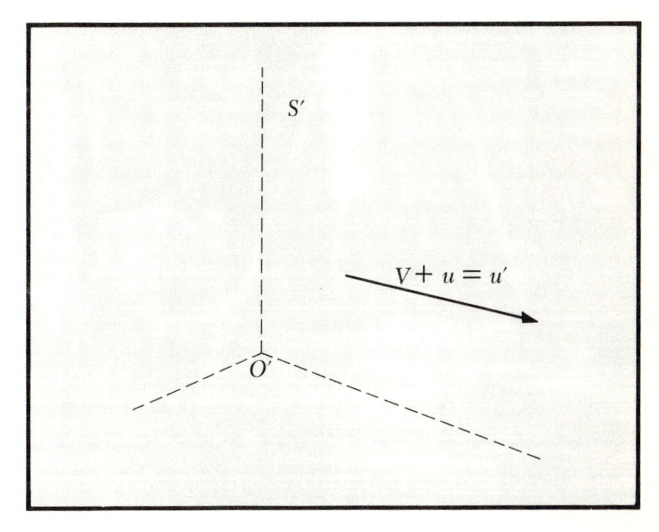

(*b*) the galilean transformation tells us that in inertial frame S′ we will observe $u' = V + u$.

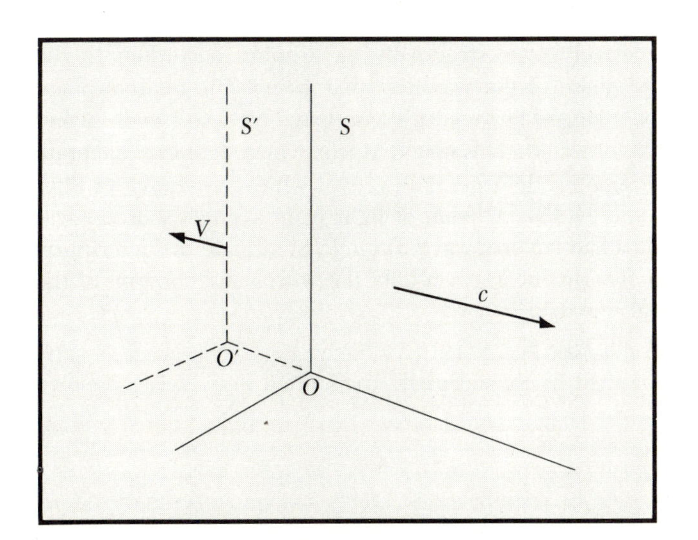

(*c*) However, experiments show that if an object has speed c in S,

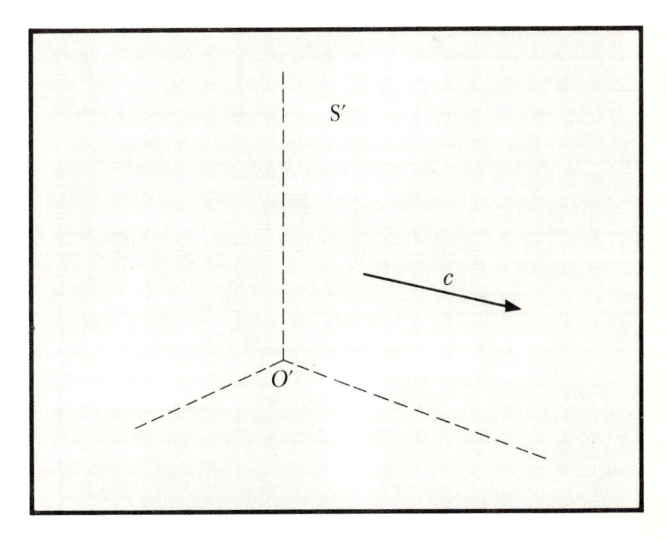

(*d*) it also has speed c in S′.

in which the propagation takes place, and if we admit this medium as an hypothesis, I think it ought to occupy a prominent place in our investigations, and that we ought to endeavour to construct a mental representation of all the details of its action, and this has been my constant aim in this treatise.

The obvious direct experiment to test the possible dependence of the velocity of light on the motion of the earth is to time accurately the one-way passage of a pulse of light over a measured path. This would be done separately in both directions on a north-south line, and then on an east-west line, and finally over again after 6 months, when the velocity of the earth about the sun has a reversed direction. With the development of lasers, sufficiently accurate clocks exist to permit such a direct experiment; the limiting technological factor at present appears to be the rise time of a pulse. At 10^{-9} s this introduces an effective error of $10^{-9}\,c = 30$ cm in the length of the path. The clocks in such an experiment would have to be synchronized at one spot and then separated slowly to their final positions.

A number of experiments have been performed to test Eq. (10.4), that is, to detect *ether drift* (see Fig. 10.13). All have failed to show a movement of the earth through the ether; very important and conceptually straightforward were those carried out by Michelson and Morley.[1]

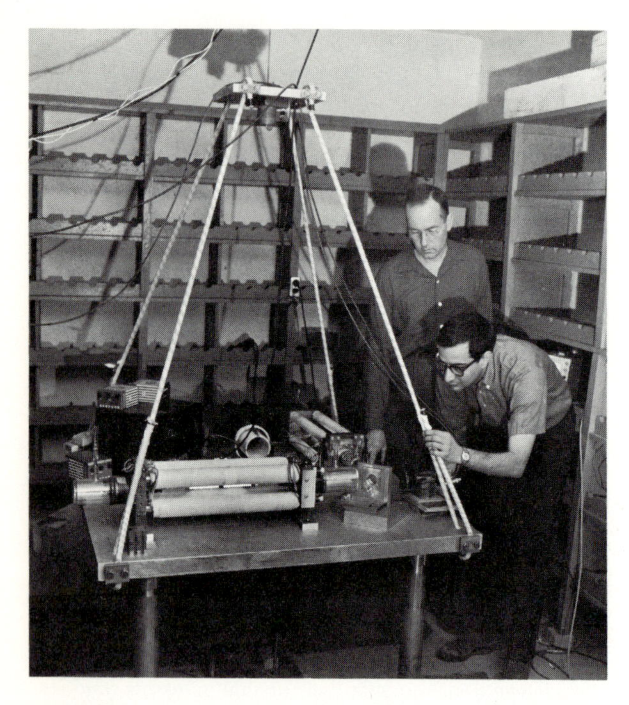

FIG. 10.13 A precision apparatus for a relativistic optical experiment using two gas lasers. The site is a former wine cellar in Round Hill, Mass. The workers are Charles H. Townes and Ali Javan.

Michelson–Morley Experiments Two sets of light waves derived from a common monochromatic source may interfere constructively or destructively at a point, according to the relative phase of the waves at that point. The relative phase may be changed by requiring one wave train to travel farther than the other. Michelson and Morley constructed an elaborate interferometer, the essential parts of which are shown in Figs. 10.14 and 10.15*a*. A beam of light from a single source *s* was split by a half-silvered mirror at *a*. We continue the description of the experiment in essentially the words and notation of the original workers:[2]

Let *sa* [see Fig. 10.15*a* to *h*] be a ray of light which is partly reflected in *ab*, and partly transmitted in *ac*, being returned

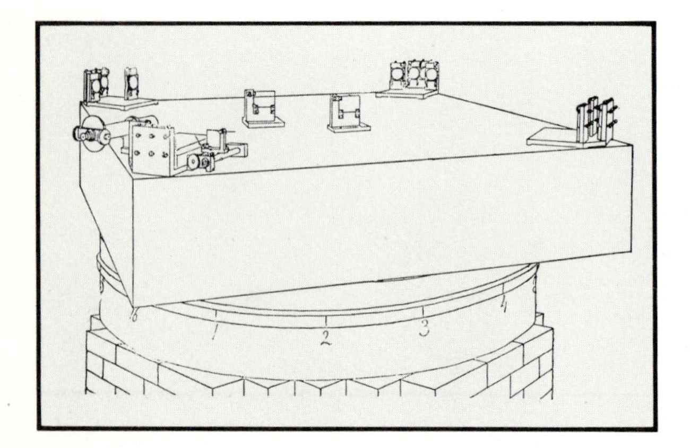

FIG. 10.14 Perspective of the apparatus described by Michelson and Morley in their 1887 paper.

[1]The influence of this experiment on Einstein in his work is discussed in an interesting article by Holton, *Am. J. Phys.*, **37**:968 (1969).

[2]A. A. Michelson and E. W. Morley, *Am. J. Sci.*, **34**:333 (1887). This was one of the most remarkable experiments of the nineteenth century. Simple in principle, the experiment led to a scientific revolution with far-reaching consequences. Note that the ratio of the speed of the earth in its orbit to the speed of light is about 10^{-4}. In reproducing the excerpt, we have written c for their V, and V for their v; interpolated remarks are enclosed in brackets.

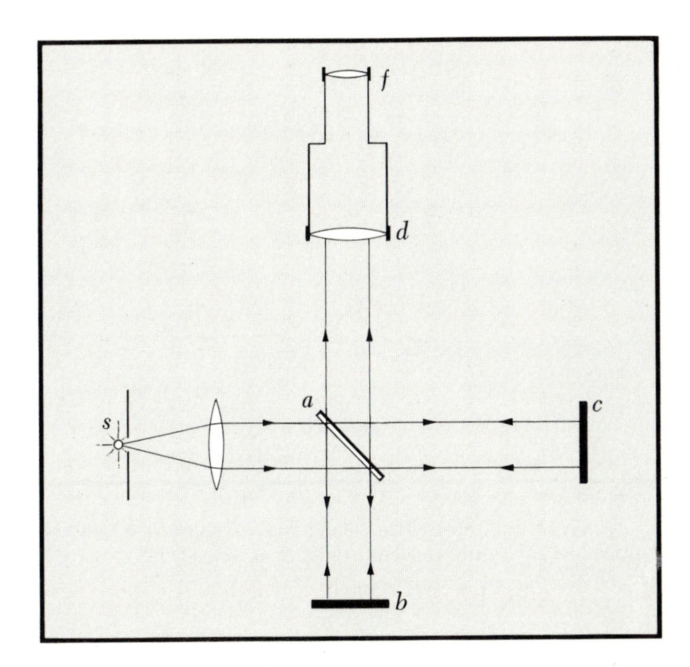

FIG. 10.15 (*a*) The Michelson–Morley experimental interferometer consists of a light source s, half-silvered mirror a, mirrors b and c, and a telescope detector d; f represents the focus of the telescope.

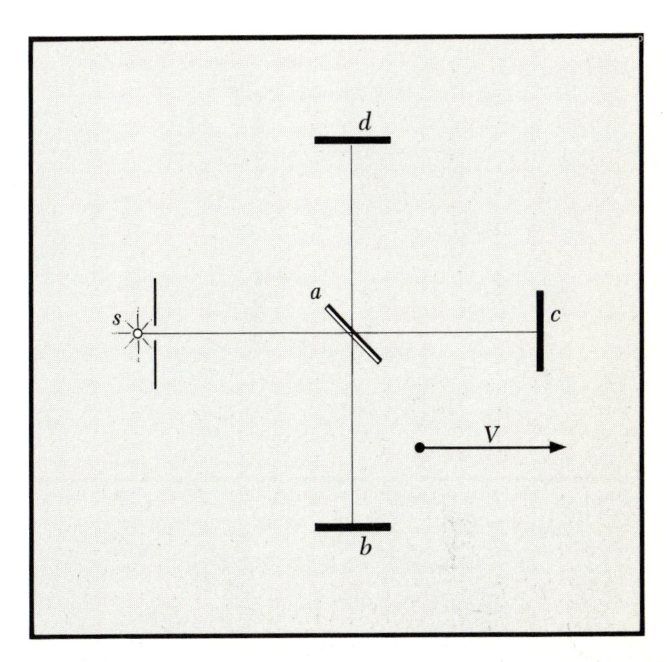

(*b*) If the interferometer is at rest in the ether, an interference pattern between the beams aba and aca is observed at d. If the apparatus (and earth) have velocity V with respect to the hypothetical ether, we would expect the interference pattern to change at d, since the times to traverse aba, aca would now change by different amounts.

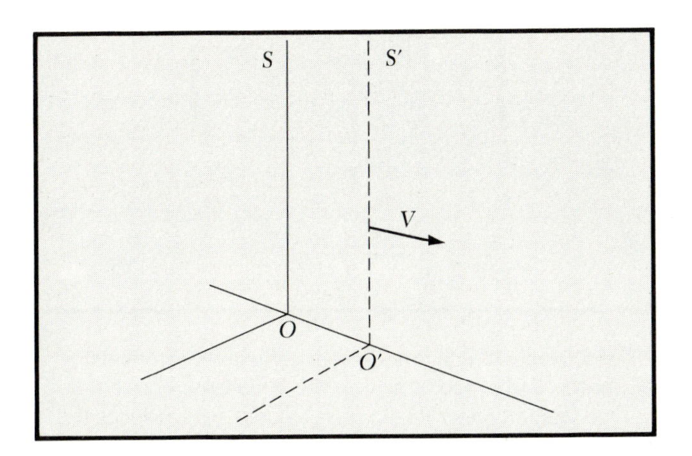

(*c*) To see this consider a galilean frame S' moving with earth and interferometer. S is a galilean frame at rest in the ether.

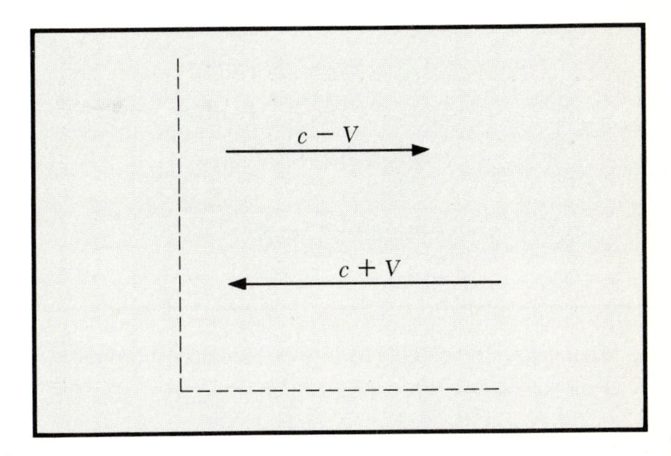

(*d*) According to the galilean transformation, light moving to right has speed $c - V$ in S'; light moving to left has speed $c + V$ in S'.

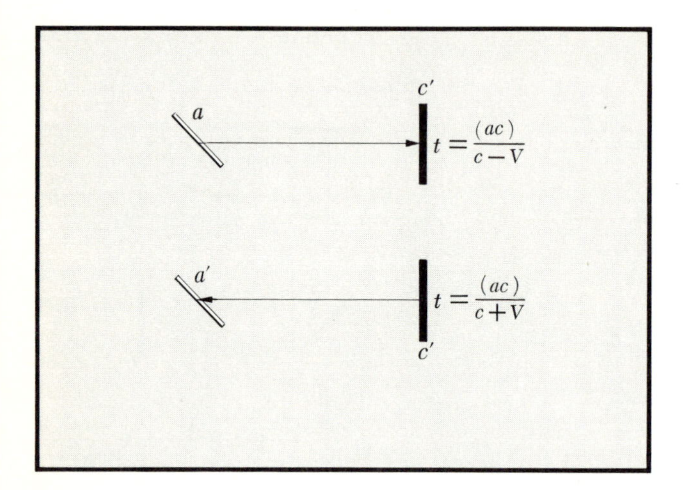

FIG. 10.15 (*cont'd*) (*e*) Thus the time to go from a to c' and back to a' is

$$\Delta t(ac'a') = \frac{(ac')}{c - V} + \frac{(ac')}{c + V}$$

where (ac') denotes the distance between a and c'.

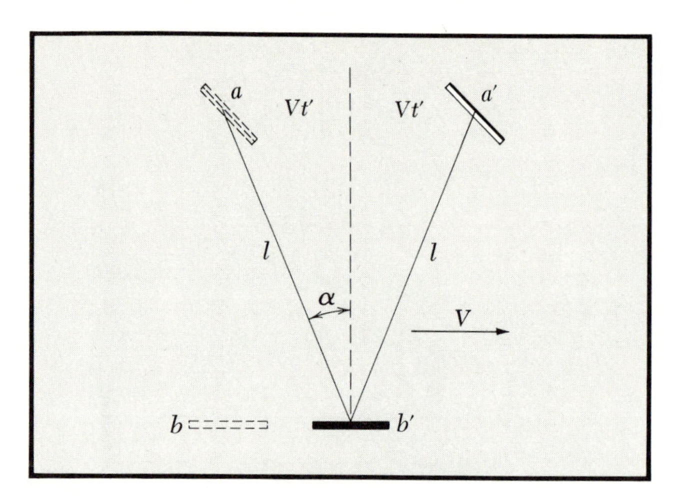

(*f*) What is the time $\Delta t(ab'a') = 2t'$ to go from a to b' and back to a'? In the galilean frame **S** at rest in the ether, the interferometer has velocity V to the right; light has speed c.

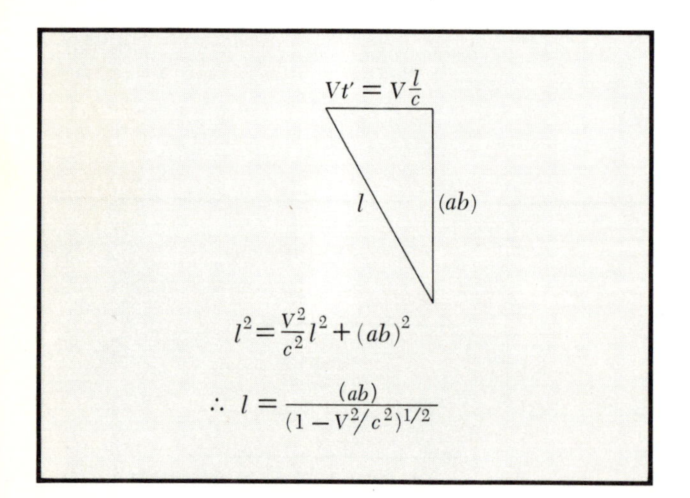

(*g*) $\Delta t(ab'a') = 2t' = 2(ab)/\sqrt{c^2 - V^2}$. To terms of the order V^2/c^2, this time is the same as

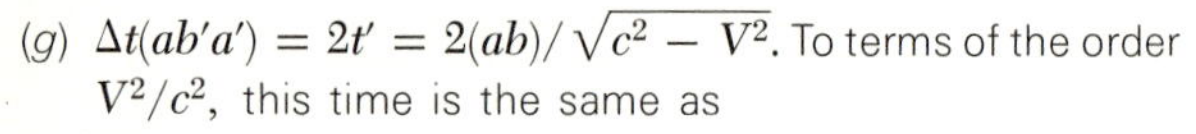

$$\frac{2(ab)\sqrt{1 + \frac{V^2}{c^2}}}{c}$$

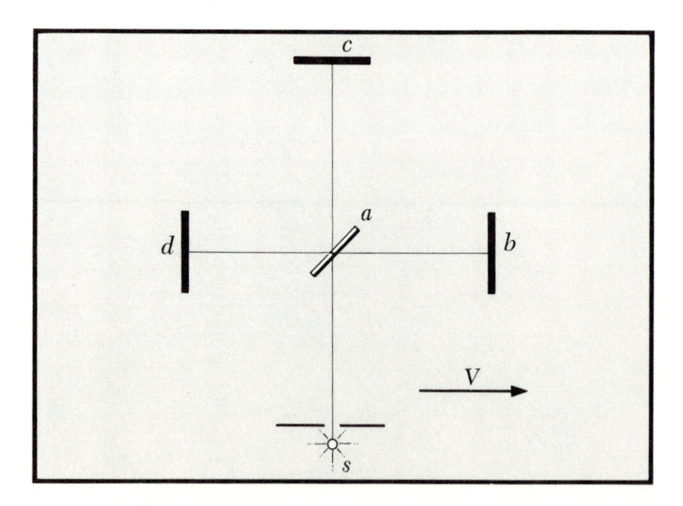

(*h*) Thus even if $(ab) = (ac)$, the galilean transformation leads us to expect a shift in the interference pattern if the interferometer changes its velocity with respect to the ether. None was observed. Here the apparatus is shown turned through 90° to repeat the test with the motion parallel to *ab* instead of *ac*.

by the mirrors b and c, along ba and ca. ba is partly transmitted along ad, and ca is partly reflected along ad. If then the paths ab and ac are equal, the two rays interfere along ad. Suppose now, the ether being at rest, that the whole apparatus moves in the direction sc with the velocity of the earth in its orbit; the directions and distances traversed by the rays will be altered thus:—The ray sa is reflected along ab' [as in Fig. 10.15*f*]; it is returned along $b'a'$, where the angle $ab'a'$ is twice the aberration angle, or 2α, and goes to the focus of the telescope, whose direction is unaltered. The transmitted ray goes along ac', is returned along $c'a'$ [as in Fig. 10.15*e*] and is reflected at a', making $c'a'd'$ [not shown] equal $90° - \alpha$, and therefore still coinciding with the first ray. It may be remarked that the rays $b'a'$ and $c'a'$ do not now meet exactly in the same point a', though the difference is of the second order; this does not affect the validity of the reasoning. Let it now be required to find the difference in the two paths $ab'a'$ and $ac'a'$.

Let c = velocity of light
V = velocity of the earth in its orbit
D = distance ab or ac
T = time light occupies to pass from a to c'
T' = time light occupies to return from c' to a'

Then

$$T = \frac{D}{c - V} \qquad T' = \frac{D}{c + V}$$

The whole time of going and coming is

$$T + T' = 2D\frac{c}{c^2 - V^2}$$

and the distance traveled in this time is

$$2D\frac{c^2}{c^2 - V^2} \approx 2D\left(1 + \frac{V^2}{c^2}\right)$$

neglecting terms of the fourth order. The length of the other path is evidently

$$2D\sqrt{1 + \frac{V^2}{c^2}}$$

or to the same degree of accuracy,

$$2D\left(1 + \frac{V^2}{2c^2}\right)$$

The difference is therefore

$$D\frac{V^2}{c^2}$$

If now the whole apparatus be turned through 90°, the difference will be in the opposite direction, hence the displacement of the interference fringes should be $2D(V^2/c^2)$. Considering only the velocity of the earth in its orbit, this would be $2D \times 10^{-8}$. If, as was the case in the first experiment, $D = 2 \times 10^6$ waves of yellow light, the displacement to be expected would be 0.04 of the distance between the interference fringes.

In the first experiment one of the principal difficulties encountered was that of revolving the apparatus without producing distortion; and another was its extreme sensitiveness to vibration. This was so great that it was impossible to see the interference fringes except at brief intervals when working in the city, even at two o'clock in the morning. Finally, as before remarked, the quantity to be observed, namely, a displacement of something less than a twentieth of the distance between the interference fringes may have been too small to be detected when masked by experimental errors.

The first named difficulties were entirely overcome [in the second experiment] by mounting the apparatus on a massive stone floating on mercury; and the second by increasing, by repeated reflection, the path of the light to about ten times its former value.

. . . Considering the motion of the earth in its orbit only, this displacement should be

$$2D\frac{V^2}{c^2} = 2D \times 10^{-8}$$

The distance D was about eleven meters, or 2×10^7 wavelengths of yellow light; hence the displacement to be expected was 0.4 fringe [if the earth were traveling through an ether]. The actual displacement was certainly less than the twentieth part of this, and probably less than the fortieth part [see Fig. 10.16]. But since the displacement is proportional to the square of the velocity, the relative velocity of the earth and the ether is probably less than one-sixth the earth's orbital velocity, and certainly less than one-fourth.

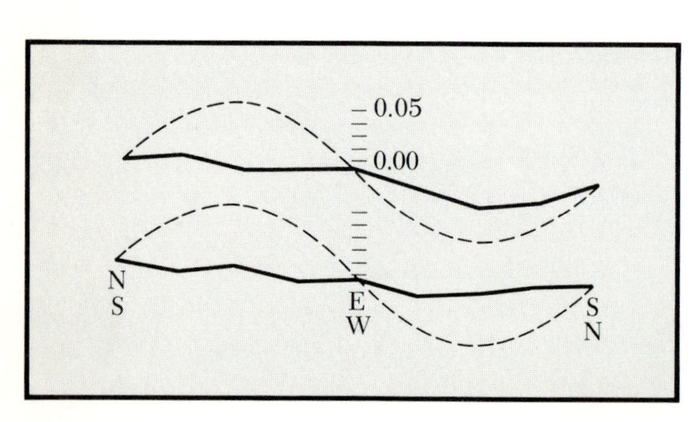

FIG. 10.16 "The results of the observations are expressed graphically [in the figure]. The upper is the curve for the observations at noon, and the lower that for the evening observations. The dotted curves represent *one-eighth* of the theoretical displacements. It seems fair to conclude from the figure that if there is any displacement due to the relative motion of the earth and the luminiferous ether, this cannot be much greater than 0.01 of the distance between the fringes." [A. A. Michelson and E. W. Morley, *Am. J. Sci.*, 34:333 (1887).] The vertical axis is the displacement of the fringes; the horizontal axis refers to the orientation of the interferometer relative to an east-west line.

The experimental results of Michelson and Morley were contrary to what we would expect, based on the galilean transformation. The experiments have since been repeated (with variations) with different wavelengths of light, with starlight, with extremely monochromatic light from a modern laser, at high altitudes, under the earth's surface, on different continents, and at different seasons over a period of some 80 yr. We can say that the change in c (the ether drift) is zero to a precision which is best expressed by saying that the speeds of light upstream and downstream are equal within a variation of less than 10^3 cm/s, or of 1 part in 1000 of the earth's orbital velocity about the sun.

Invariance of c The null result of the Michelson–Morley experiment suggests that the effects of the ether are undetectable. The result also suggests that the speed of light is independent of the motion of the source or of the observer. The experimental evidence on the latter point is quite good, but could be improved. The work by Sadeh quoted in Chap. 11 shows that the velocity of γ-rays is constant within ± 10 percent, independent of the velocity of the source, for source velocities of the order of $\frac{1}{2}c$. We conclude from all the experimental evidence that *a spherical wave front of light emitted from a point source in one inertial frame will appear as spherical to an observer in any other inertial frame.*

We noted in an earlier section that the speed of electromagnetic waves is independent of frequency over the range 10^8 to 10^{22} cps. Careful measurements also show that c is independent of the intensity of the light and also of the presence of other electric and magnetic fields. Our discussions have been limited entirely to electromagnetic waves traveling in free space.

DOPPLER EFFECT

The doppler effect or doppler shift relates the measured frequency of a wave to the relative velocities of the source, the medium, and the receiver. It is familiar, for sound, to anyone who has listened to an automobile approaching and then receding; or to those "older" people who have stood on a railroad platform and listened while a whistling train passed by. When the source is approaching, the number of waves emitted in 1 s will reach the receiver in less than 1 s because the source is closer when the last wave is emitted than when the first. Therefore, the frequency is higher. Vice versa, when the source is receding the frequency is lower. The same type of argument applies to a fixed source and moving receiver. The relations for sound are given by

$$\nu_R = \nu_T \frac{1 + v_R/\mathcal{V}}{1 - v_S/\mathcal{V}} \tag{10.6}$$

where $\mathcal{V}$ is the velocity of the sound wave in the medium, e.g., air, considered at rest, v_S is the velocity of the source considered positive when it is moving toward the receiver, v_R is the velocity of the receiver considered positive when it is moving toward the source, ν_T is the frequency of the source (transmitter) measured by an observer at rest with respect to the source, and ν_R is the frequency measured by the receiver.

Note that if $v_S \ll \mathcal{V}$ (assume $v_R = 0$),

$$\nu_R = \nu_T\left(1 + \frac{v_S}{\mathcal{V}}\right) \tag{10.7}$$

and

$$\frac{\nu_R - \nu_T}{\nu_T} = \frac{\Delta\nu}{\nu} = \frac{v_S}{\mathcal{V}} \tag{10.8}$$

In the case of light similar effects are present though we shall see some essential differences. In explaining and analyzing the doppler effect for sound we must consider the medium bearing the sound waves and the motion of source or receiver relative to the medium. In the case of light we must not understand the doppler effect in this way since the Michelson–Morley experiment result does not permit us to consider a medium (i.e., the ether). The doppler effect provides some interesting tests of special relativity and also some important results, particularly for astronomy. We shall treat the doppler effect correctly for light in Chap. 11.

EXAMPLE

The Recessional Red Shift Spectrographic analysis of light received from distant galaxies shows that certain prominent spectral lines identified in spectroscopic studies in the laboratory are shifted very significantly toward the red, or low-frequency, end of the visible spectrum. This shift may be interpreted as a doppler shift arising from the velocity of recession of the source. It is also known that the velocities calculated from these doppler shifts are directly proportional to the distances of the sources from us determined by independent means.

This is an extraordinary and provocative observational fact. The simplest nonrelativistic explanation of the distance-velocity relation is known as the "big-bang" theory, according to which the universe was formed from an explosion about 10^{10} yr ago. The fastest-moving products of the original explosion now form the outermost regions of the universe. Thus the greater the radial velocity of matter (relative to us), the farther it is from us and the greater is its red shift. There also are more sophisticated explanations of the recessional red shift. None is proved (see Fig. 10.17).

A pair of easily recognizable absorption lines in the spectrum of potassium (the K and H lines) are prominent in the spectra of many stars. These lines occur near wavelength[1] 3950 Å in laboratories on earth. We assume that laboratory observers moving in the rest frame of any star would measure the same wavelength. In light coming from a nebula in the constellation Boötes we observe these same lines at a wavelength of 4470 Å, a shift toward the red of $4470 - 3950 = 520$ Å. This is a relative shift of

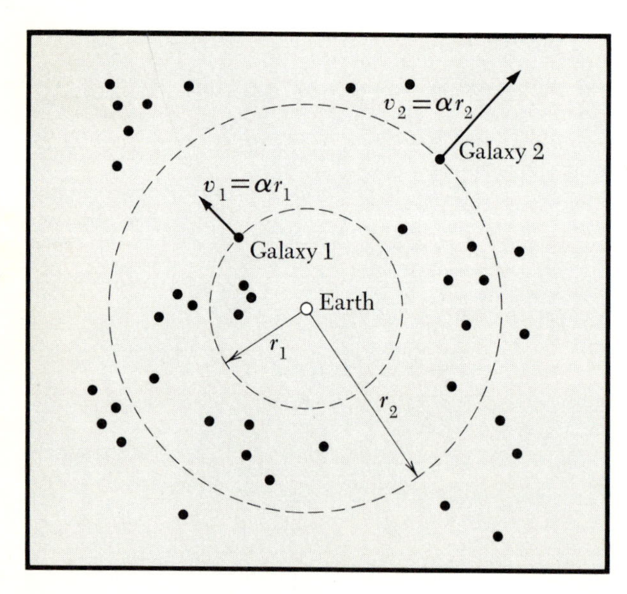

FIG. 10.17 The doppler effect observed in light from distant stars indicates that the galaxies are receding from us with a velocity proportional to their distance from earth. Galaxies 1 and 2 are assumed to have their distances r_1 and r_2 measured by other means, their velocities v_1 and v_2 by the doppler effect.

[1] angstrom $\equiv 10^{-8}$ cm $\equiv 1$ Å

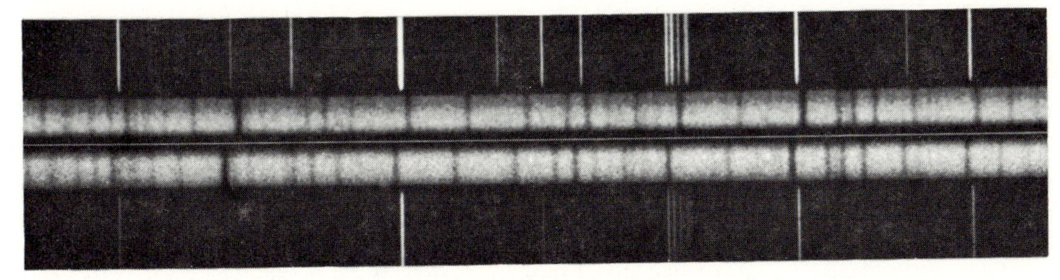

FIG. 10.18 Two spectrograms (taken at different times) of the binary star α^1 Geminorum. Only one of the two stars in this binary emits enough light to be detected. Notice that the spectral lines from the star are shifted, with respect to the laboratory reference lines, in different directions corresponding to two phases of motion of the star. In one phase the star is moving toward the earth and the frequency of the light is increased; in the other phase the star is moving away from the earth and the frequency is decreased. (*Lick Observatory photograph*)

$$\frac{\Delta\lambda}{\lambda} = \frac{520}{3950} = 0.13$$

We observe that, by using Eq. (10.8) with $\mathcal{V}$ equal to c (as will be justified for light waves in Chap. 11), and differentiating[1] $\nu = c/\lambda$ with c constant,

$$\frac{\Delta\nu}{\nu} = -\frac{\Delta\lambda}{\lambda} \qquad \text{or} \qquad \frac{\Delta\lambda}{\lambda} \approx \frac{v_S}{c} \tag{10.9}$$

We infer from Eqs. (10.8) and (10.9) that the nebula is receding from us with a relative speed $|v| \approx 0.13c$, which is really quite fast. For *higher speeds* we need to use one or another relation for the doppler shift as modified by the theory of relativistic models of the universe.[2] Also, the approximate expressions in Eqs. (10.8) and (10.9) suitable for either sound or light at speeds low compared to the speeds of sound or light, respectively, must be replaced by the correct expressions for light.

Similar observations on large numbers of galaxies can be combined with independent estimates of their distances to obtain an amazing empirical result: The relative velocity of a galaxy at distance r from us may be represented by the relation

$$v = \alpha r \tag{10.10}$$

where the constant α is empirically determined to be about $1.6 \times 10^{-18}\ \text{s}^{-1}$. (The estimation of galactic distances is a complex subject for which an astronomy text must be consulted.) The reciprocal of α has the dimensions of time:

$$\frac{1}{\alpha} \approx 6 \times 10^{17}\ \text{sec} \approx 2 \times 10^{10}\ \text{yr} \tag{10.11}$$

It is the time, beginning with the "big bang," taken by the star to

[1]Note a little computational trick: Suppose that $y = Ax^n$, where A, n are constants, and we want to find dy/y in terms of dx/x. We take the natural logarithm of both sides to form $\log y = \log A + n \log x$. We then take differentials of both sides to obtain $dy/y = n\,dx/x$. Here we have used the relation $d \log x/dx = 1/x$.

[2]See G. C. McVitties, *Physics Today*, p. 70 (July 1964).

reach its present distance. When we multiply $1/\alpha$ by c, we obtain a length:

$$\frac{c}{\alpha} \approx (3 \times 10^{10})(6 \times 10^{17}) \approx 2 \times 10^{28} \text{ cm} \qquad (10.12)$$

The time [Eq. (10.11)] is loosely called the *age of the universe;* the length [Eq. (10.12)] is loosely called the *radius of the universe.* The real significance of these quantities is not known at present, although several different cosmological models have been proposed to account for the form of the relations.

THE ULTIMATE SPEED

We have seen that electromagnetic waves in free space can only travel with the speed c. Can the speed of anything exceed the speed limit c?

Consider the motion of charged particles in an accelerator. Can particles be accelerated to travel faster than c? We have not as yet in this course encountered directly any principle which prevents the acceleration of charged particles to arbitrarily high velocities (see Fig. 10.19).

The following experiment[1] illustrates the proposition that a particle cannot be accelerated to a speed greater than c. Pulses of electrons are accelerated by successively larger electrostatic fields in a Van de Graaff accelerator, after which the electrons drift with constant velocity through a field-free region. Their time of flight, and hence their velocity over a measured distance AB, is measured directly, and the kinetic energy (which is turned to heat at the target at the end of the path) is measured by means of a calibrated thermocouple.

In the experiment the accelerating potential Φ is known with good precision. The kinetic energy of an electron is

$$K = eEL = e\Phi$$

where L is the distance over which acceleration occurs and $\Phi = EL$ is the difference in electric potential between the ends of the accelerating path. If $\Phi = 10^6$ V, the electron after acceleration has an energy of 1×10^6 eV (1 MeV). Now $10^6 \text{ V} \approx 10^6/300$ statvolts, so that the kinetic energy acquired by an electron is

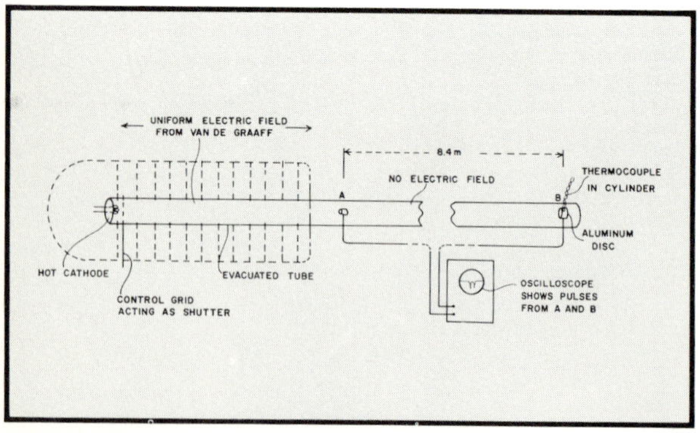

FIG. 10.19 The general arrangement of the ultimate speed experiment. The electrons are accelerated in a uniform field on the left and timed between A and B by the oscilloscope.

[1]This experiment was performed by W. Bertozzi in connection with the PSSC film "The Ultimate Speed." Our account draws directly from chap. A-3 of the PSSC Advanced Topics Program. See *Am. J. Phys.*, **32**:551 (1964).

$$\frac{(4.80 \times 10^{-10})(10^6)}{300} \approx 1.60 \times 10^{-6} \text{ erg}$$

If N electrons per second travel in the beam, the power delivered to the aluminum target at the end of the beam should be $1.60 \times 10^{-6}\, N$ erg/s. This agrees exactly with the direct thermocouple determination of the power absorbed by the target. This result confirms that the electrons deliver to the target the kinetic energy acquired during their acceleration. Further, on the basis of nonrelativistic mechanics, we expect that

$$K = \tfrac{1}{2}mv^2 \tag{10.13}$$

so that a graph of v^2 against the kinetic energy K should be a straight line. For energies greater than about 10^5 eV, however, the linear relation between v^2 and K does *not* hold experimentally. Instead, the velocity is observed to approach the limiting value 3×10^{10} cm/s at higher energies. So when the measured velocity is compared with the velocity calculated from Eq. (10.13) it is found to be less than Eq. (10.13) predicts. In fact the graph of v^2 against K bends over as shown in Fig. 10.20, approaching the value 9×10^{20} cm^2/s^2. The experimental results may be summarized: The electrons absorb the expected energy from the accelerating field, but their velocity does not increase without limit. Our only recourse in understanding this fact is to assume that m in Eq. (10.13) is not constant as K becomes large. We shall deal with this problem in Chap. 12. Many other experiments suggest, as this one does, that c is the upper limit to the velocity of particles. Thus we believe firmly that c is the maximum signaling speed with either particles or electromagnetic waves: *c is the ultimate speed.*

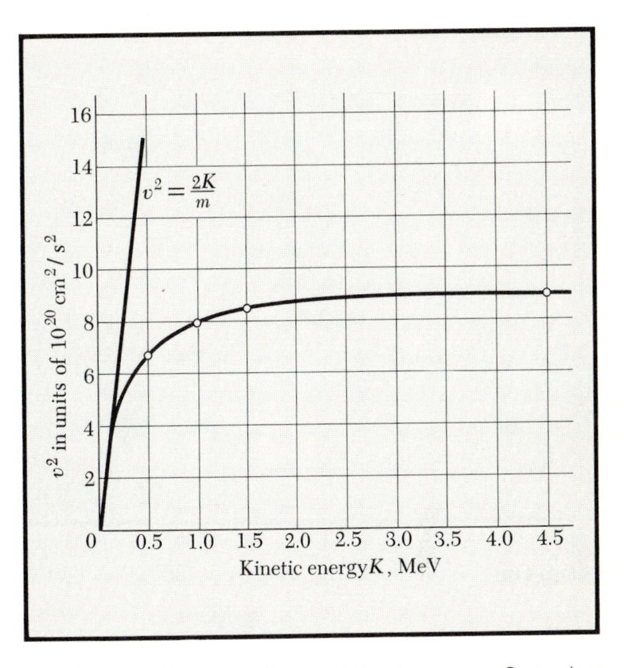

FIG. 10.20 Graph of v^2 against kinetic energy. Open dots are experimental points.

CONCLUSIONS

We are now prepared to study special relativity in Chap. 11, with the knowledge from experiment that

1. c is invariant among inertial frames, that is, frames of reference moving with uniform velocity with respect to each other.
2. c is the maximum speed at which energy can be transmitted.
3. The absolute velocity of a frame of reference has no meaning. Only relative velocities can be experimentally determined.

4 Simple galilean transformations do not provide satisfactory explanations of phenomena involving high speeds.

5 The newtonian formula for kinetic energy, $\frac{1}{2}mv^2$, fails when v approaches c.

We have reviewed only a very small fraction of the experiments that support the special theory of relativity, which is now very solidly established. Physicists place as much reliance on this theory as any other part of physics. Our next endeavor must be to formulate it precisely and to understand some of its major consequences.

PROBLEMS

1. *Doppler shift.* A space navigator wishes to determine his velocity of approach as he nears the moon. He sends a radio signal of frequency $\nu = 5000$ Mc/s and compares this frequency with its echo, observing a difference of 86 kc/s. Calculate the velocity of the space vehicle relative to the moon. (The nonrelativistic expression for the doppler effect is sufficiently accurate for many purposes.) *Ans.* 2.6×10^5 cm/s.

2. *Recessional red shift.* A spectral line appearing at a wavelength of 5000 Å in the laboratory is observed at 5200 Å in the spectrum of light coming from a distant galaxy.
   (*a*) What is the recessional velocity of the galaxy? *Ans.* 1.2×10^9 cm/s.
   (*b*) How far away is the galaxy? *Ans.* 8×10^{26} cm.

3. *Speed of light.* In Michelson's celebrated measurement of the speed of light, an octagonal reflecting prism rotating about the axis of the prism reflected a beam of light from a distant light source and back to an observer near the source. The timing provides that the transit time of the light equal one-eighth of the period of rotation of the octagonal prism. The one-way distance was $L = 35.410 \pm 0.003$ km and the frequency of rotation of the prism was $\nu = 529$ cps to an accuracy of 3×10^{-5} cps.
   (*a*) Calculate the speed of light from these data. (A fractional correction of the order of 10^{-5} for atmospheric effects had to be applied.)
   (*b*) The angle between any two adjacent prism faces was $135° \pm 0.1''$. Estimate the overall precision of the measurement of c.

4. *Eclipses of Io.* Jupiter's satellite Io moves in an orbit of radius 4.21×10^{10} cm with an average period of 42.5 h. Roemer observed that the period varied regularly during the year, with a period of variation of about 1 yr. The maximum deviation of the period from the average was 15 s, at times approximately 6 months apart. Neglect the orbital travel of Jupiter.
   (*a*) Estimate the distance the earth travels in one period of Io's motion about Jupiter. *Ans.* 4.5×10^{11} cm.
   (*b*) When does Io's period appear to be greatest?
   (*c*) Use the preceding result and the data provided to estimate the velocity of light.
   (*d*) Estimate the accumulated delay in the 6 months following the point of zero delay when the earth is closest to Jupiter.

5. *Stellar parallax and aberration.* Stellar parallax was predicted by Aristarchus of Samos (ca. 200 B.C.) and it was finally observed for certain by Bessel in 1838. A notably unsuccessful attempt was made by Bradley, who discovered instead the aberration of starlight. During the course of a year the apparent position of a star shifts between extremes by approximately 40″ of arc due to aberration.
   (*a*) What would be the distance in parsecs of a star with a parallax of 20″? The nearest known star is α Centauri at a distance of about 1.3 parsecs. *Ans.* 0.05 parsec.
   (*b*) Show that the apparent annual motion from aberration of stars near the ecliptic is a straight line whose ends

subtend a 40″ angle. The ecliptic is the plane of the earth's orbit.

6. *Rotation of galaxies.* In 1916, before the great distances of the nebulae (galaxies) were known, the spiral M101 was reported to rotate like a solid body with a period of 85,000 yr. The observed angular diameter is 22′. Calculate the maximum possible distance of the nebula if the above period is correct, supposing that the extremities of the nebula are not to move faster than c. (Recent measurements of stars in M101 place it at a distance of 8.5×10^{24} cm. It is apparent that the rotation period reported in 1916 was underestimated.)

7. *Variable stars.* The 200-in. Mt. Palomar telescope can barely resolve individual stars in galaxies at a distance of 3×10^{25} cm. One method for calibrating distances of this order of magnitude involves observation of the periods in the luminosity of certain Cepheid-type variable stars. A Cepheid-type star is a gravitationally unstable star that exhibits periodic pulsations in which its radius may change by perhaps 5 to 10 percent. The period of a Cepheid is related to its average luminosity. The temperature of the star changes with the same period as the radius, so that one observes periodic variations in brightness. Periods as short as a few hours have been found. A Cepheid whose intrinsic luminosity is 2×10^4 times that of the sun has a period of 50 days in our galaxy.

(*a*) Estimate from the distance-velocity relation [Eq. (10.10)] the radial velocity for a galaxy at a distance of 3×10^{25} cm.

(*b*) What would we expect to observe for the period of this Cepheid in a galaxy at the distance cited above?

Ans. 50.08 days.

8. *Novae.* Occasionally a star is seen to experience an explosion in which a portion of its outer layers is thrown out with high velocity. Such a star is called a *nova.* A recent nova was observed visually to have a surrounding shell after its outburst. The angular diameter of the shell was found to increase by 0.3″/yr. The spectrum of the nova is a normal stellar spectrum with superimposed broad emission lines, the widths (in wavelengths) of which remain constant at 10 Å (in the vicinity of a wavelength of 5000 Å), though the lines are dimming. The width is to be interpreted as a measure of the doppler shift between the parts of the shell advancing toward us and receding from us. Estimate the distance to the nova, if the shell is optically thin (so that we receive as much light from the far hemisphere as from the near).

Ans. 1.2×10^{21} cm.

9. *Velocities of galaxies.* Measured radial velocities of galaxies relative to the earth are not isotropic over the sky. Nonisotropy results from the motion of the sun (orbital velocity) with respect to the center of our galaxy, and from our galaxy's own motion with respect to the local extragalactic standard of rest. Let us examine all galaxies at a particular distance, say, 3.26×10^7 light yr.

(*a*) What is the mean radial velocity of these galaxies?

Ans. The mean velocity of the galaxies as calculated from the velocity-distance relation is 930 km/s.

(*b*) Where in their spectra will be the average location of the Hα line of hydrogen? (In the laboratory, $\lambda_{H\alpha} = 6.563 \times 10^{-5}$ cm.)

Ans. The Hα line will be, on the average, at 6.584×10^{-5} cm.

In our sample we find that in a certain direction the velocities are 300 km/s larger than the average and in just the opposite direction they are this much too small.

(*c*) What is the velocity of the sun in this frame of reference?

Ans. 300 km/s.

(*d*) Is that necessarily the orbital velocity of the sun around the center of our galaxy?

Ans. No, for it can include any motion of our galaxy as a whole in this reference frame.

(*e*) Assuming that this is the orbital velocity, estimate the mass of our galaxy, taking all the mass to be at its center and the orbit of the sun to be circular (the distance to the center of the galaxy is 3500 light yr). Compare with the mass of 8×10^{44} g quoted for the mass of the galaxy and explain the difference.

Ans. 4.5×10^{43} g. This is less than that usually quoted because much of the mass of our galaxy is not at the center—in fact, much mass lies exterior to the sun, where it would not affect the sun's motion or be detectable in this way.

10. *Rotation of stars.* The sun is seen from its surface features to rotate slowly, with a period of 25 days at the equator. Some stars, however, rotate far faster. How can this be determined in view of the fact that the stars are too distant to be seen except as points of light?

FURTHER READING

HPP, "Project Physics Course," chaps. 16 (sec. 6) and 20 (sec. 1), Holt, Rinehart and Winston, New York, 1970.

A. A. Michelson, "Studies in Optics," The University of Chicago Press, Chicago, 1927; paperback reprint, 1962.

CONTENTS

Special Relativity: The Lorentz Transformation

BASIC ASSUMPTIONS

The null result of the Michelson–Morley experiment to detect the drift of the earth through an ether and the other results discussed in Chap. 10 can only be understood by making a revolutionary change in our thinking; the new principle we need is simple and clear:

> The speed of light is independent of the motion of the light source or receiver.

That is, the speed of light is the same in all reference frames in uniform motion with respect to the source. To this new assumption must be added our earlier assumption:

> Space is isotropic and uniform. The fundamental laws of physics are identical for any two observers in uniform relative motion.

All the vast consequences of the special theory of relativity follow from these assumptions.

Electromagnetic waves or photons are not unique in having a velocity independent of the motion of the source. Physicists believe, with strong evidence, that there are other particles, notably neutrinos and antineutrinos, that have velocities equal to c. We shall, however, discuss photons because it is easier to carry out experiments with them.

Consider first a light wave spreading out from a point source. The wave front (surface of equal phase) will be a sphere if viewed in the reference frame in which the source is at rest. But according to our new principle the wave front must also be a sphere when viewed in a reference frame in uniform motion with respect to the source; otherwise we could tell from the shape of the wave front that the source is moving. The fundamental assumption that the speed of light is independent of the motion of the source demands that we be unable to tell from the shape of the wave front whether or not the source is in uniform motion.

LORENTZ TRANSFORMATION

In Chap. 4 we introduced the galilean transformation in order to understand how phenomena would look from two different points of view. We shall use the same ideas here with two different frames of reference S and S', moving with uniform velocity V with respect to each other. We wish to find a transformation of coordinates, and possibly of the time also, such as the galilean transformation [Eq. (4.14)] relating the coordinates and time in one frame of reference to the coordinates and time in another frame of reference in such a way as to be consistent with the relativity assumptions. If we assume that in the frame S a light source is at the origin, the equation of a spherical wave front emitted at $t = 0$ is

$$x^2 + y^2 + z^2 = c^2t^2 \tag{11.1}$$

In the frame of reference S′ in which the coordinates are x', y', z', and t', the equation of the spherical wave front must be

$$x'^2 + y'^2 + z'^2 = c^2t'^2 \tag{11.2}$$

The speed of light c is the same in both Eqs. (11.1) and (11.2).

We can try the galilean transformation to see whether it gives results in agreement with Eqs. (11.1) and (11.2).

$$x' = x - Vt \qquad y' = y \qquad z' = z \qquad t' = t \tag{11.3}$$

When we substitute Eq. (11.3) in Eq. (11.2) we obtain directly

$$x^2 - 2xVt + V^2t^2 + y^2 + z^2 = c^2t^2$$

This result is certainly not in agreement with Eq. (11.1). Thus the galilean transformation fails, and we must attempt to find some other transformation. It must reduce to the galilean transformation when the velocity V becomes very small compared with the velocity of light c.

Let us try

$$x' = \alpha x + \epsilon t \qquad y' = y \qquad z' = z \qquad t' = \delta x + \eta t$$

We know that for $x' = 0$, $dx/dt = V$; and for $x = 0$, $dx'/dt' = -V$. The algebra leads to

$$V = -\frac{\epsilon}{\alpha} \qquad -V = \frac{\epsilon}{\eta}$$

or

$$\alpha = \eta$$

When we write, repeating Eq. (11.2),

$$x'^2 + y'^2 + z'^2 = c^2t'^2$$

we get

$$\alpha^2x^2 + 2\alpha\epsilon xt + \epsilon^2t^2 + y^2 + z^2 = c^2(\delta^2x^2 + 2\delta\alpha xt + \alpha^2t^2)$$

This is to be compared to Eq. (11.1), and we see that consistency is possible if

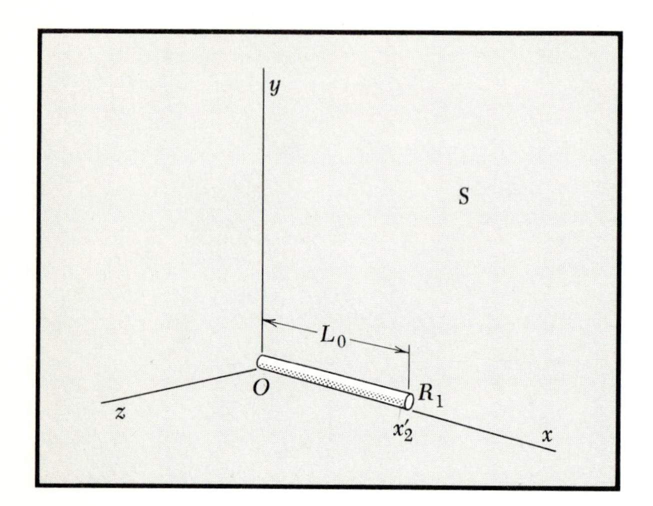

FIG. 11.1 (*a*) Consider a rigid rod R_1 of length L_0 in its rest frame S.

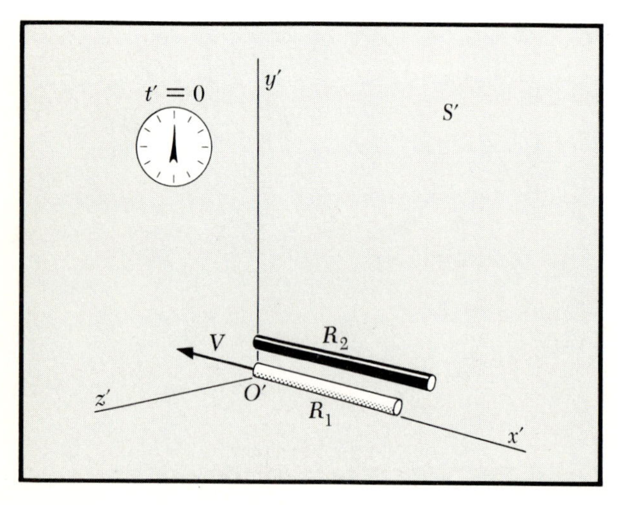

(*b*) The Lorentz transformation tells us that R_1, which has speed V in S', will be measured to have length $L' = L_0\sqrt{1 - V^2/c^2}$ in S'. Note that $x_1 = x_1' = 0$ in the figure.

$$2\alpha\epsilon = 2c^2\delta\alpha$$
$$\alpha^2 - c^2\delta^2 = 1$$

and

$$c^2\alpha^2 - \epsilon^2 = c^2$$

Eliminating ϵ by using $\epsilon = -V\alpha$, we get

$$\alpha = \frac{1}{(1 - V^2/c^2)^{\frac{1}{2}}} \qquad \epsilon = \frac{-V}{(1 - V^2/c^2)^{\frac{1}{2}}}$$
$$\delta = \frac{-V/c^2}{(1 - V^2/c^2)^{\frac{1}{2}}} \qquad \eta = \frac{1}{(1 - V^2/c^2)^{\frac{1}{2}}}$$

Our transformation is then

$$x' = \frac{x - Vt}{(1 - V^2/c^2)^{\frac{1}{2}}} \qquad y' = y \qquad z' = z$$
$$t' = \frac{t - (V/c^2)x}{(1 - V^2/c^2)^{\frac{1}{2}}} \tag{11.4}$$

This is the *Lorentz transformation*.[1] It is linear in x and t; it reduces to the galilean transformation for $V/c \to 0$; when substituted in Eq. (11.2) it gives

$$x^2 + y^2 + z^2 = c^2t^2$$

exactly as required. That is,

$$x'^2 + y'^2 + z'^2 = c^2t'^2$$

is *invariant* under a Lorentz transformation. The form of the equation describing the wave front is the same in all frames moving with uniform relative velocity. Equation (11.4) is the unique solution to all our difficulties. It is a good shorthand way to remember many important results in relativity. We shall discuss several of them below with the help of the Lorentz transformation.

It is usually convenient to make use of the standard notation used in relativity:

$$\beta \equiv \frac{V}{c} \tag{11.5}$$

That is, β (Greek beta) is the velocity measured in a natural

[1] This transformation has a long history. It was first used by J. Larmor to explain the null result of the Michelson–Morley experiment, in his "Aether and Matter," pp. 174–176, Cambridge University Press, New York, 1900. Larmor claims accuracy only to order v^2/c^2; in fact, his results are exact.

system of units in which $c = 1$. It is also convenient to introduce γ (Greek gamma):

$$\gamma \equiv \frac{1}{(1 - \beta^2)^{\frac{1}{2}}} \equiv \frac{1}{(1 - V^2/c^2)^{\frac{1}{2}}} \tag{11.6}$$

Note that $\gamma \geq 1$. The Lorentz transformation Eq. (11.4) then becomes

$$x' = \gamma(x - \beta ct) \qquad y' = y \qquad z' = z \qquad t' = \gamma\left(t - \frac{\beta x}{c}\right) \tag{11.7}$$

and the reader can prove (Prob. 2) that the inverse transformation is

$$x = \gamma(x' + \beta ct') \qquad y = y' \qquad z = z' \qquad t = \gamma\left(t' + \frac{\beta x'}{c}\right) \tag{11.8}$$

Length Contraction Consider a rod (see Fig. 11.1*a*) lying along the x axis and at rest in reference frame S. Because the rod is at rest in S, the position coordinates of its ends x_1 and x_2 are independent of time. Thus

$$L_0 = x_2 - x_1$$

is called the *rest length* or *proper length* of the rod. Also consider a rod (see Fig. 11.2*a*) lying along the x' axis and at rest in reference frame S′. For the same reason

$$L_0 = x_2' - x_1'$$

is called the *rest length* or *proper length* of the rod in S′.

We now wish to determine the lengths of these rods when viewed from a moving reference frame. First look at the rod in Fig. 11.1*a* from the reference frame S′ which moves with velocity $V\hat{\mathbf{x}}$ with respect to the rod at rest in S. (See Fig. 11.1*b* and note that the rod R_2 from Fig. 11.2*a* is at rest in S′.) We determine the length of the rod as viewed from S′ by determining at a given time t' the positions x_1' and x_2' that coincide with the ends of the rod. The important point here is that the time t' is the same for x_1' and x_2'. To say this another way, the distance between positions x_1' and x_2' in S′ which coincide

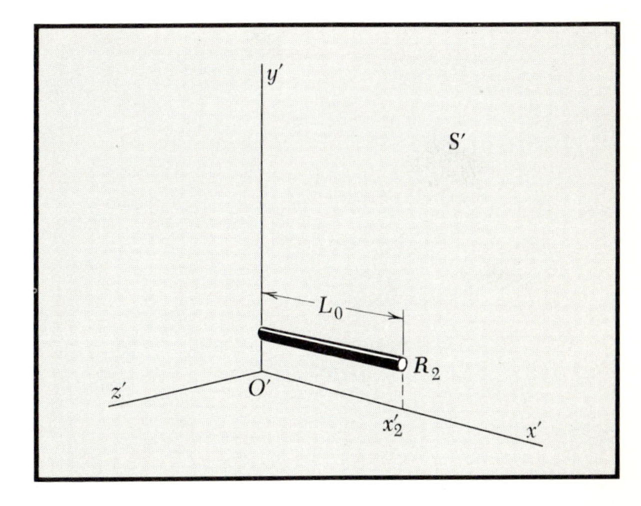

FIG. 11.2 (*a*) Consider a similar rigid rod R_2 of length L_0, as measured in its rest frame S'.

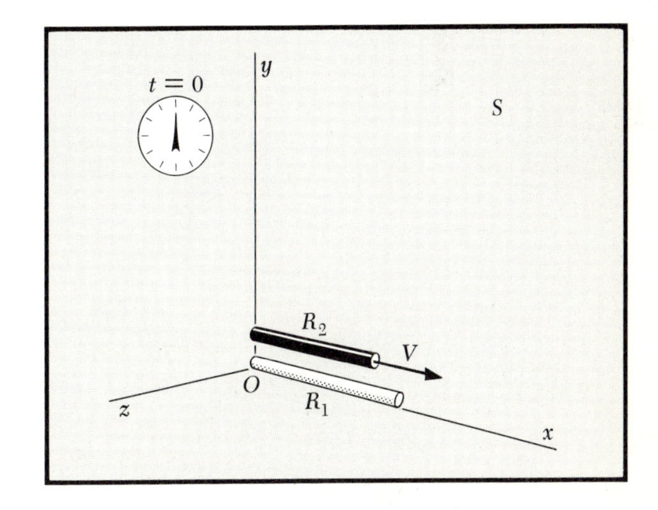

(*b*) The Lorentz transformation also tells us that R_2, which has speed V in S, will be measured to have length $L = L_0\sqrt{1 - V^2/c^2}$ in S. Note that $x_1' = x_1 = 0$ in the figure.

simultaneously (in S') with the endpoints of the rod is the natural definition of the length L in the moving frame S'.

From the Lorentz transformation, Eq. (11.8), we have

$$x_1 = \gamma(x'_1 + Vt'_1)$$
$$x_2 = \gamma(x'_2 + Vt'_2)$$
$$x_2 - x_1 = L_0 = \gamma(x'_2 - x'_1) + \gamma V(t'_2 - t'_1)$$

Now letting $t'_2 = t'_1$ as we saw was necessary for the measurement in S', we get

$$L_0 = \gamma(x'_2 - x'_1) = \gamma L$$

or

$$L = \frac{L_0}{\gamma} = L_0(1 - \beta^2)^{\frac{1}{2}} \qquad (11.9)$$

by using our definition $\gamma = (1 - \beta^2)^{-\frac{1}{2}}$. In other words, the measurement in the moving frame gives a shorter length than the measurement in the stationary frame.

Alternatively we look at the rod in Fig. 11.2*a* (at rest in S') from the reference frame S which moves with velocity $-V\hat{\mathbf{x}}'$ with respect to the rod at rest in S'. (See Fig. 11.2*b* and note that the rod R_1 from Fig. 11.1*a* is at rest in S.) The procedure is the same, but now the time t is the same for the determination of the endpoints x_1 and x_2. From the Lorentz transformation, Eq. (11.7), we have

$$x'_1 = \gamma(x_1 - Vt_1)$$
$$x'_2 = \gamma(x_2 - Vt_2)$$
$$x'_2 - x'_1 = L_0 = \gamma(x_2 - x_1) - \gamma V(t_2 - t_1)$$

and letting $t_2 = t_1$, we get

$$L_0 = \gamma(x_2 - x_1) = \gamma L$$
$$L = L_0(1 - \beta^2)^{\frac{1}{2}}$$

The measurement of the moving rod again gives a length shorter than the measurement of the stationary rod.

This is the famous Lorentz–Fitzgerald contraction of a rod moving parallel to its length with respect to the observer. One may worry at this point whether the rod has "actually contracted." Of course nothing physical has happened to the rod, but the process of measurement in the moving frame has given a different result. For a discussion of the figures of rapidly moving objects as photographed with a camera, see the excellent review by Weisskopf.[1] It has been shown, for example,

[1] V. F. Weisskopf, *Physics Today*, 13:24–27 (Sept. 1960).

by calculation of trajectories that a moving sphere will photograph as a sphere and not as an ellipsoid.

In the foregoing discussion we have emphasized that the observer makes his measurement of length by *simultaneously* recording the positions of the ends of the rod in his own reference frame. That is what we required of the observer in the moving frame S′ when he measured the length of the rod stationary in S with the result L_0/γ [Eq. (11.9)]. It is essential that we recognize that this act of simultaneously registering the endpoints at time t' in S′ does *not* transform into simultaneous events at the endpoints x_1 and x_2 in S; on the contrary the Lorentz equations indicate a time interval

$$t_2 - t_1 = \frac{\beta(x_2 - x_1)}{c}$$

in S for the registering of the two endpoints that was done simultaneously in S′. We will presently see that for a rod lying on the y axis, the question of simultaneity does not arise, but for a rod along the x axis, the matter of simultaneity is all important.[1]

This is illustrated by a different example. We can easily synchronize a series of clocks in S, the frame in which the rod is at rest. Let the clocks at $x = 0$ and $x = L_0$ (at each end of the meter stick) each emit at $t = 0$ a directional flash of light in the y direction. These two flashes are received in S′ by two of a series of counters spaced along the x' axis. How far apart are the two counters which were triggered? From Eq. (11.7) we have, for the location of the two counters,

$$x_1' = 0 \cdot \gamma - c \cdot 0 \cdot \beta\gamma = 0$$
$$x_2' = L_0\gamma - c \cdot 0 \cdot \beta\gamma = L_0\gamma$$

so that their distance apart is

$$x_2' - x_1' = L_0\gamma = \frac{L_0}{(1 - \beta^2)^{\frac{1}{2}}} \tag{11.10}$$

This does not agree with Eq. (11.9)! But we have done a *different* experiment and obtained a different result. Our earlier experiment was based on the natural definition of length in S′, using the requirement of simultaneity in S′. That earlier experiment involved comparing $\Delta x'$ with Δx when $\Delta t' = 0$,

[1]The reader is referred to Taylor and Wheeler, "Space-Time Physics—An Introduction," pp. 64–66, W. H. Freeman and Company, San Francisco, 1965.

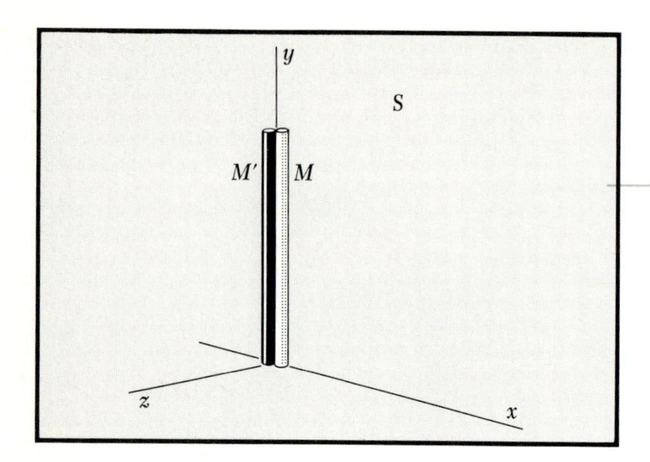

FIG. 11.3 (*a*) Suppose we have two identical rods M' and M at rest in S.

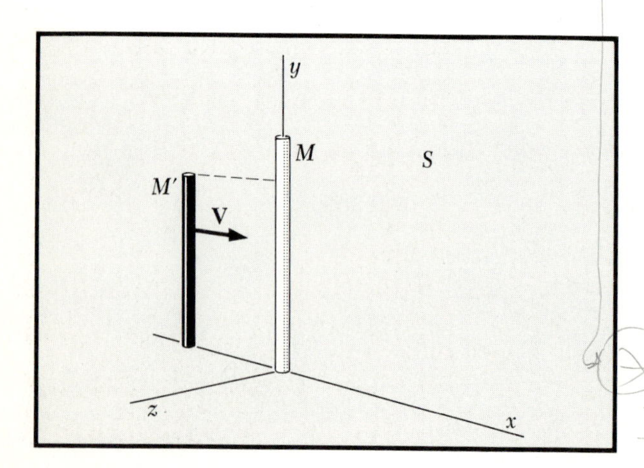

(*b*) Suppose M' appears shorter to observer in S when it moves relative to S.

whereas the second experiment involved comparing $\Delta x'$ with Δx when $\Delta t = 0$.

We have learned indirectly from the result in Eq. (11.10) of the second experiment that two events simultaneous in S are not, in general, simultaneous in S′. Thus from Eq. (11.7) we see that two events *simultaneous* ($\Delta t = 0$) in S, which are separated by Δx in space, will be separated in S′ in both space and time:

$$\Delta x' = \gamma\,\Delta x \qquad c\,\Delta t' = -\beta\gamma\,\Delta x$$

Measurement of Length Perpendicular to Relative Velocity

Contrary to the measurement of the distance in the direction of the relative velocity, we see from the Lorentz transformation, Eq. (11.7), that

$$y' = y \qquad z' = z$$

These relations are equivalent to the statement that the measurement of the length of a meter stick is independent of its velocity *if* the meter stick moves perpendicular to its length.

How would we verify this statement experimentally? We can take a meter stick and move it at uniform velocity past another meter stick which is at rest. There is no problem in making the origins of both meter sticks cross exactly. Then the 1-m mark of each will also cross exactly, or, if the motion changes the length, we can arrange for the 1-m mark of the shorter stick to make a scratch on the longer stick (see Fig. 11.3*a* to *c*). This provides a definite physical record of the length.

Let S be the rest frame of one meter stick and S′ the rest frame of the other. Suppose the motion does change the apparent length. Then if the laws of physics are to remain the same for an observer on S as for an observer on S′, it is necessary that the stick which appeared the shorter to an observer on S should appear the longer to an observer on S′. But this reversal of the roles is incompatible with our physical record that one meter stick is shorter than the other. Therefore, the lengths must be equal when viewed from S and S′ (see Fig. 11.3*d* and *e*). This discussion merely confirms that $y = y'$ and $z = z'$.

These results concerning the measurements of lengths parallel and perpendicular to the relative velocity imply that the measurements of angles involving x coordinates will be different in the two frames. This is true, and the reader can work out for himself the relations between the trigonometric functions of the angles in the two frames. (See Prob. 5 at the end

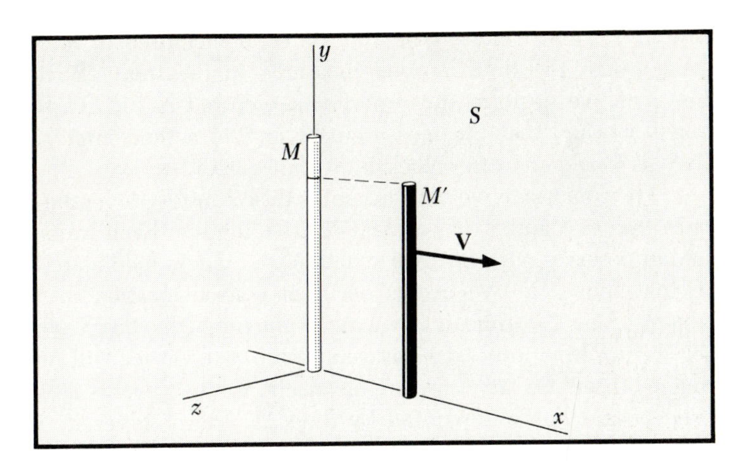

FIG. 11.3 (*cont'd*) (*c*) Then we could arrange it so that the end of M' leaves a scratch on M as it passes by.

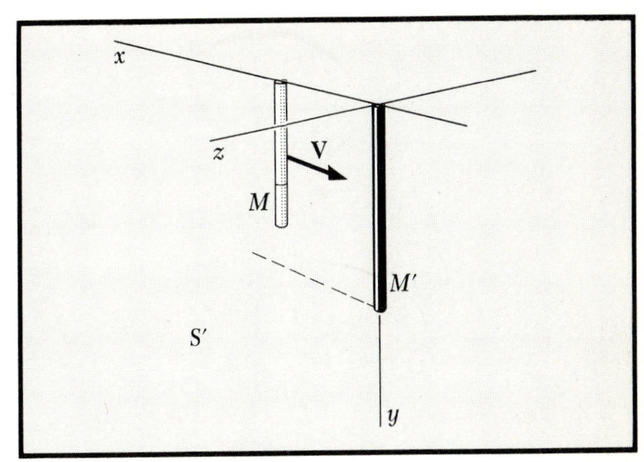

(*d*) The scratch is a physical result of an experiment, and it must be observed in another frame, for example, upside down in the rest frame of M'. But now M must appear *shorter* than M', since M is moving and M' is at rest.

of the chapter.) Remember that the important point here is to determine in which frame the measurements of the ends of the length are simultaneous.

Time Dilation of Moving Clocks Used in the ordinary sense, the word *dilate* means *enlarge beyond normal size;* in connection with a clock, it means to lengthen an interval of time. We now consider a clock which is at rest in reference frame S.

The result of the measurement of a time interval in the frame in which the clock is *at rest* is denoted as

$$\tau = t_2 - t_1$$

and is called the *proper time*. Then using the Lorentz transformation [Eq. (11.7)], we get

$$t_2' = \gamma\left(t_2 - \frac{\beta x_2}{c}\right) \qquad t_1' = \gamma\left(t_1 - \frac{\beta x_1}{c}\right)$$

or

$$\boxed{t_2' - t_1' = \gamma(t_2 - t_1) = \gamma\tau = \frac{\tau}{(1-\beta^2)^{\frac{1}{2}}}} \tag{11.11}$$

where we have set $x_2 - x_1 = 0$; the clock stays at the same place in S. This is the time interval measured by a clock at rest in S' moving with velocity $V\hat{\mathbf{x}}$ with respect to the frame

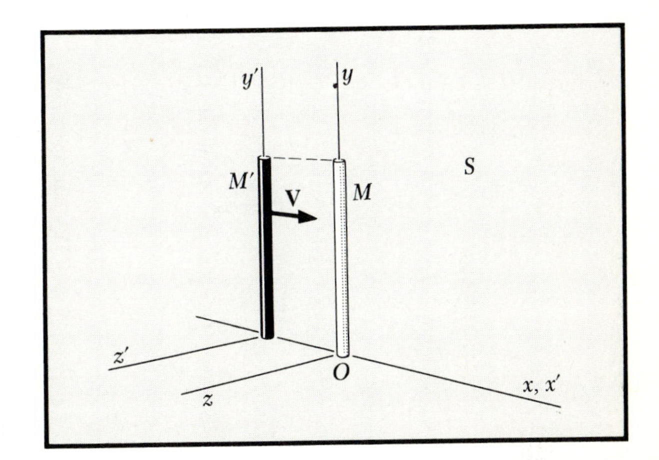

(*e*) Thus we have a contradiction, which is resolved only if M' and M have the same length even when one of them is moving. Thus $y' = y$. By a similar argument $z' = z$.

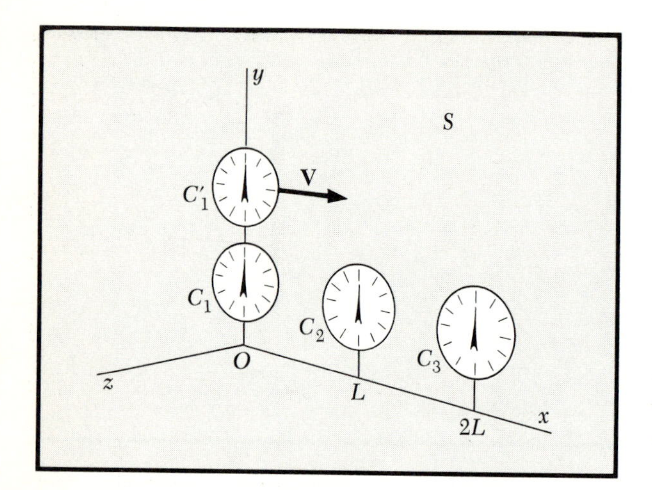

FIG. 11.4 (*a*) Clocks C_1, C_2, C_3 are at rest in S, spaced at equal intervals L along the x axis, and all synchronized. Clock C_1' has velocity V with respect to S. Suppose $t' = 0$ when $t = 0$, as shown.

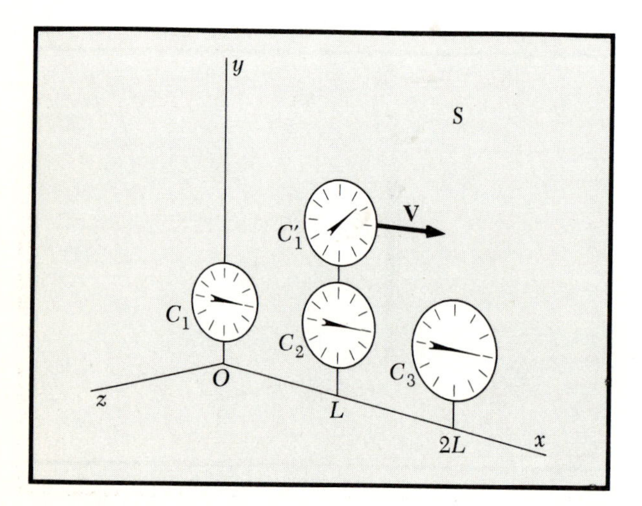

(*b*) The Lorentz transformation yields $t' = (t - xV/c^2)\,\gamma = t\sqrt{1 - V^2/c^2}$ since $x = L = Vt$. To the observer in S, the moving clock C_1' runs slow.

S of the original clock. The time interval measured in S′ is longer than the time interval measured in the frame S. If, however, we perform the experiment pictured in Fig. 11.4*a* and *b* we find that the measurement in S of a time interval in S′ is longer than the clock in S′ shows it to be.

The conclusion we must accept is this: Consider two reference frames S and S′ in constant relative motion. Each frame has an observer with his own synchronized clocks held at rest in that frame. If two events occur at a fixed location in S separated by the time interval Δt as measured by the S observer, the time interval measured by the S′ observer will be longer; it will be $\Delta t' = \gamma\,\Delta t$. Conversely, for two events at a fixed position in S′ separated by time $\Delta t'$, the observer in S will measure a longer interval; he will measure $\Delta t = \gamma\,\Delta t'$ (see Fig. 11.5*a* and *b*).

This effect is called *time dilation*. Moving clocks appear to advance more slowly than clocks at rest. This is not easy to understand in an intuitive way and it may take you a long time to feel content with time dilation. The root of the apparent paradox is the invariance of c, and a straightforward problem illustrates how time dilation is forced upon us by this constancy of the speed of light. Let us construct a standard clock in the reference frame S (see Fig. 11.6). The clock can be used to measure the time τ needed for a light pulse to travel a fixed distance L from a source at rest to a mirror at rest, and back again. The light path is along the y axis. Thus

$$\tau = \frac{2L}{c} \tag{11.12}$$

This time can be read on a dial or it can be printed out on a piece of paper. Observers in any frame can look at the printed record of the flight time of the pulse and they will all agree that a clock in the rest frame S recorded the time τ. But what do their own clocks, not in S, record?

An observer in a frame S′ (moving uniformly in the x direction with respect to S) (see Fig. 11.6) can also time the light-reflection experiment while it is carried out in S. The observer in S′ will do this by using a set of synchronized clocks at rest in S′. We start two clocks at rest in S′ at the same time (synchronized) by flashing a light source located midway between them; each starts from zero at the instant when the flash reaches it. The procedure may be extended to other clocks. We can also synchronize any number of clocks in one reference frame by synchronizing them when they are close together in

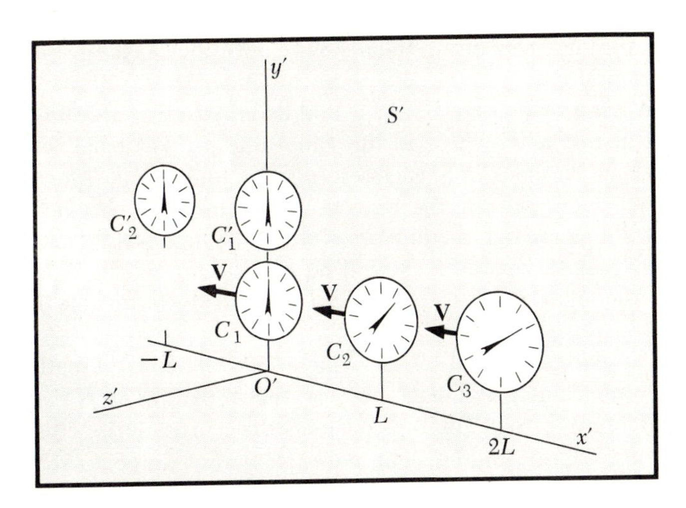

FIG. 11.5 (*a*) In S' clocks C'_1, C'_2, etc., are at rest, separated by distance L, and synchronized. To the observer in S', clocks C_1, C_2, C_3 are *not synchronized!* What do they read?

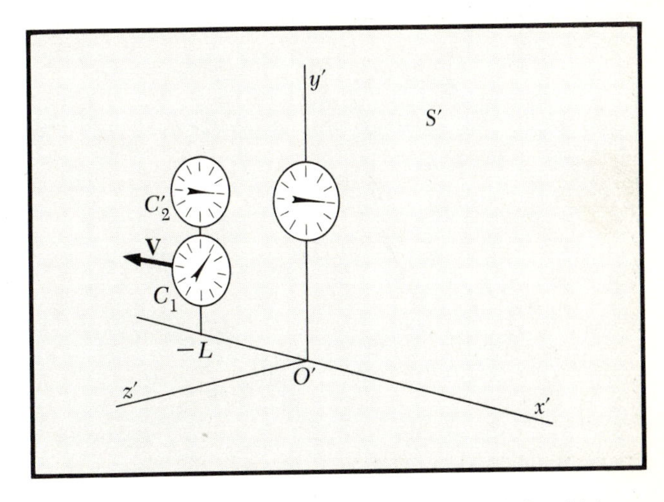

(*b*) To the observer in S', it is the *moving clock* C_1 which runs slow! Where are clocks C_2, C_3 and what do they read at this instant?

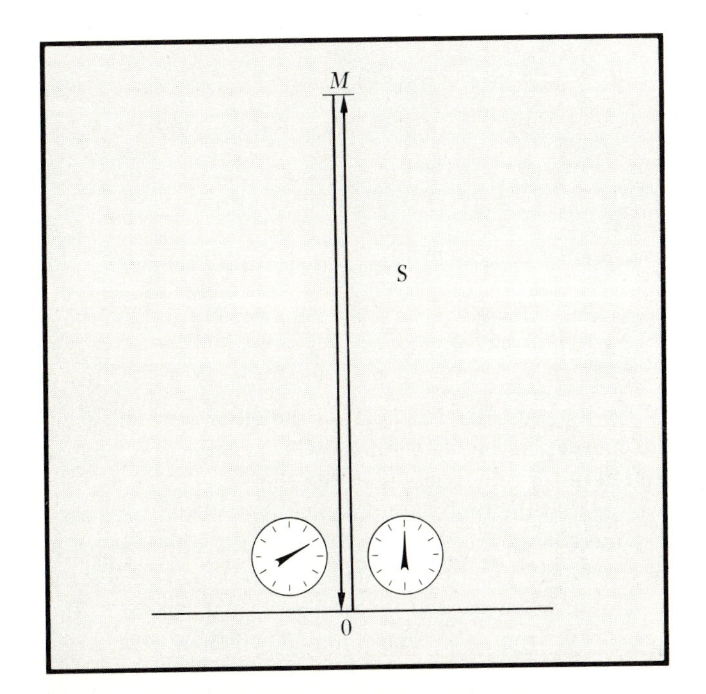

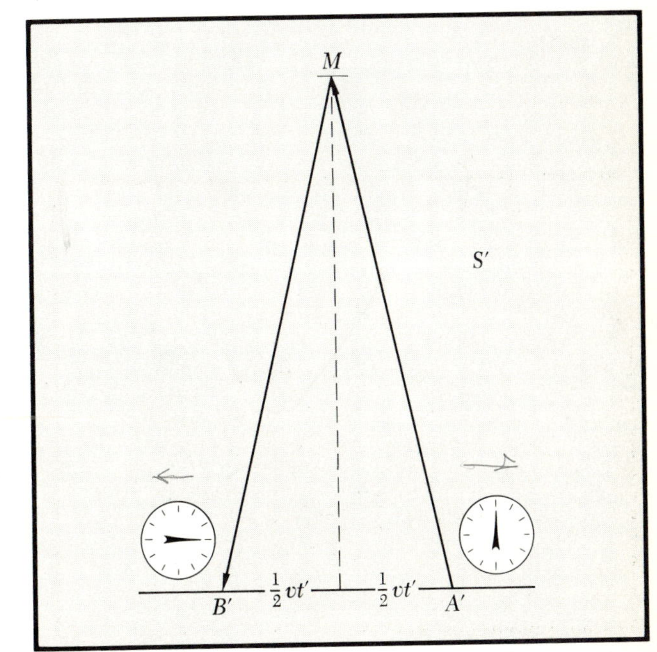

FIG. 11.6 View of path of light in frames S and S'. Point A' is coincident with 0 at time light is emitted. In S', light travels from A' to mirror M to B'.

space and then separating them slowly until they take up the desired positions.

We can read any clock in S' and be certain that all other clocks at rest in S' will read the same time. In particular we read whatever clock in S' is closest in space to the single clock in S which is used for the reflection experiment. One clock in S' will be closest and will be read when the light pulse starts out in S; another clock in S' will be closest and will be read when the light pulse returns and is recorded by the clock in S.

The path traversed by the light in S is $2L$. But the path as viewed from S' is longer because the apparatus in S has moved relative to S' by $V \cdot \frac{1}{2}t'$ along the x axis during the outbound passage of the light pulse from the source to the mirror and by another $V \cdot \frac{1}{2}t'$ during the inbound passage (see Fig. 11.6). Here t' is the time as observed in S'. The distance in S' traveled by the pulse is

$$2[L^2 + (\tfrac{1}{2}Vt')^2]^{\frac{1}{2}}$$

because the pulse travels always with the speed c, this distance must equal ct'. Thus

$$(ct')^2 = 4L^2 + (Vt')^2$$

or

$$t' = \frac{2L}{(c^2 - V^2)^{\frac{1}{2}}} = \frac{2L}{c} \frac{1}{(1 - \beta^2)^{\frac{1}{2}}}$$

or, by reference to Eq. (11.12),

$$t' = \frac{\tau}{(1 - \beta^2)^{\frac{1}{2}}} \tag{11.13}$$

exactly the same as Eq. (11.11). Thus the clock in S will seem to the timekeepers in S' to run slowly, because the S clock has printed out a time τ less than the time t'.

We see that the time-dilation effect does not involve mysterious processes in the interior of atoms; the effect arises in the measurement process. The clock at rest in S reads the proper time τ when viewed by an observer at rest in S. But when we view from S' a time interval which is τ in S, we see a longer time t' because of the longer light path. Any kind of clock will behave in the same way. In particular if τ is the decay half-life of mesons or of radioactive matter as measured in the frame S in which the particles are at rest, then

$$t' = \frac{\tau}{(1 - \beta^2)^{\frac{1}{2}}} \tag{11.14}$$

is the decay half-life observed in the frame S′ in which the particles are moving with velocity β. This is illustrated in Fig. 11.7*a* to g, which refers to the following example.

EXAMPLE

Lifetime of π^+ Mesons It is known that a π^+ meson decays into a μ^+ meson and a neutrino. The π^+ meson in a frame in which it is at rest has a mean life before decaying of about 2.5×10^{-8} s.† If a beam of π^+ mesons is produced with a velocity $\beta \approx 0.9$, what is the lifetime of the beam as viewed from the laboratory reference frame? A π^+ meson is a positively charged unstable particle with mass about $273m$, where m is the mass of the electron. The μ^+ meson has a mass of about $207m$; the neutrino has zero rest mass.

The proper lifetime τ of the π^+ meson is 2.5×10^{-8} s. If $\beta \approx 0.90$, then $\beta^2 \approx 0.81$ and the expected lifetime in the laboratory frame will be, from Eq. (11.14),

$$t' \approx \frac{2.5 \times 10^{-8}}{(1 - 0.81)^{\frac{1}{2}}} \approx 5.7 \times 10^{-8} \text{ s}$$

Thus on the average, before decaying, the particle will travel over twice as far as we would expect nonrelativistically from the product of the velocity times the proper lifetime.

Experiments on the lifetime of π^+ mesons (positive pions) are reported by R. P. Durbin, H. H. Loar, and W. W. Havens, Jr., *Phys. Rev.*, **88**:179 (1952). The results are in good agreement with the predicted time dilation for the appropriate velocity. Beams of π^+ mesons have been produced with

$$\beta = 1 - (5 \times 10^{-5})$$

their mean life in the beam is 2.5×10^{-6} s, or 100 times the proper lifetime of π^+ mesons at rest.

Consider a beam of π^+ mesons traveling with a velocity nearly equal to c. If the relativistic time-dilation effect did not exist, they would traverse a mean distance equal to $(2.5 \times 10^{-8}\text{ s})(3 \times 10^{10}\text{ cm/s}) \approx 700$ cm before decaying. They actually travel much farther than this, because of time dilation. The hydrogen bubble chamber at the Lawrence Berkeley Laboratory was about 100 m from the pion source in the Bevatron. The distance the pions travel before decay is of the order of $(2.5 \times 10^{-6})(3 \times 10^{10}) \approx 10^5$ cm, or about 100 times the distance they would travel before decay without the time-dilation effect. The design of apparatus for high-energy experiments in particle physics takes advantage of the long

† If N_0 is the number of radioactive particles present at time $t = 0$, the number left after time t is $N_0e^{-\lambda t}$. The mean life is $1/\lambda$; λ is the decay constant.

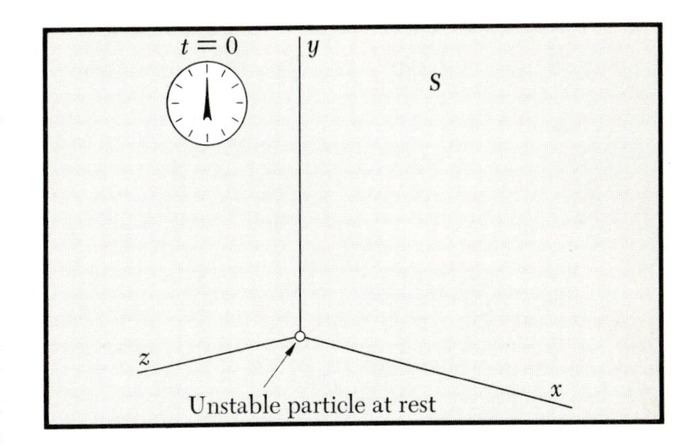

FIG. 11.7 (*a*) Another example of time dilation: An unstable particle is at rest in S. We begin to observe it at $t = 0$.

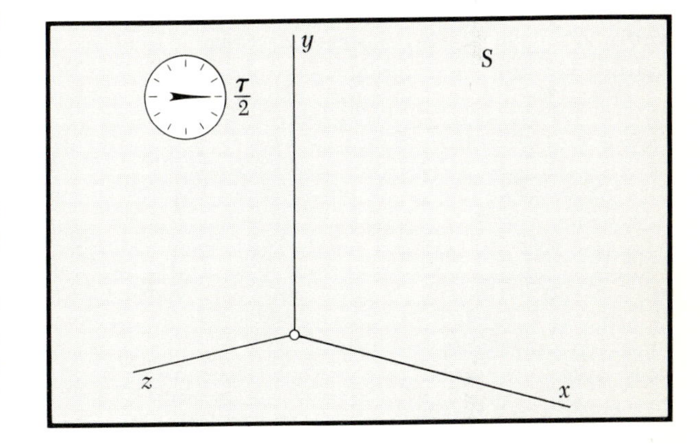

(*b*) Time elapses.

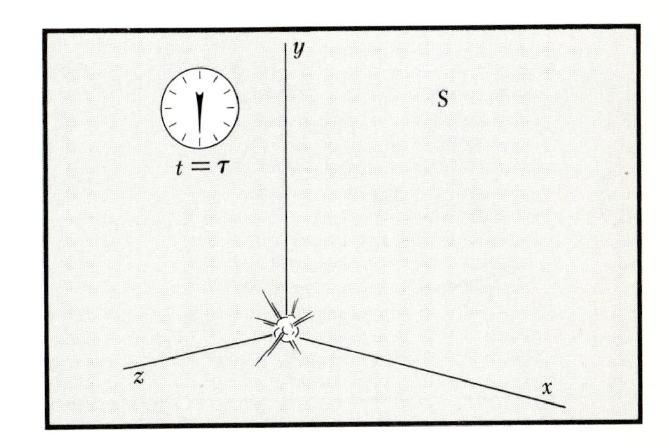

(*c*) The particle decays at time $t = \tau$.

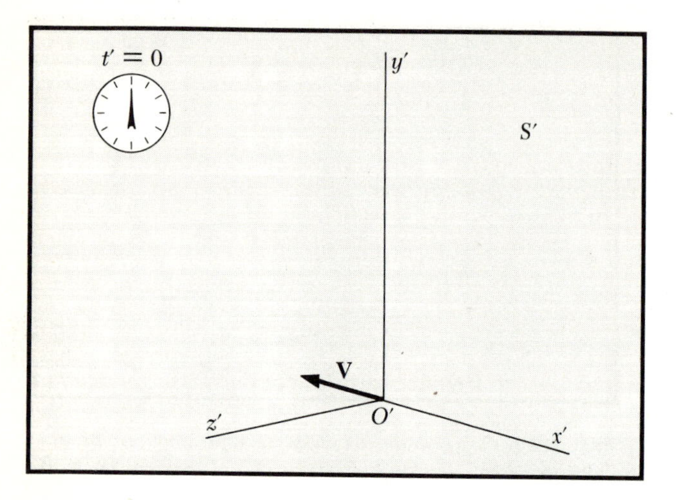

FIG. 11.7 (*cont'd*) (*d*) The same phenomenon observed from S'. Now the particle has speed V. We begin to observe it at $t' = 0 = t$.

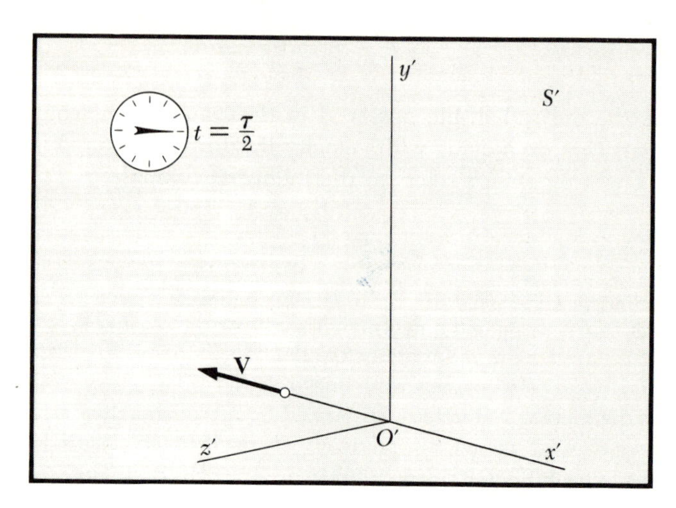

(*e*) Time elapses.

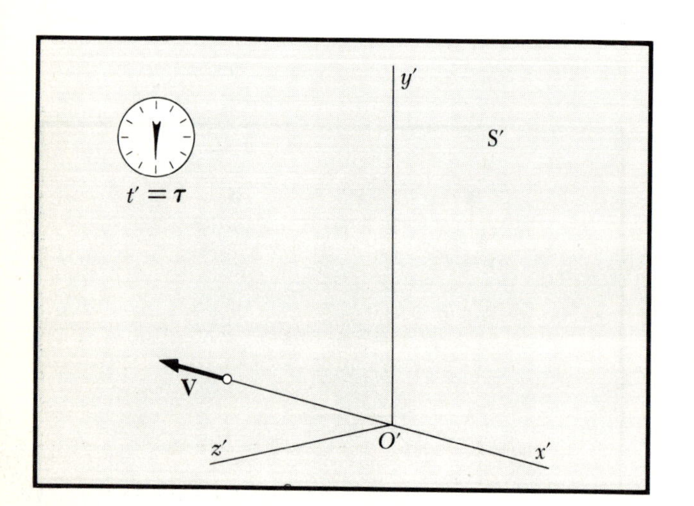

(*f*) But at $t' = \tau$ the particle has not yet decayed!

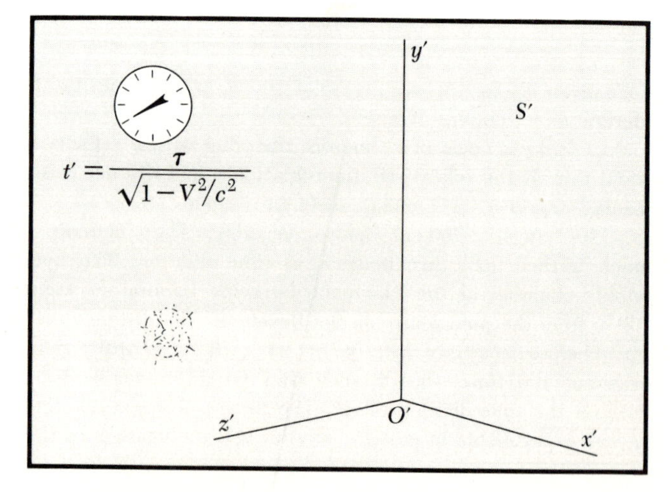

(*g*) The particle decays at $t' = \tau(1 - V^2/c^2)^{-1/2}$ according to an observer in S'.

decay distance due to relativity. It has been said that almost every high-energy physicist tests special relativity every day. He uses the Lorentz transformation with the same confidence that physicists in the nineteenth century used Newton's laws.

We repeat that there is nothing mysterious about the clocks. If there is anything mysterious about special relativity, it is the constancy of the speed of light. Granted that, everything else follows directly and fairly simply. Every new situation must be analyzed carefully, however. The field is rich in apparent paradoxes. Perhaps the most famous of these is the twin paradox.[1]

These two effects, length contraction and time dilation, are the most famous effects predicted by special relativity and verified by experiment. However, there are many more effects that have been thoroughly verified by experiment and we give some of them below. First we shall discuss transformations of velocities. In the galilean transformation we saw that velocities in the x direction simply add, and so we might expect that when the velocities approach the speed of light they would also add. However, we have seen in Chap. 10 that the speed of light is the greatest possible speed, and therefore we must change our conception derived from the galilean transformation of how velocities add.

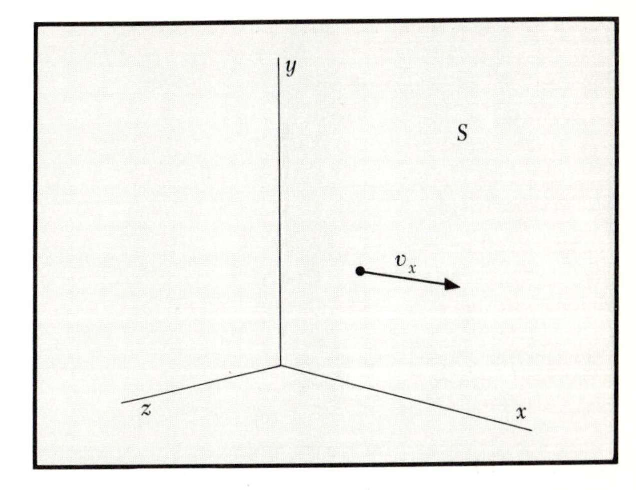

FIG. 11.8 (*a*) Suppose a particle has velocity v_x in S.

Velocity Transformation Suppose the S′ reference frame moves with uniform velocity $V\hat{\mathbf{x}}$ relative to the S reference frame. A particle moves with uniform velocity components v_x, v_y, v_z relative to the S frame. What are the velocity components v'_x, v'_y, v'_z of the particle relative to the S′ frame (see Fig. 11.8*a* and *b*)?

From Eq. (11.7) we have

$$x' = \gamma(x - \beta ct) \qquad t' = \gamma\left(t - \frac{\beta x}{c}\right)$$

whence

$$dx' = \gamma\, dx - \gamma\beta c\, dt \qquad dt' = \gamma\, dt - \frac{\gamma\beta\, dx}{c}$$

Thus

$$v'_x = \frac{dx'}{dt'} = \frac{\gamma\, dx - \gamma\beta c\, dt}{\gamma\, dt - \gamma\beta\, dx/c} = \frac{v_x - \beta c}{1 - v_x\beta/c}$$

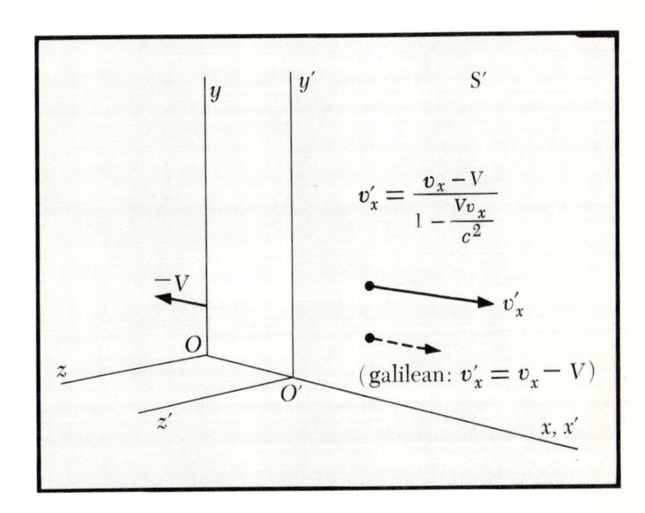

(*b*) Then in S′ the Lorentz transformation predicts $v'_x = (v_x - V)/(1 - v_xV/c^2)$. The galilean transformation would predict $v'_x = v_x - V$.

[1]This problem has recently been raised again. See M. Sachs, *Physics Today*, 24:23 (September 1971) and a group of letters in *Physics Today*, 25:9 (January 1972).

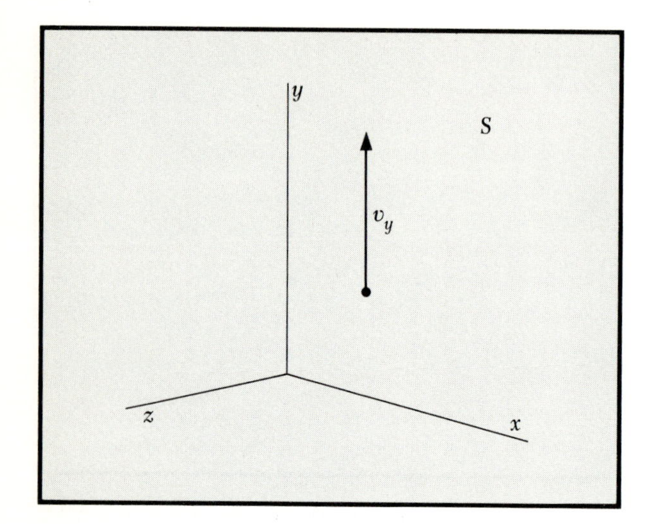

FIG. 11.9 (*a*) A particle has velocity v_y in the y direction in S.

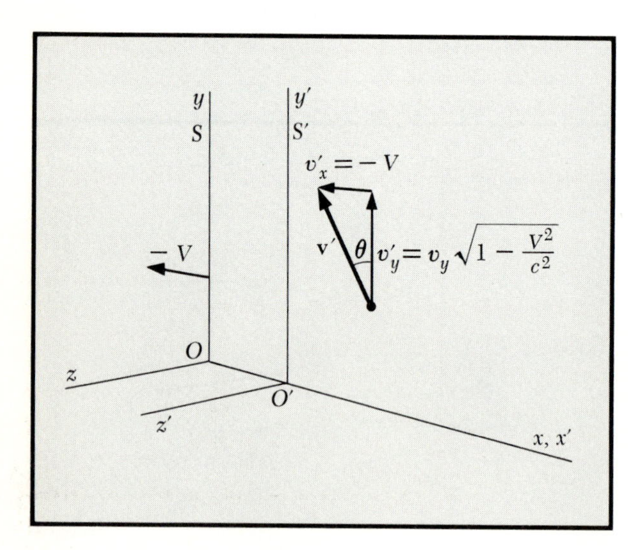

(*b*) Then it has the components shown in S', according to the Lorentz transformation

$$|\tan\theta| = \frac{V}{v_y\sqrt{1 - V^2/c^2}}$$

or

$$v'_x = \frac{v_x - V}{1 - v_x V/c^2} = \frac{v_x - V}{1 - \beta v_x/c} \tag{11.15}$$

This result may be compared with the galilean result $v'_x = v_x - V$ as in Chap. 4. Similarly, because $y = y'$ and $z = z'$ (see Fig. 11.9*a* and *b*),

$$v'_y = \frac{dy'}{dt'} = \frac{dy}{\gamma\, dt - \gamma\beta\, dx/c} = \frac{v_y}{1 - v_x V/c^2}\left(1 - \frac{V^2}{c^2}\right)^{\frac{1}{2}} = \frac{v_y}{\gamma(1 - \beta v_x/c)} \tag{11.16}$$

and

$$v'_z = \frac{v_z}{1 - v_x V/c^2}\left(1 - \frac{V^2}{c^2}\right)^{\frac{1}{2}} = \frac{v_z}{\gamma(1 - \beta v_x/c)} \tag{11.17}$$

The inverse transformations follow from Eq. (11.8) or by solving Eqs. (11.15) to (11.17) for the unprimed velocity components.

$$\begin{aligned} v_x &= \frac{v'_x + V}{1 + v'_x V/c^2} = \frac{v'_x + V}{1 + \beta v'_x/c} \\ v_y &= \frac{v'_y}{1 + v'_x V/c^2}\left(1 - \frac{V^2}{c^2}\right)^{\frac{1}{2}} = \frac{v'_y}{\gamma(1 + \beta v'_x/c)} \\ v_z &= \frac{v'_z}{1 + v'_x V/c^2}\left(1 - \frac{V^2}{c^2}\right)^{\frac{1}{2}} = \frac{v'_z}{\gamma(1 + \beta v'_x/c)} \end{aligned} \tag{11.18}$$

Note that for $V \ll c$, these reduce to the galilean transformation.

Suppose that the particle is a photon, and $v_x = c$ in S. From Eq. (11.15) we see that (Fig. 11.10)

$$v'_x = \frac{c - V}{1 - cV/c^2} = c$$

The velocity of the photon is also c in the frame S. The Lorentz transformation was designed to produce this result, and it is a reassuring check that we obtain c in both reference frames.

If $v_y = c$ and $v_x = 0$, then (see Fig. 11.11)

$$v'_x = -V \qquad \text{and} \qquad v'_y = c\left(1 - \frac{V^2}{c^2}\right)^{\frac{1}{2}}$$

so that

$$\frac{v'_x}{v'_y} = -\frac{V}{c(1 - V^2/c^2)^{\frac{1}{2}}}$$

and

$$\sqrt{v'^2_x + v'^2_y} = c$$

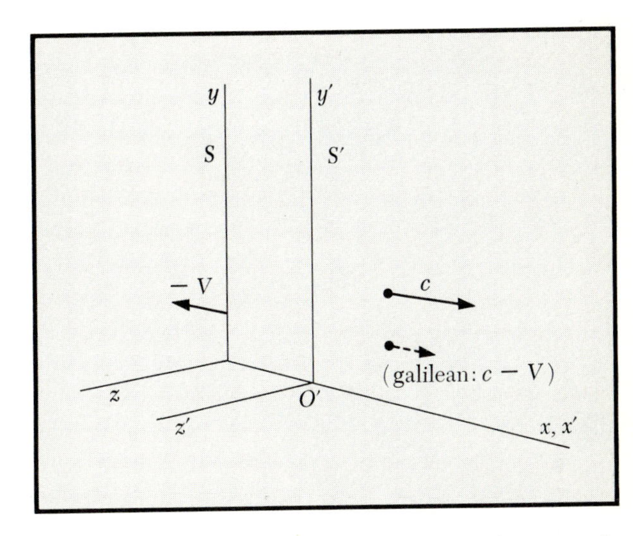

FIG. 11.10 As we know, if $v_x = c$, $v'_x = c$ also, according to the Lorentz transformation. *This* was built into our theory from the beginning.

EXAMPLE

Velocity Addition Suppose that two particles are traveling opposite to each other with velocity $v'_x = \pm 0.9c$ as observed in the S′ system. What is the velocity of one particle with respect to the other, that is, as measured by the other? To solve this problem, let S be the reference frame in which the $-0.9c$ particle is at rest. Then the velocity of S′ relative to S is $V = 0.9c$ so that the particle which in S′ has velocity $v'_x = +0.9c$ has a velocity in S [see Eq. (11.18)]

$$v_x = \frac{v'_x + V}{1 + v'_x V/c^2} \approx \frac{1.8c}{1 + (0.9)^2} = \frac{1.80}{1.81}c = 0.994c$$

Notice that the relative velocity of the two particles is less than c.

If a photon is traveling at velocity $+c$ in S′, and S′ is traveling relative to S at velocity $+c$, the photon as viewed from S is traveling only at velocity $+c$, and not at $+2c$. This result is contained in Eq. (11.18). The fact of an ultimate speed is a consequence of the structure of the velocity-addition equations which we have derived from the Lorentz transformation. Note further that there is *no* frame in which a photon (light quantum) is at rest.

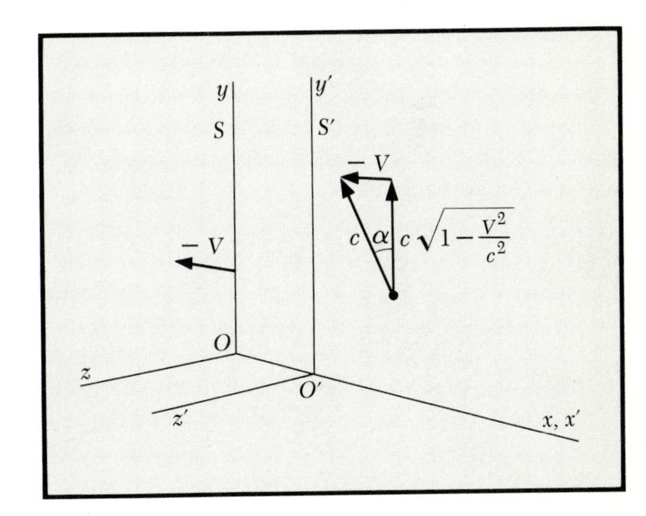

FIG. 11.11 In particular if $v_y = c$, the resultant has magnitude c in S′. Thus

$$|\tan\alpha| = \frac{V}{c\sqrt{1 - V^2/c^2}}$$

This is the relativistic theory of aberration.

D. Sadeh has carried out [*Phys. Rev. Letters*, **10**:271 (1963)] a beautiful experiment which shows that the velocity of γ-rays is constant (±10 percent), independent of the velocity of the source, for a source velocity close to $\frac{1}{2}c$ compared with a source at rest. We quote from his paper:

> In our experiments we used the annihilation in flight of positrons. In the annihilation the center-of-mass system of the positron and electron moves with a velocity close to $\frac{1}{2}c$, and two gamma rays are emitted. In the case of annihilation at rest, the two gamma rays are emitted at an angle of 180° and their velocity is c. In the case of annihilation in flight, the angle is smaller than 180° and depends on the energy of the positron. If the velocity of the gamma ray adds on to the velocity of the center of mass according to classical vector addition, and not according to the Lorentz transformation, then the gamma ray traveling with a component of motion in the direction of the positron flight will have a velocity greater than c, and that having a component in the opposite direction will have a velocity smaller than c. If it is found that the two gamma rays reach

the counters at the same time for equal distances between the counters and the point of annihilation, this would prove that even for a moving source the two gamma rays travel with the same velocity.

EXAMPLE

Aberration of Light We saw in Eq. (10.1) that for a star directly overhead (when the earth's velocity v_e is perpendicular to the line of observation) the tilt angle, or aberration, of the telescope is given by

$$\tan \alpha = \frac{v_e}{c} \tag{11.19}$$

This result was derived using a nonrelativistic argument. Now consider the problem relativistically as an exercise in the use of the Lorentz transformation.

Suppose that in reference frame S shown in Fig. 11.11 a star located at rest at O is observed by receiving light rays from it emitted along the y axis. What will be the trajectory in S′ of these rays that move along the y axis in S? In S their velocity components are $v_x = 0$, $v_y = c$, $v_z = 0$. The velocity components in S′ may be obtained by using Eqs. (11.15) to (11.17). Thus

$$v'_x = -V \qquad v'_y = \frac{c}{\gamma} \qquad v'_z = 0$$

So the direction of these rays in S′ is at the angle given by

$$\tan \alpha = \frac{-v'_x}{v'_y} = \frac{\gamma V}{c} = \beta\gamma = \frac{V/c}{\sqrt{1 - V^2/c^2}}$$

or

$$\sin \alpha = \frac{V}{c} = \beta \tag{11.20}$$

This is the correct result; it agrees within the accuracy of measurement with the nonrelativistic result of Eq. (11.19) only because V/c for the earth's motion is small, being approximately 10^{-4}.

EXAMPLE

Longitudinal Doppler Effect Consider two pulses of light sent out at $t = 0$ and $t = \tau$ by a transmitter at rest at $x = 0$ in reference frame S. Reference frame S′ moves with velocity $V\hat{\mathbf{x}}$ with respect to S. The initial pulse is received at $x' = 0$ in S′ at time $t' = 0$. The point in S′ which coincides with $x = 0$ at $t = \tau$ is given by the Lorentz transformation, Eq. (11.7),

$$x' = \frac{x - Vt}{(1 - \beta^2)^{\frac{1}{2}}} = \frac{-V\tau}{(1 - \beta^2)^{\frac{1}{2}}}$$

taking $x = 0$. The corresponding time in S′ is

$$t' = \frac{t - Vx/c^2}{(1 - \beta^2)^{\frac{1}{2}}} = \frac{\tau}{(1 - \beta^2)^{\frac{1}{2}}}$$

The time needed for the second pulse to travel in S′ from $-V\tau/(1 - \beta^2)^{\frac{1}{2}}$ to the origin is

$$\Delta t' = \frac{\tau V/c}{(1 - \beta^2)^{\frac{1}{2}}}$$

so that the total time in S′ between the reception at $x' = 0$ of the two pulses is

$$t' + \Delta t' = \tau \frac{1 + V/c}{(1 - \beta^2)^{\frac{1}{2}}} = \tau \sqrt{\frac{1 + \beta}{1 - \beta}}$$

The time between the two signals can equally well be interpreted as the elapsed time between two successive nodes of a light wave. The frequency is the reciprocal of the period of the wave, so that

$$\nu' = \nu \sqrt{\frac{1 - \beta}{1 + \beta}} \tag{11.21}$$

Here ν' is the frequency as received in S′, and ν is the frequency as transmitted in S. If the receiver is receding from the source, then $\beta = V/c$ is positive and ν' is less than ν. If the receiver is approaching the source, we take β to be negative and ν' is greater than ν. In terms of wavelength, $\lambda = c/\nu$ and $\lambda' = c/\nu'$, so that

$$\lambda' = \lambda \sqrt{\frac{1 + \beta}{1 - \beta}} \tag{11.22}$$

Equation (11.21) describes the relativistic longitudinal doppler effect for light waves in a vacuum. The frequency shift agrees to order β with the nonrelativistic result, Eq. (10.7) derived in Chap. 10.† The term of order β^2 in the series expansion of Eq. (11.21) has been confirmed experimentally by Ives and Stilwell.

H. E. Ives and G. R. Stilwell [*J. Opt. Soc. Am.*, **28**:215 (1938); **31**:369 (1941)] have carried out spectroscopic experiments on beams of hydrogen atoms in excited electronic states. The atoms were accelerated as molecular hydrogen ions H_2^+ and H_3^+ in an intense electric field. Atomic hydrogen was formed as a breakup product of the ions. The velocity of the atoms was of the order of $\beta = 0.005$. Ives and Stilwell looked for a shift in the *average* wavelength of a particular spectral line emitted by the hydrogen atoms. The average was taken over the forward and backward directions with respect to the line of flight of the atoms. From Eq. (11.22) we have, using $\beta_{\text{fwd}} = -\beta_{\text{bkwd}}$, the average wavelength

$$\frac{1}{2}(\lambda_{\text{fwd}} + \lambda_{\text{bkwd}}) = \frac{1}{2}\lambda_0\left(\sqrt{\frac{1 - \beta}{1 + \beta}} + \sqrt{\frac{1 + \beta}{1 - \beta}}\right)$$

$$= \frac{\lambda_0}{(1 - \beta^2)^{\frac{1}{2}}} \tag{11.23}$$

† The reader should show that for $\beta \ll 1$, $\sqrt{(1 + \beta)/(1 - \beta)} = 1 + \beta$.

Thus there is a shift of order β^2 in the mean position of the displaced lines, with respect to the wavelength λ_0 emitted from an atom at rest. In their 1941 paper Ives and Stilwell report an observed shift of 0.074 Å in the average wavelength, as compared with the value 0.072 Å calculated from Eq. (11.23) for a value of β deduced from the accelerating potential applied to the original ions. This is an excellent confirmation of the theory of the relativistic doppler effect.

The *transverse* doppler effect applies to observations made at right angles to the direction of travel of the light source, which is usually an atom. In the nonrelativistic approximation there is no transverse doppler effect. A transverse doppler effect for light waves is predicted by relativity theory; the frequencies must be related as the inverse of the times in Eq. (11.11), so that

$$\nu' = (1 - \beta^2)^{\frac{1}{2}}\nu$$

where ν is the frequency in the frame in which the atom is at rest, and ν' is the frequency as observed in a frame moving with velocity $V(=\beta c)$ with respect to the atom.

Accelerated Clocks The special theory of relativity describes and relates measurements which are independent of the detailed structure of real bodies. It makes no prediction about the dynamical effects of acceleration, such as the stresses induced by acceleration. If such stresses are absent or may be ignored, the theory does give us an unambiguous description of the effect of acceleration on clock rates. The result is as if at each instant an accelerated clock had a different velocity, with a rate to be calculated using in Eq. (11.11) the appropriate instantaneous velocity.

If this prediction is correct, two consequences follow:

1 If the speed is constant but the direction varies, Eq. (11.11) holds without change. The frame of the clock is noninertial.
2 If the speed is constant except for brief moments of acceleration or deceleration (moments negligibly short in comparison with the total time), then Eq. (11.11) will still describe accurately the relation between the proper time and the stationary laboratory time.

A fast charged particle in a constant magnetic field experiences an acceleration perpendicular to its motion, but the speed never changes. If the particle is unstable, the measured half-life should be exactly the same as if it moved with the same speed in a straight line with no magnetic field present. This forecast is confirmed by experiments on the μ^- meson, which decays with a proper mean life of 2.2×10^{-6} s into an electron and neutrinos. The same proper lifetime is observed for μ^- mesons which are free or spiraling in a magnetic field or allowed to come to rest. It is believed that the special theory of relativity gives a good description of the circular (accelerated) motion of particles in a magnetic field.

PROBLEMS

1. *Lorentz invariant.* Verify from Eq. (11.7) that

$$x^2 - c^2t^2 = x'^2 - c^2t'^2$$

Note that if we write $x_1 \equiv x$; $x_4 \equiv ict$, then $x^2 - c^2t^2 \equiv x_1^2 + x_4^2$. Here $i = \sqrt{-1}$.

2. *Lorentz transformation.* Given Eq. (11.7), demonstrate Eq. (11.8).

3. *Change of volume.* Show that if L_0^3 is the rest volume of a cube, then

$$L_0^3(1 - \beta^2)^{\frac{1}{2}}$$

is the volume viewed from a reference frame moving with uniform velocity β in a direction parallel to an edge of the cube.

4. *Simultaneity.* Show from the Lorentz transformation that two events simultaneous ($t_1 = t_2$) at different positions ($x_1 \neq x_2$) in reference frame S are not in general simultaneous in reference frame S′.

5. *Change of angle.* Calculate in S′ the length and angle with the x' axis of a rod of length L_0 and angle θ with the x axis in S. S′ moves with velocity $V\hat{\mathbf{x}}$ with respect to S.

6. *Addition of velocities.* Show that if in the S′ frame we have $v'_y = c \sin\theta$ and $v'_x = c\cos\theta$, then in the S frame

$$v_x^2 + v_y^2 = c^2$$

The S′ frame moves with velocity $V\hat{\mathbf{x}}$ with respect to the S frame.

7. *π^+ mesons*

(*a*) What is the mean life of a burst of π^+ mesons traveling with $\beta = 0.73$? (The proper mean lifetime τ is 2.5×10^{-8} s.) *Ans.* 3.6×10^{-8} s.

(*b*) What distance is traveled at $\beta = 0.73$ during one mean life? *Ans.* 800 cm.

(*c*) What distance would be traveled without relativistic effects? *Ans.* 550 cm.

(*d*) Answer parts (*a*) to (*c*) again, but for $\beta = 0.99$.

8. *μ mesons.* The proper mean life of the μ meson is approximately 2×10^{-6} s. Suppose that a large burst of μ mesons produced at some height in the atmosphere travels downward at $v = 0.99c$. The number of collisions in the atmosphere on the way down is small.

(*a*) If 1 percent of those in the original burst survive to reach the earth's surface, estimate the original height. [In the μ meson frame of reference the number of particles which survive to a time t is given by $N(t) = N(0)e^{-t/\tau}$.] *Ans.* 2×10^6 cm.

(*b*) Calculate this distance of travel as measured by the μ meson.

9. *Two events.* Consider two inertial frames S and S′. Let S′ move with velocity $V\hat{\mathbf{x}}$, with respect to S. At a point x'_1 an event takes place at time t'_1. At x'_2 another event takes place at time t'_2. The origins coincide at time $t = t' = 0$. Find the corresponding times and distances in S.

10. *π^+ mesons.* A burst of 10^4 π^+ mesons travels in a circular path of radius 20 m at a speed $\beta = 0.99c$. The proper mean life of the π^+ meson is 2.5×10^{-8} s.

(*a*) How many survive when the burst returns to the point of origin?

(*b*) How many mesons would be left in a burst that had remained at rest at the origin for this same period of time?

11. *Recessional velocity of galaxy.* We stated in Chap. 10 that red-shift data on distant galaxies gave a velocity of recession proportional to distance, in the nonrelativistic region:

$$V = \alpha r \qquad \alpha \approx 1.6 \times 10^{-18}\ \text{s}^{-1}$$

Calculate the recession velocity of a galaxy at a distance of 3×10^9 light yr. Is this velocity relativistic? *Ans.* 4.5×10^9 cm/s.

12. *Galactic velocities.* We observe a galaxy receding in a particular direction at a speed $V = 0.3c$, and another receding

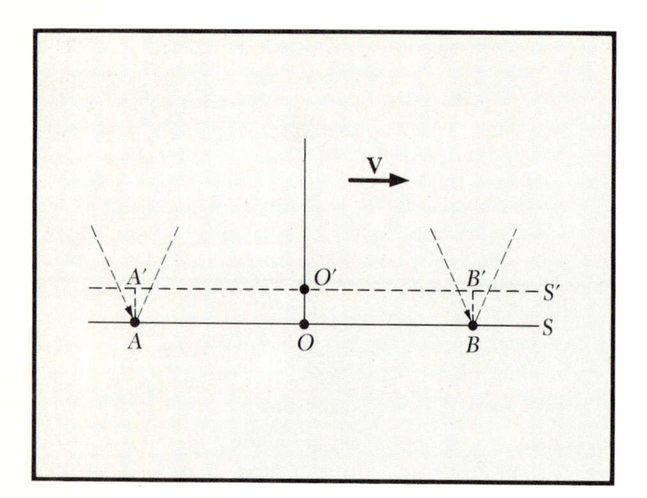

FIG. 11.12

in the opposite direction with the same speed. What speed of recession would an observer in one of these galaxies observe for the other galaxy?

13. *Simultaneity.* Consider the sources of two events to be located at rest at the points A and B, equal distances from the observer O in the frame S. Assume that at the particular instant of time (as determined by observer O in S) at which the two events occur, a second observer O' and his associated reference frame S′, moving with a velocity $V\hat{\mathbf{x}}$ with respect to S, coincide with O and his frame S (see Fig. 11.12).

(*a*) Assume $V/c = \frac{1}{3}$. Sketch the positions of the two frames and the points A, A', B, B' when the signal from B arrives at the observer O'. Has this signal arrived at the observer O? Why?

(*b*) Sketch the positions of S and S′ as both signals arrive at O.

(*c*) Sketch the positions of S and S′ as the signal from A arrives at O'.

(*d*) Assume that the two events are recorded physically at the points A', B'; for example, on photographic plates. Show under the assumptions of this problem that the distances $A'O'$ and $B'O'$ are equal.

(*e*) Show that the two events are not simultaneous as viewed by O'. The constancy of the velocity of light under all circumstances is implicitly assumed in the definition of simultaneity. To make this dependence clear consider the following. Let the two events at A and B be the simultaneous radiation of pulses of sound as observed by O, an observer at rest with respect to the medium in which the sound is propagated. Let O' be an observer moving with a velocity V one-third that of sound.

(*f*) Use the galilean transformation to show that the velocity of the sound pulses toward O' from A and B are *not* the same as observed by O'.

(*g*) Show that even though the two signals arrive at O' at different times, the fact that the pulses have traveled with different velocities compensates for this fact and that the two events are inferred to be simultaneous, even by the observer O'.

14. *Relativistic doppler shift.* Protons are accelerated through a potential of 20 kV, after which they drift with constant velocity through a region where neutralization to H atoms and associated light emission takes place. The H_β emission ($\lambda = 4861.33$ Å for an atom at rest) is observed in a spectrometer. The optical axis of the spectrometer is parallel to the motion of the ions. The spectrum is doppler-shifted because of the motion of the ions in the direction of observed emission. The apparatus also contains a mirror which is placed so as to allow superposition of the spectrum of light emitted in the reverse direction. Recall that 1 Å $\equiv 10^{-8}$ cm.

(*a*) What is the velocity of the protons after acceleration? *Ans.* 2×10^8 cm/s.

(*b*) Calculate the first-order doppler shifts, depending on v/c, appropriate to the forward and backward directions, and indicate the appearance of the relevant part of the spectrum on a diagram.

(*c*) Now consider the second-order, or v^2/c^2, effect which arises from relativistic considerations. Show that the second-order shift is $= \frac{1}{2}\lambda(v^2/c^2)$, and evaluate this numerically for this problem. Notice that it is the same for both $+\mathbf{v}$ and $-\mathbf{v}$ motions. *Ans.* 0.10 Å.

FURTHER READING

In recent years a number of books on relativity have been published, some as paperbacks. The following constitute a selection.

J. A. Wheeler and E. F. Taylor, "Space-Time Physics—An Introduction," W. H. Freeman and Company, San Francisco, 1965. Highly recommended. Not a textbook; excellent for self-study.

A. P. French, "Special Relativity," W. W. Norton and Company, Inc., New York, 1968. One of the MIT Introductory Physics Series.

C. Kacser, "Introduction to the Special Theory of Relativity," Co-Op Paperback, Prentice-Hall, Inc., Englewood Cliffs, N.J., 1967.

R. S. Shankland, "Conversations with Albert Einstein," *Am. J. Phys.*, **31**:47 (1963).

"Special Relativity Theory," selected reprints published for A.A.P.T., American Institute of Physics, 335 East 45th St., New York, 1962. This contains excellent discussion of the famous twin paradox; see especially the papers by Darwin, Crawford, and McMillan.

M. Born, "Einstein's Theory of Relativity," E. P. Dutton & Co., Inc., New York, 1924; reprint, Dover Publications, Inc., New York, 1962. A patient, full, and clear discussion of the special theory.

H. A. Lorentz, A. Einstein, H. Minkowski, and H. Weyl, "The Principle of Relativity: A Collection of Original Memoirs," translated by W. Perrett and G. B. Jeffery, Methuen & Co., Ltd., London, 1923; reprint, Dover Publications, Inc., New York, 1958.

W. Pauli, "Theory of Relativity," translated by G. Field, Pergamon Press, New York, 1958. A translation of an excellent monograph in German (Relativitätstheorie, published by B. G. Teubner, Leipzig, 1921). The contents of Part I are not difficult.

E. Whittaker, "History of the Theories of Aether and Electricity," 2 vols., Harper & Row, Publishers, Incorporated, New York, paperback reprint, 1960.

CONTENTS

Relativistic Dynamics: Momentum and Energy

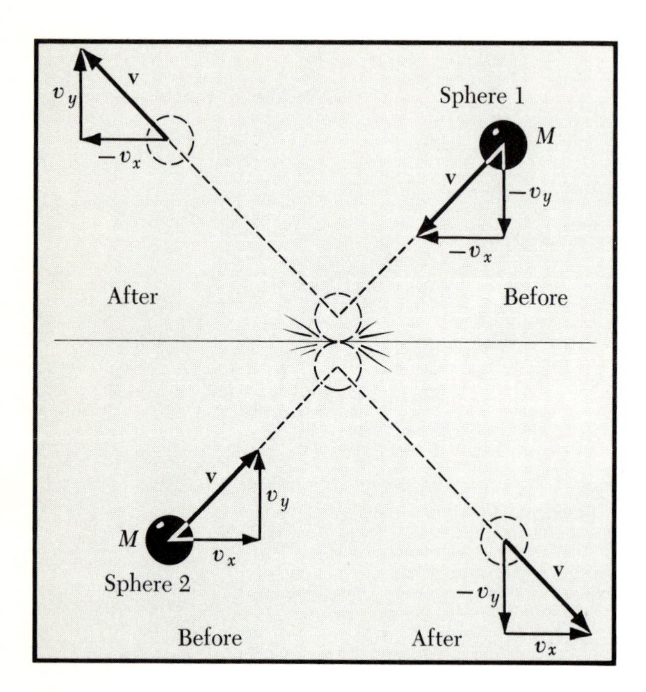

FIG. 12.1 (*a*) A collision between two spheres of mass M, taking place in the xy plane. The velocities in the x and y directions before and after collision are as shown.

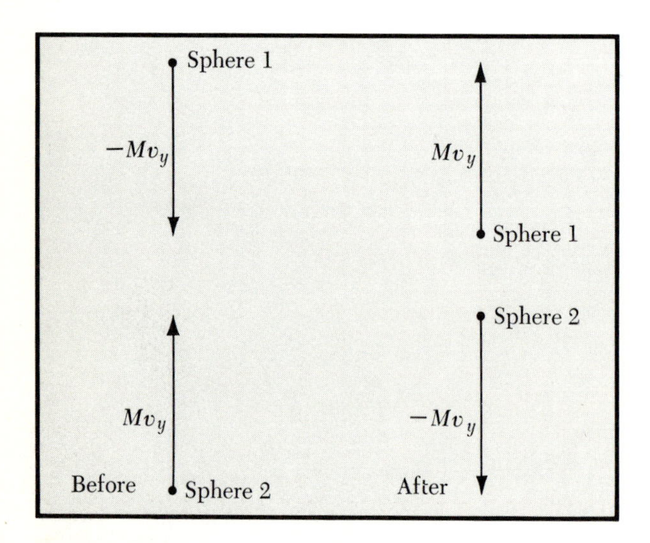

(*b*) The individual nonrelativistic momenta in the y direction are shown. The *total* momentum in the y direction is zero before *and* after the collision.

The basic change in our concepts of space and time as expressed in the Lorentz transformation deeply affects all of physics. We must reexamine now the laws of physics as developed and confirmed for low velocities ($v \ll c$) to see if they are compatible with the theory of relativity. We should not be surprised to find that the laws change when carried into new domains. The laws change in such a way that at low velocities we retrieve the newtonian forms, which we know from experience are accurate in the limit of low velocities.

As in Chap. 4, we accept as possible physical laws only those which are identical in all reference frames in unaccelerated relative motion. But instead of the galilean transformation to tell us how to transform a physical law from one reference frame to another, we now use the Lorentz transformation. The Lorentz transformation reduces to the galilean when $v/c \ll 1$. In place of insisting upon the invariance of physical laws under galilean transformations, we now insist upon their invariance under Lorentz transformations.

Two observers in different reference frames S and S′ deduce physical laws. Each expresses them in terms of lengths, times, velocities, or accelerations, as measured in his own system. The laws must be identical in form in the S frame variables and in the S′ frame variables. Thus when we use the Lorentz transformation to transform from the x, y, z, t of S to the x', y', z', t' of S′, any physical law deduced in S is translated to the language of S′ and should be unchanged in form. The meaning of this will become clear as we examine particular problems.

CONSERVATION OF MOMENTUM AND DEFINITION OF RELATIVISTIC MOMENTUM

We want to find a definition of the momentum **p** which reduces to $M\mathbf{v}$, where M is the rest mass,[1] for $v/c \ll 1$, and which assures momentum conservation in collisions regardless of the velocities of the particles relative to the reference frame. We will find the appropriate definition by consideration of a particular collision. We first show by an example that the newtonian (nonrelativistic) momentum $M\mathbf{v}$ is not conserved in collisions involving *relativistic* velocities.[2] We can already see that

[1]The rest mass is defined as the inertial mass in the nonrelativistic limit $v/c \ll 1$, in particular when $v = 0$.

[2]The question of how large v must be to be relativistic depends on the accuracy of the experimental results. We shall see that if $(v/c)^2$ can be neglected compared to 1, we can consider the velocity nonrelativistic.

Newton's Second Law can not be valid if M is constant because $\mathbf{a}$ would be $\mathbf{F}/M$ and if the force acts for a long enough time, v would be greater than c.

Consider Figs. 12.1*a* and *b* that describe a collision between particles of *equal mass*. We choose a reference frame S such that the particles approach each other with equal and opposite velocities: the y velocity component of particle 1 is $-v_y$ before the collision and $+v_y$ after the collision. In this reference frame the center of mass is at rest. The total y component of momentum must be zero by symmetry, both before and after the collision. This will be true no matter what definition we use for momentum, provided that it has opposite signs for $\pm v_y$. We therefore encounter no trouble here (whether or not the expression be correct) with the newtonian definition $\mathbf{p} = M\mathbf{v}$: the change in p_y of particle 1 is $+2Mv_y$, and the change in p_y of particle 2 is $-2Mv_y$, so that the total change in the y component of the newtonian momentum is zero.

Now consider a primed reference frame S′ moving with the particular velocity $\mathbf{V} = v_x\hat{\mathbf{x}}$ with respect to S as shown in Fig. 12.2*a*. Note that v_x is the x component of velocity of particle 2 in S, and $-v_x$ is that for particle 1. The relativistic addition of velocities is described in Eqs. (11.15) to (11.17), and by utilizing these we find that the velocity components seen in S′ will be (remembering that $V = v_x$)

$$-v_x'(1) = \frac{-v_x - V}{1 + v_xV/c^2} = \frac{-2v_x}{1 + v_x^2/c^2}$$

$$v_y'(1) = \frac{v_y}{1 + v_xV/c^2}\left(1 - \frac{V^2}{c^2}\right)^{\frac{1}{2}} = \frac{v_y}{1 + v_x^2/c^2}\left(1 - \frac{v_x^2}{c^2}\right)^{\frac{1}{2}} \tag{12.1}$$

$$v_x'(2) = \frac{v_x - V}{1 - v_xV/c^2} = 0$$

$$v_y'(2) = \frac{v_y}{1 - v_xV/c^2}\left(1 - \frac{V^2}{c^2}\right)^{\frac{1}{2}} = \frac{v_y}{(1 - v_x^2/c^2)^{\frac{1}{2}}} \tag{12.2}$$

These results are illustrated in Fig. 12.2*b*. [Note that the expressions there are in terms of V, whereas in Eqs. (12.1) and (12.2) they are in terms of v_x and V and then in terms of v_x alone.]

Clearly the magnitudes of the y components of velocity in S′ given by Eqs. (12.1) and (12.2) are not equal, even though they were equal in S. This difference in the y-component velocity magnitudes in S′ arises from the fact that the x-component velocities in S were not the same; they were the negatives of

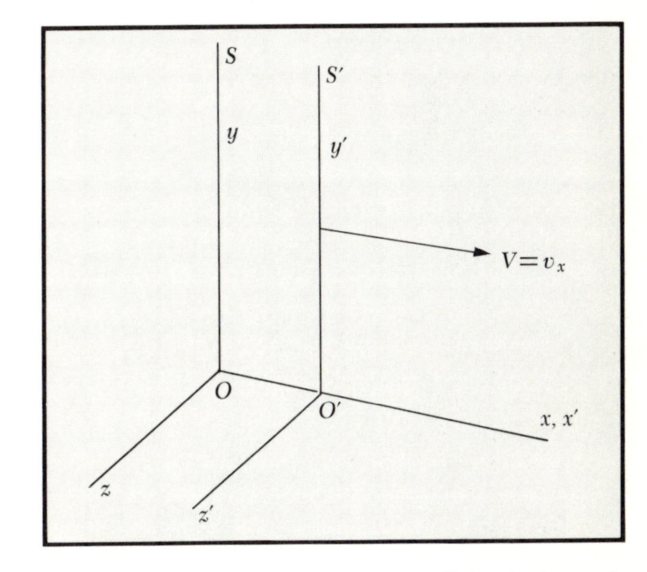

FIG. 12.2 (*a*) We have viewed the collision in frame S. What if we view it in frame S′, which has the particular velocity $V = v_x$ with respect to S, as shown?

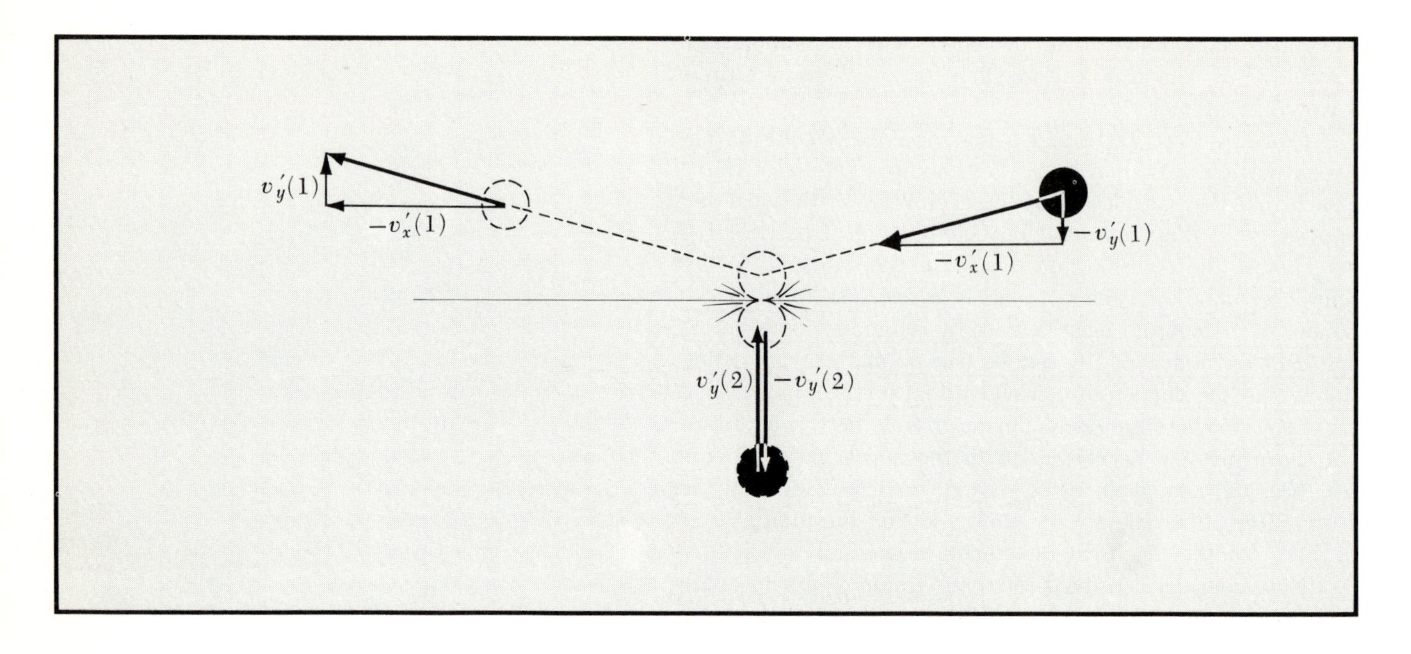

FIG. 12.2 (*cont'd*) (*b*) In S', we find (since $V = v_x$).

$$-v_x'(1) = -\frac{2V}{1 + V^2/c^2} \qquad v_x'(2) = 0$$

$$v_y'(1) = \frac{v_y}{1 + V^2/c^2}\left(1 - \frac{V^2}{c^2}\right)^{1/2}$$

and

$$v_y'(2) = \frac{v_y}{(1 - V^2/c^2)^{1/2}} > v_y'(1)$$

each other. The situation is illustrated by Fig. 12.2*c*. The nonrelativistic momentum changes $-2Mv_y'(2)$ and $2Mv_y'(1)$ would not here be equal and opposite. We see that a definition in which the momentum is directly proportional to velocity cannot assure momentum conservation in all reference systems. Either momentum conservation is incompatible with Lorentz invariance or there exists another definition of momentum such that momentum conservation is valid in all systems with constant relative velocities.

We now look for a definition of momentum which is Lorentz invariant. The definition must be such that the y component of the momentum of a particle is independent of the x component of the velocity of the reference frame in which the collision is observed. If such a definition is found, then conservation of the y component of momentum in one reference frame ensures its conservation in all reference frames. We know that under Lorentz transformations the displacement Δy in the y direction is the same for all reference frames. But the time Δt to go the distance Δy depends upon the reference frame, and thus the velocity component $v_y = \Delta y/\Delta t$ depends on the reference frame. Instead of laboratory clocks to measure Δt, we can refer to a clock carried on the particle. This clock measures the proper time interval $\Delta\tau$ for the particle. *All observers will agree on the value of* $\Delta\tau$. Thus the quantity $\Delta y/\Delta\tau$ is the same in all the reference frames.

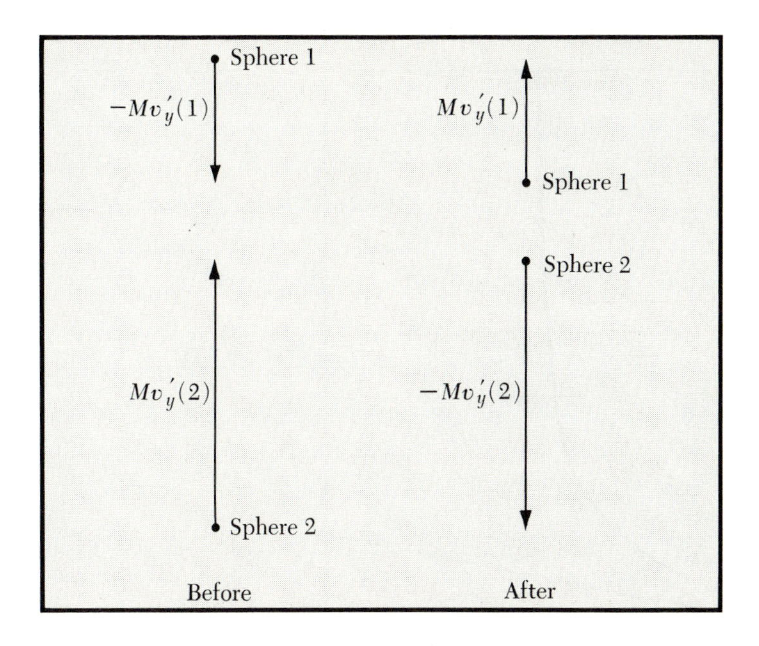

FIG. 12.2 (*cont'd*) (*c*) In the new frame S′ the nonrelativistic momentum is *not* the same in the y' direction before and after collision.

We know that Δt and $\Delta\tau$ differ by the time-dilation factor Eq. (11.11), and so we have

$$\Delta\tau = \Delta t\left(1 - \frac{v^2}{c^2}\right)^{\frac{1}{2}} \tag{12.3}$$

where v is the speed of the particle relative to the reference frame in which Δt is measured. Whence

$$\frac{\Delta y}{\Delta\tau} = \frac{\Delta y}{\Delta t}\frac{\Delta t}{\Delta\tau} = \frac{\Delta y}{\Delta t}\frac{1}{(1 - v^2/c^2)^{\frac{1}{2}}}$$

We see that the y component of $\mathbf{v}/(1 - v^2/c^2)^{\frac{1}{2}}$ will be the same in all reference frames which differ only in their x components of velocity. If we *define* the relativistic momentum by

$$\boxed{\mathbf{p} \equiv \frac{M\mathbf{v}}{(1 - v^2/c^2)^{\frac{1}{2}}}} \tag{12.4}$$

then conservation of the y component of momentum is valid in any other inertial reference frame that differs from the rest

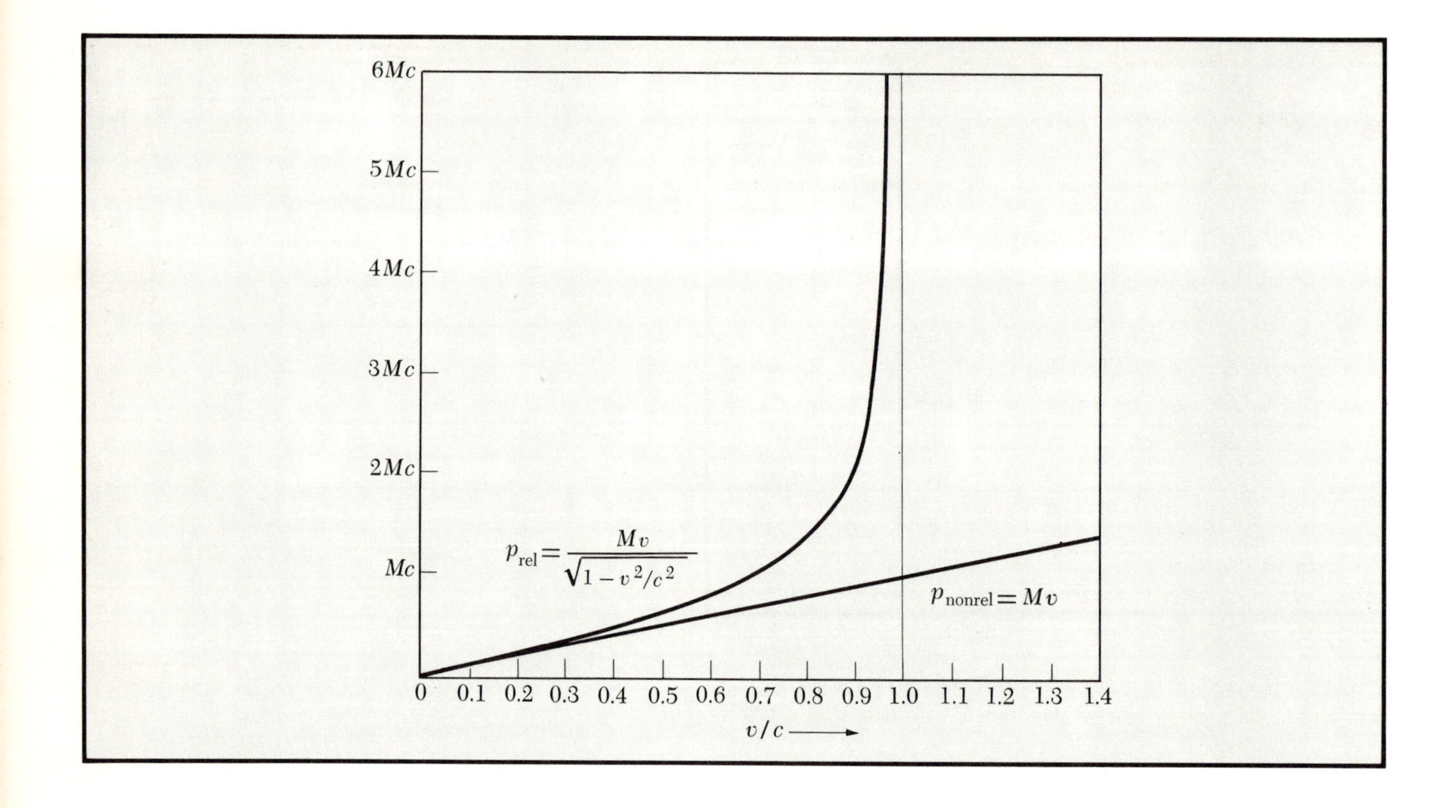

FIG. 12.3 So that momentum conservation will hold in *all* frames, we redefine $\mathbf{p}$ as follows: For a particle with velocity $\mathbf{v}$ and rest mass M,

$$\mathbf{p} = \frac{M\mathbf{v}}{\sqrt{1 - v^2/c^2}}$$

The magnitudes of both the relativistic momentum and the nonrelativistic momentum are plotted in the graph.

frame by a constant velocity in the x direction. Note that we may write for the magnitude of the momentum

$$p = Mc\beta\gamma \tag{12.5}$$

from the definitions $\beta = v/c$ and $\gamma = (1 - v^2/c^2)^{-\frac{1}{2}}$ introduced in Chap. 11. Figure 12.3 illustrates this new definition of momentum plotted against v.

For convenience in discussing this problem thus far, we have used axes arranged in a symmetrical way relative to the motions so that there is no change in the x component of velocity of either particle. Since there is no change in the magnitude of the y component either, the definition of Eq. (12.4) provides for conservation of the x component of momentum also. We shall see in the section on Transformation of Momentum and Energy (page 359) that in a general transformation the momentum in S' is related to both the momentum in S and the energy in S. The student may show that even if particle 2 has a different mass from particle 1, the above argument holds, so that we have a relativistic law of momentum conservation. For $v/c \ll 1$ the definition of momentum reduces to the nonrelativistic result $p = Mv$. It is an experimental fact that the momentum as defined by Eq. (12.4) is conserved in all collision processes.

We may write the relativistic momentum [Eq. (12.4)]

$$\mathbf{p} = M(v)\mathbf{v}$$

so that we may interpret

$$\boxed{M(v) \equiv \frac{M}{(1 - v^2/c^2)^{\frac{1}{2}}} = M\gamma} \qquad (12.6)$$

as the relativistic mass when in motion with the speed v of the particle of rest mass M. This is illustrated in Fig. 12.4. The rest mass is the mass for $v \to 0$. As $v \to c$, $M(v)/M \to \infty$. The relativistic increase of mass has been verified in various electron deflection experiments; it is also verified implicitly in the operation of every high-energy particle accelerator. An alternate formulation of Eq. (12.6) given below emphasizes directly the relation between relativistic energy and momentum, and it is often simpler to apply.

Some books use M as the variable mass and write $M = M_0/\sqrt{1 - v^2/c^2} = \gamma M_0$ where M_0 is then called the rest mass. We shall continue to denote rest mass by M and the variable relativistic mass by γM or sometimes by $M(v)$.

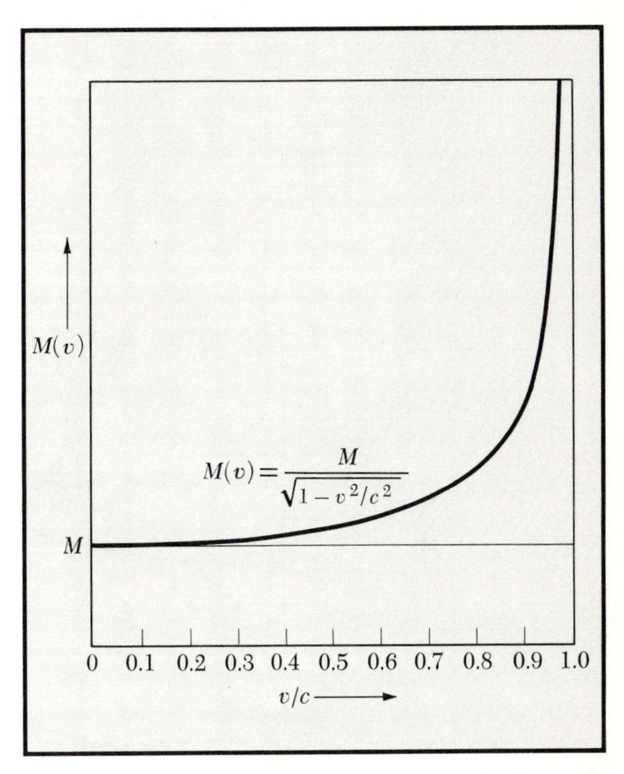

FIG. 12.4 The new definition of momentum leads to this behavior of the mass:

$$M(v) = \frac{M}{\sqrt{1 - v^2/c^2}}$$

RELATIVISTIC ENERGY

What is the relativistic kinetic energy? By what do we replace $\frac{1}{2}Mv^2$ to get a meaningful relativistic expression? First, let us remember how we defined kinetic energy in Chap. 5: The energy obtained by a free particle initially at rest when an amount of work W is done on it. We retain this definition, and we proceed by writing Newton's Second Law in the form

$$\mathbf{F} = \frac{d\mathbf{p}}{dt} = \frac{d}{dt}\frac{M\mathbf{v}}{\sqrt{1 - v^2/c^2}}$$

where the time t and the force F refer to these quantities evaluated in the laboratory frame in which the momentum p is observed. (Transformation of force from one reference frame to another is treated later in this chapter.) Let $\mathbf{F}$ be in the x direction. Then the work W is given by

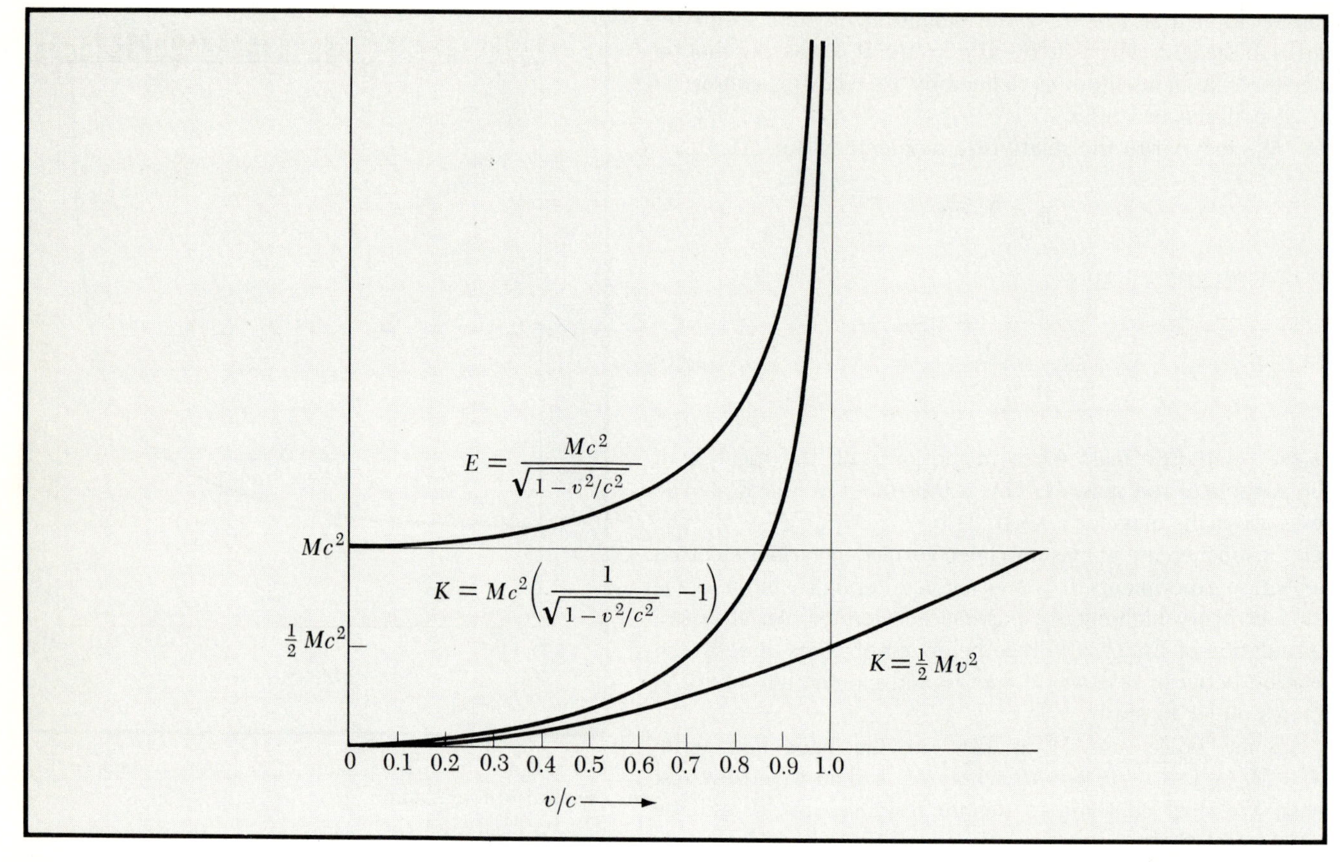

FIG. 12.5 The relativistic energy $E = Mc^2/(1 - v^2/c^2)^{1/2}$, the relativistic kinetic energy $K = Mc^2/\sqrt{1 - v^2/c^2} - Mc^2$, and the nonrelativistic kinetic energy $K = \frac{1}{2}Mv^2$ plotted vs v/c. For $v/c \ll 1$, the curves for E and K are almost identical in shape, since $Mc^2/(1 - v^2/c^2)^{1/2} \approx Mc^2 + \frac{1}{2}Mv^2$. For $v/c \sim 1$, E increases much more rapidly than $\frac{1}{2}Mv^2$.

$$
\begin{aligned}
W = \int F\,dx &= \int \frac{d}{dt}\frac{Mv}{\sqrt{1 - v^2/c^2}}\,dx \\
&= \int \frac{d}{dt}\left(\frac{Mv}{\sqrt{1 - v^2/c^2}}\right)\frac{dx}{dt}\,dt \\
&= \int \left[\frac{Mv}{\sqrt{1 - v^2/c^2}}\frac{dv}{dt} + \frac{Mv^3c^{-2}}{\sqrt{(1 - v^2/c^2)^3}}\frac{dv}{dt}\right] dt \\
&= \int \frac{Mv\,dv/dt}{\sqrt{(1 - v^2/c^2)^3}}\,dt = \int \frac{d}{dt}\left(\frac{Mc^2}{\sqrt{1 - v^2/c^2}}\right) dt
\end{aligned}
$$

where we used the fact that $dx/dt = v$.

If we assume that at the upper limit the velocity is v and at the lower limit $v = 0$, we get

$$
\boxed{W = \frac{Mc^2}{\sqrt{1 - v^2/c^2}} - Mc^2 = Mc^2(\gamma - 1)} \qquad (12.7)
$$

This will be the kinetic energy K and it is plotted against v in Fig. 12.5. Using this expression for K, we find agreement with the experimental results plotted in Fig. 10.20.

This new expression formally does not resemble $\frac{1}{2}Mv^2$ at all, so let us see what it becomes for $v/c \ll 1$.

$$\gamma = \frac{1}{\sqrt{1 - v^2/c^2}} = 1 + \frac{1}{2}\frac{v^2}{c^2} \cdots$$

Therefore

$$\gamma - 1 = \frac{1}{2}\frac{v^2}{c^2}$$

and so

$$Mc^2(\gamma - 1) \qquad \text{becomes} \qquad \frac{1}{2}Mc^2\frac{v^2}{c^2} = \frac{1}{2}Mv^2$$

for small values of v/c, thus reducing to the newtonian expression.

We now consider the relativistic energy from a formal viewpoint. From Eq. (12.5) the square of the relativistic momentum can be written as

$$p^2 = M^2c^2\beta^2\gamma^2 \tag{12.8}$$

The identity

$$\frac{1}{1 - v^2/c^2} - \frac{v^2/c^2}{1 - v^2/c^2} = 1$$

or

$$\gamma^2 - \beta^2\gamma^2 = 1$$

is a ready-made Lorentz invariant because 1 is a constant. On multiplying by M^2c^4, we have

$$M^2c^4(\gamma^2 - \beta^2\gamma^2) = M^2c^4$$

or, by use of Eq. (12.8),

$$M^2c^4\gamma^2 - p^2c^2 = M^2c^4 \tag{12.9}$$

Because the *rest mass* is a constant, we know that M^2c^4 is a constant and therefore a Lorentz invariant as required. But what physical quantity is $M^2c^4\gamma^2$? Its role in Eq. (12.9) strongly suggests that it must be an important physical quantity, for when p^2c^2 is subtracted from it, we have a number (M^2c^4) which is invariant under Lorentz transformations.

Suppose we define the *total relativistic energy* E of a free particle by the equation

$$\boxed{E \equiv Mc^2\gamma \equiv \frac{Mc^2}{(1 - v^2/c^2)^{\frac{1}{2}}}} \tag{12.10}$$

Then Eq. (12.9) tells us that

$$\boxed{E^2 - p^2c^2 = M^2c^4} \tag{12.11}$$

which is a Lorentz invariant. If we transform from one reference frame to another, with $p \to p'$ and $E \to E'$, then the invariance of Eq. (12.11) means that

$$E'^2 - p'^2c^2 = E^2 - p^2c^2 = M^2c^4$$

This is what we mean when we say that Eq. (12.11) is a Lorentz invariant. We emphasize that M denotes the rest mass of the particle and is a number invariant under a Lorentz transformation. Note that from Eq. (12.11)

$$E = \sqrt{p^2c^2 + M^2c^4} \tag{12.12}$$

If $pc \ll Mc^2$, then

$$E = Mc^2 \sqrt{1 + \frac{p^2c^2}{M^2c^4}} = Mc^2\left(1 + \frac{1}{2}\frac{p^2c^2}{M^2c^4} \cdots\right)$$

$$= Mc^2 + \frac{1}{2}\frac{p^2}{M}$$

and $K = \frac{1}{2}p^2/M$ is the nonrelativistic result. However if $pc \gg Mc^2$, then

$$E = pc$$

This is an approximation made often by high-energy physicists. We shall see later (page 365) that it is valid for light quanta where $M = 0$.

In between these two limits, there is no simple relation between E and p or between the kinetic energy K and p (or v). Note that K, as given in Eq. (12.7), now becomes, with the use of $E = \gamma Mc^2$,

$$K = E - Mc^2 \qquad \text{or} \qquad E = Mc^2 + K \tag{12.13}$$

It is important here to note that if $v = 0$, $E = Mc^2$. In other words, the mass M has energy even when at rest. This energy is naturally called the *rest energy*, and we shall see some examples of its importance. The difference between the energy E (in case $v > 0$) and the rest energy is the kinetic energy K.

Note also that $E = \gamma Mc^2$ [Eq. (12.10)] and γM is just the relativistic mass, so that E is just the relativistic mass times c^2. Mass and energy are just different names for the same quantity. It makes no particular sense to ask: Does a particle have more mass because it has kinetic energy or does it have kinetic energy because it has more mass? "More mass" and "kinetic energy" *must* go together.

The conservation of energy for particles in collision assumes the form

$$\sum_{i}^{n} E_i = \text{const}$$

the same before and after a collision where E_i is the relativistic energy of the ith particle as given in Eq. (12.12). The conservation of relativistic energy holds even for what we have called inelastic collisions because the loss of kinetic energy (into internal excitation of the particles) appears in an increase in particle mass. The conservation of momentum assumes the form

$$\sum_{i=1}^{n} \mathbf{p}_i = \text{const}$$

the same before and after a collision.

TRANSFORMATION OF MOMENTUM AND ENERGY

We now write Eq. (12.4) in components, using Eq. (12.3),

$$p_x = M\frac{dx}{d\tau} \qquad p_y = M\frac{dy}{d\tau} \qquad p_z = M\frac{dz}{d\tau} \tag{12.14}$$

Similarly it follows from Eqs. (12.10) and (12.3) that we can write E as

$$E = Mc^2\frac{dt}{d\tau} \tag{12.15}$$

Because M and τ are Lorentz invariants, it follows from Eqs. (12.14) and (12.15) that p_x, p_y, p_z, and E/c^2 must transform under a Lorentz transformation exactly as x, y, z, and t transform. Because we know how the latter transform, Eq. (12.16) follows simply. Using the transformations given in Chap. 11, we have *the transformation relations* for momentum and energy:

$$\boxed{\begin{gathered} p'_x = \gamma\left(p_x - \frac{\beta E}{c}\right) \qquad p'_y = p_y \qquad p'_z = p_z \\ E' = \gamma(E - p_x c\beta) \end{gathered}} \tag{12.16}$$

The inverse transformations follow on changing $-\beta$ into $+\beta$ and interchanging primed and unprimed quantities:

$$p_x = \gamma\left(p'_x + \frac{\beta E'}{c}\right) \qquad p_y = p'_y \qquad p_z = p'_z$$
$$E = \gamma(E' + p'_x c\beta) \tag{12.17}$$

We can determine the velocity of the particle from its momentum and energy, using Eqs. (12.14) and (12.15):

$$v_x = \frac{dx}{dt} = \frac{dx}{d\tau}\frac{d\tau}{dt} = \frac{p_x}{M}\frac{Mc^2}{E} = \frac{c^2 p_x}{E}$$

or

$$\boxed{\mathbf{p} = \mathbf{v}\frac{E}{c^2}} \tag{12.18}$$

EXAMPLE

Inelastic Collision Suppose that two identical particles 1 and 2 collide and stick together to make a third particle 3. In the reference frame S in which the center of mass is at rest we have (by definition of the center of mass)

$$\mathbf{p}_1 + \mathbf{p}_2 = 0$$

The product particle must be at rest. In another reference frame S′ we have

$$\mathbf{p}'_1 + \mathbf{p}'_2 = \mathbf{p}'_3$$

We may express this in terms of quantities observed in S by means of the transformation equation (12.16):

$$p'_{x1} + p'_{x2} = \gamma(p_{x1} + p_{x2}) - \frac{\gamma\beta(E_1 + E_2)}{c}$$
$$= p'_{x3} = \gamma p_{x3} - \frac{\gamma\beta E_3}{c} \tag{12.19}$$

Here E_1 and E_2 are the energies in S of the initial particles; E_3 is the energy in S of the product particle. But $p_{x3} = 0$ and $p_{x1} + p_{x2} = 0$, so that Eq. (12.19) reduces to

$$E_3 = E_1 + E_2$$

This result tells us that the relativistic energy is conserved in the collision. This discussion will remind you of our discussion of energy and momentum conservation in Chap. 4.

Now because the particles are identical $E_1 = E_2$; using Eq. (12.10) for E_1, we have for E_3 in frame S,

$$M_3c^2 = \frac{2Mc^2}{(1 - v^2/c^2)^{\frac{1}{2}}} \tag{12.20}$$

Here M_3 is the *rest* mass of the product particle, and $\mathbf{v}$ is the initial velocity of particle 1 or 2 in the frame S. The rest mass M_3 of the product particle is greater in this example than the sum $2M$ of the rest masses of the initial particles. The kinetic energy of the initial particles has been converted into a contribution to the rest mass of the product particle.

In the consideration of the most general collisions it is found that momentum can be conserved only if the sum

$$\sum_i \frac{M_i c^2}{(1 - v_i^2/c^2)^{\frac{1}{2}}} = \sum_i E_i \tag{12.21}$$

extended over the incoming particles is equal to the same sum extended over the outgoing particles.[1] We saw an example of this in Eq. (12.20). That is, momentum can be conserved in a relativistic collision only if the relativistic energy is also conserved.

The new rest mass M_3 is greater than the sum $2M$ of the initial rest masses. For $\beta \ll 1$ we can describe this increase partly in terms of nonrelativistic concepts. Because

$$\frac{1}{(1 - v^2/c^2)^{\frac{1}{2}}} \approx 1 + \frac{v^2}{2c^2} + \cdots$$

we have from Eq. (12.20) that

$$\begin{aligned} M_3 &\approx 2\left(M + \frac{1}{2}M\frac{v^2}{c^2} + \cdots\right) \\ &\approx 2\left(M + \frac{\text{kinetic energy}}{c^2}\right) \end{aligned} \tag{12.22}$$

Thus the rest mass M_3 is composed not only of the sum of the rest masses of the incoming particles but also of a contribution proportional to their kinetic energy. This example of an inelastic collision shows that there has been a conversion of kinetic energy to mass. [We wrote Eq. (12.22) for small β only because this makes it easier to appreciate the mass-energy conversion, but the conversion is valid for all β.] From Eq. (12.22) the relation between the mass increase

$$\Delta M = M_3 - 2M \tag{12.23}$$

and the kinetic energy which has disappeared is

$$\text{Kinetic energy} = c^2\,\Delta M \tag{12.24}$$

From the definition of kinetic energy given in Eq. (12.7), namely, $K = (\gamma - 1)Mc^2$, the results [Eqs. (12.22) to (12.24)] can be seen to be true in general, not only for small β.

[1] We cannot apply Eq. (12.21) directly if photons are involved in the collision because $v = c$ for a photon. In Eqs. (12.26) and (12.27) we shall show how to handle the problem for photons and for other particles having zero rest mass.

Equivalence of Mass and Energy The possibility of an interchange between rest mass and energy (and the quantitative relation between them) was considered by Einstein to be the most significant contribution of the theory of relativity. As long as particles never acquire velocities that are significant relative to c, we may use the nonrelativistic definition of kinetic energy, from which we conclude that in any collision among particles (even if the numbers incident and outgoing are different) any net loss or gain in rest mass times c^2 equals the net gain or loss in kinetic energy. Conversely, in an inelastic collision in which there is a loss of kinetic energy, there must be an increase in the rest masses of the outgoing particles.

We see from Eqs. (12.6) and (12.10) that we can write $E = M(v)c^2$. Thus the natural definition of energy in relativity theory is such that the statement [Eq. (12.24)] is exactly valid for the total energy without restriction to $v/c \ll 1$:

$$\Delta E = c^2 \, \Delta M$$

(An exact derivation is given in the Historical Note at the end of this chapter.) The mass change ΔM associated with conversion of kinetic energy into rest mass is generally very small in everyday processes because c is so much greater than common velocities.

Because mass is equivalent to energy, a system with total relativistic energy E has associated with it an inertial mass $M = E/c^2$. Consider a massless box which contains N particles at rest, each of mass M. The box displays an inertial mass NM when we try to accelerate it. The momentum is $NM\mathbf{V}$ if the box is given velocity $\mathbf{V}$. But if each of the particles has a velocity $\mathbf{v}$ and a kinetic energy $\frac{1}{2}Mv^2$ in the box, then the inertial mass of the box is $N(M + Mv^2/2c^2)$ and its momentum is $N\mathbf{V}(M + Mv^2/2c^2)$. In these expressions the velocities V and v have been assumed to be much less than c.

Similarly, a compressed spring has a greater mass than an uncompressed one, by the work needed to compress it divided by c^2. If the compressed spring is completely dissolved in acid, the reaction products are slightly (but too small to measure) more massive than if the spring had not been under stress. This would reveal itself as a slightly elevated temperature of the solution, if the elevation could be measured.

EXAMPLE

Mass-energy Conversions

(1) If two 1-g masses with equal and opposite velocities of 10^5 cm/s collide and stick together, the additional rest mass of the joined pair is

TABLE 12.1 Comparison of Calculated and Observed Energies of Disintegration*

	Decrease of mass, u	Energy released, MeV	
		$\Delta M\,c^2$	ΔE
$Be^9 + H^1 \rightarrow Li^6 + He^4$	0.00242	2.25	2.28
$Li^6 + H^2 \rightarrow He^4 + He^4$	0.02381	22.17	22.20
$B^{10} + H^2 \rightarrow C^{11} + n^1$	0.00685	6.38	6.08
$N^{14} + H^2 \rightarrow C^{12} + He^4$	0.01436	13.37	13.40
$N^{14} + He^4 \rightarrow O^{17} + H^1$	−0.00124	−1.15	−1.16
$Si^{28} + He^4 \rightarrow P^{31} + H^1$	−0.00242	−2.25	−2.23

*S. Dushman, *General Electric Review*, **47**: 6–13 (October 1944).

$$\Delta M = \frac{\Delta E}{c^2} \approx 2\left(\frac{1}{2}M\frac{v^2}{c^2}\right) \approx 1 \times 10^{-11}\ \text{g}$$

This is less than the precision with which a mass of 1 g can be measured.

(2) A hydrogen atom consists of an electron bound by the force of an electrical attraction to a proton: its rest mass M_H is lighter than the sum of the rest mass m of the free electron and M_p of the proton. The extra mass of the free particles is equal to the ionization (binding) energy divided by c^2. The mass M_H of an H atom is 1.6736×10^{-24} g. The binding energy of the electron to the proton is known to be 13.6 eV or 22×10^{-12} erg, so that

$$M_p + m - M_H = \frac{22 \times 10^{-12}}{c^2} \approx 2.4 \times 10^{-32}\ \text{g}$$

which is 1 part in 10^8 of the H atom mass, again too small to be measured.[1]

(3) The sum of the rest masses of the proton and neutron is

$$\begin{aligned} M_p + M_n &= (1.67265 + 1.67496) \times 10^{-24} \\ &= 3.34761 \times 10^{-24}\ \text{g} \end{aligned}$$

However the mass of the deuteron is 3.34365×10^{-24} g. The difference, 0.00396×10^{-24} g, is equal to 3.56×10^{-6} erg or 2.23 MeV which is just the energy necessary to dissociate the deuteron into a free neutron and proton and is called the binding energy of the deuteron (Figure 12.6 shows the binding energy of nuclei plotted against the mass number.) Actually these data provide one method of getting the neutron mass. The disintegration of the neutron into proton, electron, and neutrino provides another and the agreement is very good.

(4) In Table 12.1 the observed energy release ΔE is compared with the observed mass change ΔM for several nuclear reactions.

[1] Present measurements are 10 to 100 times less accurate. The effects of electronic binding have been observed in a nuclear reaction.

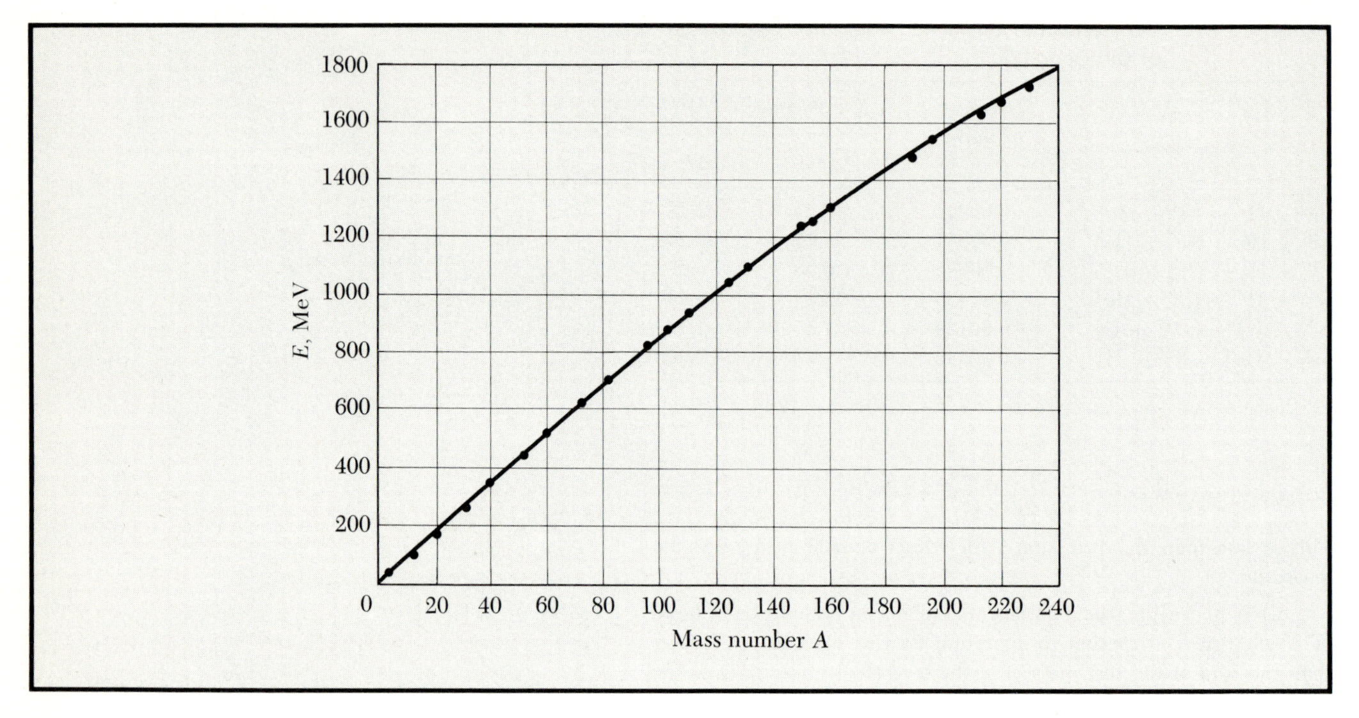

FIG. 12.6 Binding energy of nuclei, in MeV, as a function of the mass number A of the nucleus. Recall that 1 MeV is equivalent to a mass of 1.76×10^{-27} g. Not all nuclei are represented in the graph.

One unified atomic mass unit (u) is equal to one-twelfth the mass of one atom of C^{12}.

EXAMPLE

Stellar Energy Reactions The most important source of energy in the sun and in most stars arises from the nuclear burning of protons to form helium.

The energy release per helium atom (see Fig. 12.7) formed can be calculated from the net mass change in the reaction:

$$\begin{aligned} 4M(H^1) - M(He^4) &= 4(1.6736 \times 10^{-24})\ g - 6.6466 \times 10^{-24})\ g \\ &\approx 0.0478 \times 10^{-24}\ g \\ &\approx 52\ m \end{aligned} \tag{12.25}$$

where m denotes the mass of the electron. The result is equivalent to 26.7 MeV. Atomic masses as tabulated include the mass of the normal number of atomic electrons. The positrons in the reactions below annihilate with electrons to give γ-rays.

The temperature at the center of the sun is $\sim 2 \times 10^{7\circ}$ K. At this temperature the nuclear processes are believed to be dominated by the following set of reactions (illustrated in Fig. 12.8, p. 366):

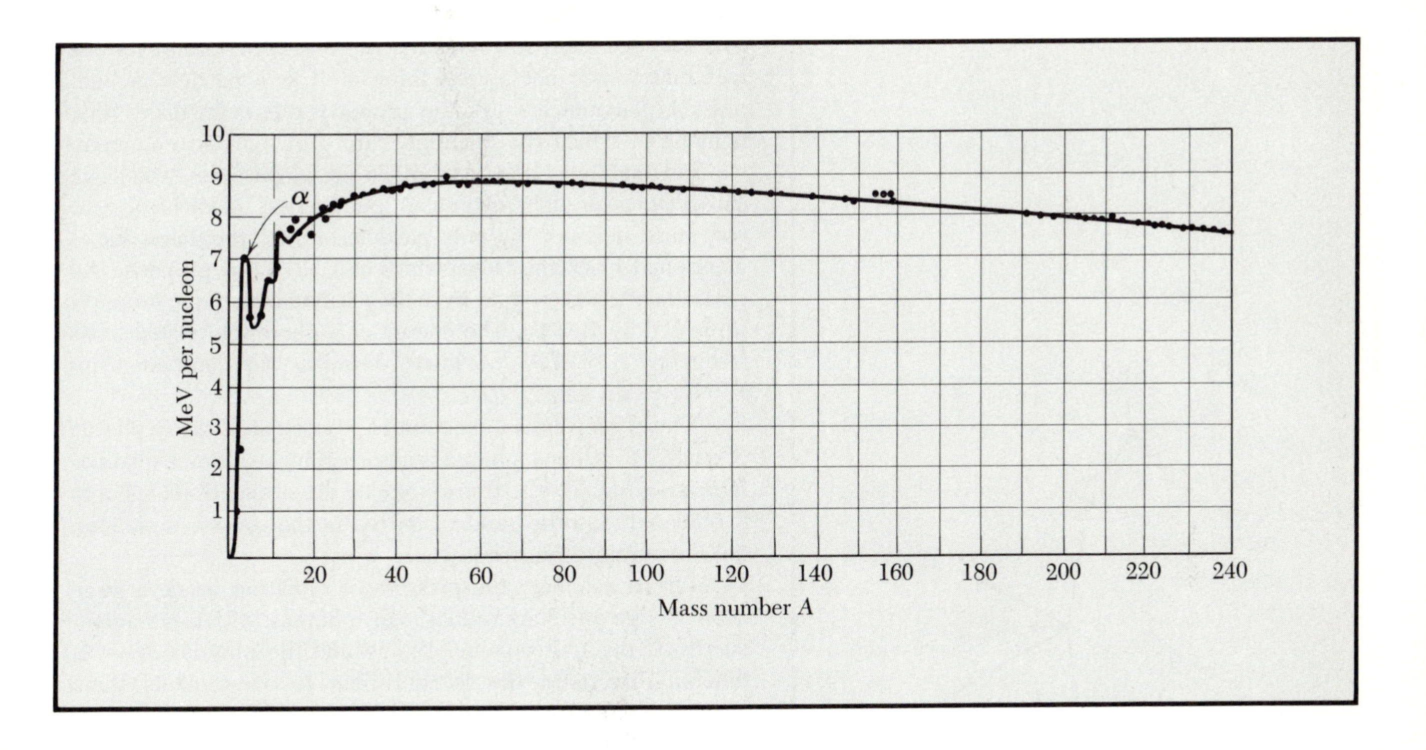

FIG. 12.7 The binding energy per nucleon in MeV per nucleon, as a function of the mass number A. The point labeled α corresponds to He^4, which has a relatively large binding energy.

$$H^1 + p = H^2 + e^+ + \text{neutrino}$$
$$H^2 + H^1 = He^3 + \gamma$$
$$He^3 + He^3 = He^4 + 2H^1$$

The net effect is to burn hydrogen to produce He^4. Note that a neutrino (massless neutral particle) is given out in the first step, so that the sun is a powerful source of neutrinos. Neutrinos interact very weakly with matter; thus nearly all the neutrinos produced in nuclear reactions in stars escape into space. They may carry off up to 10 percent of the energy emitted from the sun.[1]

Particles with Zero Rest Mass When $M = 0$ in Eq. (12.11) we have

$$E = pc \tag{12.26}$$

so that Eq. (12.18) becomes

$$v = c \tag{12.27}$$

and we see that a particle with zero rest mass always moves with the speed of light. It has this same speed for any observer

[1]For an excellent discussion of the origin of the elements, see William A. Fowler, *Proc. Nat. Acad. Sci.*, **52**:524–548 (1964).

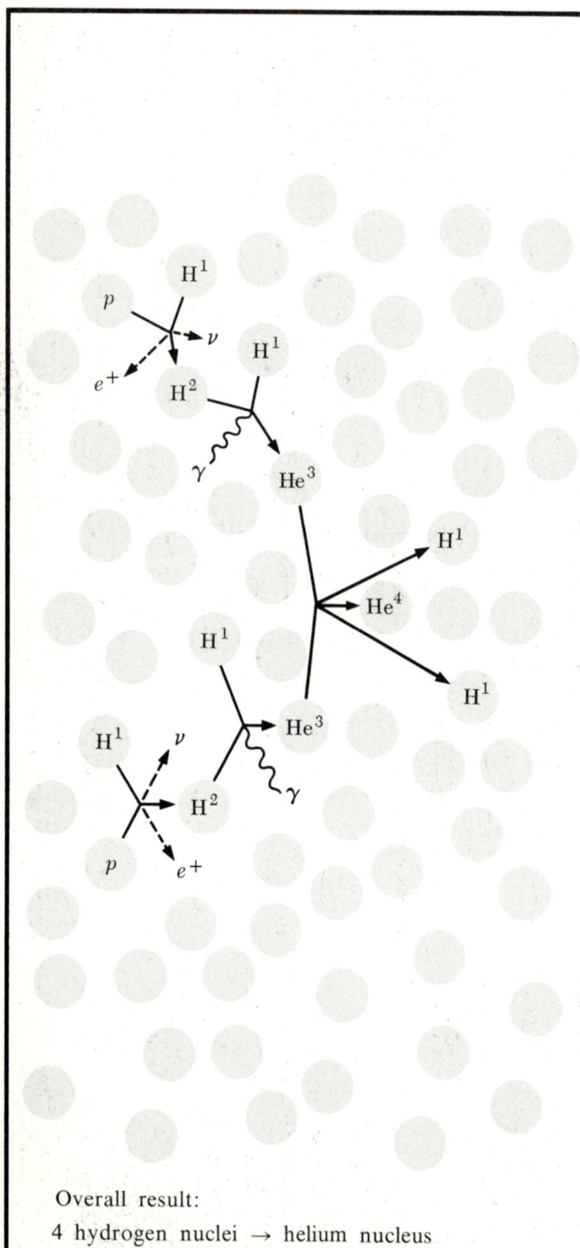

FIG. 12.8 Diagram of the fusion of hydrogen into helium by the *p-p* chain which occurs in mainsequence stars of one solar mass or less. Density: 10^2 g/cm^3. Temperature: 10^7°K. (*After W. A. Fowler*)

and the same zero rest mass for any observer. Except for the fact that we do not always think of it as a particle, a light pulse in vacuum has just the property $v \equiv c$. In many phenomena in which the quantum nature of light is prominent, we find that light acts as if made up of particles which we call *photons* or light quanta. A photon is a particle of zero rest mass; it is not the only particle of zero rest mass, for as mentioned in Chap. 11 neutrinos also have this property. All particles of zero rest mass have the particularly simple property expressed by $E = pc$. The energy of a photon is related to its frequency ν by $E = h\nu$, where h is Planck's constant. Thus $E = h\nu = pc$, or $p = h\nu/c$.

There is always a momentum E/c associated with a photon of energy E. When a photon is absorbed by an atom, a quantity of momentum E/c is transferred to the atom. If the photon is reflected (absorbed and reemitted in the reverse direction), the momentum transfer is $2E/c$.

Let us calculate the pressure of radiation inside a large cube of edge L which contains many photons, with total radiant energy U per unit volume. We assume the photons move in random directions; this is equivalent to one-third of them moving parallel to any one edge of the cube. In unit time a given cube face sustains $\left(\frac{N}{3}\right)\left(\frac{c}{2L}\right)$ collisions, where N is the total number of photons in the box. The momentum change per collision is $2E/c$. The time-average force on the face is

$$F = \text{(collisions per unit time)(momentum change per collision)}$$
$$= N\left(\frac{c}{6L}\right)\left(\frac{2E}{c}\right) = N\frac{E}{3L}$$

If n is the number of photons per unit volume, then $N = nL^3$, and we have

$$F = nL^2\frac{E}{3} \qquad \text{or} \qquad P = \frac{1}{3}U\dagger$$

† For nonrelativistic particles (kinetic theory of gases) the relation is

$$P = \frac{Nmv^2}{3L^3} = \frac{2}{3}\left(\frac{1}{2}\frac{Nmv^2}{L^3}\right) = \frac{2}{3}\text{ kinetic energy density}$$

The transition from $P = \frac{2}{3}$ kinetic density density to $P = \frac{1}{3}$ energy density (which the pressure will be if $v \approx c$) corresponds to the transition from

$$K = \frac{p^2}{2m} \qquad \text{to} \qquad K = E = pc$$

for the radiation pressure, where $P = F/L^2$ and $U = nE$.

It is easy to derive the expression for the doppler effect from the expression for p and E and the Lorentz transformation [Eq. (12.16)]. If in frame S, $E = h\nu$, $p_x = h\nu/c$, what are E' and p' in frame S′?

$$p'_x = \gamma\left(\frac{h\nu}{c} - \beta\frac{h\nu}{c}\right) \qquad E' = \gamma\left(h\nu - \beta c\frac{h\nu}{c}\right)$$

Therefore,

$$p'_x = \frac{h\nu}{c}(1-\beta)\gamma \qquad E' = h\nu(1-\beta)\gamma$$

$$p'_x = \frac{h\nu}{c}\sqrt{\frac{1-\beta}{1+\beta}} \qquad E' = h\nu' = h\nu\sqrt{\frac{1-\beta}{1+\beta}} \tag{12.28}$$

E' is, of course, equal to $p'c$.

Sunlight deposits on the earth's surface as radiant energy about 10^6 ergs/cm^2-s. If all the incident energy is absorbed, the resulting pressure is $(10^6/c)$ dyn/cm$^2 \approx 3 \times 10^{-5}$ dyn/cm^2. The pressure is twice as much if the energy is all reflected. This is an extremely small pressure and has an entirely negligible effect on the motion of the earth. On the very diffuse tail of a comet or on Echo satellites, the surface area is large for a given mass and the cumulative effect of such a pressure may not be negligible. Figure 12.9 is an illustration. Bombardment of the comet tail by material particles from the sun may be more important, however. Inside a very hot low-density star the radiation pressure can become enormously significant.

Any particle with sufficiently high energy so that $E \gg Mc^2$ will have approximately the same momentum and energy relations as a photon. For a particle we can always postulate a moving observer for whom the particle is at rest. But the photon, although its energy and momentum differ for different observers, will always have $v = c = E/p$ and it can never be brought to rest by changing reference frames.

Consider a single hydrogen atom at rest but in an excited electronic state. The hydrogen atom emits a light quantum of energy E and momentum $(E/c)\hat{\mathbf{x}}$. The atom recoils with momentum $-(E/c)\hat{\mathbf{x}}$. In consequence of the recoil, the center of mass of the system (atom plus light quantum) cannot remain

FIG. 12.9 The Mkros comet, August 27, 1957. (*Mount Wilson and Palomar Observatories photograph*).

at rest unless the light quantum possesses a mass, say M_γ. To find M_γ we set

$$\dot{\mathbf{R}}_{\text{c.m.}} \equiv \frac{M_{\text{H}}\dot{\mathbf{r}}_{\text{H}} + M_\gamma \dot{\mathbf{r}}_\gamma}{M_{\text{H}} + M_\gamma} = 0$$

Now $M_{\text{H}}\dot{\mathbf{r}}_{\text{H}} = -(E/c)\hat{\mathbf{x}}$ and $\dot{\mathbf{r}}_\gamma = c\hat{\mathbf{x}}$, so that

$$-\frac{E}{c} + M_\gamma c = 0 \qquad M_\gamma = \frac{E}{c^2}$$

This mass is just that given by the Einstein relation. The mass of the light quantum is not a rest mass; it is the mass equivalent of the energy E. The rest mass of a light quantum is zero.

TRANSFORMATION OF THE RATE OF CHANGE OF MOMENTUM

We are interested in Newton's Second Law

$$\mathbf{F} = \frac{d\mathbf{p}}{dt} = M\frac{d}{dt}\frac{\mathbf{v}}{\sqrt{1 - v^2/c^2}}$$

and how it transforms. It is apparent that

$$\frac{d\mathbf{p}}{dt} \neq \frac{d\mathbf{p}'}{dt'}$$

Let us consider the frame S' as that in which the mass M is instantaneously at rest. Then S' moves with velocity $v\hat{\mathbf{x}}$ with respect to S. From Eq. (12.16)

$$\Delta p_y = \Delta p'_y \qquad \Delta p_z = \Delta p'_z$$

and from Eq. (12.3)

$$\Delta t' = \Delta\tau = \sqrt{1 - \frac{v^2}{c^2}}\,\Delta t$$

where $\Delta\tau$ is the proper time. Therefore

$$\frac{\Delta p_y}{\Delta t} = \frac{\Delta p'_y \sqrt{1 - v^2/c^2}}{\Delta t'} = \frac{1}{\gamma}\frac{\Delta p'_y}{\Delta t'}$$

Since

$$F_y = \frac{\Delta p_y}{\Delta t} \qquad \text{and} \qquad F'_y = \frac{\Delta p'_y}{\Delta t'}$$

we see that

$$F_y = \frac{1}{\gamma}F'_y \qquad \text{and} \qquad F_z = \frac{1}{\gamma}F'_z$$

The x components of $\Delta\mathbf{p}/\Delta t$ are not so simple.

$$p_x = \gamma\left(p'_x + \frac{vE'}{c^2}\right)$$

$$\Delta p_x = \gamma\,\Delta p'_x + \gamma v\frac{\Delta E'}{c^2} \tag{12.29}$$

We wish to evaluate $\Delta E'$ in terms of $\Delta p'_x$

$$E' = (M_0{}^2c^4 + c^2p'^2)^{\frac{1}{2}}$$

$$\Delta E' = \frac{c^2p'\,\Delta p'}{\sqrt{M_0{}^2c^4 + c^2p'^2}}$$

But p'_x, p'_y, and p'_z are all equal to zero. So $\Delta E' = 0$, and returning to Eq. (12.29),

$$\frac{\Delta p_x}{\Delta t} = \frac{\gamma\,\Delta p'_x}{\Delta t'}\frac{\Delta t'}{\Delta t} = \frac{\gamma\,\Delta p'_x}{\Delta t'}\frac{1}{\gamma} = \frac{\Delta p'_x}{\Delta t'}$$

or

$$\frac{dp_x}{dt} = \frac{dp'_x}{dt'}$$

and

$$F_x = F'_x \tag{12.30}$$

These equations play a fundamental role in Volume 2, Chap. 5. They are, of course, special cases of more general results.

CONSTANCY OF CHARGE

The law of motion $q\mathbf{E} = \dot{\mathbf{p}}$ of a particle of charge q in an electric field $\mathbf{E}$ is incomplete unless we know the dependence of the charge on the speed and acceleration of the particle whose momentum is $\mathbf{p}$. The best experimental evidence that the charge on a proton or electron is very accurately constant is the observation that beams of hydrogen atoms or molecules suffer no deflection in uniform electric fields perpendicular to the beam. The hydrogen atom consists of an electron e and a proton p. The H_2 molecule consists of two electrons and two protons. Even when the protons are moving very slowly, the electrons move around the protons with an average velocity of about $10^{-2}c$.† An undeviated molecule has constant momentum, so that the experimental result tells us that $\dot{\mathbf{p}}_p + \dot{\mathbf{p}}_e = 0 = (e_p + e_e)\mathbf{E}$. Thus it follows from experiment that in the atom or molecule $e_e = -e_p$, despite the fact that the electron has a high velocity and the proton low velocity, and even

† The simple Bohr theory of the atom gives for the ground state $v = c/137$.

though the average velocity of the electron differs in the atom and the molecule. Quantitatively, the electron charge is known to be independent of velocity and equal to that of the proton to at least one part in 10^9 up to an electron velocity of $10^{-2}c$. In addition it is known that charge occurs only in multiples of the electronic charge, so that total charge can be determined by the simple process of counting, which is independent of the frame of reference.

The experimental situation is discussed in Volume 2. The experimental result is that the charge is observed to be independent of the velocity of the particle or the observer. Thus charge and mass transform in different ways when the frame of reference is changed.

PROBLEMS

1. *Relativistic momentum.* What is the momentum of a proton having kinetic energy[1] 1 BeV? (If E is measured in BeV, we may measure p in BeV/c.) *Ans.* 1.7 BeV/c.

2. *Relativistic momentum.* What is the momentum of an electron having kinetic energy 1 BeV? *Ans.* 1.0005 BeV/c.

3. *Photon momentum.* What is the momentum of a photon of energy 1 BeV?

4. *Energy and momentum of fast proton.* Given a proton whose $\beta = 0.995$ measured in the laboratory; what are the corresponding relativistic total energy and momentum? What is the kinetic energy?

5. *Energetic cosmic-ray particles.* It is known that cosmic-ray particles have energies up to 10^{19} eV, and perhaps higher.
   (*a*) What is the apparent mass of such a particle (approximately)? *Ans.* 1.8×10^{-14} g.
   (*b*) What is the momentum (approximately)? *Ans.* 5×10^{-4} g-cm/s.

6. *Transformation of energy and momentum.* A proton has $\beta = 0.999$ in the laboratory. Find the energy and momentum as observed in a frame traveling in the same direction, with $\beta' = 0.990$ with respect to the laboratory.

[1]The word *billion* has different meanings in different countries. In the United States it means 10^9, but in British usage it means 10^{12}. The prefix *giga* (abbreviated G) is often used to denote 10^9. Thus 1 BeV (American) $\equiv$ 1 GeV $\equiv 10^9$ eV.

7. *Energy of fast electron.* An electron has $\beta = 0.99$. What is its kinetic energy? *Ans.* 3.1 MeV.

8. *Recoil in γ-ray emission.* What is the recoil momentum in the laboratory of an Fe^{57} nucleus recoiling due to the emission of a 14-keV photon? Is the momentum of the nucleus relativistic? *Ans.* 7.5×10^{-19} g-cm/s.

9. Consider a γ-ray of energy E_γ directed toward a proton at rest in the laboratory.
   (*a*) In the laboratory frame what is the momentum of the γ-ray?
   (*b*) Show that the velocity V of the center of mass in the laboratory frame is given by

$$\frac{V}{c} = \frac{E_\gamma}{E_\gamma + M_p c^2}$$

   (*c*) What is the γ-ray energy in the center-of-mass frame? Also the proton energy in the center-of-mass frame?

10. *Neutron decay.* Use values given in Chap. 12 to calculate the amount of energy released as a neutron decays into a proton and an electron. *Ans.* 0.79 MeV.

11. *Lorentz invariance in two-particle system.* Let the total momentum and energy for a two-particle system be $\mathbf{p} = \mathbf{p}_1 + \mathbf{p}_2$ and $E = E_1 + E_2$, respectively. Show explicitly that the Lorentz transformations on $\mathbf{p}$ and E are consistent with the invariance of the quantity $E^2 - p^2c^2$.

12. *Transformation to rest frame from center-of-mass reference frame.* Two protons travel in opposite directions from a common point with velocities $\beta = 0.5$.

(*a*) What are the energy and momentum of one proton relative to the common point?

(*b*) Use the Lorentz transformation to find the energy and momentum of one proton in the rest frame of the other. (In problems of this sort, it is usually convenient to express the energy as a multiple of some rest mass energy.)

13. *Radiation of mass by radio transmitter.* What is the mass equivalent of the energy from an antenna radiating 1000 W of radio energy for 24 h? ($1 \text{ W} \equiv 10^7$ ergs/s.)

14. *Solar energy.* The *solar constant* is the flux of solar energy per square centimeter per second at the distance of the earth from the sun. By measurement it is found that the value of the constant is 1.4×10^6 ergs/s-cm^2.

(*a*) Show that the total energy generation of the sun is $\approx 4 \times 10^{33}$ ergs/s.

(*b*) Show that the average rate of energy generation per gram of matter on the sun is ≈ 2 ergs/g-s $\approx 6 \times 10^7$ ergs/g-yr.

(*c*) Show that the energy equivalent of 1 g of hydrogen burned to produce He^4 is $\approx 6 \times 10^{18}$ ergs.

(*d*) Show that if the mass of the sun were one-third hydrogen and the nuclear burning process continued without change, then the sun could continue to radiate at its present rate for 3×10^{10} yr.

15. *Radiation propulsion.* One possible means of propulsion in space is a large reflecting metallic sheet connected to a small vehicle. Make reasonable estimates of the accelerations which might result for some typical vehicle at a distance of 1 AU from the sun.

16. *Momentum of laser pulse.* A large laser can produce a pulse of light having an energy of 2000 J. ($1 \text{ J} \equiv 10^7$ ergs.)

(*a*) Show that the momentum of the pulse is of the order of 1 g-cm/s.

(*b*) Discuss how you might detect this momentum. The duration of the pulse might be 1 ms (10^{-3} s).

ADVANCED TOPIC

Recoilless Emission of Gamma Rays A nucleus in an excited energy state may emit a photon (γ-ray) in making a transition to the ground, or unexcited, state of the nucleus. The inverse process also may occur: A nucleus in its ground state may absorb a photon, leaving the nucleus in an excited state (see Fig. 12.10).

Suppose we prepare a source containing excited nuclei. In the course of time the source will emit photons. We allow

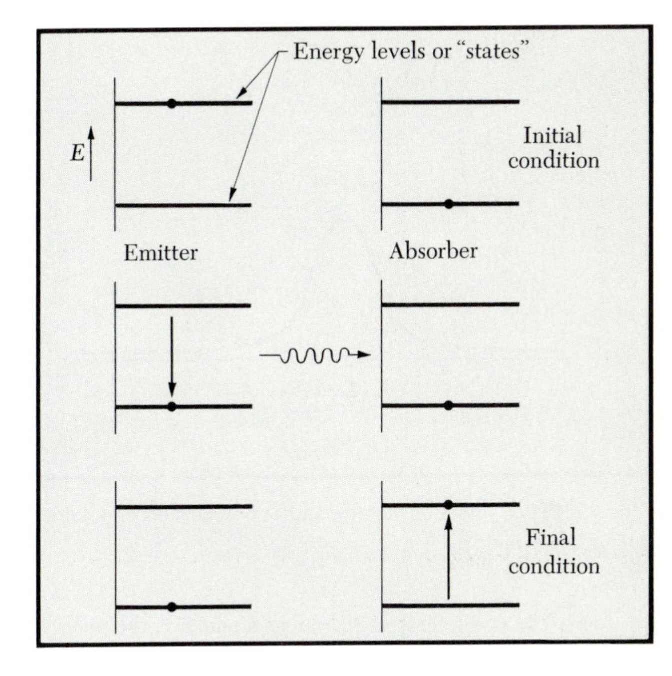

FIG. 12.10 Nuclear energy level changes in the emission and absorption of radiation.

the photons to strike an absorber which contains similar nuclei in their ground state. These nuclei will absorb the incident photons and will then reemit photons. The phenomenon of absorption and reemission is known as *nuclear fluorescence.* The photons emitted (by both the source and the absorber) will have a range of energy of approximate width Γ, as shown in Fig. 12.11. Here Γ is the Greek character capital gamma.

A good example is the nucleus Fe^{57}. This is formed in an excited state as the product of the radioactive decay of Co^{57}. The excited state of Fe^{57} emits a photon of energy 14.4 keV, leaving the Fe^{57} nucleus in its ground state.

Consider an Fe^{57} nucleus in an excited state, and suppose that the nucleus is initially at rest in free space. When the photon is emitted, the nucleus will recoil in the direction opposite to the photon.

(*a*) What is the frequency ν of a photon of energy 14.4 keV? Recall that $E = h\nu$, where h is Planck's constant and E is the energy. *Ans.* 3.5×10^{18} cps.

(*b*) The momentum of the photon is $h\nu/c$. What is the recoil momentum of the nucleus? *Ans.* 7.7×10^{-19} g-cm/s.

(*c*) Show that the recoil energy R of the nucleus is

$$R = \frac{E^2}{2Mc^2}$$

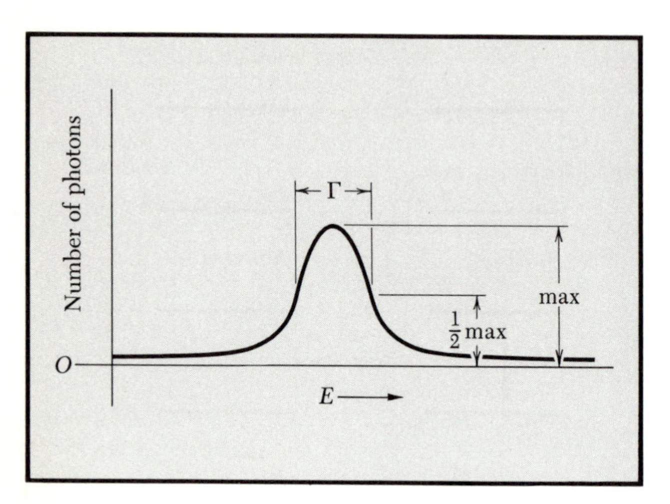

FIG. 12.11 Gamma-ray energy distribution caused by nuclear energy level width.

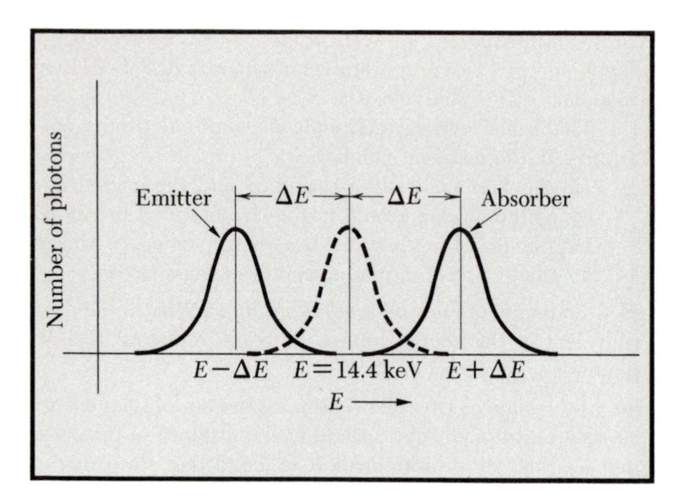

FIG. 12.12 Shifts in gamma-ray energy distribution for nuclear emission and nuclear absorption.

where M is the mass of the nucleus and E is the energy of the photon. Evaluate R in electron volts for Fe^{57}.

Ans. 2×10^{-3} eV.

Nuclear energy levels are not perfectly sharp but have a width Γ according to the uncertainty principle

$$\Gamma\tau \approx \frac{h}{2\pi}$$

where τ is the mean life of the state. For low-energy γ-rays like those from Fe^{57}, the spread in energy of the nuclear energy levels may be much less than the recoil energy R. In this situation the emitted γ-ray cannot normally be reabsorbed by a nucleus in the ground state because the frequency is not right (see Figs. 12.11 and 12.12).

One method of bringing emitter and absorber frequencies effectively into tune is to give the source a velocity relative to the absorber.

(*d*) What is the required magnitude of this velocity for Fe^{57}?

(*e*) Mössbauer observed that in some of the emissions from certain crystals, the recoil momentum is taken by the crystal as a whole rather than by the individual nucleus. At room temperature, about 70 percent of the photons from an Fe crystal are almost recoilless in this sense. Calculate R for a recoilless photon if the mass of the crystal of Fe is 1 g.

Ans. 2×10^{-25} eV, which is entirely negligible.

HISTORICAL NOTE

Mass-energy Relation Einstein's first paper on the special theory of relativity, titled "On the electrodynamics of moving bodies," appeared in *Annalen der Physik*, **17**:891–921 (1905). This volume of the *Annalen* contains three classic papers by Einstein: one on the quantum interpretation of the photoelectric effect (pp. 132–148); one on the theory of Brownian motion (pp. 549–560); and that on special relativity cited above (many of the results in this paper had been anticipated by Larmor, Lorentz, and others). In the same year a short paper by Einstein appeared in vol. 18, pp. 639–641; this was entitled "Does the inertia of a body depend on its energy content?" We paraphrase here Einstein's argument:

Consider (as in Einstein's paper on electrodynamics) a packet or group of plane waves of light. Let the packet possess the energy ϵ and move in the positive x direction in the frame S. As viewed in the frame S', which moves with velocity $V\hat{\mathbf{x}}$ relative to S, the wave packet has the energy

$$\epsilon' = \epsilon\left(\frac{1-\beta}{1+\beta}\right)^{\frac{1}{2}} \qquad \beta = \frac{V}{c} \tag{12.31}$$

This result was derived in Einstein's paper on electrodynamics, without reference to the notion of a photon. It follows most directly from another argument: We note from the result [Eq. (12.28)] of the longitudinal doppler effect that the frequencies seen by observers at rest in S' and S are related by

$$\nu' = \nu\left(\frac{1-\beta}{1+\beta}\right)^{\frac{1}{2}} \tag{12.32}$$

According to the quantum picture a light pulse may be viewed as made up of an integral number of light quanta or *photons*, each of energy $h\nu$ (as viewed in S), where h is Planck's constant. When we view the pulse from S′, the number of photons remains unchanged, but the energy of a photon becomes $h\nu'$. (We suppose that the value of h is the same in S′ as in S.) It follows that the energy ϵ' of the light pulse is proportional to ν', whence Eq. (12.31) follows from Eq. (12.32).

Now let there be a stationary body in the system S, and let its initial energy be E_0 in S and E_0' in S′. We suppose that the body emits a light pulse of energy $\frac{1}{2}\epsilon$ in the positive x direction and a similar pulse of energy $\frac{1}{2}\epsilon$ in the negative x direction. The body will remain at rest in S. Let E_1, E_1' denote in S, S′ the energy of the body after emission of the two light pulses. Then by conservation of energy,

$$E_0 = E_1 + \tfrac{1}{2}\epsilon + \tfrac{1}{2}\epsilon \tag{12.33}$$

$$E_0' = E_1' + \frac{1}{2}\epsilon\left(\frac{1-\beta}{1+\beta}\right)^{\frac{1}{2}} + \frac{1}{2}\epsilon\left(\frac{1+\beta}{1-\beta}\right)^{\frac{1}{2}}$$

$$= E_1' + \frac{\epsilon}{(1-\beta^2)^{\frac{1}{2}}} \tag{12.34}$$

from which, on subtraction of Eq. (12.34) from Eq. (12.33), we have

$$E_0 - E_0' = E_1 - E_1' + \epsilon - \frac{\epsilon}{(1-\beta^2)^{\frac{1}{2}}} \tag{12.35}$$

Now the difference in energy $E_0' - E_0$ must be just the initial kinetic energy K_0 of the body as viewed in S′, because the body is initially at rest in S. Similarly, $E_1' - E_1$ is the final kinetic energy K_1 as viewed in S′. Thus Eq. (12.35) may be written as

$$K_0 - K_1 = \epsilon\left(\frac{1}{(1-\beta^2)^{\frac{1}{2}}} - 1\right)$$

We see that the kinetic energy of the body diminishes as a result of the emission of light. The amount of diminution is independent of the properties of the body. If $\beta \ll 1$

$$K_0 - K_1 \approx \frac{1}{2}\epsilon\beta^2 = \frac{1}{2}\frac{\epsilon}{c^2}V^2$$

so that the rest mass of the body decreases by

$$\Delta M = \frac{\epsilon}{c^2}$$

From this relation Einstein concluded:

If a body gives off the energy ϵ *in the form of radiation, its mass diminishes by* ϵ/c^2. The fact that the energy withdrawn from the body becomes energy of radiation evidently makes no difference, so that we are led to the more general conclusion that

The mass of a body is a measure of its energy content; if the energy changes by ϵ, the mass changes in the same sense by $\epsilon/(9 \times 10^{20})$, the energy being measured in ergs, and the mass in grams.

It is not impossible that with bodies whose energy content is variable to a high degree (e.g. with radium salts), the theory may be successfully put to the test.

If the theory corresponds to the facts, radiation conveys inertia between the emitting and absorbing bodies.

FURTHER READING

M. Born, "Einstein's Theory of Relativity," chap. 6, secs. 7–9, (reprint) Dover Publications, Inc., New York, 1962.

A. P. French, "Special Relativity," chap. 7, W. W. Norton and Company, Inc., New York, 1968.

C. Kacser, "Introduction to the Special Theory of Relativity," chaps. 5–7, Co-Op Paperback, Prentice-Hall, Inc., Englewood Cliffs, N.J., 1967.

CONTENTS

Problems in Relativistic Dynamics

In Chap. 3 we discussed a number of problems involving the nonrelativistic motion of particles in electric and magnetic fields. In Chaps. 3 and 4 and again in Chap. 6, we discussed elastic and inelastic collisions of two nonrelativistic particles. Now we extend several of the earlier solutions to the relativistic region. Often the solutions present no special difficulty, and several of them are of very great importance to the worlds of high-energy particle physics and to astrophysics.

It is well known that in collisions of high-energy particles new particles not present before the collision may be created by conversion of energy into matter. In the center-of-mass system, all the kinetic energy of incident particles can go into production of new particles and into possible internal energy states of particles. For any reaction leading to a particular array of new particles or excited states there is a threshold energy necessary to produce them. If the reaction involves relativistic energies, the fraction of the laboratory energy represented by this center-of-mass threshold energy is less than if the situation is nonrelativistic. This is an important consideration in high-energy-particle physics experiments. An example is given on page 387.

The fact that the momentum of an accelerated relativistic particle can increase indefinitely, even though the velocity changes little when close to c, is at the basis of the great accelerators and of the momentum analysis of high-energy particles by means of a deflecting magnetic field. Magnetic deflection methods are widely used in research with high-energy particles.

We treat first the acceleration of a relativistic particle by an electric field in order to gain familiarity with the use of certain standard manipulations.

ACCELERATION OF A CHARGED PARTICLE BY CONSTANT LONGITUDINAL ELECTRIC FIELD[1]

The equation of motion of a particle of charge q and rest mass M in a uniform constant electric field[2] $\mathcal{E}\hat{\mathbf{x}}$ is

$$\dot{p}\hat{\mathbf{x}} = q\mathcal{E}\hat{\mathbf{x}} \tag{13.1}$$

or, with $\mathbf{p} = M\mathbf{v}/(1 - v^2/c^2)^{\frac{1}{2}}$

[1] The student who is concerned with whether we are correct in Eq. (13.1) should remember Eq. (12.30). A detailed discussion is given in Volume 2, pp. 168–171.

[2] Here we denote the electric field intensity by $\mathcal{E}$ to avoid confusion with the energy E.

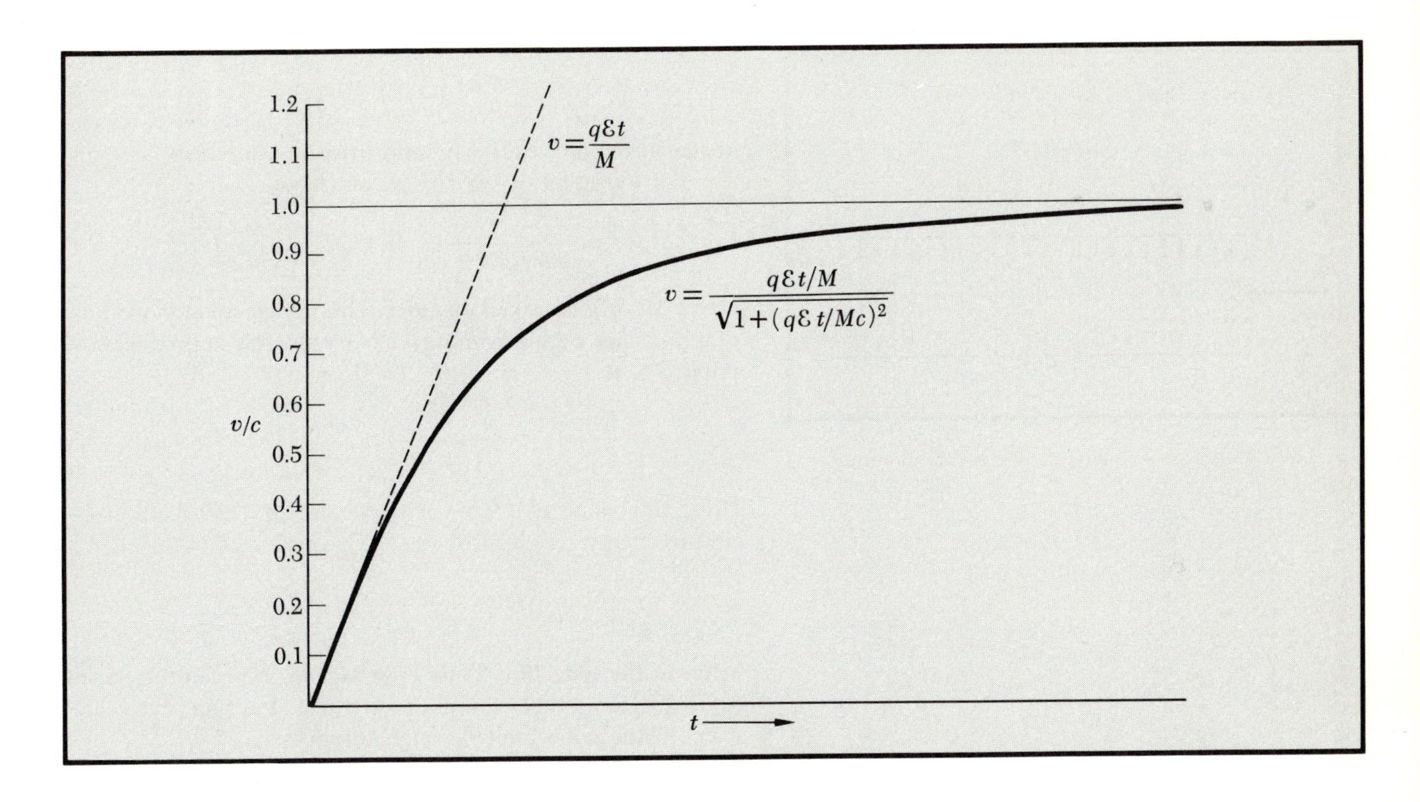

FIG. 13.1 The velocity v of a charge q of rest mass M, accelerated from rest by a uniform electric field $\mathcal{E}$, is plotted vs time. The velocity v approaches the limit c for $t \gg 0$. The dotted line represents the velocity of the charge as predicted by nonrelativistic mechanics.

$$M\frac{d}{dt}\frac{v}{(1 - v^2/c^2)^{\frac{1}{2}}} = q\mathcal{E} \tag{13.2}$$

where we assume $v_y = v_z = 0$, as required for acceleration from rest in the x direction. On integrating Eq. (13.2) with respect to the time, we see directly that

$$p = M\frac{v}{(1 - v^2/c^2)^{\frac{1}{2}}} = q\mathcal{E}t$$

with $v(0) = 0$. After squaring both sides and rearranging, or using Eqs. (12.11) and (12.18), we obtain

$$v^2 = \frac{(q\mathcal{E}t/Mc)^2}{1 + (q\mathcal{E}t/Mc)^2}c^2 \tag{13.3}$$

v/c is plotted against t in Fig. 13.1. For short times[1] $t < Mc/q\mathcal{E}$, the denominator in Eq. (13.3) may be replaced by unity and we have

[1]With $\mathcal{E} = 1$ statvolt/cm, we have for an electron

$$\frac{mc}{e\mathcal{E}} \approx \frac{(10^{-27})(3 \times 10^{10})}{5 \times 10^{-10}} \approx 10^{-7}\text{ s}$$

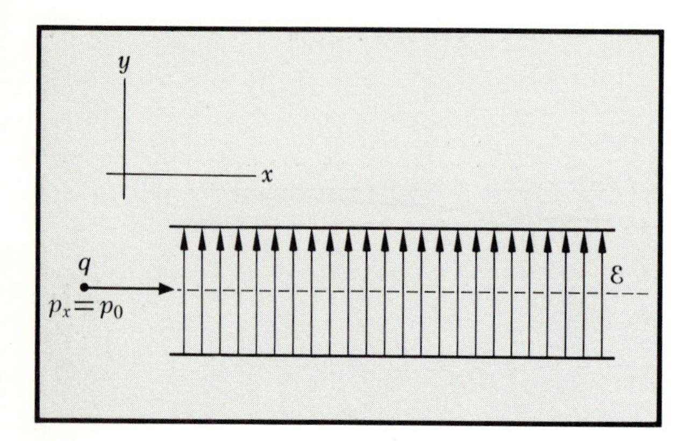

FIG. 13.2 Suppose a charge q with initial momentum p_x enters a transverse field $\mathcal{E}$.

$$v^2 \approx \left(\frac{q\mathcal{E}}{M}t\right)^2 = \frac{p^2}{M^2}$$

just as in the nonrelativistic approximation in Chap. 3.

For long times $t \gg Mc/q\mathcal{E}$, we have

$$v^2 = \frac{1}{(Mc/q\mathcal{E}t)^2 + 1}c^2 \approx \left[1 - \left(\frac{Mc}{q\mathcal{E}t}\right)^2\right]c^2$$

where $Mc/q\mathcal{E}t$ is a small quantity. This last equation shows how v approaches c as a limiting velocity. In this approximation, using $\beta \equiv v/c$

$$\frac{1}{(1-\beta^2)^{\frac{1}{2}}} \approx \frac{q\mathcal{E}t}{Mc}$$

Using this value of $1/(1-\beta^2)^{\frac{1}{2}} = \gamma$ in Eq. (12.10), the relativistic energy[1] is given by

$$E = \frac{Mc^2}{(1-\beta^2)^{\frac{1}{2}}} \approx q\mathcal{E}ct$$

again in the long time limit $t \gg Mc/q\mathcal{E}$. The limiting result is just the force times the distance traveled in time t at velocity c. In the same limit the momentum is

$$p \approx q\mathcal{E}t \approx \frac{Mc}{(1-\beta^2)^{\frac{1}{2}}} = \frac{E}{c}$$

Note the characteristic relativistic feature that the momentum and the energy may go on increasing even after the velocity has for all practical purposes leveled off and that the energy is proportional to p rather than to p^2.

The displacement x is found from the square root of Eq. (13.3), writing dx/dt for v:

$$dx = \frac{(q\mathcal{E}/Mc)t}{\sqrt{1 + (q^2\mathcal{E}^2t^2/M^2c^2)}}c\,dt \tag{13.4}$$

After integrating between 0 and t, we obtain the displacement

$$x = \frac{Mc^2}{q\mathcal{E}}\left\{\left[1 + \left(\frac{q\mathcal{E}t}{Mc}\right)^2\right]^{\frac{1}{2}} - 1\right\} \tag{13.5}$$

We have assumed $x = 0$ and $v = 0$ at $t = 0$. Note that if $q\mathcal{E}t/Mc \gg 1$, $x \approx ct$, whereas if $q\mathcal{E}t/Mc \ll 1$, $x \approx \frac{1}{2}(q\mathcal{E}/M)t^2$, which is the nonrelativistic case.

[1] The general expression for E is obtained from Eq. (12.11)

$$E^2 = M^2c^4 + q^2\mathcal{E}^2t^2c^2$$

ACCELERATION BY A TRANSVERSE ELECTRIC FIELD

We consider a charged particle that moves along the x axis with a high momentum p_0 and enters a region of length L in which there is a transverse electric field $\mathcal{E}\hat{\mathbf{y}}$. Find the angle through which the particle is deflected by the electric field (see Fig. 13.2).

The equations of motion are

$$\frac{dp_x}{dt} = 0 \qquad \frac{dp_y}{dt} = q\mathcal{E}$$

from which

$$p_x = p_0 \qquad p_y = q\mathcal{E}t$$

as shown in Fig. 13.3. We want to find the velocity $\mathbf{v}$. If we can find the energy E, then we can find the velocity from the momentum by using the relation $\mathbf{v} = \mathbf{p}c^2/E$ derived in Eq. (12.18).

The energy is given by

$$\begin{aligned} E^2 &= M^2c^4 + p^2c^2 = M^2c^4 + p_0{}^2c^2 + (q\mathcal{E}tc)^2 \\ &= E_0{}^2 + (q\mathcal{E}tc)^2 \end{aligned} \tag{13.6}$$

where E_0 is the initial energy. Therefore from Eq. (13.6) and the velocity-momentum relation we have

$$v_x = \frac{p_0c^2}{[E_0{}^2 + (q\mathcal{E}tc)^2]^{\frac{1}{2}}} \tag{13.7}$$

$$v_y = \frac{q\mathcal{E}tc^2}{[E_0{}^2 + (q\mathcal{E}tc)^2]^{\frac{1}{2}}} \tag{13.8}$$

Note that v_x decreases as t increases (see Fig. 13.4). We see also that v_y is always less than the nonrelativistic value $q\mathcal{E}t/M$. At a time t, the angle θ the trajectory makes with the x axis is given by

$$\tan\theta(t) = \frac{v_y}{v_x} = \frac{q\mathcal{E}tc^2}{p_0c^2} = \frac{q\mathcal{E}t}{p_0}$$

The time t_L required to traverse the distance L is found on integrating Eq. (13.7)

$$\int_0^L dx = p_0c^2 \int_0^{t_L} \frac{dt}{[E_0{}^2 + (q\mathcal{E}tc)^2]^{\frac{1}{2}}}$$

or, by Dwight 728.1,

$$L = \frac{p_0c}{q\mathcal{E}} \sinh^{-1} \frac{q\mathcal{E}t_Lc}{E_0}$$

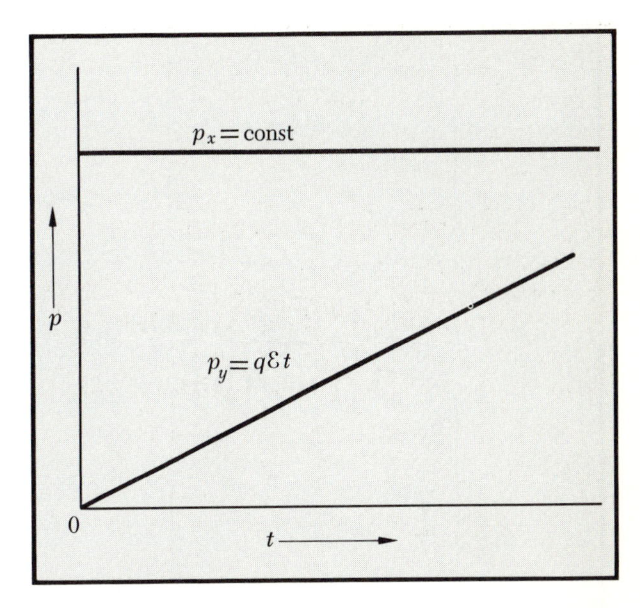

FIG. 13.3 The force in the y direction is $q\mathcal{E}$, so $p_y = q\mathcal{E}t$, while p_x remains constant. The energy

$$E = c\sqrt{(p_x{}^2 + p_y{}^2) + M^2c^2}$$

increases.

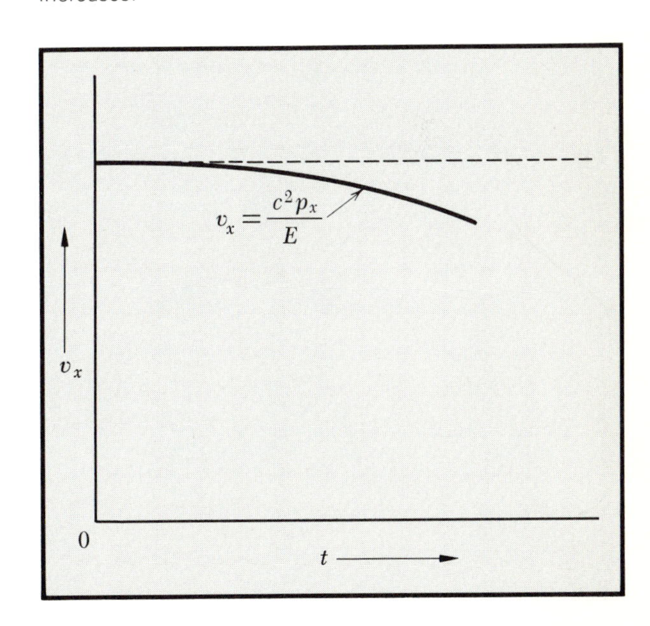

FIG. 13.4 Since $v_x = c^2p_x/E$, v_x actually *decreases* when the particle is accelerated in the y direction. Nonrelativistic mechanics would, of course, predict v_x = const.

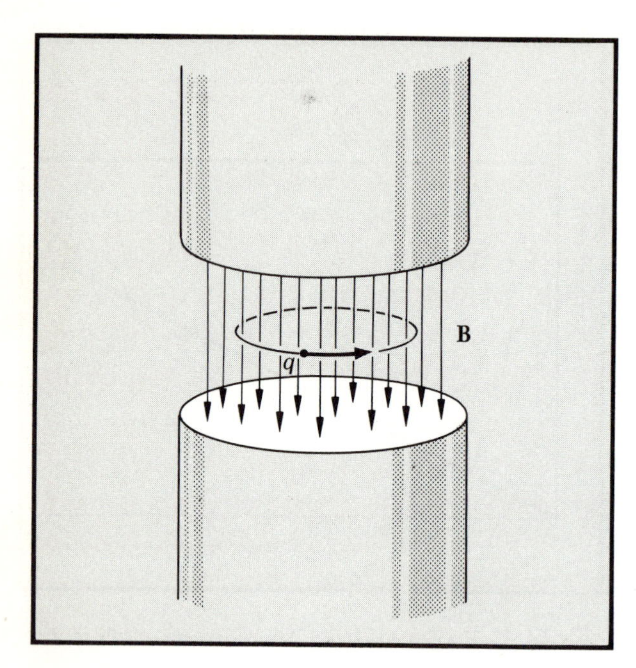

FIG. 13.5 A charge with velocity **v** perpendicular to a uniform magnetic field describes circular motion, with radius $\rho = pc/qB$.

so that

$$t_L = \frac{E_0}{q\mathcal{E}c} \sinh \frac{q\mathcal{E}L}{p_0c} \dagger$$

CHARGED PARTICLE IN A MAGNETIC FIELD

We next consider an important practical problem dealing with the motion of a particle of charge q in a uniform constant magnetic field **B.** The equation of motion is, by reference to Eq. (3.23),

$$\frac{d\mathbf{p}}{dt} = \frac{q}{c}\mathbf{v} \times \mathbf{B} \tag{13.9}$$

Just as in the nonrelativistic problem (Chap. 3, page 76), we have $dp^2/dt = 0$ because

$$\frac{d}{dt}p^2 = 2\mathbf{p} \cdot \frac{d\mathbf{p}}{dt} = 2\frac{q}{c}\mathbf{p} \cdot \mathbf{v} \times \mathbf{B}$$

and **p** is always parallel to **v**, so that the triple product is zero. Thus the magnitude of the momentum and consequently the magnitude of the velocity of the particle are unchanged by a constant magnetic field. But if only the direction is altered by the field, then the factor

$$\frac{M}{(1 - v^2/c^2)^{\frac{1}{2}}} \tag{13.10}$$

which enters into the definition of **p**, is constant.

The equation of motion [Eq. (13.9)] can now be written as.

$$\frac{d\mathbf{p}}{dt} = \frac{M}{(1 - v^2/c^2)^{\frac{1}{2}}} \frac{d\mathbf{v}}{dt} = \frac{q}{c}\mathbf{v} \times \mathbf{B} \tag{13.11}$$

Because of the constancy of expression (13.10), this equation has solutions in which the particle moves in a circle in a plane perpendicular to **B** (refer to Chap. 3, pages 76 to 81). Let ρ denote the radius of the circle and ω_c the angular frequency of the motion for the case **v** perpendicular to **B** (see Fig. 13.5). On substituting in Eq. (13.11) the centripetal acceleration $\omega_c{}^2\rho$ for $d\mathbf{v}/dt$ and $\omega_c\rho$ for v, we have

†$\sinh \theta = (e^{\theta} - e^{-\theta})/2$; thus for small values of θ, $\sinh \theta \approx \theta$, while for large values of θ, $\sinh \theta \approx e^{\theta}/e$.

$$\frac{M}{(1 - v^2/c^2)^{\frac{1}{2}}} \omega_c^2 \rho = \frac{q}{c} \omega_c \rho B$$

from which

$$\boxed{\omega_c = \frac{qB(1 - v^2/c^2)^{\frac{1}{2}}}{Mc}} \tag{13.12}$$

We see that the frequency of the motion is lower for fast particles than for slow particles. Thus a cyclotron can be used to accelerate particles to relativistic energies only if the frequency of the rf accelerating field (or the magnetic field intensity) is modulated to remain synchronous according to Eq. (13.12) as the energy of the particles is increased. This relationship is plotted in Fig. 13.6. For nonrelativistic particles the dependence of the frequency on the velocity may be neglected as we saw in Chap. 3.

Values for ω_c predicted from Eq. (13.12) have been confirmed experimentally in the operation of high-energy accelerators. The relation has been confirmed for electrons accelerated in a synchrotron in situations where $1/(1 - \beta^2)^{\frac{1}{2}} \approx 12{,}000$; that is, when the apparent mass of the particle is 12,000 times the rest mass. It is interesting to see what this means in terms of $c - v$, the amount by which the speed of light exceeds the speed of the particle. We have

$$(1 - \beta^2) = (1 + \beta)(1 - \beta)$$
$$\approx 2(1 - \beta) \approx (12{,}000)^{-2} \approx 7.0 \times 10^{-9} \tag{13.13}$$

Note that we have set $(1 + \beta) \approx 2$. From Eq. (13.13) we have

$$1 - \beta = \frac{c - v}{c} \approx 3.5 \times 10^{-9} \qquad c - v \approx 100 \text{ cm/s}$$

In the proton accelerator at Serpukhov, Russia, the protons are injected at an energy of 100 MeV into the circular orbit in a magnetic field and accelerated to about 80,000 MeV. This corresponds to a change in β from 0.43 to $1 - 6.8 \times 10^{-5}$.

Using Eq. (13.12) the gyroradius ρ of a relativistic particle in a magnetic field is given by

$$\rho = \frac{v}{\omega_c} = \frac{cMv}{qB(1 - \beta^2)^{\frac{1}{2}}}$$

But the right-hand side contains the momentum p, so that

$$\boxed{B\rho = \frac{cp}{q}}$$

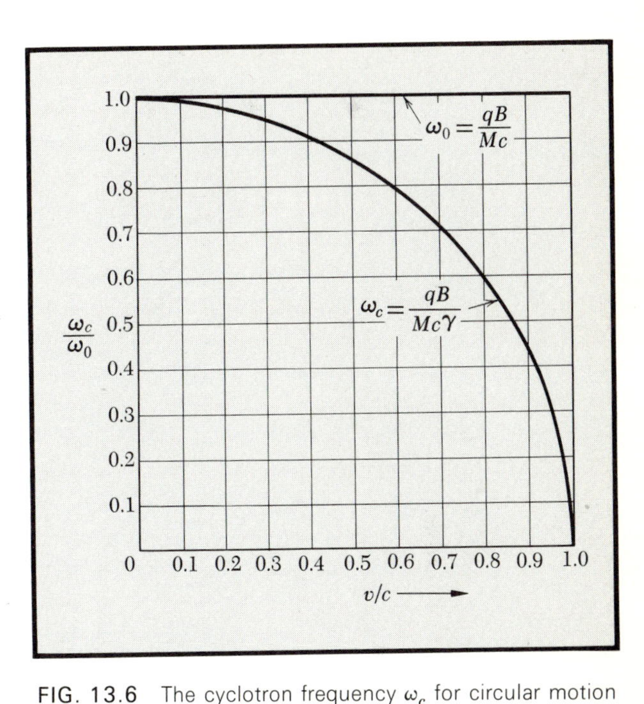

FIG. 13.6 The cyclotron frequency ω_c for circular motion of a charge q, rest mass M, in a plane perpendicular to a uniform magnetic field **B**, is plotted vs the velocity ratio v/c. The nonrelativistic cyclotron frequency ω_0 is represented by the horizontal line.

or in SI units

$$B\rho = \frac{p}{q}$$

Therefore the radius ρ of the circle traced out by a charged particle in a magnetic field is a direct measure of the relativistic momentum. This relation is the most important single method for measuring the momentum of charged relativistic particles.

CENTER–OF–MASS SYSTEM AND THRESHOLD ENERGY

Conservation of energy imposes a general limitation on the nuclear reactions or events which can take place when two particles collide. For example, a high-energy photon (γ-ray) can produce an electron-positron pair according to the reaction

$$\gamma \rightarrow e^- + e^+$$

only if the energy of the γ-ray exceeds the energy equivalent of the rest masses of the electron and positron. Thus conservation of energy alone dictates that the threshold or minimum energy for the production of an electron-positron pair is

$$E_\gamma = 2mc^2 \approx 1.02 \times 10^6 \text{ eV}$$

We recall from Chap. 9, page 292, that the rest mass of the positron is equal to the rest mass of the electron.

This reaction, however, is impossible in free space at any energy because momentum cannot be conserved. We saw in Chap. 12 that the momentum of the photon is $p_\gamma = E_\gamma/c$. We choose to view the reaction in the reference frame in which the center of mass of the electron-positron pair is at rest. In this frame the sum of the electron momentum and positron momentum is zero:

$$\mathbf{p}_{e^-} + \mathbf{p}_{e^+} = 0$$

But in this frame the momentum of the incident photon is not zero, because there is *no* reference frame in which the momentum of a photon can be made to vanish.[1] Therefore in the center-of-mass frame

$$\mathbf{p}_\gamma \neq \mathbf{p}_{e^-} + \mathbf{p}_{e^+} = 0$$

[1]We change the frequency of a photon by changing reference frames, but we can neither make it disappear nor bring it to rest in this way.

and the reaction $\gamma \rightarrow e^+ + e^-$ cannot take place because momentum is not conserved. If it cannot take place in one reference frame, it cannot take place in any reference frame.

The reaction can proceed in the vicinity of another particle, such as a nucleus of an atom, for then the nucleus can absorb the momentum change. It absorbs the change by pushing and pulling with its coulomb field on the charged particles. We can have

$$\mathbf{p}_\gamma + \mathbf{p}_{\text{nuc}} = \mathbf{p}'_{\text{nuc}} + \mathbf{p}_{e^-} + \mathbf{p}_{e^+}$$

The nucleus has its momentum changed by the reaction, but otherwise the nucleus is virtually unchanged and acts only as a catalyst of a very simple kind. The initial nuclear momentum may be zero.

A heavy particle or nucleus is a good vehicle for absorbing excess momentum without absorbing much energy. We see this from the form of the nonrelativistic kinetic energy

$$K = \tfrac{1}{2}Mv^2 = \frac{p^2}{2M}$$

The larger the mass M, the smaller the kinetic energy associated with a given momentum.

EXAMPLE

Threshold Energy for Photoproduction of π^0 Mesons The mass of the π^0 meson is 135 MeV. What is the minimum-energy γ-ray that can produce in the laboratory the reaction

$$\gamma + p \rightarrow \pi^0 + p$$

when the initial proton is at rest?

It is instructive to consider this problem from two different points of view: from the laboratory point of view and from the center-of-mass point of view.

(1) In the laboratory (see Fig. 13.7) a high-energy γ-ray strikes a proton at rest and the result at threshold is a proton and a π^0 meson traveling together at speed βc with the same momentum as the original γ-ray. We write the equations for conservation of energy and of momentum. (Notice that γ in the following equations is $\gamma = (1 - \beta^2)^{-\frac{1}{2}}$, not the symbol for a γ-ray photon as it was in the discussion immediately preceding.)

$$\text{Energy:} \quad h\nu_{\text{lab}} + M_p c^2 = \frac{(M_p + M_\pi)c^2}{\sqrt{1 - \beta^2}} = \gamma(M_p + M_\pi)c^2 \tag{13.14}$$

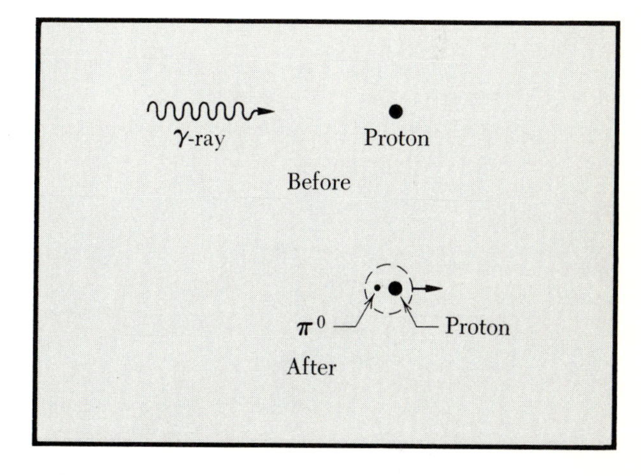

FIG. 13.7 In the laboratory a γ-ray strikes a proton at rest to give a proton and π^0 meson, which must be traveling together in the direction of the γ-ray to conserve momentum at threshold.

Momentum: $\frac{h\nu_{lab}}{c} = \frac{(M_p + M_\pi)\beta c}{\sqrt{1-\beta^2}} = \gamma(M_p + M_\pi)\beta c$

We can eliminate $h\nu_{lab}$ and solve for β and thus γ as follows:

$$h\nu_{lab} = \gamma(M_p + M_\pi)\beta c^2$$
$$\gamma(M_p + M_\pi)\beta c^2 + M_p c^2 = \gamma(M_p + M_\pi)c^2$$

Letting $M_\pi/M_p = \alpha$, we get

$$1 + \gamma\beta(1 + \alpha) = \gamma(1 + \alpha)$$

or

$$\sqrt{1-\beta^2} = (1 + \alpha)(1 - \beta)$$

which gives

$$\beta = \frac{\alpha(2 + \alpha)}{2 + 2\alpha + \alpha^2} \tag{13.15}$$

and

$$\gamma = \frac{2 + 2\alpha + \alpha^2}{2(1 + \alpha)}$$

Solving now for $h\nu_{lab}$,

$$h\nu_{lab} = \frac{M_p c^2 \alpha(1 + \alpha)(2 + \alpha)}{2(1 + \alpha)} = \frac{M_\pi c^2}{2}(2 + \alpha) \tag{13.16}$$

or

$$h\nu_{lab} = 144.7 \text{ MeV}$$

where we have used

$$\alpha = \frac{135}{938} = 0.144$$

(2) In the center-of-mass system (see Fig. 13.8) we write

Energy: $h\nu_{c.m.} + \gamma M_p c^2 = (M_p + M_\pi)c^2$

and

Momentum: $\frac{h\nu_{c.m.}}{c} = \gamma M_p \beta c$

where β and γ now refer to the initial motion of the proton in the center-of-mass frame. Using the same notation as above and eliminating $h\nu_{c.m.}$,

$$\gamma(\beta + 1) = 1 + \alpha \qquad \frac{\beta + 1}{\alpha + 1} = \sqrt{1-\beta^2}$$

and

$$\beta = \frac{\alpha(2 + \alpha)}{2 + 2\alpha + \alpha^2}$$

which is the same β as we obtained in Eq. (13.15) above as it must be since in the center-of-mass system the proton and π^0 meson are at rest in the final state. We can proceed to find

$$h\nu_{c.m.} = M_\pi c^2 \frac{2 + \alpha}{2(1 + \alpha)} = 126.5 \text{ MeV} \tag{13.17}$$

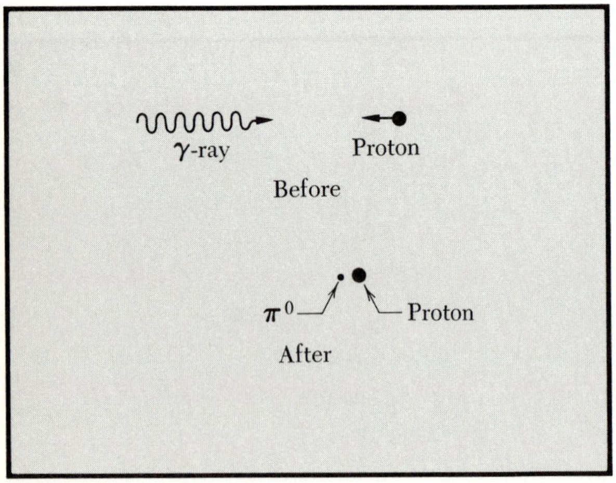

FIG. 13.8 Center-of-mass system. A γ-ray and proton travel toward each other with equal momenta. After the interaction the proton and π^0 meson are at rest if the event occurs at threshold.

We can also find $h\nu_{\text{c.m.}}$ from $h\nu_{\text{lab}}$ by the doppler effect formula [Eq. (12.28)]:

$$h\nu_{\text{c.m.}} = h\nu_{\text{lab}}\sqrt{\frac{1-\beta}{1+\beta}} = \frac{M_\pi c^2}{2}(2+\alpha)\frac{1}{(1+\alpha)}$$

which agrees with the value in Eq. (13.17).

Still a simpler way of working this problem is to remember that $E^2 - p^2c^2$ is an invariant. We write this down for the situation before and after the reaction in the laboratory to get

$$(h\nu_{\text{lab}} + M_pc^2)^2 - \left(\frac{h\nu_{\text{lab}}c}{c}\right)^2$$
$$= [\gamma(M_p + M_\pi)c^2]^2 - [\beta\gamma(M_p + M_\pi)c^2]^2 \quad (13.18)$$

which gives

$$2h\nu_{\text{lab}}M_pc^2 + M_p{}^2c^4 = (M_p + M_\pi)^2c^4$$

where we have used

$$\gamma^2 - \beta^2\gamma^2 = 1$$

Then

$$h\nu_{\text{lab}} = \frac{M_\pi{}^2c^2}{2M_p} + \frac{M_\pi M_pc^2}{M_p} = M_\pi c^2\left(1 + \frac{\alpha}{2}\right)$$

which agrees with the value above in Eq. (13.16).

In collision events in which new particles are created, the requirement of momentum conservation usually makes it impossible to convert all the initial kinetic energy in the laboratory system into rest mass of new particles formed in the collision. If there is a net momentum in the initial state before the collision, there must be an equal momentum in the final state after the collision. Therefore the particles remaining after the collision will not be at rest; some of the initial kinetic energy is transferred as kinetic energy of the final particles.

The only situation in which *all* the initial kinetic energy is available for the reaction occurs when the momentum of the initial state is zero. The momentum can always be made to appear zero by viewing the collision from a suitable reference frame, the center-of-mass reference frame.

EXAMPLE

Energy Available from a Moving Particle How much energy is available in the collision of a moving proton with a proton at rest?

Suppose first that the kinetic energy of the incident proton is much less than M_pc^2, so that the collision may be treated nonrelativistically. If the incident proton has velocity $\mathbf{v}$ in the laboratory reference frame, its kinetic energy is

$$K_{lab} = \tfrac{1}{2}M_p v^2 \tag{13.19}$$

In the center-of-mass reference frame one proton has velocity $\frac{1}{2}\mathbf{v}$ and the other has velocity $-\frac{1}{2}\mathbf{v}$. In the center-of-mass reference frame all the kinetic energy is available for the production of further particles; this kinetic energy is

$$K_{c.m.} = \tfrac{1}{2}M_p(\tfrac{1}{2}v)^2 + \tfrac{1}{2}M_p(\tfrac{1}{2}v)^2 = \tfrac{1}{4}M_p v^2 \tag{13.20}$$

From Eqs. (13.19) and (13.20) we have the nonrelativistic result

$$\frac{K_{c.m.}}{K_{lab}} = \frac{1}{2}$$

thus one-half of the energy in the laboratory system is available. If we accelerate a proton to 50 MeV, only about 25 MeV of this energy is available in a collision with another proton at rest to create other particles.

The efficiency is lower in the relativistic region, and it is straightforward to calculate it.

We can relate the total relativistic energy in the laboratory system to the total relativistic energy in the center-of-mass system by using the invariance property [Eq. (12.11), also used in Eq. (13.18) above] applied to the system of two protons:

$$\underbrace{(E_1 + E_2)^2 - (\mathbf{p}_1 + \mathbf{p}_2)^2c^2}_{\text{laboratory}} = \underbrace{(E_1 + E_2)^2 - (\mathbf{p}_1 + \mathbf{p}_2)^2c^2}_{\text{center of mass}} \tag{13.21}$$

By definition of the center-of-mass system, $(\mathbf{p}_1 + \mathbf{p}_2)_{c.m.} = 0$. If proton 2 is at rest in the laboratory system, $E_{2,lab} = M_pc^2$ and $\mathbf{p}_{2,lab} = 0$. Using

$$E_{1,lab}^2 - p_{1,lab}^2c^2 = M_p^2c^4$$

we see that Eq. (13.21) reduces to

$$2E_{1,lab}M_pc^2 + 2M_p^2c^4 = E^2_{tot,c.m.} \tag{13.22}$$

where $E_{tot,c.m.}$ denotes the sum $E_1 + E_2$ in the center-of-mass system. If we let $E_{tot,lab}$ denote the total energy $E_1 + M_pc^2$ in the laboratory system, we have from Eq. (13.22)

$$2E_{tot,lab}M_pc^2 = E^2_{tot,c.m.}$$

or

$$\boxed{\frac{E_{tot,c.m.}}{E_{tot,lab}} = \frac{2M_pc^2}{E_{tot,c.m.}}} \tag{13.23}$$

This is a measure of the "efficiency." To get a total energy of 20 BeV in the center-of-mass system, for $M_pc^2 \approx 1$ BeV, we need

$$E_{tot,lab} = \frac{E^2_{tot,c.m.}}{2M_pc^2} \approx \frac{400}{2} \approx 200 \text{ BeV}$$

In this case about 20 BeV of the kinetic energy of the 200-BeV proton in the laboratory system is available to create new particles.

Because of this low efficiency in the case of collision of a relativistic particle with a particle at rest, colliding-beam machines for electrons, in which two beams of electrons with equal and opposite momenta collide, have been built and a new ring has been built for the 28-GeV proton accelerator at CERN[1] so that colliding-beam experiments with protons can be done.

EXAMPLE

Antiproton Threshold The energy of the Bevatron at Berkeley was designed to make possible the production of antiprotons (denoted by $\bar{p}$) by bombarding stationary protons with high-energy protons. The reaction may be written

$$p + p \rightarrow p + p + (p + \bar{p})$$

that is, a proton-antiproton pair is produced. This conserves charge, because the antiproton carries charge $-e$. What is the threshold energy for the reaction?

The rest energy of a proton-antiproton pair is $2M_pc^2$ because the rest mass of the antiproton is equal to the rest mass of the proton. In the center-of-mass system the kinetic energy must therefore be at least $2M_pc^2$, which is M_pc^2 for each of the two initial protons. To this must be added the rest energy M_pc^2 for each of the initial protons, so that the minimum total energy in the center-of-mass system is

$$E_{\text{tot,c.m.}} = 4M_pc^2$$

In the laboratory system the corresponding energy is, from Eq. (13.23),

$$E_{\text{tot,lab}} = \frac{E^2{}_{\text{tot,c.m.}}}{2M_pc^2} = \frac{16}{2}M_pc^2$$

of which $2M_pc^2$ is rest energy of the two protons and $6M_pc^2$ is kinetic energy. Thus the threshold energy is

$$6M_pc^2 = 6(0.938 \text{ BeV}) \approx 5.63 \text{ BeV}$$

If the incident proton collides with a proton bound in a nucleus, the threshold energy is lower because the target proton is bound. Can you see why? The observed threshold energy for antiproton production when protons collide with nuclei is about 4.4 MeV, which is 1.2 BeV lower than calculated for free target protons at rest. This threshold is the minimum kinetic energy required of the incident proton, as seen in the laboratory system, to make the reaction go.

EXAMPLE

Compton Effect The Compton effect is one of the most convincing manifestations of the particle nature of electromagnetic

[1]Conseil Européen pour la Recherche Nucléaire, a research institution in Geneva, Switzerland, operated jointly by a number of European nations.

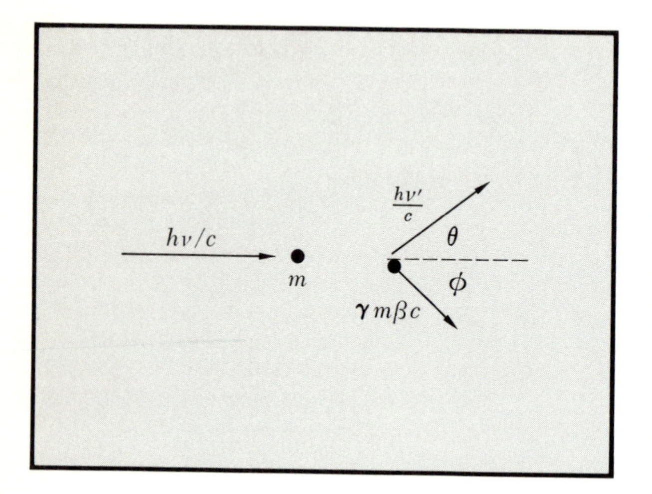

FIG. 13.9 Compton effect. Momenta before and after.

waves. We assume that students are familiar with the wave character of light such as the fact that the wavelength can be determined from interference effects. Compton in 1922 (see Volume 4, page 152ff. for a thorough discussion of the Compton effect) showed that when electromagnetic waves in the x-ray wavelength region ($\sim 10^{-8}$ cm) interact with free electrons they behave like particles in elastic collisions. We have already seen that the characteristic energy, or quantum energy, related to electromagnetic waves of frequency ν is $h\nu$; and the associated momentum is $h\nu/c$. Figure 13.9 shows the momenta in a collision with an electron in which the x-ray is scattered at angle θ with reduced frequency ν'.

Longitudinal momentum: $$\frac{h\nu}{c} = \frac{h\nu'}{c}\cos\theta + \gamma m\beta c\cos\phi$$

Transverse momentum: $$\frac{h\nu'}{c}\sin\theta = \gamma m\beta c\sin\phi$$

Energy: $$mc^2 + h\nu = h\nu' + \gamma mc^2$$

Let us find ν' in terms of θ, eliminating β and ϕ. Let

$$\frac{h\nu}{mc^2} = \alpha \qquad \frac{h\nu'}{mc^2} = \alpha'$$

$$\alpha = \alpha'\cos\theta + \gamma\beta\cos\phi$$
$$\alpha'\sin\theta = \gamma\beta\sin\phi$$

After some slightly complicated algebra we obtain

$$\alpha - \alpha' = \alpha\alpha'(1 - \cos\theta)$$
$$\frac{h\nu}{mc^2} - \frac{h\nu'}{mc^2} = \frac{h^2\nu\nu'}{m^2c^4}(1 - \cos\theta)$$
$$\frac{1}{\nu'} - \frac{1}{\nu} = \frac{h}{mc^2}(1 - \cos\theta)$$
$$\lambda' - \lambda = \frac{h}{mc}(1 - \cos\theta)$$

FIG. 13.10 Inverse Compton effect. A and B show laboratory states. The transformation referred to in the text changes A to C. $C \rightarrow D$ in the Compton effect ($\theta = \pi$) and the transformation takes D back to B.

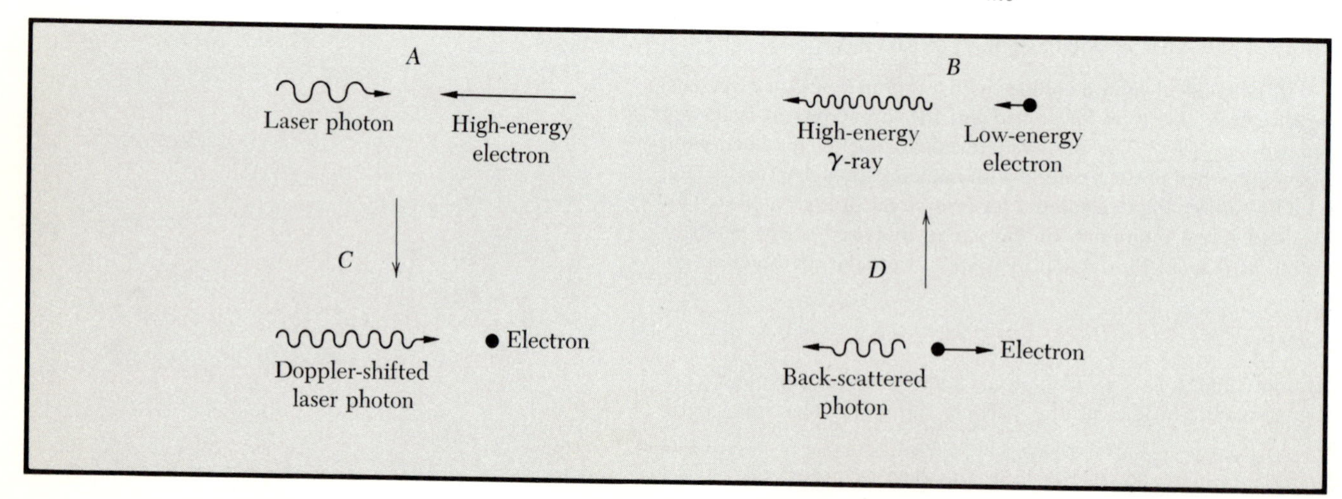

One can reach this same result by solving the problem in the center-of-mass system and then transforming back to the laboratory system, but this involves more calculation.

A recent development involving the Compton effect is the use of high-energy electron accelerators and lasers to make high-energy γ-rays by the inverse Compton effect. We look at such a collision as in Fig. 13.10. To calculate the energy of the γ-ray, we make a transformation to the system that we worked out above in which the electron is at rest.

The doppler-shifted laser photon has energy

$$h\nu_C = h\nu\sqrt{(1+\beta)/(1-\beta)}$$

[Eq. (12.28)] (we use ν_c for reference to Fig. 13.10*C*) or

$$\lambda_C = \lambda\sqrt{\frac{1-\beta}{1+\beta}}$$

where β is the v/c of the electron and is ≈ 1. Now we consider the backward scattering of the doppler-shifted photon so

$$\lambda' - \lambda_C = \frac{h}{mc}(1 - \cos\pi) = \frac{2h}{mc}$$

If

$$\beta \approx 1 \qquad \text{so that} \qquad \lambda_C \ll \frac{h}{mc}\dagger$$

then

$$\lambda' \approx \frac{2h}{mc}$$

Now we transform back to the laboratory system and λ' will again be doppler-shifted. Thus

$$\lambda_{\text{lab}} \approx \lambda'\sqrt{\frac{1-\beta}{1+\beta}} \approx \frac{2h}{mc}\frac{\sqrt{1-\beta}}{\sqrt{2}}$$

But if the initial electron energy is

$$E_{\text{lab}} \approx \gamma mc^2 = \frac{mc^2}{\sqrt{1-\beta^2}} \approx \frac{mc^2}{\sqrt{2}\sqrt{1-\beta}}$$

$$\lambda_{\text{lab}} \approx \frac{2h}{mc}\frac{1}{\sqrt{2}}\frac{mc^2}{\sqrt{2}E_{\text{lab}}} \approx \frac{h}{mc}\frac{mc^2}{E_{\text{lab}}}$$

In this approximation, then

$$\frac{c}{\lambda_{\text{lab}}} \approx \frac{mc^2}{h}\frac{E_{\text{lab}}}{mc^2} \approx \nu_{\text{lab}}$$

and

$$h\nu_{\text{lab}} \approx E_{\text{lab}}$$

where E is the electron energy; nearly all the kinetic energy goes into the photon. The inverse Compton effect is important in astrophysical considerations (see Probs. 12 and 13).

† In Prob. 13 at the end of this chapter the exact solution is asked for.

PROBLEMS

1. *Proton in magnetic field.* Calculate the gyroradius and gyrofrequency of a proton of total relativistic energy 30 BeV in a magnetic field of 15,000 G.

Ans. $\omega_c = 4.5 \times 10^6$ rad/s.

2. *Nuclear recoil.* What is the recoil energy in ergs and in electron volts of a nucleus of mass 10^{-23} g after the emission of a γ-ray of energy 1 MeV? (See Chap. 12, Advanced Topic.)

Ans. 1.4×10^{-10} erg; 90 eV.

3. *Electron-proton collision.* An electron of energy 10 BeV collides with a proton at rest.
(*a*) What is the velocity of the center-of-mass system?
(*b*) What energy is available to produce new particles? (Express in units of M_pc^2.)

4. *High-energy cyclotron frequency.* At high energies the cyclotron frequency depends on the speed of the particle which is being accelerated. In order to maintain synchronism between the cycling particle and the alternating electric field that accelerates it, the designer must require that the applied radio frequency or the magnetic field (or both) be modulated as the acceleration progresses. Show that $\omega \propto B/E$, where ω is the radio frequency, B is the magnetic field, and E is the total energy of the particle.

5. *Nonrelativistic and relativistic cyclotron frequencies.* The Berkeley 184-in. cyclotron operates at a fixed magnetic field of approximately 23,000 G.
(*a*) Calculate the nonrelativistic cyclotron frequency for protons in this field. *Ans.* 2.2×10^8 rad/s.
(*b*) Calculate the frequency appropriate for a final kinetic energy of 720 MeV.

6. *Conservation laws*
(*a*) Show that a free electron moving in a vacuum at velocity v cannot emit a single light quantum. That is, show that such an emission process would violate the conservation laws.
(*b*) A hydrogen atom in an excited electronic state can emit a light quantum. Show that such a process can satisfy the conservation laws. What is the reason for the difference between parts (*a*) and (*b*)?

7. *High-energy proton.* Calculate the momentum, total energy, and kinetic energy of a proton ($M_pc^2 = 0.94$ BeV) for which $\beta \equiv v/c = 0.99$ in the following cases:
(*a*) In the laboratory frame.

Ans. 6.58 BeV/c; 6.63 BeV; 5.69 BeV.

(*b*) In a frame moving with the particle.
(*c*) In a frame stationary with respect to the center of mass of the proton and a stationary helium nucleus ($M_{\text{He}} \approx 4M_p$).
(*d*) In the center-of-mass frame of the proton and a stationary proton.

8. *Cosmic-ray particle.* Find the radius of the orbit of a particle of charge e and of energy 10^{19} eV in a magnetic field of 10^{-6} G. (A magnetic field of 10^{-6} G is not unreasonable for a magnetic field in our galaxy.) Compare this with the diameter of our galaxy. (Particles causing "events" of this tremendous energy have been detected in cosmic rays; such particles give rise to what are called *extensive air showers* of electrons, positrons, γ-rays, and mesons.)

9. *Curvature in electric and magnetic fields*
(*a*) Calculate the radius of curvature of the path of a proton of kinetic energy 1 BeV in a transverse magnetic field of 20,000 G. *Ans.* 284 cm.
(*b*) What transverse electric field is required to produce about the same radius of curvature? Use the fact that the radius of curvature of a curve $y(x)$ is given by $\rho = [1 + (dy/dx)^2]^{\frac{3}{2}}/(d^2y/dx^2)$, and calculate ρ at a point where the proton enters the electric field, $dy/dx = 0$, and d^2y/dx^2 can be calculated from d^2y/dt^2 and $x = vt$.
Ans. 1.75×10^4 statvolts/cm $= 5.25 \times 10^8$ practical volts/m.
(*c*) Consider the magnitude of the electric field in part (*b*) and comment on the practicability of using electric fields for the deflection of relativistic particles.

10. *Deuteron disintegration.* Consider the nuclear reaction in which an incident proton of kinetic energy K_p strikes and splits a stationary deuteron according to

$$p + d \rightarrow p + p + n$$

Near threshold the two protons and the neutron move in an unbound cluster with approximately the same velocity. Write down nonrelativistic expressions for momentum and energy and show that the threshold kinetic energy $K_p{}^0$ of the incident proton is

$$K_p{}^0 = \tfrac{3}{2}E_B$$

where E_B ($\approx$2 MeV) is the binding energy of the deuteron, with respect to a free neutron and proton.

FIG. 13.11 First electron synchrotron. (*Lawrence Berkeley Laboratory photograph*)

11. *Nonrelativistic π^0 threshold.* Use the nonrelativistic expressions for the kinetic energy and momentum of the proton and π^0 meson in calculating the photothreshold [compare Eqs. (13.14)]. What is the difference of the calculated threshold from Eq. (13.16)? What is the correct kinetic energy of the proton plus π^0 meson at threshold? What is the nonrelativistic kinetic energy at threshold?

Ans. $h\nu_{\text{thresh}} = M_p c^2[1 + \alpha - (1 - \alpha^2)^{\frac{1}{2}}]$

12. *Electron-photon elastic collision.* What is the kinetic energy of an electron which will be scattered without energy loss or gain by a photon whose energy is 10,000 eV? (*Hint:* Compare with elastic scattering in the center-of-mass system.) *Ans.* 98 eV.

13. *Inverse Compton effect.* Work out the exact formula for the wavelength of a photon of wavelength λ scattered straight backward by an electron of velocity βc.

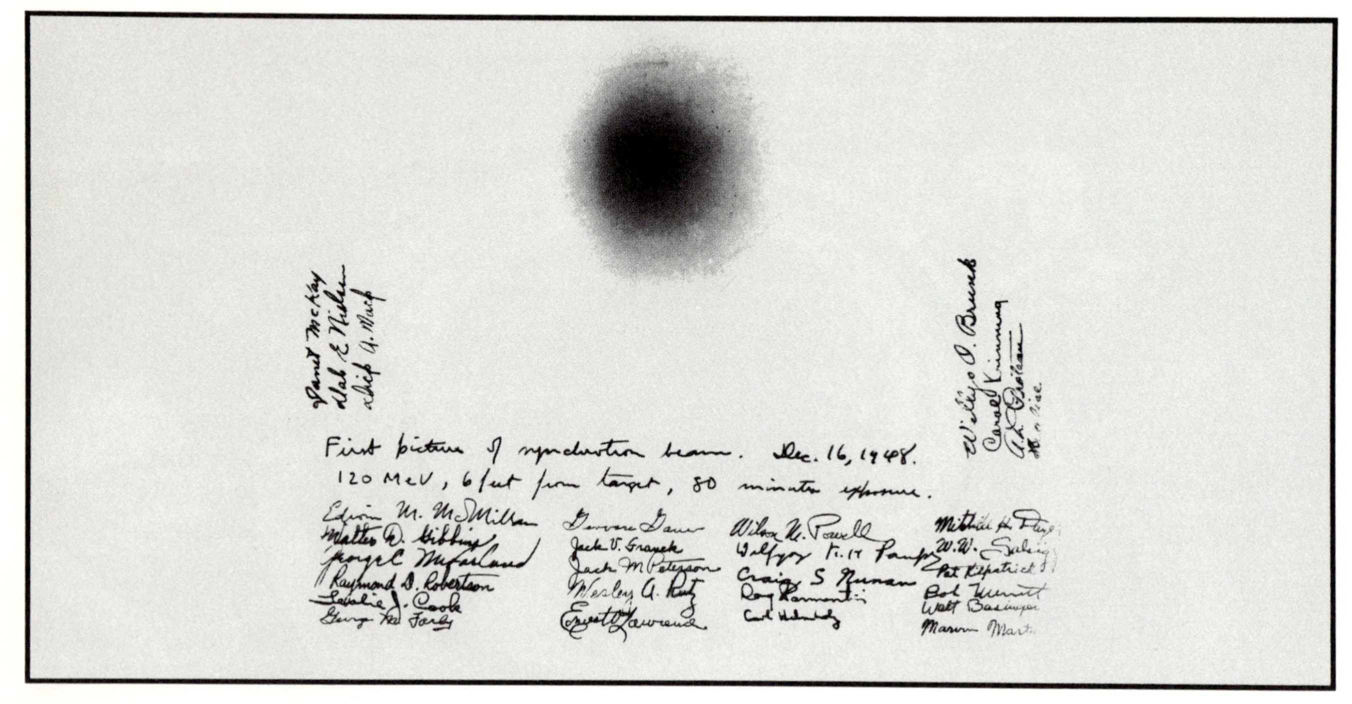

FIG. 13.12 The first picture of the synchrotron beam. (*Lawrence Berkeley Laboratory photograph*)

Apply to the case in which the incident photon has energy $(h\nu) = 3.0$ eV and the electron has total energy 1.02 MeV $(\gamma = 2)$ to find the energy of the scattered photon.

Ans. 41 eV.

HISTORICAL NOTE

Synchrotron The synchrotron principle is utilized in all high-energy accelerators in the region over 1 BeV, except for electron linear accelerators such as that at Stanford. The synchrotron is a device to accelerate particles to high energies. It is essentially a cyclotron in which either the magnetic field or the applied radio frequency is varied during the acceleration, and in which the phase of the particles with respect to the rf electric field automatically adjusts itself to the proper value for acceleration. The idea of frequency or field modulation was not new at the time; what was new was the demonstration that particle orbits could be stable during the modulation. The synchrotron principle was discovered by V. Veksler in Moscow and independently by E. M. McMillan in Berkeley. A full account of Veksler's work appeared in *Journal of Physics* (*USSR*), **9**:153–158 (1945). McMillan's account appeared in the *Physical Review*, **68**:143 (1945). We reproduce McMillan's publication. Figure 13.11 shows the first synchrotron built under McMillan's direction, and Fig. 13.12 shows the first exposure recorded from the x-ray beam.

FURTHER READING

C. Kacser, "Introduction to the Special Theory of Relativity," chap. 7, Co-Op Paperback, Prentice-Hall, Inc., Englewood Cliffs, N.J., 1967.

A. P. French, "Special Relativity," chap. 6, W. W. Norton and Company, Inc., New York, 1968.

Lawrence Radiation Laboratory, "Introduction to the Detection of Nuclear Particles in a Bubble Chamber," Ealing Press, Cambridge, Mass., 1964. A wonderful collection of bubble-chamber photographs of particle tracks (complete with stereoscopic viewer).

The Synchrotron—A Proposed High Energy Particle Accelerator

Edwin M. McMillan
University of California, Berkeley, California
September, 5, 1945

ONE of the most successful methods for accelerating charged particles to very high energies involves the repeated application of an oscillating electric field, as in the cyclotron. If a very large number of individual accelerations is required, there may be difficulty in keeping the particles in step with the electric field. In the case of the cyclotron this difficulty appears when the relativistic mass change causes an appreciable variation in the angular velocity of the particles.

The device proposed here makes use of a "phase stability" possessed by certain orbits in a cyclotron. Consider, for example, a particle whose energy is such that its angular velocity is just right to match the frequency of the electric field. This will be called the equilibrium energy. Suppose further that the particle crosses the accelerating gaps just as the electric field passes through zero, changing in such a sense that an earlier arrival of the particle would result in an acceleration. This orbit is obviously stationary. To show that it is stable, suppose that a displacement in phase is made such that the particle arrives at the gaps too early. It is then accelerated; the increase in energy causes a decrease in angular velocity, which makes the time of arrival tend to become later. A similar argument shows that a change of energy from the equilibrium value tends to correct itself. These displaced orbits will continue to oscillate, with both phase and energy varying about their equilibrium values.

In order to accelerate the particles it is now necessary to change the value of the equilibrium energy, which can be done by varying either the magnetic field or the frequency. While the equilibrium energy is changing, the phase of the motion will shift ahead just enough to provide the necessary accelerating force; the similarity of this behavior to that of a synchronous motor suggested the name of the device.

The equations describing the phase and energy variations have been derived by taking into account time variation of both magnetic field and frequency, acceleration by the "betatron effect" (rate of change of flux), variation of the latter with orbit radius during the oscillations, and energy losses by ionization or radiation. It was assumed that the period of the phase oscillations is long compared to the period of orbital motion. The charge was taken to be one electronic charge. Equation (1) defines the equilibrium energy; (2) gives the instantaneous energy in terms of the equilibrium value and the phase variation, and (3) is the "equation of motion" for the phase. Equation (4) determines the radius of the orbit.

$$E_0 = (300cH)/(2\pi f), \tag{1}$$

$$E = E_0[1-(d\phi)/(d\theta)], \tag{2}$$

$$2\pi\frac{d}{d\theta}\left(E_0\frac{d\phi}{d\theta}\right) + V\sin\phi = \left[\frac{1}{f}\frac{dE_0}{dt} - \frac{300}{c}\frac{dF_0}{dt} + L\right] + \left[\frac{E_0}{f^2}\frac{df}{dt}\right]\frac{d\phi}{d\theta}, \tag{3}$$

$$R = (E^2 - E_r^2)^{\frac{1}{2}}/300H. \tag{4}$$

The symbols are:

E = total energy of particle (kinetic plus rest energy),
E_0 = equilibrium value of E,
E_r = rest energy,
V = energy gain per turn from electric field, at most favorable phase for acceleration,
L = loss of energy per turn from ionization and radiation,
H = magnetic field at orbit,
F_0 = magnetic flux through equilibrium orbit,
ϕ = phase of particle (angular position with respect to gap when electric field = 0),
θ = angular displacement of particle,
f = frequency of electric field,
c = light velocity,
R = radius of orbit.

(Energies are in electron volts, magnetic quantities in e.m.u., angles in radians, other quantities in c.g.s. units.)

Equation (3) is seen to be identical with the equation of motion of a pendulum of unrestricted amplitude, the terms on the right representing a constant torque and a damping force. The phase variation is, therefore, oscillatory so long as the amplitude is not too great, the allowable amplitude being $\pm\pi$ when the first bracket on the right is zero, and vanishing when that bracket is equal to V. According to the adiabatic theorem, the amplitude will diminish as the inverse fourth root of E_0, since E_0 occupies the role of a slowly varying mass in the first term of the equation; if the frequency is diminished, the last term on the right furnishes additional damping.

The application of the method will depend on the type of particles to be accelerated, since the initial energy will in any case be near the rest energy. In the case of electrons, E_0 will vary during the acceleration by a large factor. It is not practical at present to vary the frequency by such a large factor, so one would choose to vary H, which has the additional advantage that the orbit approaches a constant radius. In the case of heavy particles E_0 will vary much less; for example, in the acceleration of protons to 300 Mev it changes by 30 percent. Thus it may be practical to vary the frequency for heavy particle acceleration.

A possible design for a 300 Mev electron accelerator is outlined below:

peak H = 10,000 gauss,
final radius of orbit = 100 cm,
frequency = 48 megacycles/sec.,
injection energy = 300 kv,
initial radius of orbit = 78 cm.

Since the radius expands 22 cm during the acceleration, the magnetic field needs to cover only a ring of this width, with of course some additional width to shape the field properly. The field should decrease with radius slightly in order to give radial and axial stability to the orbits. The total magnetic flux is about $\frac{1}{5}$ of what would be needed to satisfy the betatron flux condition for the same final energy.

The voltage needed on the accelerating electrodes depends on the rate of change of the magnetic field. If the magnet is excited at 60 cycles, the peak value of $(1/f)(dE_0/dt)$ is 2300 volts. (The betatron term containing dF_0/dt is about $\frac{1}{5}$ of this and will be neglected.) If we let $V=10,000$ volts, the greatest phase shift will be 13°. The number of turns per phase oscillation will vary from 22 to 440 during the acceleration. The relative variation of E_0 during one period of the phase oscillation will be 6.3 percent at the time of injection, and will then diminish. Therefore, the assumptions of slow variation during a period used in deriving the equations are valid. The energy loss by radiation is discussed in the letter following this, and is shown not to be serious in the above case.

The application to heavy particles will not be discussed in detail, but it seems probable that the best method will be the variation of frequency. Since this variation does not have to be extremely rapid, it could be accomplished by means of motor-driven mechanical tuning devices.

The synchrotron offers the possibility of reaching energies in the billion-volt range with either electrons or heavy particles; in the former case, it will accomplish this end at a smaller cost in materials and power than the betatron; in the latter, it lacks the relativistic energy limit of the cyclotron.

Construction of a 300-Mev electron accelerator using the above principle at the Radiation Laboratory of the University of California at Berkeley is now being planned.

CONTENTS

Principle of Equivalence

In this chapter we discuss some further aspects of relativity. Some of the topics bring together the general theory of relativity with the special theory discussed in Chaps. 11 to 13.

INERTIAL AND GRAVITATIONAL MASS

Newton's Second Law may be used to define the mass of an object by subjecting different masses to the same force and measuring their accelerations. Thus

$$M(1)a(1) = F = M(2)a(2)$$
$$\frac{M(2)}{M(1)} = \frac{a(1)}{a(2)}$$

and if we set $M(1) = 1$, $M(2)$ is uniquely defined. The mass determined in this way is known as the *inertial mass* and is denoted by M_i. We may also determine the mass by measuring the gravitational force exerted on it by another body, such as the earth:

$$\frac{GM_g M_E}{R_E{}^2} = F$$
$$M_g = \frac{FR_E{}^2}{GM_E} \tag{14.1}$$

The mass determined in this way is known as the *gravitational mass* and is denoted by M_g. In Eq. (14.1) the mass of the earth is M_E and R_E is its radius.

It is a remarkable fact that the inertial mass of all bodies is, within experimental accuracy, proportional to the gravitational mass. (We may consider that the constant G has been determined as in the Cavendish experiment using the definition of a force and hence reflecting the inertial mass.) The simplest experiment to check this is to see whether all bodies fall with the same acceleration. For one falling body near the surface of the earth, we have

$$M_i(1)a(1) = \frac{GM_E M_g(1)}{R_E{}^2} \tag{14.2}$$

for a second falling body,

$$M_i(2)a(2) = \frac{GM_E M_g(2)}{R_E{}^2} \tag{14.3}$$

On dividing Eq. (14.2) by (14.3), we have

$$\frac{M_i(1)a(1)}{M_i(2)a(2)} = \frac{M_g(1)}{M_g(2)}$$

$$\frac{M_i(1)}{M_g(1)} = \frac{M_i(2)}{M_g(2)} \cdot \frac{a(2)}{a(1)}$$

But falling bodies in a vacuum are always observed to fall at the same rate, so that $a(2) = a(1)$ within the experimental accuracy, and therefore we have for the ratio of inertial to gravitational mass

$$\frac{M_i(1)}{M_g(1)} = \frac{M_i(2)}{M_g(2)} \qquad (14.4)$$

As long as this ratio of masses is constant, we can always make the value of the ratios in Eq. (14.4) equal to unity by adjusting appropriately the value of G; that is, we determine as in the Cavendish experiment the force F between two masses $M_i(1)$ and $M_i(2)$, measured in the inertial system, a distance r apart and set

$$G = \frac{Fr^2}{M_i(1)M_i(2)}$$

The experimental task is to determine whether variations of the ratio M_i/M_g exist for different particles, materials, and objects.

The classical determinations were carried out by Newton, using the pendulum method of Prob. 1 at the end of this chapter. Other famous determinations include those by R. Eötvös, which he started about 1890 and continued for about 25 yr. His ingenious method can be understood by considering a pendulum suspended at the surface of the earth and at a latitude of 45°, as in Fig. 14.1. The pendulum is acted on by a gravitational force $M_g g$ directed toward the center of the earth. It is acted on also by a centrifugal force[1] $M_i\omega^2R_E/\sqrt{2}$, where $R_E/\sqrt{2}$ is the radius of the circle in which the pendulum bob moves due to the rotation of the earth. The centrifugal force is directed normal to the axis of rotation, and its horizontal component is obtained by a further multiplication by cos 45°, or $1/\sqrt{2}$. The resultant of the two forces makes an angle

$$\theta \approx \frac{M_i\omega^2R_E/2}{M_g g - \frac{1}{2}M_i\omega^2R_E} \approx \frac{M_i\omega^2R_E}{2M_g g}$$

[1]We place ourselves in a frame of reference rotating with the earth.

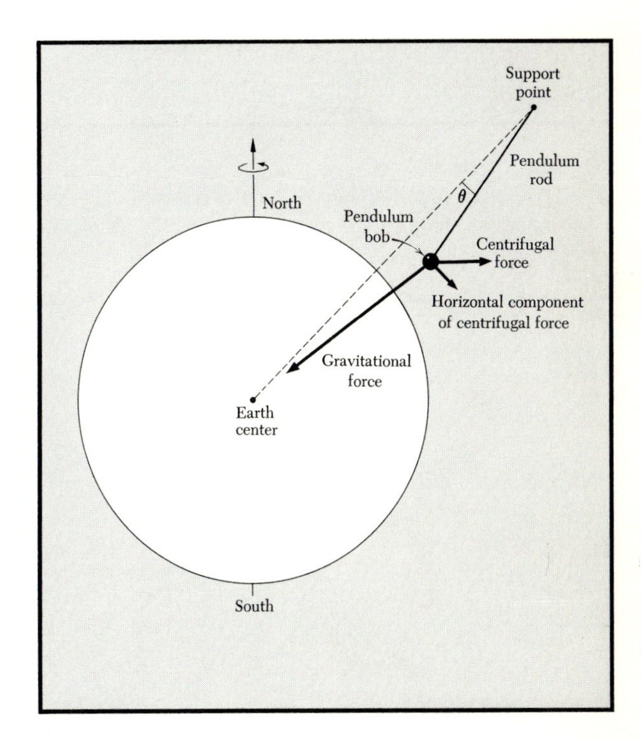

FIG. 14.1 Illustration of how a pendulum is deflected from vertical by small angle θ because of centrifugal force arising from earth's rotation. In the figure the angle θ, the distance of the bob above the surface of the earth, and the centrifugal force are all greatly exaggerated.

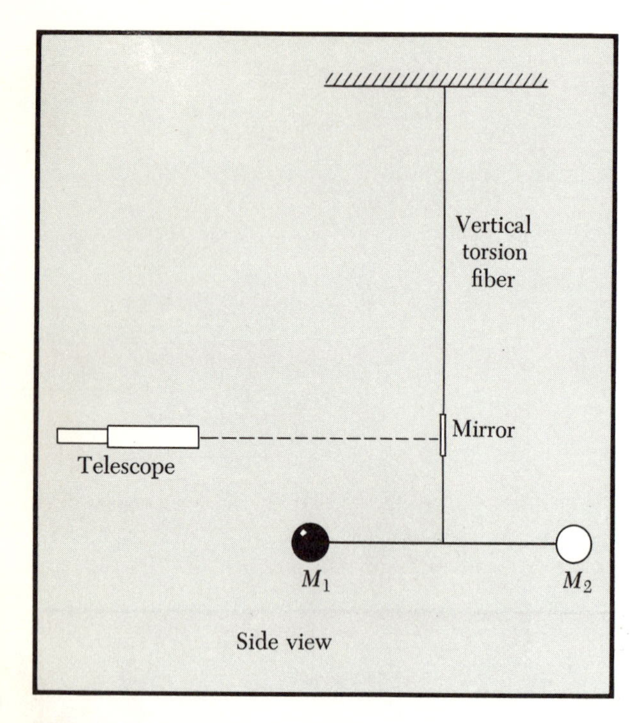

FIG. 14.2 Side view of an apparatus similar to that employed by Eötvös to determine the ratio of inertial to gravitational mass. M_1 and M_2 are two dissimilar objects of the same gravitational mass.

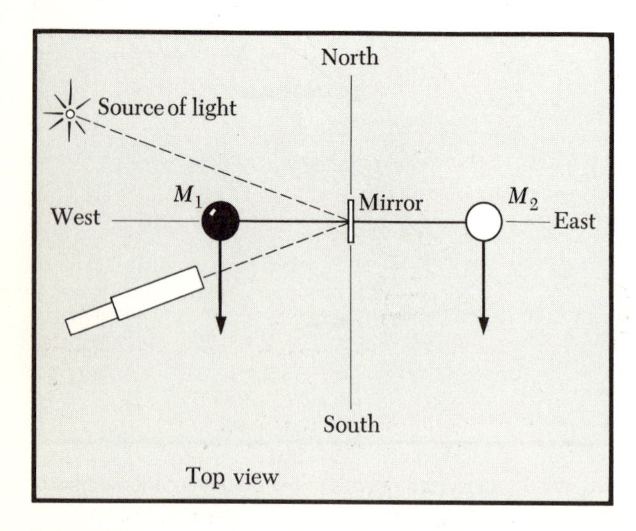

FIG. 14.3 If inertial masses of M_1, M_2 are equal, the horizontal components of centrifugal force (two arrows) are equal, and there is no net torsion on fiber.

with the direction toward the center of the earth. Here we have used the fact that the ratio $M_i\omega^2 R_E/M_g g$ is a small number and so $\tan\theta = \theta$. Using the data given at the beginning of Chap. 4, we see that the ratio has a value of about 0.003.

Now suppose that a torsion suspension is made as shown in Fig. 14.2, with the two bobs made of different materials but of equal gravitational mass, so that $M_g(1) = M_g(2)$. If $M_i(1)$ is equal to $M_i(2)$, there will be no torque tending to turn the torsion fiber; this situation is illustrated in Fig. 14.3. However, if $M_i(1)$ is greater than $M_i(2)$, the horizontal component of the centrifugal force on $M(1)$ will be greater than that on $M(2)$ and a net torque will twist the fiber, as shown in Fig. 14.4. The measurement is repeated with the apparatus turned through 180°; this helps determine the zero position of the balance. The experiment is a good example of a null experiment: an effect will be observed only if $M_i(1) \neq M_i(2)$. Eötvös compared eight different materials against platinum (Pt) as a standard. He found that

$$\frac{M_i(1)}{M_g(1)} = \frac{M_i(\mathrm{Pt})}{M_g(\mathrm{Pt})}$$

within less than 1 part in 10^8. Dicke et al.[1] have improved the experiment using the same general technique. Their results show the ratio equal to unity within 1 part in 3×10^{10}.

The present experimental situation can be summarized as follows:

If we denote the ratio M_g/M_i by Q, then

1 The value of Q for an electron plus a proton equals the value of Q for a neutron up to 1 part in 10^7. (This comparison follows directly from a comparison of light and heavy elements in the periodic table; heavy elements have a higher proportion of neutrons than light elements have.)
2 The value of Q for that part of the nuclear mass associated with nuclear binding equals one within 1 part in 10^5.
3 The value of Q for that part of the atomic mass associated with the binding of the orbital electrons equals one within 1 part in 200.
4 Q for aluminum relative to gold equals $1 \pm 3 \times 10^{-11}$.

GRAVITATIONAL MASS OF PHOTONS

We saw in Chap. 12 that a photon of energy $h\nu$, where ν is

[1] P. G. Roll, R. Krotkov, R. H. Dicke, *Ann. Phys.* (N.Y.), **26**:442 (1964).

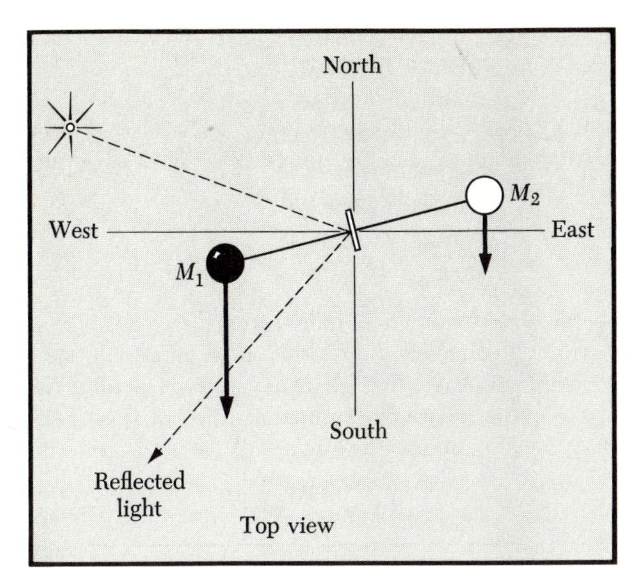

FIG. 14.4 If inertial mass of M_1 is greater than that of M_2, there will be a twist in the fiber, and the mirror will rotate.

the frequency, must have an inertial mass equal to $h\nu/c^2$. Does the photon also have a gravitational mass? Experimental evidence strongly indicates that it does, and that the gravitational mass is equal in value to the inertial mass. (The *rest* mass, of course, is zero.)

Consider a photon that, when at a height L above the surface of the earth, has frequency ν and energy $h\nu$. After falling through the distance L, it will have lost potential energy $mgL = (h\nu/c^2)gL$ and will itself have gained this much energy so that the energy of the photon will become $h\nu'$, where

$$h\nu' \approx h\nu + \frac{h\nu}{c^2}gL \tag{14.5}$$

assuming a constant mass $h\nu/c^2$ for the photon during the fall (the argument being that ν' is not much different from ν). The frequency ν' measured for the photon *after* the fall is then, from Eq. (14.5),

$$\nu' \approx \nu\left(1 + \frac{gL}{c^2}\right) \tag{14.6}$$

Figure 14.5 illustrates this effect. If $L = 20$ m, the fractional frequency shift is

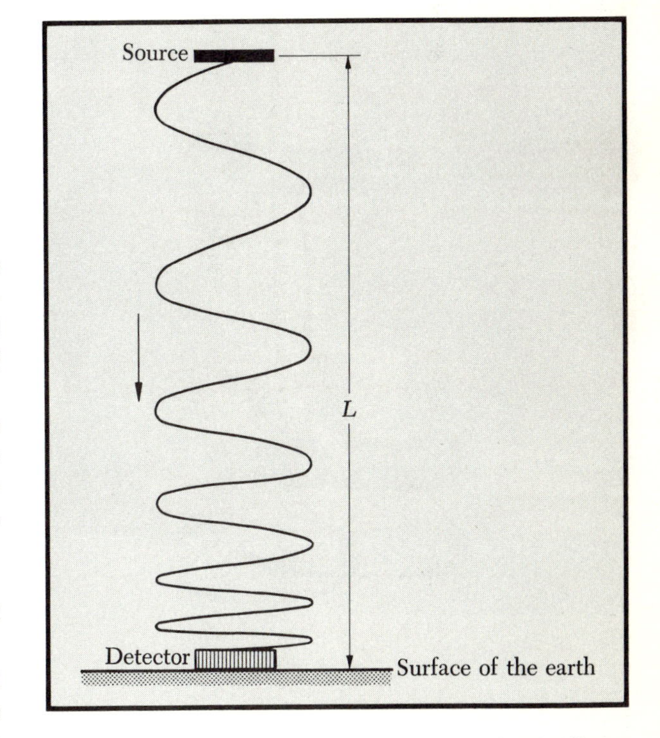

FIG. 14.5 Schematic picture of gravitational red-shift experiment. A photon emitted at the source in a direction toward the earth's center loses "potential energy" $\Delta U = (h\nu/c^2)gL$ and gains an equal amount of "kinetic energy" in falling a distance L. The photon frequency at the detector is $\nu' = \nu(1 + gL/c^2)$; the photon frequency at the source is ν. (This is a *blue shift* as here described. It would be a *red shift* if the photon moved upward.)

FIG. 14.6 Lower end of the Pound falling-photon experiment at Harvard, showing G. A. Rebka, Jr., adjusting photomultipliers on instruction from the control center. In a later version of the experiment means are provided for controlling the temperature of the source and absorber. The whole gravitational shift measured is only about 1/500 of the line width. To measure with accuracy such a small shift requires the aid of a few tricks. *(Courtesy of R. V. Pound)*

$$\frac{\Delta\nu}{\nu} = \frac{gL}{c^2} \approx \frac{(10^3)(2 \times 10^3)}{(3 \times 10^{10})^2} \approx 2 \times 10^{-15} \tag{14.7}$$

This extremely small effect has actually been observed by Pound and Rebka[1] using a γ-ray source (see Fig. 14.6). They find, with $\Delta\nu = \nu' - \nu$,

$$\frac{(\Delta\nu)_{\text{exp}}}{(\Delta\nu)_{\text{calc}}} = 1.05 \pm 0.10$$

where the calculated value is obtained from Eq. (14.6).

A photon with frequency ν emitted at an infinite distance from the earth will have the frequency ν' on reaching the surface of the earth, where, by a generalization of Eqs. (14.5) and (14.6),

$$\nu' \approx \nu\left(1 + \frac{GM_E}{R_E c^2}\right) \tag{14.8}$$

Note that the frequency shift involves the ratio of the *gravitational length* GM_E/c^2 of the earth[2] to the radius R_E of the earth. This ratio has the value 6×10^{-10}. The larger effect here is of the same kind as that considered in Eq. (14.6), but now the light source is an infinite distance from the earth.

Gravitational Red Shift As illustrated in Fig. 14.7, a photon of frequency ν that leaves a star and escapes to infinity will be observed at infinity with a frequency

$$\nu' \approx \nu\left(1 - \frac{GM_s}{R_s c^2}\right) \tag{14.9}$$

where M_s is the mass and R_s the radius of the star. This follows by a modification of Eq. (14.8); the photon now has lost energy in escaping from the gravitational field of the star. A photon in the blue region of the visible spectrum will be shifted in frequency toward the red end of the spectrum: for this reason the effect is known as the *gravitational red shift*. It must not be confused with the recessional red shift of distant stars believed to arise from their apparent radial motion away from the earth, as discussed in Chap. 10.

[1] R. V. Pound and G. A. Rebka, Jr., *Phys. Rev. Letters*, **4**:337 (1960); R. V. Pound and J. L. Snider, *Phys. Rev.*, **140**:B788 (1965).

[2] Defined by analogy with the radius of the electron (Chap. 9, page 279):

$$M_e c^2 = \frac{GM_e^2}{R} \qquad R = \frac{GM_e}{c^2}$$

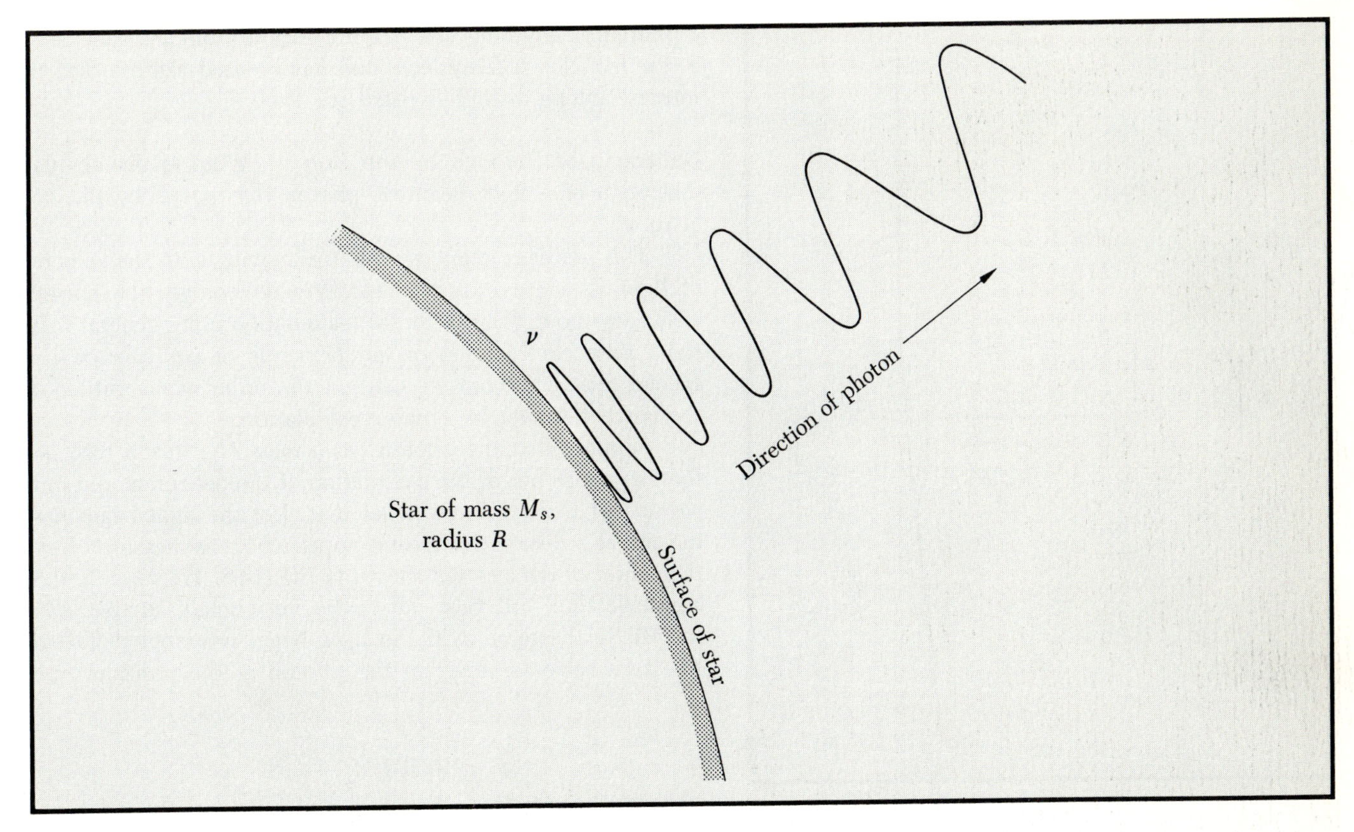

FIG. 14.7 A photon escaping to infinity from the surface of a star gains "potential energy" and loses an equal amount of "kinetic energy." If the photon frequency at the surface is ν, the photon frequency at ∞ is $\nu' = \nu(1 - GM_s/R_sc^2)$.

White dwarf stars have large values of M_s/R_s and thus have relatively large values of the gravitational red shift. For Sirius B the calculated fractional shift is

$$\frac{\Delta\nu}{\nu} \approx -5.9 \times 10^{-5}$$

and the observed value is -6.6×10^{-5}. The discrepancy is within the uncertainties in M_s and R_s.

If

$$\frac{GM_s}{R_sc^2} > 1$$

the frequency ν' in Eq. (14.9) would be negative, which is of course impossible. However, the case $GM_s/R_sc^2 \approx 1$ is a more complicated problem requiring the theory of general relativity. The result is that if

$$\frac{2GM_s}{R_sc^2} \geq 1 \tag{14.10}$$

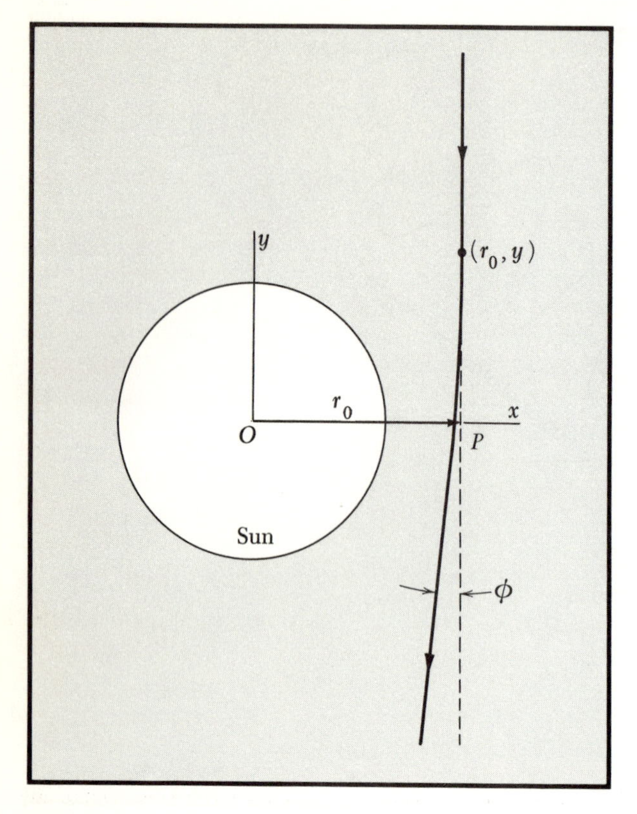

FIG. 14.8 Deflection of a photon by the gravitational field of the sun.

a photon or anything else can *not* escape from the star. Such a star is called a *black hole* and has aroused a great deal of interest among astrophysicists[1] (see Prob. 5).

Deflection of Photons by the Sun What is the angular deflection of a light beam or photon that passes by the sun at its edge?

This problem involves a photon moving with the velocity of light in a gravitational field. We do not get the correct answer without doing a careful calculation using general relativity or a combination of the principle of equivalence and special relativity;[2] but we can get the order of magnitude of the correct answer by a naïve calculation.

Suppose that the photon has a mass M_L; it will turn out that M_L drops out of the calculation of the deflection and thus we do not have to know what it is. Let the light beam pass the sun at a distance of closest approach r_0, as measured from the center of the sun and shown in Fig. 14.8. We suppose that the deflection will turn out to be very small, so that r_0 is essentially the same as if the light beam were not deflected. The transverse force F_x on the photon at the position (r_0, y) is

$$F_x = -GM_sM_L \frac{r_0}{(r_0^2 + y^2)^{\frac{3}{2}}}$$

where y is measured from the point P as in the figure.

The final value of the transverse velocity component v_x of the photon has the value given by

$$M_L v_x = \int F_x \, dt = \int F_x \frac{dy}{v_y} \approx \frac{1}{c} \int F_x \, dy$$

so that

$$v_x \approx -\frac{GM_s r_0}{c} \int_{-\infty}^{\infty} \frac{dy}{(r_0^2 + y^2)^{\frac{3}{2}}}$$

$$\approx -\frac{2GM_s r_0}{c} \int_0^{\infty} \frac{dy}{(r_0^2 + y^2)^{\frac{3}{2}}} \approx -\frac{2GM_s}{cr_0}$$

It follows that when r_0 is equal to the radius R_s of the sun, the angular deflection is (see Fig. 14.8)

[1]For example, Kip S. Thorne, *Scientific American*, **217**:5, 88 (1967); R. Ruffini and J. A. Wheeler, *Physics Today*, **24**:30 (1971).

[2]L. I. Schiff, *Am. J. Phys.*, **28**:340 (1961).

$$\tan \phi \approx \phi \approx \frac{|v_x|}{c} \approx \frac{2GM_s}{R_s c^2} \text{ rad}$$

On doing the calculation, we find $\phi = 0.87''$. The careful analysis predicts twice the value given by our argument, or 1.75″. This value has been confirmed by observation with an accuracy of perhaps 20 percent. (One still hears grumbling about the data, but the experiment is very difficult.) Figure 1.9 shows a star photographed at an eclipse when such measurements are made.

When we solve a collision problem by calculating the force on one particle as if its trajectory were a straight line, we are making what is called an *impulse approximation.* The connection between $\int F_x\,dt$ and the x component of the change in momentum is discussed in Chap. 5. The impulse approximation is often very useful, provided that the actual trajectory does not depart greatly from the straight line the particle would follow if there were no interaction.

Shapiro[1] has observed another effect predicted by Einstein's general theory of relativity. When radar signals are bounced off a planet such as Venus, the time taken for the signal to go out to Venus and return is greater if the signal passes close to the sun than if the path is far from the sun, and the observations agree with the theory.

PRECESSION OF THE PERIHELION OF MERCURY

The three classical tests of general relativity are the gravitational red shift (page 402), the deflection of light in the field of the sun (page 404), and the precession of the perihelion of Mercury. The delay in radar signals mentioned above has been referred to as a *fourth test* of general relativity.

Even at this stage of our study we can make an order-of-magnitude estimate of the precession of Mercury's perihelion. According to the calculations of Chap. 9, the line between the sun and the planet Mercury at its closest approach should remain fixed in space.[2] The actual orbit greatly exaggerated is shown in Fig. 14.9. The effect is due to the fact that v/c, or more properly v^2/c^2, is not zero. What quantity would be proportional to v^2/c^2? A reasonable possibility is the angle of advance per revolution, or the angle of advance divided by

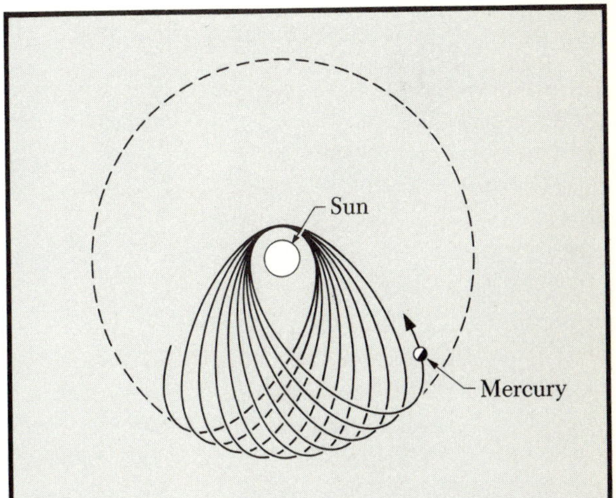

FIG. 14.9 Precession of the orbit of Mercury explained by the general theory of relativity. The plane of the orbit is in the page; the eccentricity of the orbit is greatly exaggerated for clarity. Without precession the figure would be a stationary ellipse.

[1] I. I. Shapiro, *Scientific American,* 219:1, 28 (1968).

[2] The perturbation due to other planets can be calculated and compared with experiment. The observed motion in space of this line differs from that calculated with these perturbations by 43″ per century.

FIG. 14.10 Detector for gravitational waves, a 96-cm aluminum cylinder. Its length is 151 cm and its response is centered at 1661 Hz. It has the directivity of a mass quadrupole. (*Photograph courtesy of Professor J. Weber*)

2π. We can estimate v/c from Table 9.2. We assume that the orbit is circular of radius equal to the semimajor axis. Then using the period we get

$$v = \frac{2\pi r}{\text{period}} = \frac{(2\pi \times 0.39 \times 1.5)10^{13}}{7.6 \times 10^6}$$
$$\approx 4.8 \times 10^6 \text{ cm/s}$$

$$\frac{v}{c} \approx 1.6 \times 10^{-4}$$

$$\frac{v^2}{c^2} \approx 2.6 \times 10^{-8} = \frac{\delta\theta}{2\pi}$$

$$\delta\theta(\text{in degrees}) \approx 360 \times (2.6 \times 10^{-8})$$
$$\approx 9 \times 10^{-6}$$
$$\delta\theta \approx 3 \times 10^{-2} \text{ seconds of arc/revolution}$$

The usual figure is the number of seconds per century. The period is 0.24 yr. Therefore we may expect the effect to be of the order of magnitude of

$$\delta\theta(\text{per century}) = \frac{100}{0.24}(3 \times 10^{-2}) \approx 13''$$

The experimental value is 42.9″, and the general theory of relativity predicts 43.0″, which is within the experimental error.[1]

EQUIVALENCE

The experimental result that no difference has ever been detected between the inertial mass and the gravitational mass of a body suggests that gravitation in a sense may be equivalent to acceleration. Consider an observer in an elevator which is freely falling with the acceleration g.

The equivalence principle states that to an observer in a freely falling elevator the laws of physics are the same as in the inertial frames of special relativity (at least in the immediate neighborhood of the center of the elevator). *The effects due to the accelerated motion and to the gravitational forces exactly cancel.* An observer sitting in an enclosed elevator cannot, if he observes apparent gravitational forces, tell what portion of these correspond to acceleration and what portion to actual gravitational forces. He will detect no forces at all *unless* other forces (i.e., other than gravitational forces) act on the elevator. In particular, the postulated principle of equivalence requires

[1] A careful discussion of these classic experiments is given in the first chapter of L. Witten, "Gravitation: An Introduction to Current Research," John Wiley & Sons, Inc., New York, 1962.

that the ratio of the inertial and gravitational masses be $M_i/M_g \equiv 1$. The "weightlessness" of a man in orbit in a satellite is a consequence of the equivalence principle.

Pursuit of the mathematical consequences of the principle of equivalence leads to the general theory of relativity; for further discussion, consult the references suggested at the end of Chap. 11.

GRAVITATIONAL WAVES

Just as oscillating electric charges give out electromagnetic waves, the general theory of relativity predicts that oscillating gravitational masses, such as double stars, should give out gravitational waves. Because of the small value of G these are difficult to detect, but recently Weber[1] has reported results indicating the arrival from outer space of gravitational waves. Figure 14.10 shows a detector for gravitational waves.

[1] J. Weber, *Phys. Rev. Letters*, **24**:276 (1970); *Scientific American*, **224**:5, 22 (1971).

PROBLEMS

1. *Pendulum in terms of gravitational and inertial masses.* Show that the frequency of a pendulum of length L is given by

$$\nu = \frac{\omega}{2\pi} = \frac{1}{2\pi}\left(\frac{M_g\, g}{M_i\, L}\right)^{\frac{1}{2}}$$

where M_g, M_i are the gravitational and inertial masses. (Bessel in the early days made careful pendulum observations and showed that M_g was equal to M_i within 1 part in 6×10^4.)

2. *Gravitational red shift.* Find an expression for the gravitational red shift in which you do not use the assumption that $\Delta\nu/\nu \ll 1$. (Neglect any effects associated with the curvature of space.) Start with $h\,\Delta\nu = -(h\nu/c^2)(M_sG/r^2)\;\Delta r$, and integrate over dr from R_s to infinity, and over $d\nu$ from ν to ν'.
Ans. $\nu' = \nu e^{-GM_s/R_sc^2}$.

3. *Red shift from our galaxy.* Estimate the gravitational red shift for light leaving the center of our galaxy, as observed far outside the galaxy. (Treat the distribution of mass as uniform within a sphere of radius 10,000 parsecs. The mass of the galaxy is $\sim 8 \times 10^{44}$ g.) *Ans.* $\Delta\nu/\nu = -3 \times 10^{-6}$.

4. *Radio galaxy.* In 1962 an intense extraterrestrial source of radio radiation was optically identified as a starlike object with an angular radius of approximately $\frac{1}{2}''$ of arc. It was at first thought to be a star in our galaxy giving off radio waves, but subsequently its spectrum was obtained and its spectral lines were found to be very considerably red-shifted. For instance, an atomic oxygen line with wavelength λ normally 3.727×10^{-5} cm was identified at $\lambda = 5.097 \times 10^{-5}$ cm. One explanation took it to be an exceedingly massive star with a spectrum *gravitationally red-shifted.* If this hypothetical radio star is in our galaxy, its distance must be less than 10^{22} cm from the earth.[1]
(*a*) Calculate from the angular diameter and the red shift, the mass and mean density of the star under this hypothe-

[1] For further details see J. L. Greenstein, Quasi-stellar Radio Sources, *Scientific American*, **209**:54 (December 1963).

sis, assuming the distance to be 10^{22} cm. Is this a reasonable explanation of this object?

Ans. Mass is 1.0×10^{44} g, mean density is 1.7×10^{-6} g/cm^3. This does not seem reasonable, as the mass is about 0.1 of the total mass of our galaxy (use the result of Prob. 2).

(*b*) An alternative suggestion was that it might be a peculiar "radio galaxy," with its red shift following the usual recessional red-shift relation given in Chap. 10. Calculate its distance on this second hypothesis.

Ans. 3×10^9 light yr (3×10^{27} cm).

(*c*) Does the radio-galaxy hypothesis conform with this expectation?

Ans. Yes; it has a radius of about 10^{22} cm. This is in the usual range of galactic radii.

5. *Black hole.* What would the radius of the sun have to be in order for it to be a black hole [see (Eq. 14.10)]? Compare the density it would then have to the density of a nucleus.

Ans. $\sim 3 \times 10^5$ cm.

HISTORICAL NOTE

Newton's Pendulums We quote from Newton's account in the *Principia* of his experiments with pendulums to investigate possible variations in the ratio of gravitational and inertial masses.

> But it has been long ago observed by others, that (allowance being made for the small resistance of the air) all bodies descended through equal spaces in equal times; and, by the help of pendulums, that equality of times may be distinguished to great exactness.
>
> I tried the thing in gold, silver, lead, glass, sand, common salt, wood, water, and wheat. I provided two equal wooden boxes. I filled the one with wood, and suspended an equal weight of gold (as exactly as I could) in the centre of oscillation of the other. The boxes, hung by equal threads of 11 feet, made a couple of pendulums perfectly equal in weight and figure, and equally exposed to the resistance of the air: and, placing the one by the other, I observed them to play together forwards and backwards for a long while, with equal vibrations. And therefore (by Cor. I and VI, Prop. XXIV, Book II) the quantity of matter in the gold was to the quantity of matter in the wood as the action of the motive force upon all the gold to the action of the same upon all the wood; that is, as the weight of the one to the weight of the other.
>
> And by these experiments, in bodies of the same weight, could have discovered a difference of matter less than the thousandth part of the whole.

FURTHER READING

C. Kacser, "Introduction to the Special Theory of Relativity," chap. 8, Co-Op Paperback, Prentice-Hall, Inc., Englewood Cliffs, N.J., 1967.

Appendix

The following books about mechanics are at approximately the same level as this text; the student may find them helpful for studying and for consulting while he is studying mechanics. The student should look at a number of the books before buying or spending a great deal of time on any one.

H. D. Young, "Fundamentals of Mechanics and Heat," McGraw-Hill Book Company, New York, 1964. Good text. Somewhat less advanced. Brief treatment of special relativity.

R. Resnick and D. Halliday, "Physics for Students of Science and Engineering," John Wiley & Sons, Inc., New York, 1966. Vol. I, 2d ed., or vols. I and II combined. Many examples. Good conventional text.

R. Resnick and D. Halliday, "Fundamentals of Physics," John Wiley & Sons, Inc., New York, 1970. Shortened version of their previous text.

A. P. French, "Newtonian Mechanics," W. W. Norton and Company, Inc., New York, 1971. Part of M.I.T. series. Excellent and very extensive book.

M. Alonso and E. J. Finn, "Fundamental University Physics," vol. I, "Mechanics," Addison-Wesley Publishing Company, Inc., Reading, Mass., 1967. Good, short, and concise book, but perhaps too brief for this level.

R. T. Weidner and R. L. Sells, "Elementary Classical Physics," vol. I, Allyn and Bacon, Inc., Boston, 1965. Not as advanced as other texts.

R. P. Feynman, R. B. Leighton, and M. Sands, "The Feynman Lectures on Physics," vol. I, Addison-Wesley Publishing Company, Inc., Reading, Mass., 1963. Remarkable series of lectures, with keen insights on many aspects of mechanics. Not a textbook.

In recent years a number of "problems" books have appeared. Although for most there are ample problems in the books mentioned above and in this text, some students may wish more.

J. A. Taylor, "Programmed Study Aid for Introductory Physics," part I, "Mechanics," Addison-Wesley Publishing Company, Inc., Reading, Mass., 1970. Good, with many less-advanced problems.

R. B. Leighton and E. Vogt, "Exercises in Introductory Physics," Addison-Wesley Publishing Company, Inc., Reading, Mass., 1969. Written to accompany vol. I of Feynman lectures. Excellent problems.

D. Schaum, "Theory and Problems of College Physics," Schaum Publishing Co., New York, 1961. Many problems, with solutions to about half.

Film Lists

There are many excellent films that deal with subjects in mechanics. The Resource Letter, BSPF-1, Physics Films by W. R. Riley, *Am. J. Phys.*, **36**:475 (1968), provides a good list of 16-mm films as well as information about catalogs, places where the films can be obtained, etc. Many of the comments given below are directly from this review.

In recent years there have been many film loops produced. They are very useful, particularly since they are easy to show for individual instruction. The Commission on College Physics has published a catalog, "Short Films for Physics Teaching," available from AIP, Division of Education and Manpower, Information Pool, State University of New York, Stony Brook, N.Y. 11790.

The reader will note that many films referred to below are PSSC films and so are at a more elementary level than this course. Nevertheless, these films are very well thought out and made and are helpful even to students who see them for the second time.

Recently a National Committee for Physics Demonstration Films [George Appleton, James Strickland, *Am. J. Phys.*, **38**:1945 (1970)] was formed; Strickland can be contacted at Education Development Center, Newton, Mass. 02160.

The films are listed by the chapter in which they seem to fit most coherently. Places where they may be obtained are listed at the end.

Chapter 1

The Evolution of Physical Ideas (49 min). P. A. M. Dirac; SUNY. Dirac's personal approach to theoretical physics; suggests that physicists' attempts to improve existing theories involve a search for mathematical beauty.

Measuring Large Distances (29 min). F. Watson; PSSC MLA 0103. Shows by triangulation and parallax measurements how distances to the moon and then to stars up to 500 light yr away can be measured.

Change of Scale (23 min). R. W. Williams; PSSC MLA 0106. For orientation in the ideas of estimating and scaling. Presents several nice examples of scaling stresses and of scaling where something depends on the scale of the velocity (ship design).

Chapter 2

Measurement (21 min). William Siebert; MLA. Measurement of speed of a rifle bullet, with emphasis on the relation of accuracy to the asking of appropriate questions.

Symmetry (10 min). P. Stapp, J. Bregman, R. Davisson, A. Holden; BTL. Interesting contemporary presentation of reflection, rotation, and translation symmetries and their uses.

Uniform Circular Motion (8 min). MGH. Shows changes in velocity vector, explaining that the motion is accelerated even though speed is constant. Centripetal force illustrated with several animations.

Vector Kinematics (16 min). Francis Friedman; PSSC MLA 0109. A computer traces out on a cathode-ray tube the velocity and acceleration vectors corresponding to various types of displacements of a spot: circular, simple harmonic, and free fall.

Straight Line Kinematics (34 min). E. M. Hafner; PSSC MLA. Graphs of distance, speed, and acceleration vs time are generated using special equipment in a test car; relationships among them are analyzed.

The Relation of Mathematics to Physics (57 min). Richard Feynman; EDC. Emphasizes that without having some deep understanding of mathematics honest explanation of the beauties of the laws of nature is impossible.

Chapter 3

Force, Mass and Motion (10 min). F. W. Sinden; Bell and EDC. Computer-animated film illustrating the motion of massive bodies under gravity and other forces. Orbits traced, and conservation of momentum observed.

Forces (23 min). Jerrold Zacharias; PSSC MLA 0301. Discusses forces found in nature, with experimental demonstrations. Cavendish experiment illustrated. Eight-minute version of Cavendish experiment also available.

Electrons in a Uniform Magnetic Field (11 min). Dorothy Montgomery; PSSC MLA 0412. Shows electrons in Leybold *e/m* tube.

Coulomb's Law (30 min). Eric Rogers; PSSC MLA 0403. Shows dependence of electrical forces on charge and distance.

Coulomb Force Constant (34 min). Eric Rogers; PSSC MLA 0405. Large-scale Millikan apparatus used to determine the constant of proportionality in Coulomb's law of force between electric charges.

Mass of the Electron (18 min). Eric Rogers; PSSC MLA 0413. Shows how observation of electron motion leads to mass measurement.

The Law of Gravitation, an Example of Physical Law (55 min). Richard Feynman; EDC. Excellent account of the discovery of the law and some of its consequences.

Inertia (26 min). E. M. Purcell; PSSC MLA 0302. Motion of constant-mass–dry-ice puck under no forces and under external forces.

Inertial Mass (19 min). E. M. Purcell; PSSC MLA 0303. Motion of different masses under constant force using dry-ice pucks.

Free Fall and Projectile Motion (27 min). Nathaniel Frank; PSSC MLA 0304. Study of free fall and of inertial and gravitational masses leads to projectile motion.

Chapter 4

Frames of Reference (28 min). Patterson Hume and Donald Ivey; PSSC MLA 0307. Excellent demonstrations of motion relative to inertial and accelerated frames of reference. Also available (EDC) in 6-min version on linearly accelerated frames and 7-min version on rotating reference frames.

Inertial Forces—Centripetal Acceleration ($3\frac{1}{4}$ min). Franklin Miller, Jr.; OSU 16-mm loop. Shows amusement-park rotor ride.

Inertial Forces—Translational Acceleration (2 min). Franklin Miller, Jr.; OSU 16-mm loop. Shows forces in constant velocity or accelerated motion, both up and down.

Chapter 5

Energy and Work (28 min). Dorothy Montgomery; PSSC MLA 0311. Discusses work done by constant force and by a variable force, and determination of the energy produced by such work.

Elastic Collisions and Stored Energy (28 min). James Strickland; PSSC MLA 0318. Quantitative demonstrations of the transformations between kinetic and potential energy in elastic collisions.

The Great Conservation Principles (56 min). Richard Feynman; EDC. Very interesting discussion of a number of conservation principles and their relation to physics.

Chapter 6

Vorticity (44 min). Ascher H. Shapiro; EBEC. Interesting film with references to angular momentum (see also Chap. 8).

Chapter 7

Periodic Motion (33 min). Patterson Hume and Donald Ivey; PSSC MLA 0306. Excellent film about simple harmonic motion, using frictionless puck mounted between springs.

Simple Harmonic Motion (10 min). MGH. Spring-influenced mass moving horizontally illustrates simple harmonic motion.

Tacoma Narrows Bridge Collapse (4 min, 40 sec). OSU. Spectacular pictures of the wind-excited resonant vibrations that destroyed the bridge.

The Wilberforce Pendulum (5 min). Franklin Miller, Jr.; OSU. Interesting case of resonance between torsional and translational vibrations.

Chapter 8

Angular Momentum, a Vector Quantity (27 min). Aaron Lemonick; ESI MLA 0451. Shows that angular momenta add vectorially and that torque applied to a system which already has angular momentum causes the angular momentum to precess.

Moving with the Center of Mass (26 min). Herman Branson; PSSC MLA 0320. Validity of the conservation of energy and momentum demonstrated for several magnetic-puck interactions viewed in two different reference frames.

Chapter 9

Elliptic Orbits (19 min). Albert Baez; PSSC MLA 0310. Geometric demonstration of Kepler's first two laws and the inverse-square law.

Measurement of "G"—Cavendish Experiment (4 min, 25 sec). Franklin Miller, Jr.; OSU. Short film about the Cavendish torsion pendulum.

Universal Gravitation (31 min). Patterson Hume and Donald Ivey; PSSC MLA 0309. The law of gravitation is derived on Planet X from observations on satellites and a planet.

Chapter 10

Measurement of the Speed of Light (8 min). MGH. Very good explanation of several terrestrial measurements of the speed of light, including the methods of Fizeau, Foucault, and Michelson.

Doppler Effect (8 min). MGH. Clear exposition of source in motion and observer in motion cases.

Doppler Effect and Shock Waves (8 min). James Strickland; MLA 0464. Part of a series taken with a ripple tank. Shows effects produced by a source of periodic waves moving at various speeds with respect to the wave medium.

The Ultimate Speed, an Exploration with High-energy Electrons (38 min). William Bertozzi; ESI MLA 0452. The relationship between the kinetic energy of electrons and their speed; investigated by time-of-flight and calorimetric techniques. The results indicate a limiting speed equal to c in agreement with the special theory of relativity.

Speed of Light (21 min). William Siebert; PSSC MLA. The velocity of light is measured by the time of flight of a light pulse and also by the rotating-mirror method.

Chapter 11

The Large World of Albert Einstein (60 min). Edward Teller; SUNY. Extension of ordinary time-distance relationships to the realm of relativity. Discusses impact of special relativity on physics.

Time Dilation, an Experiment with Mu-Mesons (36 min). David Frisch and James Smith; ESI MLA 0453. Using the radioactive decay of cosmic-ray mu-mesons, the dilation of time is shown in an experiment that takes place at Mt. Washington, N.H. (5300 ft) and Cambridge, Mass. (sea level). A detailed report of this experiment appears in *Am. J. Phys.* **31**:342 (1963).

BELL (Bell System): contact local Bell Telephone Company (BTL) business office or Bell Telephone Laboratories (BTL), Film Library, Murray Hill, N.J. 17971.

BTL: see BELL

EBEC: Encyclopedia Britannica Educational Corporation, 425 North Michigan Avenue, Chicago, Ill. 60611.

EDC: Education Development Center (formerly Educational Services, Inc.), Film Librarian, Education Development Center, 39 Chapel Street, Newton, Mass. 02160.

ESI MLA: ESI College Physics Films produced by Educational Services, Inc. Available from Modern Learning Aids, 1212 Avenue of the Americas, N.Y. 10036. Available for purchase, lease-to-buy, or subscription.

MGH: McGraw-Hill Book Company, Text-Film Division, 327 West 41st Street, N.Y. 10036. Sale only.

MLA: see ESI MLA

OSU: Ohio State University, Film Distribution Supervisor, Motion Picture Division, 1885 Neil Avenue, Columbus, Ohio 43210. 16-mm, loop.

PSSC MLA: Physical Sciences Study Committee–Modern Learning Aids. Rentals are handled by Modern Talking Picture Service, Inc. For purchase information contact the MLA division of Ward's Natural Science Establishment, Inc., P.O. Box 302, Rochester, N.Y. 14603.

SUNY: The State University of New York, Educational Communications Office, Room 2332, 60 East 42nd Street, N.Y. 10017.

Index